# FUNDAMENTALS OF APPLIED PHYSICS

C. THOMAS OLIVO    THOMAS P. OLIVO

*This volume is dedicated to the parents (Mr. and Mrs. Peter Olivo) of the senior author as a tribute to their drive, encouragement, and convictions about the tangible value to youth and adults of perserverance, industriousness, and the ever-present spirit of inquiry into the **Why** and **How**.*

*i*

# FUNDAMENTALS OF APPLIED PHYSICS

DELMAR PUBLISHERS
COPYRIGHT ©1978
BY LITTON EDUCATIONAL PUBLISHING, INC.

LIBRARY OF CONGRESS CATALOG CARD NUMBER: 77-79381
ISBN: 0-8273-1300-4

PRINTED IN THE UNITED STATES OF AMERICA
PUBLISHED SIMULTANEOUSLY IN CANADA BY
DELMAR PUBLISHERS, A DIVISION OF
VAN NOSTRAND REINHOLD, LTD.

**C. THOMAS OLIVO**
PROFESSOR EMERITUS
COLLEGE OF EDUCATION
TEMPLE UNIVERSITY
PHILADELPHIA, PENNSYLVANIA

**THOMAS P. OLIVO**
INSTRUCTION AND CURRICULUM
  DEVELOPMENT SPECIALIST
VOCATIONAL EDUCATION MEDIA CENTER
CLEMSON UNIVERSITY
CLEMSON, SOUTH CAROLINA

**DELMAR PUBLISHERS**
**ALBANY, NEW YORK 12205**

# Preface

The productivity, pattern of living, and culture of America depend on the extent to which the simple basic scientific truths about matter, energy, space, and time are applied to produce new substances, new parts, and ingenious mechanisms. Science, which at one time was largely an avocation of learned people, is today a planned, purposeful, intensive application of invention and technology by an individual, a group, or an institution. Each one seeks to translate and control the physical world by shaping it to meet the growing needs of today's society.

The scientist uses two powerful tools to control and use matter and energy. The first tool is that of reliable knowledge which has been verified and organized. The second tool relates to the acquiring and organizing of new knowledge and generalizations. Each scientist builds on the accumulated knowledge and work of others. The scientist establishes theories by designing scientific investigations, conducting experiments or demonstrations, making accurate observations, and interpreting and recording the results according to scientific and precise methods.

Scientific concepts revealed by investigation, exploration, and demonstration are utilized by skilled individuals who translate theory into fact. It remains for physicists, astronomers, engineers, technologists, skilled craftspersons and others to transform the generalizations into inventions, products, and processes which have widespread use.

This text is the end product of years of extensive inquiry, study, experimentation, and analysis of the fundamental understandings about science needed by all persons. It helps provide an answer to the question "To what extent and in what areas should a person study the physical sciences to prepare for a vocation, to become an intelligent consumer, and to contribute as an informed member of society?"

The units in each section were selected to provide instructional material for a well-balanced coordinated course based on life needs. At the same time, the scope of the units is such that the contents may be adapted to meet the special needs of students and to allow for progress according to varying individual abilities and desires.

Recognition is made in each unit of the value of organized laboratory work as an essential part of a physics course. Each student has opportunities to do individual exploration, to gather data and evidence through actual participation, to study and interpret results, and to apply both the findings and the principles to other practical situations. Practical problem material is included with each unit.

This edition was changed significantly to reflect accelerated scientific developments, the increased trend toward metrication, the changing concepts of consumer needs, and individual career planning and development. Some of the major changes in the text include the following:

- Each of the previous eight *Sections* under which all physical science units were grouped for study was expanded. Other *Sections* were added so there is now a

iv

total of eleven *Sections*. The unit on measurement now is included with two new units in a new *Section II* on *Systems of Measurement in Science*. Specifically, the new units relate to the *Evolving International Metric System — SI* (Unit 7) and the *International System of Units — SI Metrics* (Unit 8). An expanded *Achievement Review* is provided for this Section.

- *Section III* on *Mechanics, Machines, and Wave Motion* covers two more units than were formerly in Section II. Unit 21 is concerned with the growing importance of *Fluid Power* in pneumatic and hydraulic systems. Unit 22 is a basic unit on *Wave Motion — Transfer of Energy*.

- *Section IV* on *Magnetism and Electrical Energy* was updated and expanded to include the *Domain Theory of Magnetism. Section V* on *Heat Energy and Machines* treats gas turbines and their applications in greater detail.

- Many changes are incorporated in *Section VI* on *Light Energy,* including a thorough treatment of laser characteristics and uses.

- The Sections on *Sound Energy* (VII), *Electronics* (VIII) and *Nuclear Energy* (IX) cover new developments in each area with greater emphasis placed on nuclear fission and atomic fusion.

- *Section X* deals with new systems for the *Direct Conversion of Energy.* Fuel cells, solar energy, semiconductors such as transistors, thermionic converters, magneto-hydrodynamic energy generators, and thermoelectric generators are treated in detail.

- *Section XI* is also new. It personalizes and enhances the learning of physical science principles and applications. Unit 40 stresses the importance of *Physical Sciences Applied to Consumer Needs. Science Essential to Careers, Career Planning, and Development* is treated in Unit 41.

- The ten *Technical Reference Tables* in the previous edition were increased to seventeen tables to reflect increasing emphasis on metrication. The new tables for prefixes, symbols, derived and supplementary units of physical quantities, conversion factors, and other important problem solving data relate to SI Metrics.

- The *Glossary* was expanded considerably. It includes all of the current major measurement quantities and accepted terms in SI Metrics.

- The *Index* was changed to include all SI Metric designations and the updated physical science developments, principles, and applications treated in the text.

- Solutions to all problem materials with changes and additions are included in the *Instructor's Guide.*

One final point is in order. The measurement systems and standards that are emphasized throughout the text are those most widely used by producers, the service industries, and other occupational groups. It must be understood that although SI Metrics is accepted by legislation as the future dominant system of measurement, it will not be the sole system of standards in the United States.

Conversion to SI Metrics is to be voluntarily accepted by producers, professional bodies, and other users. SI Metrics, like all current and previous measurement systems, is a constantly evolving coherent system of measurement units. The standards are accepted over a period of years by the *General Conference on Weights and Measures*. The standards reflect compromises among scientific, engineering, technologist, and other learned groups.

In the light of daily practices in this transitional period, this edition of *Fundamentals of Applied Physics* treats measurements and standards in the context in which they are applied. Sufficient information is given, conversion tables are provided, and modern practices are suggested within the text to permit the student to solve problems in the customary British and Metric measurement units or in SI Metrics.

Albany, New York

1978

C. Thomas Olivo

Thomas P. Olivo

# Introduction to Physics

*Fundamentals of Applied Physics* serves three basic purposes:

- The content and organization provide a broad background of scientific information. This material was selected from modern technology as being important to the development of social competency in all individuals.

- The instructional units are written so as to present in a clear, functional way those principles and applications of physical science which are needed as preparation for a vocation.

- The units provide the student with fundamental understandings which are essential to the pursuit of advanced studies in related technologies.

## ORGANIZATION AND SCOPE OF THE INSTRUCTIONAL MATERIAL

*Fundamentals of Applied Physics* contains eleven sections. These are: Science and Matter (I); Systems of Measurement in Science (II); Mechanics, Machines, and Wave Motion (III); Magnetism and Electrical Energy (IV); Heat Energy and Heat Machines (V); Light Energy (VI); Sound Energy (VII); Electronics (VIII); Nuclear Energy (IX); Direct Conversion of Energy (X); and Applications of Physical Science to Consumer Needs and Career Planning and Development (XI).

## BASIC PRINCIPLE UNITS

Each Basic Principle Unit is introduced with a brief statement giving the importance and relationship of each new topic or area of scientific investigation. This is followed by a planned presentation of background information which provides a base for formulating whatever scientific laws are considered in the unit. Examples are given to show how a law or physical condition may be applied in a practical way, thus giving meaning to each new principle.

The line drawings which appear throughout the units serve as teaching-learning devices either to place emphasis on an important point or to simplify a description.

### Summary

Important new items are summarized at the end of each instructional unit for emphasis. The statements are brief and may also serve as a guide to the instructor in preparing each lesson.

## ASSIGNMENT UNITS

### Practical Experiments

Almost every Assignment Unit has a first part which identifies one or more experiments. These practical Experiments for laboratory work or demonstration provide general applications of scientific principles. Through actual participation the results show quantitative relationships. A definite pattern of organization is followed throughout the Experiment Series. The equipment and materials for each experiment are listed first. Step-by-step procedures are then given in a simple way as a guide for performing each experiment. Sample tables are shown for recording results. These suggested tables are intended to develop an orderly manner of gathering data, to provide ease of interpretation, and to instill the need for accuracy. This information is applied immediately to the problems under *Interpreting Results*. The conclusions and the solutions reached provide another technique for validating the basic principles involved.

### Practical Problems

The second part of the Assignment Unit contains a series of Practical Problems which also seek to apply each Basic Principle. For the most part, the problems are of the objective type and are arranged from the simple to the more complex. While computations are required to determine quantitative values, all such problems have been kept comparatively simple. In this form, the problems emphasize and clarify specific principles and concepts and do not consume time in lengthy mathematical processes which may not strengthen the learning of a scientific principle. The units of measurement and the standards used throughout the text include the customary British and Metric systems and the evolving SI Metrics. The use of these measurement systems is consistent with current practices.

## ACHIEVEMENT REVIEWS

Where there is more than one Basic Principle Unit in a Section, an Achievement Review is provided to serve as both test and review material. The sequence of the problem material on each Achievement Review follows the same order as the Basic Principle and companion Assignment Units in that Section. The Instructor may use the Achievement Reviews as pre-tests to measure student comprehension of certain basic concepts of science and their application in practical situations.

## RESOURCE TABLES AND GLOSSARY

The Appendices contain three items: (1) a Glossary in which many of the newer technical terms are described; (2) a series of selected Reference Tables which make the text self-contained, insofar as they provide technical data required in the solution of problems or as a basis of comparison of experiment results, and (3) an Index to assist in locating material within the text.

## SUGGESTED APPLICATIONS FOR FUNDAMENTALS OF APPLIED PHYSICS

Each teaching-learning situation requires a different emphasis and presents a different need in the use of instructional materials.

The following are ways in which the material in this text may be used effectively.

- As a science textbook in schools where the learner must develop functional skills in science and must learn how to apply these skills in practical problems that have meaning in career planning and development.

- As a basic science textbook in vocational-industrial-technical education for use either for organized class instruction for homogeneous groups, or individualized instruction for heterogeneous groups.

- As a basic practical science textbook for apprentice training and other in-plant training courses.

- As a basic science textbook for industrial cooperative work-experience classes.

- As a textbook or source book for adult programs and occupational extension classes where a sound practical working knowledge of science gives meaning to the technologies which are related to processes, tools, and materials.

- As a resource book in teacher-training classes for preparing courses of study and for studying the organizational pattern and the teaching content.

## TEACHER'S MANUAL AND ANSWER BOOK

The Instructor's Guide (which is a companion publication) contains the solutions to the objective questions in both the Assignment Units and the Achievement Reviews. This guide is intended to conserve valuable teaching time and provide uniformity to the solutions of the problem material. A Comprehensive Examination is included in this Instructor's Guide as a final integrating experience covering all of the instructional units.

* * * * * * * * * * * * * * * *

*Fundamentals of Applied Physics* incorporates a number of tested teaching-learning techniques for mastering basic scientific principles through simple, direct, and meaningful experiences. The instructional units develop an orderly working concept of the importance of science. They show how fundamental scientific laws, principles, and understandings may be applied with success to new technical developments that contribute so much to the attainment of higher standards of health, living, and welfare for all people.

# Contents

Experiments
and
Assignments

## SECTION 1  SCIENCE AND MATTER — BASIC PRINCIPLES

Unit  1  Science and the Scientific Method.................... 1   5
      2  The Science of Matter................................ 7   10
      3  Basic Properties of Solids ......................... 14   16
      4  Basic Properties of Liquids ....................... 19   22
      5  Basic Properties of Gases.......................... 27   30
         Achievement Review of Science and Matter .............   33

## SECTION 2  SYSTEMS OF MEASUREMENT IN SCIENCE

Unit  6  Customary British and Metric Systems of Measurement ..... 36   48
      7  Metrication and the Evolving International Metric
         System (SI)........................................ 55   61
      8  International System of Units — SI Metrics............. 64   75
         Achievement Review of Systems of Measurement..........   77

## SECTION 3  MECHANICS, MACHINES, AND WAVE MOTION

Unit  9  Forces and Their Effects............................. 80   85
    10  Balance and Equilibrium — Parallel and Angular Forces ..... 89   95
    11  Gravitation, Motion, and Mechanical Movements .......... 99   107
    12  Simple Machines:  Levers ......................... 112   116
    13  Simple Machines:  Inclined Plane and Wedge.............. 121   125
    14  Simple Machines:  The Wheel and Axle................. 129   132
    15  Simple Machines:  The Screw Thread ................... 136   141
    16  Simple and Compound Gear Trains..................... 145   150
    17  Simple and Compound Machines....................... 156   159
    18  Mechanical Power Transmission, Friction, and Lubrication... 164   172
    19  Mechanics of Fluids at Rest and in Motion .............. 179   184
    20  Atmospheric Pressure:  Principles and Applications ........ 190   196
    21  Fluid Power:  Principles and Applications................ 203   210
    22  Wave Motion:  Transfer of Energy...................... 214   222
        Achievement Review of Mechanics, Machines, and
        Wave Motion........................................   226

| | | | Experiments and Assignments |
|---|---|---|---|

### SECTION 4  MAGNETISM AND ELECTRICAL ENERGY

| 23 | Static Electricity and the Electron Theory | 235 | 242 |
| 24 | Magnetism | 245 | 251 |
| 25 | Electromagnetism | 255 | 259 |
| 26 | Electricity in Motion (Direct Current) | 264 | 273 |
| 27 | Electrochemical Sources of Electrical Energy | 279 | 285 |
| 28 | Generation, Transmission, and Distribution of Electricity | 289 | 300 |
| 29 | Application of Electricity to Machines and Lighting | 305 | 312 |
| | Achievement Review of Magnetism and Electrical Energy | | 318 |

### SECTION 5  HEAT ENERGY AND HEAT MACHINES — BASIC PRINCIPLES

| 30 | Heat Energy — Its Effects and Its Measurement | 323 | 331 |
| 31 | Expansion, Transfer of Heat, and Change of State | 338 | 346 |
| 32 | Heat Engines and Turbines | 352 | 365 |
| | Achievement Review of Heat Energy and Heat Machines | | 371 |

### SECTION 6  LIGHT ENERGY

| 33 | Light: Sources, Application, Control, and Measurement | 375 | 393 |
| 34 | Principles, Control, and Applications of Color | 403 | 412 |
| | Achievement Review of Light Energy | | 418 |

### SECTION 7  SOUND ENERGY

| 35 | Sound Properties, Production, and Acoustics | 421 | 433 |
| 36 | Transmitting, Recording, and Reproducing Sound | 444 | 452 |
| | Achievement Review of Sound Energy | | 460 |

### SECTION 8  ELECTRONICS

| 37 | Basic Electronics in Industry and Communications | 463 | 478 |

### SECTION 9  NUCLEAR ENERGY

| 38 | Production and Control of Nuclear Particles | 483 | 496 |

### SECTION 10  DIRECT CONVERSION OF ENERGY

| 39 | Direct Conversion of Energy Systems | 502 | 519 |

### SECTION 11  APPLICATIONS OF PHYSICAL SCIENCE TO CONSUMER NEEDS, CAREER PLANNING, AND DEVELOPMENT

| 40 | The Physical Sciences Applied to Consumer Needs | 522 | |
| | Achievement Review of Consumer Applications of Science | | 530 |
| 41 | Science Essential to Careers, Career Planning, and Development | 534 | |
| | Achievement Review of Science Essential to Careers, Career Planning, and Development | | 539 |

Glossary ........................................................... 542
Appendix .......................................................... 553
Acknowledgments.................................................. 562
Index............................................................... 563

## APPENDIX

Table I          Standard Tables of Metric Units of Measure .................... 553
Table II         Conversion of English and Metric Units of Measure.............. 554
Table III        Conversion Factors ........................................ 555
Table IV         Symbols and Derived Units of Physical Quantities .............. 556
Table V          Common Prefixes for Metric Units ........................... 557
Table VI         Units of Mass and Weight and Conversion Factors............... 557
Table VII        Comparative Densities of Gases Used in Industry ............... 557
Table VIII       Mass and Weight Densities of Common Substances ............. 558
Table IX         Effect of Altitude on Pressure............................... 558
Table X          Coefficients of Friction for Different Solids ................... 559
Table XI         Coefficients of Static and Sliding Friction for
                 Selected Materials ........................................ 559
Table XII        Heat Values of Gases, Liquids, and Solids..................... 560
Table XIII       Physical Properties of Important Pure Metals .................. 560
Table XIV        Tempering and Heat Colors................................. 561
Table XV         Sound Absorption Qualities of Common Materials .............. 561
Table XVI        Velocity of Sound at 0°C Through Common Mediums ........... 561

# Section 1
# Science and Matter-Basic Principles

## Unit 1 Science and the Scientific Method

Our technological age is being built by the peoples of all lands who seek to learn and apply fundamental scientific principles. It is this seeking out of truths and their applications which provides the scientist, the inventor, the technologist, and others with the basic knowledge required which each needs to make possible experimentation, new developments, or production.

### THE MEANING AND IMPORTANCE OF SCIENCE

The impact of science is evident everywhere. Disease has been controlled to such an extent that in the last fifty years the average life expectancy at birth has increased twenty years. Today, there is a greater abundance and variety of food; better transportation made possible by advances in design, materials, and fuels; faster communications; more comforts of life; better clothing; and better shelter. All of these improvements were made possible through new findings and gains in science.

Scientific *know-how* helps the craftsman and technologist to manufacture materials and then shape or form them into parts which are useful in the home or on the farm; or into more complex or heavier mechanisms for use in business, industry, or research.

Science provides each person with simple truths about cause and effect. Each ingenious mechanism or new substance can be traced to a scientific beginning where some basic truths, proven through experimentation, were applied.

Science forms the foundation on which new substances and materials are produced and new products are built. The development of nylon, as shown in figure 1-1, is an example of the application of scientific facts.

Once the basic scientific fact was established that a fiber such as nylon was theoretically possible, continued laboratory experimentation produced the fiber; specialists and technicians developed the required plant and production facilities; and finally, nylon became a marketable product.

Within this framework, industry's principal contribution is to bring together three different areas of human effort — fundamental scientific knowledge, invention, and technology. Each of these areas is becoming increasingly more complex and specialized.

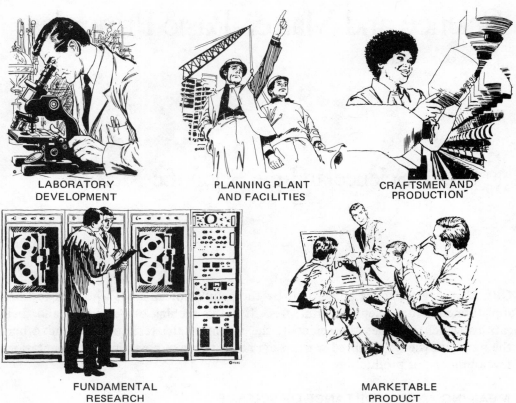

LABORATORY
DEVELOPMENT

PLANNING PLANT
AND FACILITIES

CRAFTSMEN AND
PRODUCTION

FUNDAMENTAL
RESEARCH

MARKETABLE
PRODUCT

**Fig. 1-1**

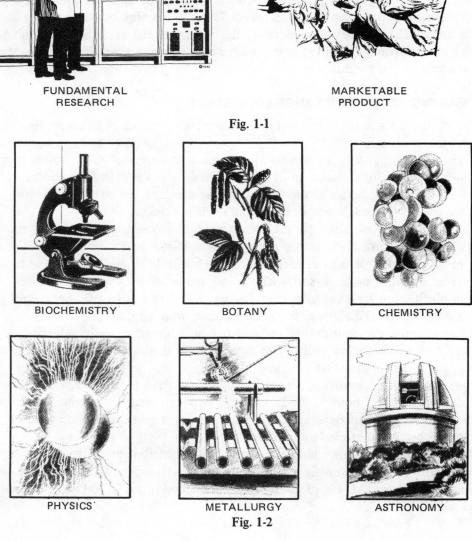

BIOCHEMISTRY

BOTANY

CHEMISTRY

PHYSICS

METALLURGY

ASTRONOMY

**Fig. 1-2**

For example, consider a single product such as a chemical weed destroyer. To produce this chemical, scientists, inventors, and technologists from many different branches of science, such as biochemistry, plant physiology, pathology, soil chemistry, and others, must work together. Such combinations of people are common and are needed to create, test, experiment, produce, and market the goods of the world. Continuous production is another example of how science is combined with experimentation and manufacturing capability.

## BRANCHES OF SCIENCE

Communication, transportation, manufacturing — physical, chemical, and biological developments — all depend on science and the application of the scientific method.

Science is *organized knowledge* derived from the use of a systematic approach to a problem and the application of the scientific method. Physical science requires planning, experimentation, observation, analysis, and problem solving under controlled conditions. Scientific principles, laws, and philosophy are based on broad generalizations. Every generalization is subject to modification. Prior deductions must be continuously assessed in the light of new evidence. Thus, science is a living and evolving body of information.

Scientific knowledge is classified into groupings such as *biology* or the *biological sciences* which deal with life and living things. *Botany* is a branch of biology dealing with plant life. *Zoology* is a scientific study of animal life.

Scientific knowledge may also be grouped around nonliving things like the sun, moon, stars, and planets. Information about these celestial bodies belongs in the area of study called *astronomy*.

*Chemistry* deals with the changes that occur in the composition of matter. *Physics* is a study of matter. (Matter is anything that takes up space and has mass and energy. Energy is the ability to produce work.) *Metallurgy* is another branch of the physical sciences dealing with the study of the properties of metals, their grain structure, and the effects of adding metals to one another.

The student should be aware of the dependence on science of engineering and technological developments, the national economy and security, standards of living, and life itself. In particular, the science of physics is considered to be the basic science whose principles and concepts are drawn upon by many other fields of science.

## THE SCIENTIFIC METHOD

The *scientific method* is a systematic procedure for discovering basic truths. There are five simple steps to the scientific method.

- Recognize the problem to be solved and the essential parts of the problem.

- State a set of hypotheses to solve the problem. Hypotheses are possible explanations or scientific conditions which are the cause of the observed results.

- Collect facts to test each possible explanation or hypothesis. These facts may be obtained by experimentation or any other method.

- Analyze and interpret the data (facts) to determine the correctness of each possible explanation (hypothesis).

- Test conclusions in enough new situations to be sure they are correct.

The scientific laws of physics, chemistry, biology, and the other sciences, which are widely used and accepted today, are the result of experimentation and tests based on the scientific method.

## SYSTEMATIC APPROACH TO PROBLEM SOLVING

Scientific investigation requires that specific procedures be followed in an orderly manner. This practice enables the investigator to analyze a problem, to gain a perspective of the magnitude and conditions of the problem, to prevent errors, and to save time. Scientific problem solving also requires the systematic processing of data to a specified degree of accuracy. The solutions are stated in a significant number of figures. Alternate computational methods are used to recheck the solutions.

The following steps are recommended for the solution of physical science problems. In addition, these steps can be applied effectively to the solution of other practical occupational problems.

1. Read the statement of the problem carefully.

2. Determine the nature and degree of accuracy required in the solution. This step is necessary to avoid extra numerals in a computation.

3. Draw an appropriate diagram, listing the given data.

4. Determine and state the physical principles which appear to be relevant to the solution of the problem.

5. Determine if all of the required data are available. Identify the sources or methods that may be used to obtain missing data.

6. When using formulas, determine if the mathematical processes are to be carried out by first using symbols and, later, substituting numerical values, or if number values are to be used immediately.

7. Check the units for each quantity to be used in the problem to insure that all units are in the same system of measurement.

8. Substitute the data obtained from the physical principles in the problem.

9. Compute the required numerical value. Record the result so that only the significant numerals (figures) are retained.

10. Determine the units in which the results are to be expressed and label the results accordingly.

11. Examine the result to see if it is a reasonable solution. Determine if an alternate method of checking may be used.

12. Recheck all values, formulas, and computations by reworking the problem using an alternate method.

This modern study of science begins with the branch of science called physics. Each unit that follows states the basic principles or laws of physics which are derived by applying the scientific method. Wherever practical, experiments are suggested to give meaning to the study of specific scientific principles. When results are obtained from performing one or more of the steps given in the procedure for each experiment, these results should be put in table form. As these pieces of data become complete, they are interpreted to identify and explain the scientific principle being demonstrated.

This opportunity to apply scientific principles, using the scientific method, should make the study of science both practical and interesting.

———————————————————— SUMMARY ————————————————————

- The seeking out of scientific truths is the foundation on which the scientist, inventor, and technologist control the physical world and shape it to meet human needs.

- Science is the gathering, testing, and arranging of organized knowledge.

- Scientific knowledge is grouped around a specific science and may deal with living things or nonliving matter.

- The fundamental findings of research in various branches of science may be combined to test, improve, finance, produce, and market the goods of the world.

- Industry's greatest contribution to mankind has resulted from bringing together three different areas of human effort: science, invention, and technology.

- Physics deals with a study of matter and energy and is the science upon which other fields of science are based.

- The scientific method, together with other organized procedures, is essential to any study, research, demonstration, and application of physical science principles and concepts.

## ASSIGNMENT UNIT 1 SCIENCE AND THE SCIENTIFIC METHOD

### PRACTICAL PROBLEMS ON THE MEANING AND IMPORTANCE OF SCIENCE

#### Science in the Nuclear Age

For statements 1 to 8, determine which are true (T) and which are false (F).

1. New production, developments, and experimentation are made possible largely through the application of fundamental scientific principles.

2. Gains in science have not affected the standard of living or the life span of an individual.

3. Scientific laws and principles help the inventor, technologist, and scientist to produce and shape materials into useful products.

4. Industry and business contribute to society and scientific progress by welding the following areas into a productive team: fundamental scientific knowledge, invention, and technology.

5. There is no dependence of one branch of science on another.

6. Scientific laws and principles are applied daily to living things and nonliving matter.

7. The scientific method has no influence on the development of scientific principles.

8. The scientist, inventor, or technologist seeks out scientific truths as the foundation for controlling the physical world and shaping it to meet human needs.

Complete statements 9 to 15 by adding a word or phrase.

9. Scientific knowledge is gained through _____ and _____.

10. The _____ sciences deal with life and living things.

11. _____ is a study of animal life.

12. _____ is that branch of science dealing with the solar system.

13. Physics is a study of _____ and _____.

14. Wherever metals are used, a physical science such as _____ deals with their properties and fabrication.

15. _____ is a study of the composition of matter.

Select the correct word to complete statements 16 to 19.

16. (Botany) (Zoology) is a study of plant life.

17. The study of animal life is called (zoology) (bacteriology) (astronomy).

18. Chemistry is a study of the (energy) (life) (composition) of matter.

19. Biology is the science of (matter) (life) (metals).

20. Identify three additional branches of science that are not mentioned in this unit.
    a. Name each branch of science.
    b. Describe briefly what each branch covers.

21. Study one new industrial product.
    a. Name the product.
    b. Identify the kind of scientific knowledge which any two branches of science contribute to the development of the product.

**The Scientific Method**

1. List the five basic steps in the scientific method. Be brief.

Add a word or phrase to complete statements 2, 3, and 4.

2. Scientific facts may be obtained by _____.

3. Facts must be _____ and _____ in the scientific method.

4. A scientific set of hypotheses means _____.

Select the correct word or phrase to complete statements 5 and 6.

5. A conclusion tested in (a single instance) (a sufficient number of new situations) is an example of a good scientific technique.

6. The (facts) (estimates), when analyzed, determine the correctness of a basic hypothesis.

7. List three values to be derived from a study of physics, chemistry, and industrial materials.

## SYSTEMATIC APPROACH TO PROBLEM SOLVING

Identify a simple physical science problem.

1. Review each of the twelve steps in the systematic approach to problem solving.

2. Analyze the problem and state how each step may be used to solve the problem.

# Unit 2 The Science of Matter

Physics is the branch of science that deals with solids (such as steel and plastics), liquids (oils and water), and gases (hydrogen and the other gases in air). The physicist calls these various types of material *matter*. Thus, matter is anything that occupies space and has weight.

Steel, oil, and air are three different states or forms of matter. The steel is a solid; the oil is a liquid; and the air is a gas. Each form or state of matter is identified by certain characteristics which are called *properties*.

This unit deals with the structure of matter and, in a general way, with the common properties of matter.

With this kind of understanding as a starting point, later units cover specific principles that apply to solids, to liquids, and to gases.

## STRUCTURE OF MATTER

All matter is made of small particles called *molecules*. Each molecule is actually the smallest particle of a material which retains the properties of the original material. For example, if a grain of table salt is divided in two, then each subsequent grain is divided again, and the process is continued as finely as possible, the smallest particle having all the properties of salt is a molecule of salt. This molecule is almost one millionth of an inch in diameter. The molecule is so small that before it can be seen in a microscope, it must be enlarged about 100 times. As small as the molecule may seem to be, nuclear physicists work with even smaller particles.

### Properties Depend on Molecular Arrangement

The molecules in a given material are all alike. However, different materials have different molecules. The characteristics and properties of different materials depend upon the nature and arrangement of molecules. For example, the differences in the weights, colors, and hardnesses of two metals all depend upon the arrangement and structure of their molecules.

When the form of a material is changed, the change is referred to as a *physical change*. This means that although a material is pressed, cut, or in some way has its shape modified, no new material is produced because the molecular structure remains the same.

## PHYSICAL PROPERTIES OF MATTER

The properties of matter are described and classified by terms that have very special meanings. Some of the common terms are: mass, weight, volume, density, porosity, cohesion, adhesion, impenetrability, and inertia. Later, in the more advanced units dealing with matter, additional terms will be introduced and described.

The terms listed in this unit provide a working understanding of matter. The mathematics used in making measurements of the properties of matter will be covered in another unit.

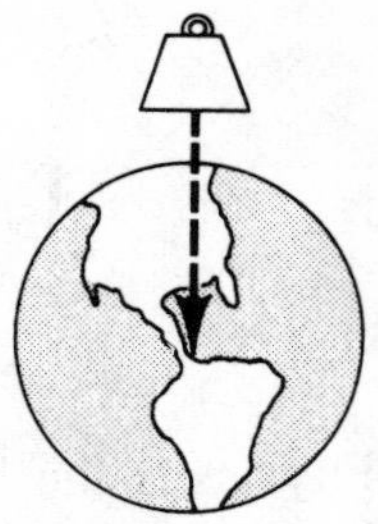

**EARTH'S PULL**

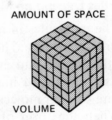

WEIGHT

**Fig. 2-1**

## Mass and Weight

The *weight* of a body is the pull of the earth on the body, figure 2-1. The weight of a body is a variable quantity which depends in part on the distance of the body from the earth's center. Since the earth is not a perfect sphere, a man weighing 160 pounds in New York will weigh 161 pounds at the North Pole. At the pole, he is about 12 miles closer to the center of the earth.

Assume that a stone weighing six pounds on the earth is taken to the surface of the moon. The moon attracts this stone with a force of about one pound. Thus, the weight of the stone on the moon is only one pound. Although it is easier to keep the stone from falling, the force that one must use to throw the stone fast is the same as it is on the earth. If someone is hit by the thrown stone, he finds that it hits just as hard as it does on the earth at the same speed.

The behavior of the thrown stone is said to be due to its *mass*. The mass of a body is the amount of material in it, as demonstrated by its opposition to a change in speed. An object always has mass; this mass is usually a constant quantity, independent of the object's position. Mass is used in scientific calculations of forces, energy, and speed changes.

- Mass is a measure of the amount of material in a body.
- Mass is not affected by the distance of the body from the earth's center.

## Volume

*Volume* is the space that a body occupies, figure 2-2. A cube of metal one foot on a side has a volume of 1728 cubic inches. This means that the total space the piece of metal occupies is a specific number of measurement units which remains the same regardless of where the metal is placed.

**AMOUNT OF SPACE**

VOLUME

**Fig. 2-2**

## Density

Materials are often said to be light or heavy. Actually, these words mean that in comparing the weights of two equal volumes, one is heavy or light in proportion to its size, figure 2-3. *Density* is the mass for a unit volume. Tables of density are available which indicate the weight for a specific volume for many types of materials.

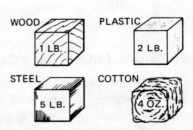

WOOD 1 LB.   PLASTIC 2 LB.

STEEL 5 LB.   COTTON 4 OZ.

**Fig. 2-3 Comparison of densities**

For example, lead weighs 700 pounds per cubic foot and cork weighs 15 pounds per cubic foot. There is a wide variation between the densities of these two materials for the same volume.

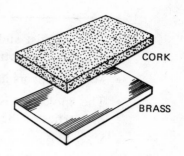

CORK

BRASS

## Porosity

*Porosity* describes the structure or arrangement of molecules in a material. The greater the spaces between the molecules of the material, the more *porous* it is said to be, figure 2-4.

**Fig. 2-4  Porosity**

LIKE MOLECULES

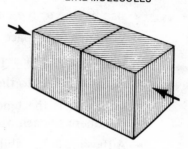

## Cohesion

The degree to which the molecules in one body attract the molecules in another body is described as *molecular attraction*. The molecules in solids do not separate but have an attraction for each other. This tendency of two pieces of the same material to stick together is called *cohesion*, figure 2-5. Without cohesion, solids and liquids could not exist. It is the property of cohesion between molecules that holds solids and liquids together in a recognizable form.

**Fig. 2-5  Cohesion**

UNLIKE MOLECULES

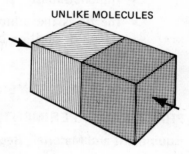

## Adhesion

Some molecules have a tendency to stick to molecules of a different material, figure 2-6. This property is called *adhesion* and is the attractive force between unlike molecules.

**Fig. 2-6  Adhesion**

## Impenetrability

All matter has another property, that of impenetrability. *Impenetrability* means that two bodies may not occupy the same space at the same time. It is apparent that two blocks of steel may not occupy the same place at the same time.

## Inertia

Force is required to start, stop, or change the direction of motion of matter. Objects such as tables or chairs remain at rest until a pushing or pulling force is applied. On the other hand, a car or a ball in motion tends to keep moving until a force changes the speed or direction of motion. The reluctance these objects show to a change is called *inertia*. The greater an object's mass, the greater is its inertia.

These, then, are the properties of matter: mass, weight, volume, density, porosity, cohesion, adhesion, impenetrability, and inertia. The more specific properties of solids, liquids, and gases are described and applied in later units.

─────────────── **SUMMARY** ───────────────

- The scientist and technician consider matter as anything that occupies space, is perceptible to the senses, and has weight and mass.
- Matter appears in any one of three forms: solid, liquid, or gas.
- The molecule is the smallest particle of a substance that retains the properties of the original substance.
- Mass is the amount of material in a body. Mass is constant and is not affected by the distance of the body from the earth.
- Weight is the pull of the earth on a body and varies with distance from the center of the earth. Depending on an object's location, weight may also be due to the pull of any celestial body.
- Volume is the space or room that a body occupies.
- Density is the mass per unit volume.
- Porosity refers to the spacing or arrangement of molecules in a material.
- Cohesion is the tendency of molecules of the same material to attract or cohere to each other.
- Adhesion is the ability of unlike molecules to attract or adhere to each other.
- Impenetrability is that property of all materials to occupy space.
- Inertia is the reluctance of a body to change its direction of motion, speed, or position of rest.

## ASSIGNMENT UNIT 2 THE SCIENCE OF MATTER

**PRACTICAL EXPERIMENTS WITH THE SPECIFIC PROPERTIES OF MATTER**

**Equipment and Materials Required**

| Experiment A Density | Experiment B Impenetrability | Experiment C Porosity | Experiment D Inertia |
|---|---|---|---|
| Small piece of steel Graduate (large enough to hold steel) Spring balance | Small piece of steel and a piece of wood Graduate | Paper filter Oilite bearing Light oil and pan Sponge | Heavy object Test stand String |

## EXPERIMENT A. DENSITY

**Procedure**

1. Weigh a small piece of steel (or other metal). This piece must be able to slide easily into a marked container called a *graduate*. Note the weight in a table such as the one shown for Recording Results.

2.  Partially fill the graduate with water.  Record the volume of the water.

3.  Lower the metal into the graduate.  Record the increased volume reading.

4.  Determine the volume of the metal piece by subtracting the volume of the water from the total volume in the graduate.

5.  Compute the density of steel:  density = mass ÷ volume.

NOTE:  If the graduate reads in fluid ounces, the readings may be converted to cubic inches by multiplying by (1.8).  The result will be in pounds per cubic inch.

### Recording Results

| Factors | Known or Computed Values |
|---|---|
| A.  Weight of metal | |
| B.  Volume of water in graduate | |
| C.  Volume of water and metal | |
| D.  Volume of steel | |
| E.  Density of steel | |

### Interpreting Results

1.  Repeat this experiment with two other materials.  Determine and record the density of each material.

2.  State two factors which must be known before the density of materials can be computed.

## EXPERIMENT B.  IMPENETRABILITY

### Procedure

1.  Partially fill a graduate with water.  Mark or record the level of the water.

2.  Add a small, irregularly shaped piece of metal.  Note the rise in the water level.

3.  Add a piece of wood and again note the rise in the water level.

### Recording Results

| Condition | Height in Graduate |
|---|---|
| A.  Level of water | |
| B.  Level of water and metal | |
| C.  Level of water, metal and wood | |

### Interpreting Results

1.  Explain briefly why the water rises in the graduate each time a new object is added.

2.  What term is used to describe this property of matter?

## EXPERIMENT C. POROSITY

### Procedure

1. Pour water into a cone-shaped paper filter. Note the action that takes place.
2. Cap or seal one end of an *oilite* bearing. Pour a very light oil into the open end. Note what happens on the outside of the bearing.
3. Squeeze a sponge and fill it with water. Note the action.

### Interpreting Results

1. Which of the three materials tested is the most porous: the paper filter, the oilite bearing, or the sponge?
2. In which one of these materials are the molecules closest together?
3. What is the advantage of an oilite bearing over a solid metal bearing?

## EXPERIMENT D. INERTIA

### Procedure

1. Suspend a weight or a heavy object.
2. Attach a light string to the weight.
3. With the weight at rest, quickly pull the string. Note what happens.
4. Swing the weight and try to stop it quickly by jerking the string. Note what happens.

### Interpreting Results

1. Does this simple experiment prove or disprove the statement about the property of inertia? Explain briefly.

## PRACTICAL PROBLEMS ON THE STRUCTURE AND PHYSICAL PROPERTIES OF MATTER

### The Structure of Matter

For statements 1 to 5, determine which are true (T) and which are false (F).

1. Matter may exist in any one of three states: solid, liquid, or gas.
2. All of the molecules in a given material are not alike.
3. The arrangement and structure of molecules determine the properties of different materials.
4. The physical working of a material produces a physical and not a chemical change.
5. The smallest particles of matter that scientists experiment with are molecules.

### Physical Properties of Matter

1. Define briefly what is meant by mass.
2. Which of the two properties, mass or weight, may vary? Give reasons.

Select the correct word or phrase to complete statements 3 to 6.

3. The space that a body occupies refers to its (volume) (density).

4. The arrangement of the molecules in a body determines its (inertia) (adhesion) (porosity).

5. The ability of the molecules of the same material to attract each other is (adhesion) (cohesion) (impenetrability).

6. (Density) (Adhesion) refers to the ability of unlike molecules to attract each other.

Indicate the correct physical term to complete statements 7 to 10.

7. _____ is the amount of material in a body and is not affected by distance from the earth's center.

8. The resistance of a body to change in motion is known as _____.

9. Matter has the property of _____, which means that two bodies cannot occupy the same space at the same time.

10. A material that is less porous than cork is _____.

11. On a separate sheet, match each term in Column I with the correct physical property described in Column II.

| Column I | Column II |
|---|---|
| 1. Impenetrability | a. A comparison of the weight of equal volumes of matter. |
| 2. Cohesion | b. The amount of matter in a body. |
| 3. Adhesion | c. The pull of the earth on a body. |
| 4. Mass | d. The property of matter to occupy space. |
| 5. Density | e. The ability of unlike molecules to attract each other. |
| | f. The ability of like molecules to attract each other. |

12. Select three common materials and compare some of their physical properties.

   a. Prepare a table like the one illustrated.
   b. List the three materials.
   c. Identify and record the color of each material.
   d. Compare the porosity of the three materials and record the results using a 1, 2, 3 scale.
   e. Compare and record the density of the three materials, using the same scale.
   f. Estimate and record how the three materials compare for cohesive properties.

| | Materials | Comparison of Physical Properties | | | |
|---|---|---|---|---|---|
| | | Color | *Porosity | *Density | *Cohesion |
| A | | | | | |
| B | | | | | |
| C | | | | | |

*Use a 1, 2, 3 rating scale: (1) to indicate the greatest or most; (2) next; (3) the least.

# Unit 3  Basic Properties of Solids

Matter exists in three states: solid, liquid, or gas. In each state, matter has specific properties. These properties are used to identify matter and determine its use. The properties to be studied in this unit are those of solid matter.

A solid consists of molecules arranged so that they cohere; that is, they attract each other so strongly that they stay fairly close together. This cohesion gives the solid definite size and shape and affects such properties as hardness, elasticity, and machinability. These and other properties of solids will be analyzed and defined. Thus, as a solid is considered from this point on, it will be understood in terms of its properties.

## HARDNESS, TOUGHNESS, MALLEABILITY

*Hardness* is the ability of a material to resist forces that tend to push the molecules apart. Hard materials resist being scratched, worn away, penetrated, or indented.

**Examples:** Industrial diamonds, which are harder than rocks, are used in well drilling.

Abrasive wheels are used to cut away metals, plastics, and nonmetallic materials; cutting of this type is possible because the individual abrasive particles are harder than the material to be cut.

*Toughness* is the property of a material that enables it to withstand a permanent change.

**Example:** Laminated wood is used on building trusses because of its toughness. This property helps prevent collapse when the structural member is loaded.

*Malleability* is the property of a material which deals with its ability to withstand mechanical processing such as hammering and rolling.

**Example:** When an iron casting is chilled in cooling, it is brittle. This same casting can be heat treated until it is malleable enough to be hammered into another shape without breaking.

**Fig. 3-1**

## DUCTILITY, ELASTICITY, TENACITY

*Ductility* refers to the ability of a material to be drawn into shape without losing other mechanical properties, figure 3-2.

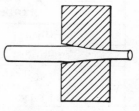

**Example:**  In wire drawing, a metal rod is drawn through progressively smaller openings until the desired size is reached. While it is being drawn, however, the wire retains its physical properties.

**Fig. 3-2 Ductility**

*Elasticity* is the tendency of a material to return to its original form when a deforming force is removed. Once a pattern of molecules is formed in a solid, these molecules seek to return to this pattern after the force tending to push them out of the pattern is removed.

**Example:**  A structural steel girder must be elastic so that it will return to its original shape after a load is removed.

*Tenacity* is the cohesive ability of a material to resist forces which tend to pull the material apart.

**Example:**  The tenacity of the fine strands of wire in a large bridge cable enables them to withstand the tremendous loads encountered in supporting a bridge.

## MACHINABILITY, FUSIBILITY, CONDUCTIVITY

*Machinability* refers to the ease with which a material can be shaped by cutting tools.

**Example:**  The machinability of aluminum is greater than that of tool steel. Machining operations such as drilling and turning can be performed more easily and quickly on aluminum than on tool steel.

*Fusibility* is the ability of a solid to change from the solid to a liquid state when heated. This term also applies to the ease with which a solid may be united or fused with another material.

**Example:**  The fusibility of welding rod with certain metals makes the joining of parts by welding possible.

*Conductivity* refers either to *heat conductivity* or *electrical conductivity*. Heat conductivity is the ability of a material to permit the flow of heat through it; electrical conductivity is the ability of a material to permit the flow of electricity.

**Examples:**  Copper is a good conductor of heat and electricity.

Wood is a poor conductor of both heat and electricity and is used as an insulation against them.

───────────────────── SUMMARY ─────────────────────

- Every solid consists of molecules held together by cohesion.
- The cohesive force gives each solid its shape and size.
- The abilities of the molecules in a solid to resist physical and electrical changes are described as the properties of the solid.

| Property | Characteristics of Solid Due to Property | |
|---|---|---|
| Hardness | Resist scratch and wear | |
| Toughness | Withstand permanent change | |
| Malleability | Take mechanical processing | |
| Elasticity | Return to original form | |
| Ductility | Be pulled to shape | |
| Machinability | Be shaped by cutting tools | |
| Tenacity | Resist forces pulling material apart | |
| Fusibility | Change shape and unite | |
| Conductivity | Heat | Permit flow of heat |
| | Electricity | Permit flow of electricity |

## ASSIGNMENT UNIT 3 BASIC PROPERTIES OF SOLIDS

## PRACTICAL EXPERIMENTS WITH PROPERTIES OF SOLIDS

### Equipment and Materials Required

| Experiment A Hardness | Experiment B Malleability | Experiment C Ductility |
|---|---|---|
| Small pieces of four common materials<br><br>Scriber or other marking device | Hardened steel part<br>Chilled cast iron plate<br>Mechanic's hammer<br>Protective shield<br>Tempered steel part<br>Malleable cast iron plate | Two lengths of fine copper wire<br>Two spools or round pieces of wood<br>Heat source<br>Yardstick |

## EXPERIMENT A. HARDNESS

### Procedure

1. Select any four common materials, such as a small piece of wood, metal, chalk, and glass.
2. Scratch each one with a hardened metal scriber or other pointed tool. Observe what happens in each case.

### Interpreting Results

1. Describe briefly what the term hardness means.
2. State which of the four materials tested is the hardest.
3. Arrange the four materials in order from the hardest to the softest.

## EXPERIMENT B. MALLEABILITY

### Procedure

1. Secure a thin sample of either a hardened steel plate or a chilled cast iron plate; also obtain a sample of both materials, tempered and heat-treated.

2.  Place the hardened steel plate and the chilled cast iron plate on a solid base with a protecting shield.

3.  Tap both pieces sharply with a mechanic's hammer. Note what happens to each piece.

4.  Repeat steps 2 and 3 with the tempered steel plate and the heat-treated cast iron plate.

### Interpreting Results

1.  What happens when the hardened steel plate and the chilled cast iron plate are hammered?

2.  Do the tempered steel plate and the malleable cast iron plate shatter when hammered? Explain.

3.  List the four materials in the order of malleability from the least malleable to the most malleable material.

## EXPERIMENT C. DUCTILITY

### Procedure

1.  Cut two pieces of fine copper wire of the same diameter to the same length.

2.  Wind each end of one piece of wire around a separate spool or round pieces of wood for ease and safety in pulling.

3.  Pull on the spools until the wire breaks. Measure the stretched length of wire.

4.  Heat the other wire over a flame until it is red-hot over its entire length.

    CAUTION:  Allow the wire to cool before handling.

5.  Wrap each end of the heat-treated wire around separate spools and pull the wire until it breaks. Measure the length of the stretched wire.

### Interpreting Results

1.  Determine which of the two wires is more ductile.

2.  Explain why it is possible to pull one wire more than the other.

## PRACTICAL PROBLEMS ON PROPERTIES OF SOLIDS

### Hardness, Toughness, Malleability

Select the correct word or phrase to complete statements 1 to 4.

1.  A steel saw is used to cut wood because the saw is (harder) (softer) than wood.

2.  (Machinability) (Toughness) is the ability to withstand permanent change.

3.  A brittle casting is (more) (less) malleable than a soft lead plate.

4.  Heat treating a hardened steel part or chilled cast iron casting makes it (brittle) (tougher).

5.  List three metals in the order of their hardness.

### Elasticity, Ductility, Tenacity

For statements 1 to 4, determine which are true (T) and which are false (F).

1. Ductile materials may be drawn into shape easily.
2. Elasticity is the ability to withstand hammering and rolling.
3. Elasticity is the ability to resist forces that tend to push molecules apart.
4. A ductile material retains its physical properties even when drawn to a smaller size.
5. Describe (a) elasticity  (b) tenacity.

### Machinability, Fusibility, Conductivity

1. On a separate sheet, match the properties in Column I with the material in Column II that has the required properties.

| Column I | Column II |
|----------|-----------|
| 1. Machinable | a. Abrasive grinding wheel |
| 2. Fusible | b. Bronze rod |
| 3. Conductor | c. Silver wire |
| | d. Solder |

2. Describe:  (a) conductivity     (b) fusibility
3. List three materials that are good heat conductors.

# Unit 4  Basic Properties of Liquids

A different set of properties is used to describe the characteristics of matter in the liquid form. Terms such as surface tension, solvent, buoyancy, specific gravity, and capillary action are associated with liquids. This unit identifies and illustrates those properties which the student must understand to be able to work with liquids.

## COHESION, ADHESION, SURFACE TENSION

### Cohesion

A liquid has the properties of cohesion and adhesion, but to a lesser degree than solids. The ability of like molecules in a liquid to cohere is the property that holds the liquid particles together.

### Adhesion

*Adhesion* is the tendency of different molecules to stick together. For example, water "wets" some metals because the unlike molecules of the water and the metal stick together. There is less wetting action between water and oil because the molecules of these liquids are less adhesive. If water did not have the property of adhesion, it would not "wet" a surface.

### Surface Tension

The molecules on the surface of a liquid are pulled downward by cohesion. Thus, the surface tends to contract. Surface tension refers to the tendency of the surface of a liquid to shrink and become as small as possible. Surface tension also makes the surface of a liquid seem tougher than the interior.

### Capillary Action

The three properties of cohesion, adhesion, and surface tension in combination produce capillary action. *Capillary action* generally describes how a liquid behaves when its surface is in contact with a solid.

### Liquids with Greater Adhesion than Cohesion Properties

Capillary action can be explained with the aid of a simple demonstration. If a glass tube with a small bore is placed in a container of water, the water will rise in the tube, figure 4-1. This action results from the fact that the cohesion and adhesion of a liquid are seldom equal and the surface

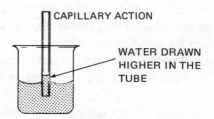

Fig. 4-1  Liquid with greater adhesion than cohesion

tension tends to shrink the surface. In this case, the adhesion of water to the glass is greater than the cohesion between the molecules of water. Thus, the water wets the glass surface and moves up the walls of the tube.

As the water crawls up the walls of the tube, the surface of the water becomes curved. Surface tension (which is trying to contract the surface and keep it as flat as possible) pulls the water up under it.

The water in the tube rises until the weight of the water under the surface film is equal to the combined adhesive and surface tension forces. The smaller the surface area of the water in the tube, the greater are the adhesive and surface tension forces and the higher the water will be lifted before its weight equals the forces.

### Liquids with Greater Cohesion than Adhesion Properties

Not all liquids show the same capillary action. For example, the cohesion of mercury is greater than its adhesion. As a result, instead of crawling up the walls of a tube or container, mercury seems to crawl downward, figure 4-2. Since the surface tension and cohesion of the mercury molecules are greater than the adhesive force, the molecules tend to stick to one another rather than to a different substance. The result is that the mercury in the tube drops to a lower level than the mercury in the container. The surface of the mercury both in the tube and the container is mounded (convex).

## LIQUIDS AND EMULSIONS

The properties of adhesion and cohesion are important in industrial applications. High production is possible because *coolants* may be used to reduce the heat produced in a machining operation. The coolant wets the work and clings to it to remove heat.

Water is a good coolant because it is cheap and has the ability to remove heat effectively. However, water tends to produce rust. A mixture called an *emulsion* is produced by adding oil to water. This oil contains a special soap that allows the oil to mix with the water. This type of oil is said to be *miscible*. The milky-appearing water and oil emulsion combines the cooling properties of water with the rust prevention properties of oil to make an ideal coolant.

In draining the crankcase of an automobile, the sludge that is removed is an emulsion of oil, water, and dirt particles.

## VISCOSITY

*Viscosity* refers to the resistance of a fluid to the motion of bodies through the fluid or to currents in the fluid. The viscosity of fluids is important because the friction that must be overcome due to viscous drag requires the expenditure of a great amount of energy. In the design of a system, viscosity must be considered so that materials, parts, and mechanisms can be selected and shaped to move in or through a fluid with a minimum of resistance. When the design involves liquids, the pressures exerted on the liquid and the operating temperature are other design factors. For example, automotive and aircraft parts require lubricants of different viscosities and properties. There are variations of viscosity even with the same liquid. At a constant temperature,

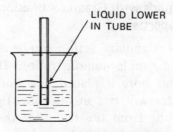

LIQUID LOWER IN TUBE

Fig. 4-2 **Liquid with greater cohesion than adhesion**

a No. 10 grade oil offers less resistance to flow than a heavier oil. The important consideration in selecting a lubricant is that the viscosity of the oil must insure that contacting surfaces are always separated by an oil film.

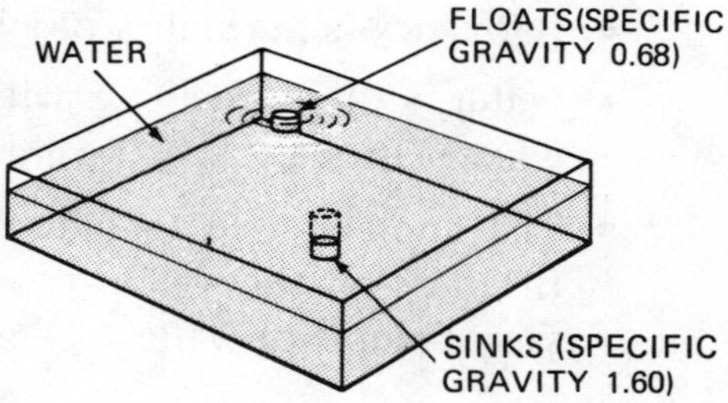

## BUOYANCY AND SPECIFIC GRAVITY

Fig. 4-3

### Buoyancy

Liquids exert a force against materials which are placed in them. Because of the impenetrability of materials, an object placed in water pushes the water aside and begins to sink. When the weight of the water displaced (pushed aside) by the object equals the weight of the object, the object floats. When the weight of the object is greater than the weight of the water displaced, the object sinks.

The tendency of the liquid to produce an upward force on an object in the liquid is known as *buoyancy*. The amount of this buoyant force equals the weight of the liquid displaced.

### Specific Gravity

Water is taken as the standard of comparison when the buoyancy of an object is to be determined. A cube of water twelve inches on a side weighs 62.4 pounds. An object with a weight per volume equal to that of water will float. This relationship is known as the *specific gravity* of the material. For water, the specific gravity is 1.0. An object with a specific gravity greater than 1.0 will sink; one whose specific gravity is less than 1.0 will float, figure 4-3.

Specific gravity, then, is stated as a number (a ratio) which indicates the relationship between the weight of a given volume of material and the weight of an equal volume of water. For example, if a cubic foot of one liquid weighs 93.6 pounds and a cubic foot of water weighs 62.4 pounds, the specific gravity of the liquid is the ratio 93.6/62.4 or 1.5. This value means that the liquid is 1.5 or 1 1/2 times heavier than water.

─────────── SUMMARY ───────────

- Molecules in liquids have cohesive properties.
- The property of adhesion causes many liquids to wet the surfaces of solids.
- Surface tension produces a surface that appears tougher than the interior and is shrunken in size.
- Capillary action depends on cohesion, adhesion, and surface tension.
- Capillary action is the result of the interaction of these three properties. This action continues until the column height of the liquid is equal to the combined adhesive and surface tension forces.
- Practical applications of cohesive properties are found in emulsions which are widely used in manufacturing, agriculture, business, and the home.
- Viscosity is that property of the molecules of a fluid which resists the motion of bodies or currents through or in the fluid.

- Buoyancy is the ability of a liquid to exert force and support an object.
- Water, with a specific gravity of 1.0, is used as a standard of comparison when stating the specific gravity of a material.
- The number used to indicate the specific gravity of a material expresses the ratio of the weight of a volume of the material to the weight of the same volume of water.

## ASSIGNMENT UNIT 4 BASIC PROPERTIES OF LIQUIDS

### PRACTICAL EXPERIMENTS WITH PROPERTIES OF LIQUIDS

**Equipment and Materials Required**

| Experiment A<br>Cohesion and Adhesion | Experiment B<br>Surface Tension | Experiment C<br>Capillary Action |
|---|---|---|
| Glass container (beaker)<br><br>Small quantity of water, alcohol, mercury | Glass container (beaker)<br><br>Small quantity of water<br><br>Needle<br><br>Razor blade | Glass container (beaker)<br><br>Small quantity of water, alcohol, mercury<br><br>Glass tubes (two sizes) |

| Experiment D<br>Buoyancy | | Experiment E<br>Specific Gravity | |
|---|---|---|---|
| One-inch cubes of steel, wood, aluminum, plastic<br><br>Rubber band or string<br><br>Spring balance | | One-inch cubes of steel, wood, aluminum, plastic<br><br>Rubber band or string<br><br>Spring balance | |

### EXPERIMENT A. COHESION AND ADHESION

**Procedure**

1. Place a small quantity of water in a glass container (beaker). Prepare a table for Recording Results as illustrated and sketch the shape of the liquid surface.
2. Replace the water with alcohol and sketch the shape of the liquid surface.
3. Replace the alcohol with mercury. Note any change in the shape of the liquid surface. Sketch the liquid surface.

**Recording Results**

| Liquid | Shape of Liquid Surface |
|---|---|
| 1. Water | |
| 2. Alcohol | |
| 3. Mercury | |

**Interpreting Results**

1. State which of the liquids wet the surface of the container. Why? Explain briefly.
2. Explain which property in mercury is greater: cohesion or adhesion.

## EXPERIMENT B.  SURFACE TENSION

### Procedure

1. Float a needle on water and observe what happens.
2. Repeat the step with a razor blade placed flat on the surface of the water.
3. Determine what property of liquids produces the resulting action.

## EXPERIMENT C.  CAPILLARY ACTION

### Procedure

1. Place some alcohol in a beaker.
2. Immerse one size of glass tube in the alcohol. Note the shape of the liquid surface and the height of liquid in the container and tube. Make a simple sketch of the liquid surfaces.
3. Immerse a smaller glass tube in the alcohol. Observe the shape and height of the liquid and make a sketch of the surfaces.
4. Replace the alcohol with mercury.
5. Immerse the two sizes of glass tube in the mercury.
6. Note and sketch the shape and height of the mercury in the container and the two tubes.
7. Repeat the procedure with any other available liquid.

### Recording Results

| | Liquid | Shape and Height of Liquid Surface |
|---|---|---|
| A | Alcohol | |
| B | Mercury | |
| C | Any other liquid | |

**Interpreting Results**

1. Name the property of the liquid which is visibly identified by the shape of its surface.

2. What effect does decreasing the size of the tube have on the capillary action of the liquid?

## EXPERIMENT D. BUOYANCY

### Procedure

1. Weigh a one-inch aluminum cube on a spring scale. Note the weight of the cube in ounces in a table prepared for Recording Results.

2. Lower the aluminum cube into a container of water. Record its weight in water in the table.

3. Repeat steps 1 and 2 with a one-inch cube of steel, of wood, and of a plastic material. Note the original and the changed weights in water.

### Recording Results

|   | Material | Weight in Air | Weight in Water |
|---|----------|---------------|-----------------|
| A | Aluminum |               |                 |
| B | Steel    |               |                 |
| C | Wood     |               |                 |
| D | Plastic  |               |                 |

### Interpreting Results

1. Explain briefly the effect of the buoyancy of water on the four materials.

2. Was the loss of weight of any one of the materials in water any greater than the others? Explain.

## EXPERIMENT E. SPECIFIC GRAVITY

### Procedure

1. Prepare a table for Recording Results, as illustrated. Record the weight of the one-cubic-inch blocks of aluminum, steel, wood, and plastic.

2. Record the weight of one cubic inch of water in ounces. Compare the weight in ounces of each cube with the weight of one cubic inch of water.

3. Compute the specific gravity of each material. The values are to be accurate to one decimal place.

   NOTE:  To find the specific gravity, divide the weight in ounces of one cubic inch of the material by the weight in ounces of 1 cubic inch of water.

### Recording Results

|   | Material | Weight of 1 cu. in. (in ounces) | Specific Gravity of Material |
|---|----------|---------------------------------|------------------------------|
| A | Aluminum |                                 |                              |
| B | Steel    |                                 |                              |
| C | Wood     |                                 |                              |
| D | Plastic  |                                 |                              |
| E | Water    |                                 |                              |

### Interpreting Results

1. Name the materials, if any, that have a specific gravity greater than that of water (1.0).

2. List the materials, if any, with a specific gravity less than (1.0).

## PRACTICAL PROBLEMS WITH PROPERTIES OF LIQUIDS

### Cohesion, Adhesion, Surface Tension

1. Describe briefly what is meant by (a) adhesion, (b) cohesion, and (c) surface tension.

2. Name four common liquids used in industry or business.

3. For each of the four liquids listed in problem 2, state whether the property of cohesion or adhesion is greater.

### Capillary Action

1. Using a simple cross-sectional sketch, show the capillary action of a liquid that has greater properties of cohesion than adhesion. Label the sketch.

### Solutions, Emulsions and Viscosity

Complete statements 1 to 4 by adding the correct term for each.

1. Heat may be carried away in an industrial process by a _____ .

2. A mixture of water and oil is referred to as a (an) _____ .

3. _____ refers to the ability of the molecules of a liquid to resist a change in shape.

4. The heavier the grade of an oil, the greater is the _____ offered to flow.

5. Describe briefly what viscosity means.

6. Give one example each of an important application of viscosity in an automobile and in an airplane.

### Buoyancy

1. Indicate which of the following materials will float in gasoline.

    a. Cork          d. Lead        g. Bakelite
    b. Bronze        e. Copper      h. Slate
    c. Aluminum      f. Magnesium   i. Wood

2. Name three industrial materials that sink in water.

3. List three nonmetallic materials that float in oil.

### Specific Gravity

For statements 1 to 5, determine which are true (T) and which are false (F).

1. Mercury is the standard on which the specific gravity of materials is based.

2. Specific gravity is the relationship between the weight per volume of the material and water.

3. A material with a smaller specific gravity than water floats.

4. A material with a specific gravity of 0.75 sinks in water.

5. The specific gravity of water is (1.0).

6. List five materials that have a specific gravity greater than water. A listing of materials can be found in a handbook table on Specific Densities of Materials. The specific gravity can also be determined by laboratory tests.

# Unit 5  Basic Properties of Gases

Gases have many properties which are different from those of liquids or solids. The student should have an understanding of the common properties of gases such as weight, fluidity, pressure, and density.

## MOLECULAR MOTION IN GASES

### Fluidity of Gases

Gases have the ability to expand indefinitely or to be compressed into a smaller volume. When heated, a gas expands and becomes less dense. This action is explained by the fact that the molecules of gases move farther apart when heated than they do when cool. Heated gases are lighter than cooler gases and so they tend to rise. The *fluidity* of air (its ability to flow in a manner similar to a liquid) is brought about by changes in pressure and variations in temperature. This fluidity causes the common storm.

The molecules of a gas may move freely in any direction. Single molecules move at high speeds. Although the movement of one molecule produces little effect, large numbers of molecules acting on a container may produce a pressure.

This principle can be explained by noting the increase in pressure when a closed container of air is heated. As the container is heated, the speed and distance between the molecules increase and a pressure builds up. As the source of heat is removed, the molecules slow down and move closer together again. If any air has escaped from the container, the air pressure outside the container exerts a force against the container and causes it to collapse. When the pressures inside and outside a container are equal, no further action takes place.

## COMPRESSIBILITY AND PRESSURE OF GASES

When a gas is compressed to take up a smaller space, the molecules are packed tighter together. Since the molecules in motion (at the same temperature) now strike

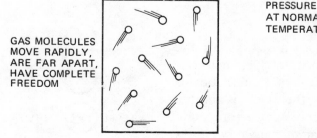

GAS MOLECULES
MOVE RAPIDLY,
ARE FAR APART,
HAVE COMPLETE
FREEDOM

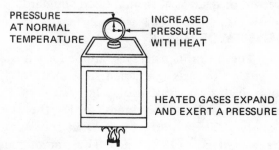

PRESSURE
AT NORMAL
TEMPERATURE

INCREASED
PRESSURE
WITH HEAT

HEATED GASES EXPAND
AND EXERT A PRESSURE

**Fig. 5-1**

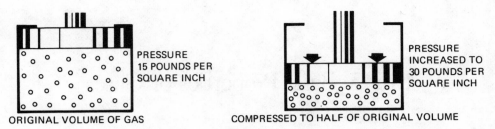

PRESSURE
15 POUNDS PER
SQUARE INCH

ORIGINAL VOLUME OF GAS

PRESSURE
INCREASED TO
30 POUNDS PER
SQUARE INCH

COMPRESSED TO HALF OF ORIGINAL VOLUME

**Fig. 5-2**

the sides of the container more frequently, the pressure rises. Tests made with gases prove that the pressure of a gas varies inversely with its volume (when the temperature of the gas remains constant). Thus, if a gas is forced (compressed) to occupy a space one-half as large as its original volume and there is no increase in temperature, the pressure is twice the original value, figure 5-2.

### Temperature Rise Affects Pressure

The pressure of a confined gas increases as the temperature of the gas rises. This fact is explained by the movement of the gas molecules.

An understanding of pressure is very important. Explosions often result because people overlook the simple facts of the expansion of gases and increased pressures resulting from the application of heat.

### Buoyancy of Gases

All gases have weight. Air, which is a mixture of nitrogen, oxygen, and other gases, weighs almost 1 1/4 ounces per cubic foot. The weight of air is taken as a standard against which other gases are compared. It is the weight of air that creates a buoyant force and, at the same time, applies pressure on everything.

### Atmospheric Pressure

At sea level there is a pressure of almost fifteen pounds per square inch. This pressure is the weight of air pressing down on objects. The higher an object rises above the earth's surface, the less is the pressure on the object. This fact is used by designers, engineers, and others who design guided missiles, rockets, aircraft, and other mechanisms where it is necessary to compensate for changes in pressure. The pressure of the air is called *atmospheric pressure.* This pressure is measured by an instrument known as a barometer. The barometer indicates whether the pressure of the atmosphere has increased or decreased from a fixed standard.

### DENSITY OF GASES

The term *density,* as applied to a gas, means the weight of the gas contained in a unit volume, such as a cubic inch or a cubic foot. The weight of a cubic foot of ammonia gas is 0.045 pounds; for air, the weight is approximately 0.075 pounds; for bottled cooking gas (propane), the weight is 0.117 pounds. Thus, ammonia gas, which is lighter than air, rises because of the buoyant effect of air; propane gas sinks.

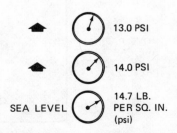

13.0 PSI

14.0 PSI

SEA LEVEL

14.7 LB.
PER SQ. IN.
(psi)

**Fig. 5-3 Atmospheric pressure decreases with altitude**

## SPECIFIC WEIGHT OF GASES

*Specific weight* is a numerical comparison between the weight per volume of a gas and the weight of the same volume of air. A comparative table of *Densities of Gases Used in Industry* is included in the Appendix. For example, the table shows that the specific weight of ammonia is 0.60, that of air is 1.00, and propane gas is 1.56. Note that the standard specific weight of air is 1.00.

## INDUSTRIAL APPLICATIONS OF GASES

A knowledge of the specific properties of gases is important to insure that they are handled safely in industry and in other daily uses. Carbon dioxide is often compressed and used in fire extinguishers because of the cooling effect of the gas as the pressure is relieved.

The compressibility of gases is important to welding in that the compressed gases can be transported safely to the job for cutting and fusing operations. Modern food processing includes the packaging of foods in pressure containers. These containers are based on the gas laws and have limitless applications for the packaging of paints, medicines, and cosmetics.

## CHANGE OF STATE

### Changing from a Gas to a Liquid or a Solid State

When the temperature of a gas is decreased to an extremely low temperature, the gas may change to a solid form. Dry ice is compressed carbon dioxide gas. The compressed carbon dioxide is changed to a solid when its temperature is lowered to a point 78 Celsius degrees less than the freezing temperature of water ($0°$ Celsius). Under such conditions, it is possible to change the state of many gases to another form.

### Evaporation of Liquids to Form Gases

Many liquids may change their state to become gases. Turpentine, alcohol, gasoline, and paint removers are examples of common liquids that evaporate quickly to form vapors. These liquids are said to be volatile. The gases or vapors of many volatile liquids are explosive, highly flammable, and dangerous when mixed with air. This is one reason why containers of liquids which evaporate into or give off volatile vapors should be sealed and kept away from flame.

───────────────────── **SUMMARY** ─────────────────────

- Gases expand and become less dense when heated.
- Gas pressure in a closed container increases with an increase of temperature.
- The pressure of a compressed gas increases inversely as its volume is decreased.
- The weight of air creates a buoyant effect.
- The air pressure at sea level is almost 15 pounds per square inch.
- Air pressure decreases as the altitude increases.

- The density of a gas is the weight per volume; the density of air is 0.075 pounds per cubic foot.
- Specific weight is a number which compares the weight per volume of a gas with the weight of the same volume of air. When these quantities are divided (gas by air), the numerical result is the specific weight.
- Volatile liquids evaporate and change into the gaseous state.

## ASSIGNMENT UNIT 5 BASIC PROPERTIES OF GASES

### PRACTICAL EXPERIMENTS WITH GASES
#### Equipment and Materials Required

| Experiment A<br>Molecular Motion and Weight | Experiment B<br>Compressibility - Pressure |
|---|---|
| Balloon and string<br>Yardstick<br>Pivot | Metal container<br>Asbestos gloves<br>Heat source |

### EXPERIMENT A. MOLECULAR MOTION IN GASES (Fluidity and Weight)

#### Procedure

1. Inflate a balloon and attach it near one end of a yardstick, figure 5-4.
2. Balance the yardstick and note the dimension of the balance point.
3. Detach the balloon and release about half of the air.
4. Attach the balloon to the yardstick and set the yardstick at the dimension noted in step 2. Note the result.
5. Remove the balloon and place it over a heated radiator or some other heat source. Note what happens to the balloon size.

#### Interpreting Results

1. Indicate whether or not the inflated balloon lost weight as some air was released.
2. State three conclusions about air and weight that can be drawn from this experiment.

### EXPERIMENT B. COMPRESSIBILITY AND PRESSURE OF GASES

#### Procedure

1. Wash a metal container thoroughly to remove any remaining liquids or gases.
2. Add a few tablespoons of water and place a cover on the can.

**Fig. 5-4**

3. Heat the can slightly and record the action in a table prepared for Recording Results.

    CAUTION:   Use asbestos gloves to hold the container. Remove the container from the heat quickly as soon as the container is heated and the walls expand slightly.

4. Remove the cover from the container. Heat the container until a vapor appears.

5. Remove the heat source and cover the container.

6. Allow the container to cool. Note what happens to the container and record the result in the table.

**Recording Results**

| Condition | Action |
|---|---|
| A  Closed container, heat applied | |
| B  Open container, heat applied | |
| C  Container closed after heating, then cooled | |

**Interpreting Results**

1. Describe briefly what happens to the molecules of a gas when they are heated.

## PRACTICAL PROBLEMS ON PROPERTIES OF GASES

### Molecular Motion of Gases

Select the correct word to complete statements 1 to 5.

1. Molecules in gases move freely in (one) (any) direction.
2. Gases (contract) (expand) when heated.
3. The speeds of the molecules in a gas (increase) (decrease) when heated.
4. A (heated) (cooled) gas is lighter than the gas it displaces.
5. (Small) (Large) numbers of gas molecules that are moving quickly build up a pressure in a container.

### Compressibility and Pressure of Gases

For statements 1 to 4, determine which are true (T) and which are false (F).

1. As a gas is compressed, the volume decreases and the pressure increases.
2. The atmospheric pressure at sea level is almost 15 pounds per square inch.
3. Air pressure increases as the distance traveled from the earth's surface increases.
4. A barometer is used to measure temperature.

### Density and Specific Weight of Gases

Refer to the table of *Densities of Gases Used in Industry* in the Appendix.

1. Determine the specific weights of (a) hydrogen, (b) carbon monoxide, (c) oxygen, and (d) butane.

2. Check the table for the density of (a) acetylene, (b) nitrogen, (c) oxygen, and (d) gasoline vapors.

3. Prepare a simple table to show which of the seven different gases in Problems 1 and 2 are (a) less dense than air and (b) more dense than air.

## Industrial Applications of Gases

1. List three gases that are lighter than air and give a practical industrial application of each.

2. Name three gases that are heavier than air. Give a common use of each gas.

## Change of State

1. Indicate which of the following are volatile liquids.

   a. Motor oil
   b. Ether
   c. Benzine
   d. Carbon tetrachloride
   e. Water
   f. Kerosene
   g. Paints (rubber base)
   h. Paints (water base)
   i. Lacquer
   j. Gasoline

2. State two safety precautions that must be observed with volatile fluids.

3. Describe briefly two conditions which produce a change in the state of matter.

# Achievement Review
## BASIC PRINCIPLES OF SCIENCE AND MATTER

### SCIENCE AND THE SCIENTIFIC METHOD

Complete statements 1 to 5 by adding the correct word or phrase.

1. The scientist, inventor, and technologist seek out and apply _____ .

2. Scientific knowledge is derived from _____ and _____ .

3. _____ is that branch of the physical sciences dealing with the structure and properties of metals.

4. The *scientific method* is a _____ of arriving at fundamental truths.

5. Before a scientific fact is valid it must be _____ .

6. Name one new important industrial mechanism or product.

    a. Identify two materials used in its construction.

    b. List at least three different branches of science which contributed to the development or manufacture of one of the materials.

### THE SCIENCE OF MATTER

Match each term in Column I with the correct description in Column II.

| Column I | Column II |
|---|---|
| 1. Porosity | a. The reluctance of a body to change its position of rest, speed, or direction |
| 2. Volume | |
| 3. Inertia | b. The smallest particle of a substance retaining the original properties |
| 4. Impenetrability | |
| | c. The mass for a unit volume |
| | d. The arrangement or spacing of molecules in a material |
| | e. The ability of matter to occupy space |
| | f. The space occupied by a body |

For statements 5 to 9, determine which are true (T) and which are false (F).

5. Steel, oil, and air are all in the same state of matter.

6. The arrangement and structure of molecules in a given material determine its physical characteristics.

7. The mass of a body varies while the weight remains constant.

8. The weight of a body is not affected by its density.

9. Density, cohesion, mass, weight, and porosity are some of the properties which are used to identify matter.

10. a. Prepare a table in which hardwood, cork, plastic, and aluminum are listed vertically and the three properties of weight, density, and porosity are listed horizontally.

    b. Rate the properties of each material, using a scale of (1) to indicate the most or heaviest, (2) next, (3) next, and (4) least.

## BASIC PROPERTIES OF SOLIDS

Select the correct word or words to complete statements 1 to 5.

1. When one material cuts another, the material doing the cutting is the (harder) (softer) of the two.
2. (Conductivity) (Elasticity) (Fusibility) is the ability of matter to change state and be welded together.
3. A (ductile) (hardened) (tough) metal may be drawn easily into a different shape.
4. A (hardened) (malleable) metal part has the greater ability to resist scratching.
5. (Toughness) (Elasticity) is the tendency of a material to return to its original shape when a force is removed.

Identify which of the properties (a to h) is the main property of each material (6 to 15).

6. Industrial diamond                    a. Hard
7. Welding rod                           b. Tough
8. Spring steel                          c. Malleable
9. Glass (ordinary)                      d. Elastic
10. Rubber                               e. Ductile
11. Abrasive grains                      f. Machinable
12. Aluminum wire                        g. Fusible
13. Asbestos                             h. Heat-resistant
14. Silver
15. Wood (oak)

## BASIC PROPERTIES OF LIQUIDS

Indicate the term which each statement (1 to 6) describes.

1. _____ is that property of liquids to wet the surface of a solid.
2. _____ refers to the ability of the molecules of a liquid to resist change.
3. _____ is the ability of a liquid to support an object by exerting a force.
4. _____ is the tendency of the surface of a liquid to become as small and as tough as possible.
5. _____ depends on cohesion, adhesion, and surface tension.
6. _____ is the ratio of a given volume of a material to an equal volume of water.
7. Name three common metals that are heavier than water.
8. Identify three common materials that float in oil.

## BASIC PROPERTIES OF GASES

Complete statements 1 to 6 by adding the correct word or phrase.

1. When heated, gases _____ and become _____ .
2. The gas pressure in a closed container _____ as the temperature increases.

3. The ratio of the weight per volume of gas to air is the _____ of the gas.

4. The _____ of a gas is the weight per volume.

5. A _____ liquid is one which evaporates and changes to the gaseous state.

6. _____ refers to the pressure of the air pushing down on an object.

7. a. List two industrial gases that are lighter than air.

   b. Give a practical application of each one.

8. a. Name two common gases that are lighter than air.

   b. Give two examples of where each one is used.

# Section 2
# Systems of Measurement in Science

## Unit 6 Customary British and Metric Systems of Measurement

Scientific progress depends on the degree to which extremely fine particles of matter can be identified and measured. When a measurement is taken directly with the use of tools, instruments, or other calibrated measuring devices, it is known as a *direct measurement*. When the measurement must be determined using a formula (computation), the measurement is called an *indirect* or *computed measurement*.

The most important measurements considered in this and other units are those dealing with length, mass, and time. Measurements related to heat, light, and electrical energy are treated in later sections. These measurements are applied as each new scientific principle is studied.

## STANDARDIZING BASIC UNITS OF MEASURE

The starting point in any study of measurement is an understanding of the need for a universal standardization of units. In primitive times, parts of the human body were used as standards of measure. Due to progress, these rough, inaccurate measurements were discarded for newer and more accurate systems of measurement. Increased world trade also creates a need for universal standards for units of measure. It is essential that all countries have the opportunity to check their units of measure against established standards.

### British, Metric, and Le Système International d'Unites (SI) Metric Measurements

Measurements are used internationaly as a technical language in which precise relationships are stated for physical quantities. There are two parts to every measurement: (1) quantity and (2) a designation of the nature of the quantity. The magnitude of a physical quantity is specified by a numerical value and a unit.

Of the many systems of measurement in common use today, three are used in this text: (1) the British gravitational system (BGS), (2) the metric (MKSA) system, and (3) the Système International d'Unites (SI). The SI metric system is a newly evolving, modernized version of the MKSA system.

There are seven basic measurement units in the SI system. These units relate to length, mass, time, electric current, thermodynamic temperature, luminous intensity, and the substance of a system. For the present, the first four basic units will be considered. The remaining basic, supplementary, and derived units will be covered in later sections of this text.

Every unit of area, speed, force, and electrical charge is expressed in terms of a combination of the basic units of measurement. For example, every unit of force (stated in *newtons*) is the product of a *length* unit and a *mass* unit divided by the square of a *time* unit.

In the British gravitational system (BGS), the units of length, mass, time, and electric current are expressed in standard measurements of the foot, slug, second, and ampere, respectively. These same units in the MKSA system are stated in terms of the meter (length), kilogram (mass), second (time), and ampere (electrical current). The MKSA (Meter, Kilogram, Second, Ampere) system is an internationally accepted system of units that is independent of gravity. The advantage of the MKSA system over the British system is that calculations are simplified and the possibility of error is reduced because of the use of decimals. The MKSA system also allows the more precise scientific measurements to be expressed simply by using conventional prefixes or by a "power of ten" system of notation.

### Standards as Constant Measurements

The need for a standard like the meter to be absolutely constant in magnitude has caused scientists, engineers, and other measurement specialists to fix each value to an invariant quantity in nature. An *invariant quantity in nature* is one that is regarded as a constant (without variation) under all circumstances; in other words, it is unchanging with time.

Until recently, the standard measurement of length was based on a specific platinum-iridium bar kept in Sevres, France. Today the meter is specified as 1 650 763.73 wavelengths of one of the spectral lines of a krypton isotope ($^{86}$Kr). Established in this manner, a precise meter measurement can be reproduced in any laboratory in the world. Having the standard meter (m) measurement, the "yard" as the standard unit of measurement in the British Gravitational System (BGS) is 0.914 m. The standard meter, in terms of BCS inches, is 39.37.

### Conversion of Measurement Units

The magnitude of a physical quantity may be represented in a set of measurement units different from those that are required. The dimensions of physical quantities such as distance (length) and speed (a ratio of length to a time) must be expressed in the same units.

| British and Metric Systems of Measurement | | | | |
|---|---|---|---|---|
| **Dimension** | **British Systems** | | **Metric Systems** | |
| | *Absolute | Gravitational (BGS) | *Absolute (MKS) | Gravitational |
| Length | foot (ft.) | foot (ft.) | meter (m) | meter (m) |
| Mass | pound (lb.) | slug (sl.) | kilogram (kg) | kg·sec$^2$/m |
| Time | second (sec) | sec | sec | sec |
| Energy | ft. poundal | ft. lb. | joule (J) | kg·m |
| Force | poundal | lb. | newton (N) | kg |

*The British and metric absolute systems are independent of gravity.

Fig. 6-1 British and Metric Systems of Measurement

When a quantity is given in terms of a certain unit and this quantity must be expressed in another unit of the same kind, the original value must be changed to the new unit. This process is called *conversion*. The factors that are applied to the quantity to convert the units are called *conversion factors*.

Two major steps must be followed in the conversion process:

Step 1   Determine the value of one unit in the given system in relation to its equivalent conversion factor in the desired unit.

Step 2   Perform any mathematical processes required by substituting the conversion factor in the algebraic equation(s) and formula.

## EXAMPLE

To convert 60 km/hr. (kilometers per hour) to ft./sec. (feet per second), three common factors are needed.

(1)   1 kilometer (km)   = 0.621 mile (mi.)   = $\dfrac{0.621 \text{ mi.}}{1 \text{ km}}$

(2)   1 mile (mi.)           = 5,280 feet (ft.)   = $\dfrac{5,280 \text{ ft.}}{1 \text{ mi.}}$

(3)   1 hour (hr.)           = 3,600 sec.           = $\dfrac{1 \text{ hr.}}{3,600 \text{ sec.}}$

$$\frac{60 \text{ km}}{1 \text{ hr.}} \times \frac{0.621 \text{ mi.}}{1 \text{ km}} \times \frac{5,280 \text{ ft.}}{1 \text{ mi.}} \times \frac{1 \text{ hr.}}{3,600 \text{ sec.}} = 41 \text{ ft./sec.}$$

NOTE:   (1)  the manner in which each conversion factor is expressed, and
(2) that the units designating the original values and the conversion factors cancel out, except for the required units of feet and second (ft./sec.).

## MEASUREMENT OF LENGTH

The measurement of length is often called *linear measure* or the measure of straight line distances between two points, lines, or surfaces. These measurements may be expressed in either the metric system or the British system (BGS). Of the two systems, the British system of linear measure is more widely used in the United States. While the foot is the standard unit of length in the British system, the inch is the smallest unit of measure.

The tools most commonly used in the laboratory for measuring length are the yardstick (meter stick in the metric system), steel rule, measuring tape, micrometer, vernier caliper, vernier micrometer, and gauge blocks. The measurements taken with these tools vary in accuracy from about 1/32 inch with the yardstick to 0.000 1 inch with the vernier micrometer.

### Yardsticks, Tapes, and Rules

The inch, which is the smallest unit of linear measure in the British system, is subdivided into smaller fractional parts. The fractional divisions of an inch which are most commonly used on yardsticks and steel tapes represent halves, quarters, eighths, sixteenths, and thirty-seconds of an inch.

The steel rule has a greater degree of accuracy than either the yardstick or the steel tape. There are two standard methods of graduating steel rules: (1) into fractional divisions, and (2) into decimal divisions, figure 6-2.

The fractional divisions represent halves, quarters, eighths, sixteenths, thirty-seconds, and sixty-fourths of an inch. When smaller units of measure are needed, steel rules graduated in halves, tenths, fiftieths and hundredths of an inch are used. The decimal rule is convenient to use for accurate measurements because the possibility of error in changing from common fractions to decimals is reduced.

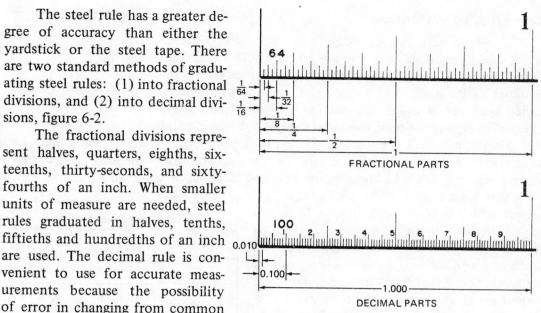

**Fig. 6-2 Enlarged views of subdivisions of an inch**

## PRECISION MEASUREMENTS

*The Micrometer (0.001 inch).* Laboratory measurements of length are often required to an accuracy of one ten-thousandth part of an inch or more. For most practical purposes, an instrument called a *micrometer* is used, figure 6-3.

Direct readings on the standard micrometer are obtained by turning a graduated thimble which moves a spindle toward or away from a fixed anvil. The spindle is carefully advanced to the workpiece until the space between the spindle and the work is just large enough to permit the micrometer to be moved over the work. The measurement is read directly from the graduations on the barrel and the thimble.

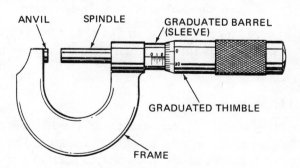

**Fig. 6-3 The micrometer**

## MEASUREMENT OF AREA AND VOLUME

The measurement of the space occupied by a body having the three linear dimensions of length, height, and depth is expressed as cubic or volume measure. When the three dimensions can be taken directly from a solid shape such as a cube, a rectangular solid, or a cylinder, the volume can be computed. The volume of irregularly shaped objects can be determined by the displacement method which will be described later.

## Concept of Square Measure

Common activities of daily living and work often require the determination of the space occupied by a flat surface. The space occupied by a rectangular plot of land is the product of its length and width. Mathematically, this computed quantity is called the area of the object and is expressed in square units of the same kind as the linear units.

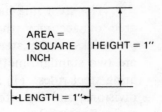

Fig. 6-4

One square inch is the area occupied by a square object which is one linear inch long and one linear inch high, figure 6-4. One square foot is the area occupied by a square object which is 12 linear inches long and 12 linear inches high.

One square yard is the area of a square object which is 36 inches long and 36 inches high. The area of such a square is also 1,296 square inches or 9 square feet, figure 6-5. The units in which the area is expressed depend on whether the dimensions of the length and height are given in inches, feet, or yards. Figures 6-5 and 6-6 compare the square inch, square foot, and square yard.

| Table of Surface Measure | |
|---|---|
| Unit of Surface Measure | = 1 sq. in. |
| 144 sq. in. | = 1 sq. ft. |
| 9 sq. ft. | = 1 sq. yd. |

Fig. 6-5

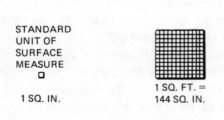

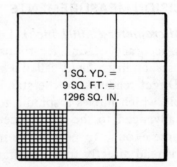

Fig. 6-6  Comparison of units of area measure

Area measure is used to compute the size of square, rectangular, round, and irregularly shaped surfaces. A solid is formed when a third dimension of depth is added to these surfaces.

## Concept of Volume Measure

Volume measure refers to the measurement of the space occupied by a body having three linear dimensions of length, height, and depth. The principles of volume measure are applied in this unit to common shapes such as the cube, the rectangular solid, the cylinder, and various combinations of these shapes, and to objects that are irregularly shaped. For these objects, the volume may be determined more accurately by methods other than computation.

## Cubes

The volume of a solid is the product of three linear dimensions. Each dimension must be in the same unit before they can be multiplied. For example, if the length, height, and depth of a cube are given in feet, then the volume will be in cubic feet.

Figure 6-7 compares the volume of cubes one inch, one foot, and one yard on a side respectively. Note that one cubic foot = 1728 cubic inches and one cubic yard = 27 cubic feet. The measurement of the volume of a cube is equal to its length x depth x height.

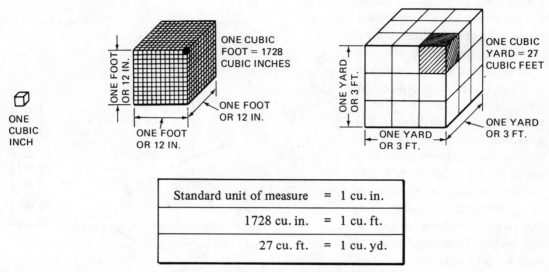

| Standard unit of measure | = 1 cu. in. |
| 1728 cu. in. | = 1 cu. ft. |
| 27 cu. ft. | = 1 cu. yd. |

Fig. 6-7 Comparison of units of volume measure

## Rectangular Solids

When the surface of a flat rectangle is extended in a third direction, a rectangular solid is formed, figure 6-8. The space occupied by this solid is measured in terms of the number of cubic units which it contains. The volume measure of a rectangular solid is equal to the product of the length, depth and height.

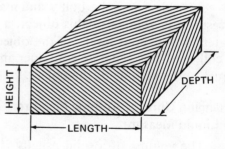

Fig. 6-8

## Cylindrical Solids

The volume of a cylinder is a measure of the cubic units which it contains. The volume is equal to the area of the circular base multiplied by the length or height of the cylinder, figure 6-9.

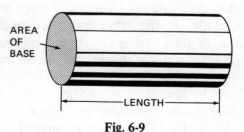

Fig. 6-9

## Displacement Method of Measuring Volume

There are two types of irregularly shaped objects. The first type consists of a combination of regular solid shapes. For this object, the volume can be determined easily by computation. The second type of object may be so irregular in shape (a lump of coal, for example) that its volume must be determined by a simpler and more accurate method than computation.

The *displacement method* is based on the principle of the impenetrability of matter. This principle states that two bodies cannot occupy the same space at the same time. As a result, when a solid is immersed in a liquid, the volume of the liquid is noted before and after the body is placed into it. In this manner, the volume of the body or solid can be determined quickly.

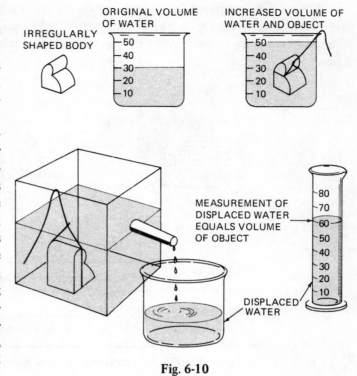

In figure 6-10, the volume of water in the container is 30 cubic inches. An irregularly shaped metal part is placed in the container. The new volume of the water and the object immersed in it is equal to 50 cubic inches. The volume of the metal part can be determined by subtracting the original volume of 30 cubic inches from the new volume of 50 cubic inches. Thus, the volume of the metal part is 20 cubic inches.

Fig. 6-10

An object may be bulky and may not fit conveniently into a graduated vessel. To measure the volume of this object, a tank fitted with an overflow spout is filled with water to the spout level. As the object is lowered into the tank, the overflow of water is collected in a separate container and measured. The volume of the displaced water is equal to the volume of the object.

### Changing Units of Volume Measure to Liquid Measure

The volume of a solid usually is given in terms of cubic units of measure. A cube one inch on a side has a volume of one cubic inch. Liquids, on the other hand, are measured in the English system by liquid measure. The standard units in this system are the cup, pint, quart, gallon, and barrel, figure 6-11.

| Table of Liquid Measure | |
|---|---|
| 2 cups | = 1 pint (pt.) |
| 2 pints (pt.) | = 1 quart (qt.) |
| 4 quarts (qt.) | = 1 gallon (gal.) |
| | = 231 cu. in. |
| 31 1/2 gallons | = 1 standard barrel (bbl.) |

Fig. 6-11

By law, the United States gallon contains 231 cubic inches of liquid. This relationship makes it possible to change a unit from cubic measure to volume measure by dividing by 231, or from volume to cubic measure by multiplying by 231.

## MEASUREMENT OF WEIGHT AND TIME

### Weight of a Body

The basic units of weight in the English system of measurement are the ounce, pound, hundredweight, and ton. Small objects and mechanisms may be weighed directly with some form of spring balance. The weight of simple shapes can be computed by multiplying the volume of the shape by the density of the material.

The weight of a body may vary when it is moved from place to place because of variations in its distance from the earth's center. However, for all practical purposes, a spring balance is accurate enough for the measurement of weight.

### Measurement of Time

The units of time measurement are standardized around the world. The standard day is the interval during which the earth rotates once around its axis with respect to the sun. The day is divided into 24 parts called hours. The hour is divided into 60 parts (minutes) and each minute is divided into 60 parts (seconds).

The second in this system of standard time units is 1/86 400th of the mean solar day. The solar day represents the average time required for a single rotation of the earth around its axis in relation to the sun.

Scientists today need time units divided as finely as billionths of a second. This need for smaller and smaller units of time measurement led to a new standard for the second based on the periodic time intervals stimulated in a beam of cesium ($^{133}$Cs) atoms.  The *cesium-clock system* defines the standard second as the time of 9 192 631 770 vibrations in a beam of $^{133}$Cs atoms.  An important property of this natural time unit is the fact that it is reproducible anywhere in the world.

### THE METRIC SYSTEM OF MEASUREMENT

Scientists, physicists, chemists, engineers, and others throughout the world use a second system of linear, surface, volume, and weight measure. This system is called the *metric sytem.*

The standard unit of linear measure in the metric system is the meter. The prefixes micro, milli, centi, deci, deka, hecto, and kilo are combined with the word meter to indicate how a measurement is related to the meter.  For instance, figure 6-12 indicates that the prefix deci means one-tenth.  A decimeter is one-tenth of a meter; similarly, a centimeter is one-hundredth of a meter; and a kilometer is one thousand meters.  The metric system has the advantage that, as a decimal system, it is convenient.   In the English system, the meter (on which the other measurements are based) is 39.37 inches, or 3.37 inches larger than the English yard.

| Prefix | Relation to Meter | Abbrev. | Metric Units of Linear Measure | | Unit Value (English) |
|--------|-------------------|---------|------------------|-------------------|----------------------|
| | | | Linear Units | Unit Value (Metric) | |
| | | m | 1 meter | | 39.37           in. |
| deci | 1/10 | dm | 1 decimeter | 0.1        meter | 3.937           in. |
| centi | 1/100 | cm | 1 centimeter | 0.01       meter | 0.393 7         in. |
| milli | 1/1 000 | mm | 1 millimeter | 0.001      meter | 0.039 4         in. |
| micro | 1/1 000 000 | $\mu$m | 1 micrometer | 0.000 001  meter | 0.000 039 37 in. |
| deka | 10 | dam | 1 decameter | 10.        meters | 32.80           ft. |
| hecto | 100 | hm | 1 hectometer | 100.       meters | 328.            ft. |
| kilo | 1000 | km | 1 kilometer | 1000.      meters | 3280. ft.   0.621 mile |

**Fig. 6-12**

## CHANGING UNITS OF MEASURE FROM ONE SYSTEM TO THE OTHER

The units of linear measure in the English system may be expressed in more than one kind of unit, such as 3 yards, 2 feet, and 9 inches. The metric system, however, is based on multiples of ten; therefore, a linear dimension of 2 meters, 5 decimeters, and 3 millimeters may be expressed as a decimal (2.503 m). The metric system provides a simplified way of writing such dimensions.

| English Units | Equivalent in Metric Units (Approximate) | |
|---|---|---|
| One inch (in.) | 2.54<br>25.4 | centimeters (cm) or<br>millimeters (mm) |
| One foot (ft.) | 0.3048 | meter (m) |
| One yard (yd.) | 0.9144 | meter (m) |
| One statute mile (mi.) | 1.61 | kilometers (km) |

**Fig. 6-13 Comparison of English and Metric Units of Linear Measure**

Tables comparing the English and metric systems are given in science and mathematics handbooks. The most widely used linear units in both of these systems are shown in figure 6-13.

To convert from one system to the other, the equivalent in the required system must be known. The desired unit may then be computed by either multiplying or dividing by a conversion factor.

## COMPUTING AREAS OF SURFACES IN THE METRIC SYSTEM

The area of a surface in the metric system refers to the number of square units of measure contained in the surface. Commonly used metric area units are the square meter, the square centimeter, the square decimeter, and the square kilometer.

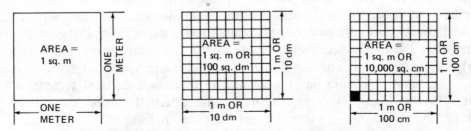

**Fig. 6-14 Relationship of the meter to a decimeter and a centimeter**

Surface area is computed by multiplying the linear dimensions of height and length in the metric units of measure.

Figure 6-15 gives the values of common units of square measure in the metric system and the equivalent values in the English system.

The principles presented using English units of measure for computing surface area also apply in the metric system. The only differences are in the name of the unit and its size. Thus, the areas of squares, rectangles, parallelograms, circles, and parts of circles are computed the same way.

### Computing Volumes in the Metric System

The volume of a solid can be determined using the displacement method. The linear dimensions of a solid in the metric system are expressed in meters, decimeters, centimeters, or a combination of any of the metric units.

| Values of Metric Units of Square Measure | | |
|---|---|---|
| Metric Unit of Surface Measure | Value of Unit in Metric System | Equivalent Value in English System |
| 1 square meter (sq. m; m²) | Standard Unit of Measure | 1550 sq. in. (in.²) |
| 1 sq. decimeter (sq. dm; dm²) | 0.01 square meter (sq. m) | 15.50 sq. in. (in.²) |
| 1 sq. centimeter (sq. cm; cm²) | 0.000 1 sq. m | 0.155 sq. in. (in.²) |
| 1 sq. millimeter (sq. mm; mm²) | 0.000 001 sq. m | 0.001 55 sq. in. (in.²) |

| Comparison of English and Metric Units of Surface Measure | |
|---|---|
| English Units | Equivalent in Metric Units (Approximate) |
| 1 sq. in. | 6.452 sq. cm (cm²) |
| 1 sq. ft. | 0.092 9 sq. m (m²) |
| 1 sq. yd. | 0.836 sq. m (m²) |

**Fig. 6-15**

The volume of a solid is a measure of the cubic units contained by the solid. Thus, the volume in metric units for a solid is usually given as cubic meters (cu. m or m³), or cubic centimeters (cu. cm or cm³), or cubic millimeters (cu. mm or mm³). A comparison of the cubic meter and the cubic decimeter is shown in figure 6-16.

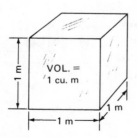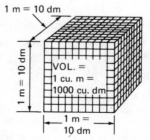

**Fig. 6-16**

One of the most familiar measurements in science is the cubic centimeter (cu. cm or cm³). (An older method of expressing cubic centimeters is cc; however, this method is no longer recommended because of the necessity of standardized metric notation. The preferred abbreviation is cm³.) This volume represents the space occupied by a cube one centimeter on a side. The tables in figure 6-17 compare common metric units of volume with the corresponding units in the English system.

| Values of Metric Units of Volume Measure | | |
|---|---|---|
| Metric Unit of Volume Measure | Value of Unit in Metric System | Equivalent Value in British System (Approximate) |
| 1 cu. meter (cu. m or m³) | 1000 cu. decimeters (cu. dm or dm³) | 35.314 cu. ft. or 1.308 cu. yd. |
| 1 cu. decimeter (cu. dm or dm³) | 1000 cu. centimeters (cu. cm or cm³) | 61.023 cu. in. |
| 1 cu. centimeter (cu. cm or cm³) | 1000 cu. millimeters (cu. mm or mm³) | 0.061 cu. in. |

| Comparison of English and Metric Units of Volume Measure | |
|---|---|
| English Units | Equivalent in Metric Units (Approximate) |
| 1 cu. in. | 16.387 cu. cm |
| 1 cu. ft. | { 0.028 3 cu. m<br>{ 28.3 liters |
| 1 cu. yd. | 0.764 6 cu. m |

**Fig. 6-17**

## Computing Volumes in Liquid Measure

The most widely used standard unit of liquid measure in the metric system is the liter. The *liter* is equal to a volume of 1000 cubic centimeters.

For accurate work, a *milliliter* (ml), which is one-thousandth of a liter, is used instead of the cubic centimeter. The value of the liter and other metric units of liquid measure are compared with corresponding units in the English system in figure 6-19.

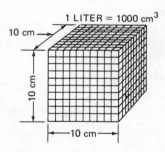

**Fig. 6-18**

## METRIC UNITS OF MASS MEASUREMENT

Mass is the measurement of the earth's attractive force on a material. In the metric system, mass is measured in terms of the gram (g), decigram (0.1 g), centigram (0.01g), milligram (0.001 g), kilogram (1000 g), and the metric ton (1000 kg). The basic unit, the gram, is the mass of one cubic centimeter of water at 4° Celsius. One thousand grams equals a kilogram. The weight of a kilogram in the English system is 2.204 6 pounds. Thus, 454 grams in the metric system are approximately equal to one pound in the English system, figure 6-20.

| Value of the Liter | | |
|---|---|---|
| Metric Unit of Liquid Measure | Value of Unit in Metric System | Equivalent Value in British System (Approximate) |
| 1 liter | Standard Unit of Liquid Measure | 1.057 qt. |
| | 1000 cu. cm 1 cu. dm | 61.023 cu. in. |

| Comparison of English and Metric Units of Liquid Measure | |
|---|---|
| English Units | Equivalent in Metric Units (Approximate) |
| 1 qt. | 0.946 liter |
| 1 gal. | 3.785 liters |

Fig. 6-19

| Metric Units of Mass with Corresponding Values in the English System | | |
|---|---|---|
| Metric Unit | Value in Other Metric Units | Value in English Units |
| 1 gram (g) | Mass of 1 $cm^3$ of water at 4°C | 1/28 oz. approx. |
| 454 grams (g) | | 1 lb. approx. |
| 1000 grams (g) | 1 kilogram (kg) | 2.204 6 lb. 0.068 5 slug |
| 1 kilogram (kg) | 1000 grams (g) | |
| 14.6 kilograms (kg) | | 32.2 lb. (1 slug) |

Fig. 6-20

## SUMMARY

- The National Bureau of Standards is authorized by law to develop, establish, and maintain standard units of measure.

- The measurement of any quantity requires a comparison with a standard quantity (unit) of the same kind.

- Common measuring tools such as rules and tapes are used to take measurements of length to an accuracy of 1/100 inch. Precision measuring instruments yield linear measurements with accuracies ranging from one-thousandth to two-millionths of an inch.

- The displacement method of measuring volume is widely used for irregularly shaped solids. The volume of liquid displaced when an object is immersed is equal to the volume or mass of the object.

- Tables of comparison between metric, English, and SI units of measure furnish values which may be used to convert from one system to another.
- Volume (cubic) measurements are usually expressed in cubic centimeters, cubic meters, or in any other metric units of measurement.
- The liter is the standard unit of liquid measure and is equal to 1000 cm$^3$ or 1.057 qt.
- The volume of a body may be computed in both the metric system and the English system. The volume expressed in linear measure can be converted to a unit of liquid measure.
- The metric units of weight are based on the gram which is the mass of 1 cm$^3$ of water at 4°C. One kilogram (1000 grams), at the earth's surface, is equal to 2.204 6 pounds.
- The BGS, MKSA, and SI are common systems of measurement in which the basic units of length, mass, time, and electric current are standardized. All other units in each system are derived from the basic units in that system.
- In the BGS, the units of length, mass, time, and electric current are expressed in foot, slug, second, and ampere measurements, respectively.
- In the MKSA and SI systems, the standard units are the meter (length), kilogram (mass), second (time), and ampere (electric current).
- Conversion is the process of determining a given quantity of a certain unit in terms of another unit of the same kind.

## ASSIGNMENT UNIT 6 CUSTOMARY BRITISH AND METRIC SYSTEMS OF MEASUREMENT

## PRACTICAL EXPERIMENTS WITH MEASUREMENT

### Equipment and Materials Required

| Experiment A<br>Measuring Volume | Experiment B<br>Units of Liquid Measure | Experiment C<br>Units of Linear Measure |
|---|---|---|
| Three irregularly shaped objects<br>Graduated beaker or vessel<br>Vessel with run-off spout<br>Container for displaced water | Beaker graduated in metric units (500 cm$^3$ capacity)<br>Quart bottle or container | Yardstick<br>Meterstick |

## EXPERIMENT A. DISPLACEMENT METHOD OF MEASURING VOLUME

### Procedure A

1. Obtain three irregularly shaped objects.
2. Pour water into a graduated beaker until it is half full. Note the quantity of water added by referring to the graduated reading on the beaker.

3. Lower the first object gently into the beaker. Note the new volume of the water and the object.

4. Determine the volume of the object by subtracting the original volume of the water from the new volume. Prepare a table for Recording Results and record this volume in Column A.

5. Remove the first object and repeat steps 2-4 with each of the remaining two objects. Record the volumes.

6. Repeat this experiment using a container with a run-off spout.

   a. Fill the container to the level of the spout and until a few drops run off.

   b. Place another container under the spout to collect the water displaced.

   c. Immerse the first object, collect the displaced water, and measure it in a graduated beaker.

   d. Record the amount of displaced water in the table in Column B.

7. Repeat steps 6a–6d with each of the two remaining objects.

**Recording Results**

|  | A<br>Graduated Vessel | B<br>Run-off Method | C<br>Converted Units |
|---|---|---|---|
| Object 1 |  |  |  |
| Object 2 |  |  |  |
| Object 3 |  |  |  |

**Interpreting Results**

1. What happens to the level of the water as a body is immersed in it? Why does this happen?

2. Regardless of the method used for the one object, is the amount of water displaced always the same?

3. Does the weight of an object (heavier than water) or its volume determine the amount of water displaced by the object? Why?

**Procedure B. Changing Units of Volume Measure**

1. Refer to a Table of Liquid Measure to determine the equivalents in volume and liquid measure.

2. Change the volume of water displaced for Object 1 from one system to the other. If the graduations on the beaker or other glass container are given in cubic units of measure, such as cubic inches, change the units to units of liquid measure and vice versa. Record the results in Column C in the table.

3. Repeat step 2 for each of the two remaining objects.

## EXPERIMENT B. COMPARISON OF UNITS OF LIQUID MEASURE

### Procedure

1. Secure a beaker graduated in metric units.
2. Fill the graduated beaker with water to the 500 cm³ mark.
3. Pour the measured water into a quart bottle.
4. Refill the graduated beaker to the same 500 cm³ level.
5. Pour some of this water into the quart container until the quart is filled.
6. On the graduated scale of the beaker, read how much water is left in the beaker.

   NOTE: 1000 cm³ of water equals one liter. If a liter graduate is available, it may be used in place of the graduated beaker.

### Interpreting Results

1. Determine the approximate number of cubic centimeters of water contained in one quart.
2. Compare the amount of water in one quart and in one liter. Is a quart liquid measure (a) smaller than, (b) the same amount as, or (c) larger than one liter? Explain briefly.
3. Check your findings against a table of comparisons between English and metric units of liquid measure.

## EXPERIMENT C. COMPARISON OF LINEAR METRIC AND ENGLISH UNITS OF MEASUREMENT

### Procedure

1. Place a yardstick graduated in inches beside a meterstick so that the ends rest on the same surface.
2. Prepare a table for Recording Results.
3. Compare the difference in lengths between the yardstick and the meterstick and record the findings.
4. Determine the equivalent gradation in the metric system of the one-foot gradation. Record the approximate reading.
5. Record the approximate equivalent reading on the meterstick for the one-inch gradation of the yardstick.

### Recording Results

|   | Reading in English Units of Measurement | Approximate Equivalent in Metric Units |
|---|---|---|
| A | One yard | |
| B | One foot | |
| C | One inch | |

### Interpreting Results

1. Based on the actual comparisons recorded in the table (checked by a table of values for metric and British units), give the unit in the metric system in Column II that corresponds with each dimension in Column I.

| Column I | Column II |
|---|---|
| a. One foot | 1. 39.37 inches |
| b. One yard | 2. 2.54 centimeters |
| c. One inch | 3. 0.4 inch (approx.) |
| d. One meter | 4. 30.48 centimeters |
| e. One centimeter | 5. 9.144 decimeters |

## PRACTICAL PROBLEMS WITH ENGLISH AND METRIC UNITS OF MEASUREMENT

### Nonprecision Linear Measurements with Yardstick, Tape, and Rule

1. Determine the reading of each measurement indicated on the yardstick in figure 6-21.

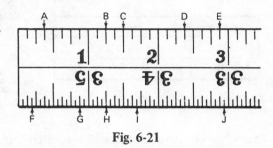

Fig. 6-21

2. Determine the reading of each measurement indicated on the steel rules in figure 6-22. The rule on the left is graduated in decimal parts; the rule on the right is graduated in fractional parts of an inch.

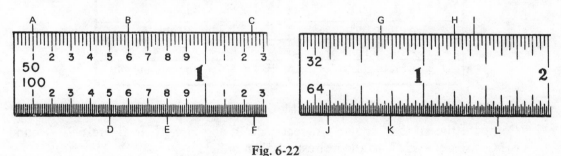

Fig. 6-22

### Precision Linear Measurements with the Standard Micrometer

1. Determine the linear dimensions indicated at A, B, C, D, and E on the standard micrometer in figure 6-23.

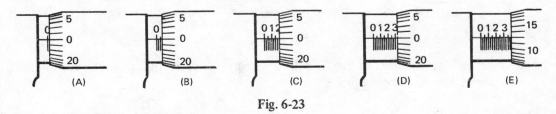

Fig. 6-23

### Volumes of Solids (British System)

1. Determine the volumes of cubes A, B, and C and the rectangular solids D, E, and F in figure 6-24, page 52.

| | A | B | C | D | E | F |
|---|---|---|---|---|---|---|
| Length | 10″ | 2 1/2″ | 3 1/2″ | 5″ | 2 ft. | 1 yd. |
| Depth | 10″ | 2 1/2″ | 3 1/2″ | 2″ | 1 ft. | 1 yd. 1 ft. |
| Height | 10″ | 2 1/2″ | 3 1/2″ | 4″ | 8 in. | 1 yd. 8 in. |

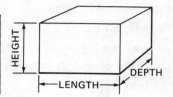

**Fig. 6-24**

2. In figure 6-25, find the volumes of the cylinders A, B, and C and the cored cylinders D and E, correct to the two decimal places.

| | A | B | C |
|---|---|---|---|
| Diameter | 2″ | 6″ | |
| Radius | | | 12.2″ |
| Length | 4″ | 4 1/2″ | 3.4″ |

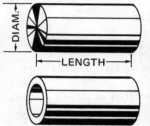

| | D | E |
|---|---|---|
| Outside Diameter | 2″ | 5″ |
| Inside Diameter | 1″ | 3″ |
| Length | 10″ | 1′-4″ |

**Fig. 6-25**

## Changing Units of Volume Measure to Liquid Measure (British System)

1. Change the quantities given in the table from one unit of liquid measure to the one stated in each case.

| A | 12 qt. to pints | E | 2 bbl. to gallons |
|---|---|---|---|
| B | 3 gal. to quarts | F | 3 1/2 bbl. to quarts |
| C | 2 gal. to pints | G | 3 gal. 5 qt. to quarts |
| D | 8 pt. to gills | H | 6 3/4 pt. to pints and gills |

## Applying Units of Square Measurement (Metric System)

1. Compute the areas of square A, rectangle B, and circle C in figure 6-26. Round off C, using $\pi = 22/7$, to the nearest whole number.

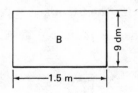

**Fig. 6-26**

2. Change each metric area in Problem 1 to its equivalent in English units. Round off all answers to one decimal place.

## Applying Volume Measurements

1. Determine the volumes of rectangular solid A in cubic meters and cube B in cubic centimeters (correct to two decimal places).

| | Length | Depth | Height |
|---|---|---|---|
| A | 2 m | 12 dm | 8.6 dm |
| B | 6.8 cm | 6.8 cm | 6.8 cm |

2.  Compute the volume of cylinders A, B, and C (correct to two decimal places). Use $\pi = 22/7$.

|   | Diameter | Length | Find Volume in |
|---|----------|--------|----------------|
| A | 8.4 cm | 50 cm | $cm^3$ |
| B | 14.4 dm | 2.1 dm | cu. dm ($dm^3$) |
| C | 5 dm 6 cm (5.6 dm) | 10 dm 5 cm 2 mm (10.52 dm) | $cm^3$ |

## Applying Liquid Measurements (Metric System)

1.  Find the volume of liquid (in liters) required to fill the square, rectangular, and round containers to the levels indicated.

|   | Shape | Inside Dimensions of Container | Height of Liquid Level |
|---|-------|-------------------------------|------------------------|
| A | Square | 4 meters (m) | 5  dm |
| B | Rectangular | L = 9.2 dm<br>W = 4.4 dm | 10  dm |
| C | Round | Radius = 2.1 dm | 5.55 dm |

## Applying Units of Weight Measurement (Metric System)

1.  In figure 6-27, determine the weight of water in the rectangular tank (A), and the round tank (B) to the nearest kilogram. Use $\pi = 22/7$.

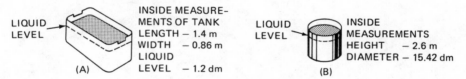

Fig. 6-27

2.  Change the kilogram weights of water in the rectangular and round tanks of problem 1 to pounds. Round off answers to the nearest pound.

## Conversion of Measurement Units (BGS and MKSA Systems)

Refer to a Conversion Table which gives equivalent values for BGS and MKSA units of length, mass, and time.

1.  Insert the equivalent values of quantities A through F in either the BGS or MKSA system, as required.

| Quantity | Measurement System | |
|----------|------|------|
|          | BGS | MKSA |
| A | 39.37 in. | m |
| B | in. | 1 m |
| C | lb. | 1 kg |
| D | 0.068 5 sl. | kg |
| E | sl. | 14.6 kg |
| F | hr. | 1/3600 hr. or 1 sec |

2. Convert each of the BGS quantities of mass (A, B, C, and D) to the equivalent MKSA units.

| Mass | Quantity | |
|---|---|---|
| | BGS | MKSA |
| A | 22.046 lb. | kg |
| B | 25 sl | kg |
| C | 36.4 sl | kg |
| D | 0.025 sl | kg |

# Unit 7 Metrication and the Evolving International Metric System (SI)

There is a slow but persistent effort toward the adoption of the International System of Units (SI) by the United States and other nations which use measurement standards based on the British gravitational system and other metric systems. The SI evolved from a consortium of over fifty industrialized nations in the International Standards Organization (ISO). The term *metrication* is used during these changing times in connection with any program or process of conversion to the International System of Units (SI).

## ORGANIZATION AND FUNCTIONS OF THE GOVERNMENT AND THE PRIVATE SECTOR IN METRICATION

### The Metric Conversion Act

In December of 1975, the President of the United States signed the Metric Conversion Act of 1975. This event gives official federal sanction to a movement which has been gaining in acceptance for several years. The Metric Conversion Act establishes a United States Metric Board which is charged with the tasks of ". . . devising and carrying out a broad program of planning, coordination, and public education, consistent with other national policy and interests, with the aim of implementing the policy set forth in the Act."

The United States Metric Board will work closely with the organizations described as follows to implement metrication in the United States.

### The American National Standards Institute (ANSI) and the American National Metric Council (ANMC)

The American National Standards Institute (ANSI) represents the United States at the International Standards Organization. In this role, the ANSI is the national coordinating agency that is involved in the compromises which must be made if acceptable international standards of weights and measures are to be developed. The ANSI is a voluntary national organization which serves all interests and segments of society. Its functions are advisory and it operates under the principles of *nonlegislative consensus*. The ANSI provides the nucleus for planning and coordinating activities for organizations and industries converting to SI units of measurement. To assist in the coordination of planning for metrication in the private sector, the ANSI established the American National Metric Council (ANMC).

The major functions of the ANMC are as follows:

- to promote the consistent application of metric units and metric practices, guided by the principle of voluntary membership.

- to assist existing standards-making bodies by identifying needs for SI metric units, evaluating priorities, and making recommendations for developing standards.

- to coordinate standards development efforts and activities with the National Metric Conversion Board.

- to maintain a close working relationship with the metric conversion groups in Canada, the United Kingdom, Australia, New Zealand, and other countries. These nations can and do provide considerable expertise and advice on the many problems of metrication.

**Organizational Chart of the American National Metric Council**

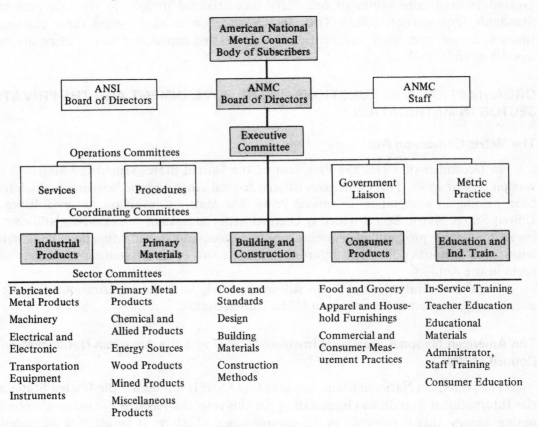

## ANMC Committee Structure

There are two basic categories of committees performing the functions of the ANMC: the operations committees and the coordinating committees. The operations committees deal with services, procedures, government liaison, and metric practices.

The coordinating committees are concerned directly with materials, goods, products, and services. Coordinating committees have been established in the areas of primary materials, industrial products, building and construction, consumer products, and educational and industrial training. Within each coordinating committee, there are additional sector committees.

The sector committees deal with the status of metrication, issues confronting industry, the nature and need for codes and standards, and the problems of applying new

standards and units of measurement. The sector con
Industrial Training Coordinating Committee work with co
educational materials, and consumer education. The secto
divided into task forces. These groups are concerned with major
egories of activities.

## Government Responsibilities in Metrication

The enactment of the Metric Conversion Act means that:

- there is now an official national policy of voluntary changeover to a mod
  (SI) metric system. The International Metric System will require additional u
  of measurement to make SI usable in all sectors of society;

- a conversion time period is necessary to establish the International Metric System
  as the predominant, but not the sole, system of weights and measures;

- a United States Metric Board is established. The board consists of 17 members
  including the chairman who is appointed by the President and approved by the
  Senate.

## The National Bureau of Standards

The responsibilities of the federal government in the task of metrication within
the United States will be fulfilled through the National Bureau of Standards (NBS).
The National Bureau of Standards is expected to:

- provide technical advice to the United States Metric Board and to the nation
  regarding SI and the extension of the system to meet various needs;

- consult with the United States Metric Board (USMB) on the basic principles on
  which metrication should be based;

- supply information to the USMB and to the nation resulting from the National
  Metric Study of the National Bureau of Standards. The goal of these efforts is to
  encourage changes within society to promote the use of SI metric measurements
  while making minimal changes or additions of units to meet the needs within all
  sectors; and

- work toward a coordinated effort among the states to change commercial weights
  and measures to SI metric units. This function is the responsibility of individual
  states and, thus, is controlled by each state.

## BASE TEN SCIENTIFIC NOTATION SYSTEM

The International System of Units (SI) uses a *base ten scientific notation system*
which simplifies the designation of quantities and mathematical processes and reduces
errors in calculations. Part A of table 7-1 (page 58) shows standard prefixes and symbols and the equivalent value of each prefix as a power of ten. The values for the positive powers of ten are expressed by an exponent. For instance $10^2$ indicates a quantitative value of $10 \cdot 10$ or 100; $10^3 = 10 \cdot 10 \cdot 10$ or 1000.

Submultiples of the base units are illustrated in Part B of table 7-1. Each prefix
is identified by a different symbol. The value of each prefix is shown as a negative
power of ten. A negative exponent indicates that the value is less than (1). If there is
no negative sign (-) in front of the exponent, then the exponent is positive.

mittees of the Education and
nversion training programs,
committees are further
training groups or cat-

rnized
nits

| | Multiplication Factor |
|---|---|
| | 10 |
| | 100 |
| | 1 000 |
| | 1 000 000 |
| | 1 000 000 000 |
| | 1 000 000 000 000 |

| | en | Multiplication Factor |
|---|---|---|
| | | 0.1 |
| | | 0.01 |
| milli | m | $10^{-3}$ | 0.001 |
| micro | $\mu$ | $10^{-6}$ | 0.000 001 |
| nano | n | $10^{-9}$ | 0.000 000 001 |
| pico | p | $10^{-12}$ | 0.000 000 000 001 |
| femto | f | $10^{-15}$ | 0.000 000 000 000 001 |
| atto | a | $10^{-18}$ | 0.000 000 000 000 000 001 |

(B)

Table 7-1  SI unit prefixes, symbols, and power of ten
multiple and submultiple values

### Performing Mathematical Processes in the Base Ten System

There are two basic steps in solving multiplication and division problems using values in whole numbers or decimals in the base ten system. First, the whole numbers or decimals are multiplied. Secondly, the exponents are added or subtracted depending on the (+) or (–) value of the exponent. The value of the answer is expressed as a positive or negative quantity.

**Example 1.** Multiply 4.4 to the $10^{-7}$ power by 3.1 to the $10^{-9}$ power.

- Multiply the decimal values $(4.4) \cdot (3.1) = 13.64$
- Add the exponents $(10^{-7}) + (10^{-9}) = 10^{-16}$
  Answer: **13.64 ($10^{-16}$ )**

All scientific problems deal with quantities and units of measurement. Thus, each solution must include an identification of the quantity and unit of measurement.

**Example 2.** Multiply 4.4 to the $10^{-7}$ power by 3.1 to the $10^{9}$ power.

- Multiply the decimal values $(4.4) \cdot (3.1) = 13.64$
- Add the exponents $(10^{-7}) + (10^{9}) = 10^{2}$
  Answer: **13.64 ($10^{2}$ ) or 1364**

For scientific problems involving the division of whole numbers or decimal values which are expressed in powers of ten, the division of the values is completed first. Then the sign of the exponent in the denominator is changed and added to the exponent in the numerator.

**Example 3.** Divide $4.4 (10^3)$ by $2.2 (10^{-6})$

- Divide 4.4 by 2.2 = 2.0

- Change the sign of the exponent in the denominator $(10^{-6})$ to $(10^6)$

- Add the values of the exponents $(10^3) + (10^6) = 10^9$

  Answer: $2 \cdot (10^9)$

  This is a simplified method of writing the value 2 000 000 000.

**Example 4.** Divide $4.4 \cdot 10^3$ by $2.2 \cdot 10^4$

- Divide 4.4 by 2.2 = 2.0

- Change the sign of the exponent $(10^4)$ to $(10^{-4})$ and add it to the numerator $(10^3)$

  $(10^{-4}) + (10^3) = 10^{-1}$

  Answer: $2.0 \cdot 10^{-1} = 0.2$

## GUIDELINES FOR WRITING AND USE OF INTERNATIONAL METRIC UNITS

Basic rules govern the method of expressing values and units in the SI metric system. The following rules promote uniformity in writing and accuracy in interpretation of the units.

- Base units, supplementary units, derived units, and combinations of these units, with multiple and submultiple quantities, are used.

- Insignificant digits and decimals are eliminated.

- Powers of ten are used for computation.

- Prefixes express the order of magnitude of a value. For example, 56.5 km is a precise definition of a specific measurement.

- Prefixes indicating values of 1000 are preferred in the SI metric system. For example: milli $(10^{-3})$, kilo $(10^3)$, and mega $(10^6)$.

- Symbols are capitalized in the SI metric system when they are derived from a proper name: Henry (H), Siemens (S), and Testa (T).

- The numerical prefixes of tera (T), giga (G), and mega (M) are also capitalized.

- The SI metric system symbols are written in singular form when abbreviated. For instance, ten meters is written as 10 m; 16.2 kilograms as 16.2 kg.

- Number values are written in groups of three digits without commas. For example, 2755932.6 Hz is written 2 755 932.6 Hz. Similarly, the value of 8.37250 W is written 8.372 50 W. However, the space is not recommended for four-digit numbers, unless such numbers are grouped in a column with numbers of five digits or more.

- A zero is used before a decimal quantity: 0.2 kg.

- A heavy dot (·) is used in mathematical equations to indicate multiplication. A slash line (/) indicates division.

- Values may be rounded off to a stated degree of accuracy after all calculations are complete. The process of rounding off may be started before this stage if the conversion factors contain more digits in the decimal values than are necessary for the computation.

  **Example 5.** 9.435 V, when rounded off to the nearest 1/100 V, equals 9.44 V. Since the third decimal digit is followed by at least one digit other than zero and the value of this digit is more than 5, the last digit to be retained (hundreds) is increased by 1.

  If the original value is changed to 9.435 0 V and is rounded off to two decimal places, its value is 9.43 V. In this case, the value of the last two digits is not more than 5.

## CONVERTING MEASUREMENTS USING CONVERSION FACTORS AND PROCESSES

The system of measurement in the United States is in a transitional period between the use of the conventional British and metric systems and the use of the SI metric system. During this period, it is necessary to convert values from one system to the other. To do this, two basic arithmetical processes normally are involved, using conversion factors. As stated in unit 6, a *conversion factor* is a numerical value. It represents the degree to which a quantity of a specific measurement in one system is equivalent to a similar quantity measurement in the required system.

  **Example 6.** One meter is equal to 39.370 08 inches. To determine how many meters are contained in 59.055 12 inches, this value is divided by the conversion factor of 39.370 08. The answer is 1 1/2 meters, expressed as 1 1/2 m.

  **Example 7.** If an object is 1 1/2 m long, its equivalent value in inches is found by multiplying 1 1/2 m · 39.370 08 in./m = 59.055 12 in.

Equivalent values are given in many conversion tables to six or more decimal places. In general, ordinary measurements do not require this degree of accuracy. The table values should be rounded off depending on the precision required in the result. The conversion factor used in a computation usually contains one more decimal place than is needed. The result is then rounded off.

────────────────────────── **SUMMARY** ──────────────────────────

- Metrication relates to any program or process of converting to the International System of Units (SI).

- The major task of the National Bureau of Standards, as the federal agency involved in metrication, is to provide

  - Technical assistance to the United States Metric Board, within a conversion time period.

- Consultation services regarding basic principles on which metrication is to be based, and
- Coordinating materials and experiences to the Board which the NBS obtained during the National Metric Study.

- The American National Metric Council (ANMC) was established by the American National Standards Institute (ANSI) to represent the private sector in metrication. The ANMC is to:

  - Promote metrication consistent with the principle of voluntary participation,
  - Assist standards-making bodies, identify needs, evaluate priorities, and recommend standards, and
  - Coordinate standards development efforts with the United States Metric Board.

- All functions within the ANMC are conducted through operations committees and coordinating committees. The coordinating committees, in turn, include sector committees and other task forces.

- The base ten notation system is used in the International Metric System (SI). Multiples (+) and submultiples (-) of the powers of ten simplify the mathematical processes involved in solving scientific problems and simplify the statement of values having large numbers of digits.

- Rules adopted for writing and expressing values and processes in SI cover the base, supplementary, and derived units of measure. Some of the major rules are:

  - insignificant digits are eliminated.

  - prefixes express the order of magnitude of a value.

  - symbols derived from proper names and the numerical prefixes of giga (G), tera (T), and mega (M) are capitalized.

  - number values are placed in groups of three digits and are written in this manner, without commas.

  - values may be rounded off by increasing the last required digit by 1 if the value of the digits beyond this position is more than 5.

- Scientific measurements in one system may be converted to equivalent values in another system by the use of conversion factors and simple mathematical processes.

- Tables usually give conversion factors to six or more decimal places. The degree of accuracy required in the solution determines the actual number of decimal places used in computations and for rounding off the final result.

## ASSIGNMENT UNIT 7  METRICATION AND THE EVOLVING INTERNATIONAL METRIC SYSTEM (SI)

### ORGANIZATION AND FUNCTIONS OF THE GOVERNMENT AND PRIVATE SECTOR IN METRICATION

1. Describe the function of the United States Metric Board.

2. State briefly how the American National Metric Council relates to metrication in this nation.

3. Identify three major functions of the ANMC.

4. List three complementary functions of the national government in metrication that are carried on by the National Bureau of Standards.

5. Select any three of the sector committees of the ANMC coordinating committee.

   a. Name the three sector committees.

   b. State two issues or problems within any of the sectors that must be resolved before metrication can be implemented in that sector.

## BASE TEN SCIENTIFIC NOTATION SYSTEM

1. Four prefixes that are used with SI metric units and positive factors and four symbols used with negative factors are given in the following table.

   a. Give the appropriate symbol, the base ten exponent, and the multiplication factor for items A through D.

   b. Complete the table for items E through H by giving the prefix, base ten exponent, and multiplication factor.

| | Prefix | Symbol | Base Ten Exponent | Multiplication Factor |
|---|---|---|---|---|
| A | giga | | | |
| B | mega | | | |
| C | kilo | | | |
| D | hecto | | | |
| E | | d | | |
| F | | m | | |
| G | | $\mu$ | | |
| H | | p | | |

2. Use the measurement units and exponential factors given in the table for problems A through F.

   a. Solve each problem. Give the computed numerical value and base ten factor.

   b. Restate the answer in terms of its numerical value.

| Problem | Quantity | Solution Quantity and Base Ten Exponent | Numerical Value |
|---|---|---|---|
| A | $5.4 \text{ g} (10^2) + 6.2 \text{ g} (10^2)$ | | |
| B | $5.4 \text{ m} (10^2) \times 6.2 \text{ m} (10^2)$ | | |
| C | $(5.4 \text{ cm}^2 \cdot 10^4) \cdot (0.62 \text{ cm} \cdot 10^{-2})$ | | |
| D | $(5.4 \text{ MV} \cdot 10^{-4}) \cdot (6.2 \text{ MW} \cdot 10^2)$ | | |
| E | $\dfrac{5.4 \text{ kg} \cdot 10^6}{2.7 \text{ kg} \cdot 10^4}$ | | |
| F | $\dfrac{5.4 \text{ Hz} \cdot 10^6}{2.7 \text{ Hz} \cdot 10^{-2}}$ | | |

**Guide for Style and Usage: International Metric System (SI)**

1. Complete the following table as indicated.

   a. Write each computed value (A through E) according to the rules for SI metric measurements.

   b. Round off each of the values (A-E) to the degree of accuracy indicated in each case. State each rounded off value as an SI metric number.

   NOTE: The answer is given for Part A to show what is required.

| | Computed Value | Stated as SI Metric Value | Accuracy (Decimal Places) | SI Metric Numerical Value |
|---|---|---|---|---|
| A | 1004.25 | 1 004.25 | 1 | 1 004.2 |
| B | 100004.2543 | | 2 | |
| C | 1004.249655 | | 1 | |
| D | 1000042.2512722 | | 4 | |
| E | 100004.25672 | | 2 | |

# Unit 8 International System of Units-SI Metrics

The International Metric Convention established the International Bureau of Weights and Measures in 1875. In that same year, a General Conference on Weights and Measures (CGPM) was organized to deal with international issues concerning the metric system. The General Conference continues to be the controlling body of the International Bureau of Weights and Measures. The Bureau preserves metric standards, compares other national standards with the metric standards, and conducts research to establish new standards. The United States representative to the CGPM is the National Bureau of Standards.

Historically, the regulation of weights and measures was recognized in the United States Constitution in 1787. However, it was not until 1866 that the metric system was legalized in this country. Later, in 1893, the meter and kilogram became the fundamental standards of length and mass for both the metric system and the customary United States units of weight and measure. For almost 200 years, the United States has played an active role in developing and establishing the international units of measurement which are so important to agriculture, commerce, industry, science, and to the consumer.

The United States is rapidly moving into the SI metric system through voluntary participation. Since metrication is an evolving process requiring the design, testing, and evaluation of new standards in SI metrics, these standards must be laboratory tested against the engineering, scientific, production, and other standards of the United States. These standards were established as functional values and continue to provide high factors of safety with excellent productivity (efficiency/effectiveness of input/output).

## CGPM DEVELOPMENTS LEADING TO THE INTERNATIONAL SYSTEM OF UNITS (SI)

In the years since its organization, the CGPM has tried to provide a coherent set of metric measurement units. The original metric system of the 1790s contained two fundamental units: the meter as a unit of length and the kilogram as a unit of mass. These units were adequate for the measurement of length, area, volume, capacity, and mass. A unit of time (second) was added in 1881 and the *centimeter-gram-second* (cgs) system was formed.

Scientific, industrial, and commercial progress made it necessary to develop other fundamental and derived units and other systems. Around 1900, it became practical to base measurements on the units of meters-kilograms-seconds; thus, the *MKS* system was formed. Experimentation in the field of electromagnetism led to the adoption of the ampere as the fourth base unit. In 1950 CGPM accepted the ampere unit and the *MKSA* system was established.

At the time of the tenth CGPM in 1954, two more base units were added to the four MKSA units. The *degree Kelvin* (°K) was recognized as the unit of temperature and the candela (cd) as the unit of luminous intensity. The twelfth CGPM (1964) and thirteenth CGPM (1967) brought other refinements: the second was redefined; the unit of temperature was renamed Kelvin (K); and the definition of candela was revised.

The seventh base unit, the *mole* (mol) was added during the fourteenth CGPM in 1971. In addition, the 1971 conference accepted the following special terms: the *pascal* as a unit of pressure and stress, the *newton*, and the *siemens* (S) as a unit of electrical conductance. The years since 1971 have brought worldwide acceptance of SI for the many advantages the system provides.

## ADVANTAGES OF SI METRICS

SI is recognized in all languages as the abbreviation of the International System of Units. The full title and abbreviation were accepted in 1960 by 36 countries, including the United States. Some of the more important advantages of SI are as follows:

- SI provides only one base unit for each physical quantity: meter (length), kilogram (mass), second (time), Kelvin (temperature), ampere (electrical current), candela (luminous intensity), and mole (substance of a system). Additional units are derived from these base units. The derived units are defined by simple equations.

- SI provides a unique and well-defined set of symbols and abbreviations, each of which relates to a specific phenomenon or condition.

- SI retains the decimal relationship between multiples (+) and submultiples (–) of the base unit for each physical quantity. Prefixes are used to simplify long multiple and submultiple values.

- SI is a *coherent system*. It has seven base units with established names, symbols, and precise definitions. Coherence also means that a product or quotient of any two quantities is a unit of the resulting quantity. For example, a unit area is the product of a unit of length and another unit of length. In a coherent system where the meter is the unit of length, the square meter is the unit of area.

- The base units of SI are accurately defined in terms of physical measurements. With the exception of the kilogram mass that is preserved at the International Bureau of Weights and Measures, the measurement for each base unit can be produced in laboratories.

- Other units that are not a part of SI, but are associated with it, are related to the base units of the system by powers of ten.

## SEVEN BASE UNITS OF SI METRICS

The term *SI metrics* is used in this text as a modification of the accepted SI abbreviation of the International System of Units. Although there is variation among the nations on the spelling of scientific terms, the spelling of the SI terms in this and subsequent units follows the common usage in the United States.

There are seven *base units,* two *supplementary units,* and other additional units that are *derived* from the base units in SI metrics.

### Unit of Length

The unit of length in SI metrics is the meter (m). A *meter* is defined as a length that is equal to 1 650 763.73 wavelengths, in a vacuum, of the radiation corresponding to the transition of the krypton$_{86}$ atom between the $(2p_{10})$ energy level and the $(5d_5)$ energy level. A definition of this type allows laboratory technologists and scientists anywhere in the world to duplicate the meter with extreme accuracy.

## Unit of Mass

The unit of mass is the kilogram (kg). The *kilogram* is equal to 2.2 pounds. Thus, the mass of a body weighing 220 pounds is 100 kilograms. Another body weighing 200 pounds has a mass of 90.909 kg.

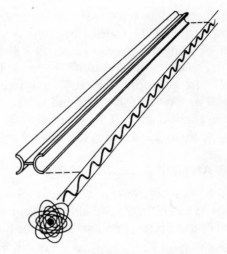

Fig. 8-1

## Unit of Time

The unit of time is the second (s). A *second* is defined as the duration of 9 192 631 770 periods of radiation of the cesium$_{133}$ atom corresponding to the transition between the two hyperfine levels of the ground state.

The duration, or time interval of one second, is still 1/86 400 of the mean solar day. The advantage of the new definition is that it permits a standard to be set. Since the second is based on two energy levels of an atom, the measurement may be reproduced in any properly equipped laboratory.

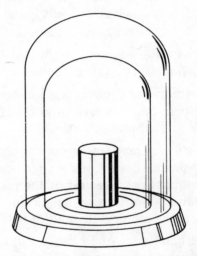

Fig. 8-2 U.S. kilogram mass prototype #20

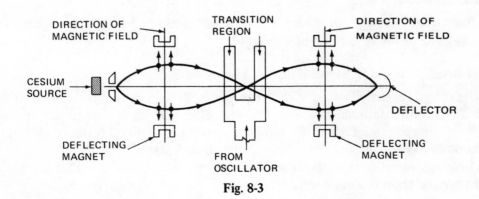

Fig. 8-3

## Unit of Electrical Current

The SI unit of electrical current is the ampere (A). The ampere is defined as ". . . that constant which, if maintained in two straight parallel conductors of indefinite length (of negligible cross section) and placed one meter apart in a vacuum, will produce a force between these conductors equal to 2 x 10⁻⁷ newton/meter of length."

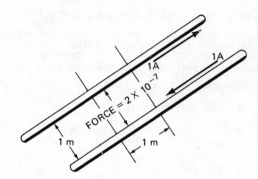

Fig. 8-4

### Unit of Temperature

The unit of thermodynamic temperature in SI is the Kelvin (K). The *Kelvin* is equal to 1/273.16th of the thermodynamic temperature of the triple point of water. On the Kelvin scale, water boils at 373.15 K degrees and freezes at 273.15 K degrees.

A more practical and commonly used unit of temperature measurement is the *Celsius degree* ($C°$). Water freezes at 0° and boils at 100° on the Celsius scale (formerly called the centigrade scale). To convert from a °C to a Kelvin degree, add 273.15 to the Celsius measurement: $K = °C + 273.15$.

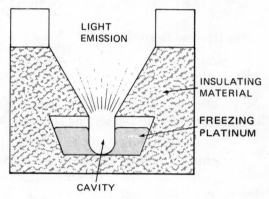

Fig. 8-5

### Unit of Luminous Intensity

The unit of luminous intensity is the candela (cd). The *candela* represents the luminous intensity (in the perpendicular direction) of a surface 1/600 000th of a square meter of a blackbody at the temperature of freezing platinum under a pressure of 101 325 newtons per square meter.

### Unit of Substance of a System

The seventh base unit in SI metrics is the mole (mol). The *mole* represents the amount of substance of a system which contains as many elementary entities as there are atoms in 0.012 kilogram of carbon-12. Elementary entities such as atoms, molecules, ions, electrons, and other particles must be specified when the mole is used.

## SUPPLEMENTARY UNITS OF ANGULAR MEASURE

SI Metrics has two supplementary units which deal with two types of angles: plane and solid. The *plane angle* or *radian* (rad), figure 8-6, is the unit of measure of an angle which has its vertex at the center of a circle and is subtended by an arc that is equal in length to the radius.

$$1 \text{ rad} = 180°/\pi = 57.259\ 78°$$

The *solid angle* or steradian (sr), figure 8-7 (page 68), has its vertex at the center of a sphere and encloses an area of the spherical surface. This area is equal to that of a square whose sides are equal in length to the radius of the sphere.

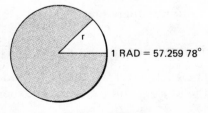

Fig. 8-6 Plane angle

## DERIVED UNITS OF MEASURE

### Derived Area and Volume Units

The following units of area are derived from the meter as a base unit.

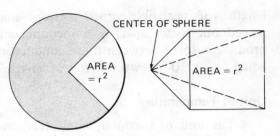

Fig. 8-7 Solid angle

- The square meter ($m^2$)

- The *hectare* (h) which is equal to 10 000 $m^2$

- The cubic meter ($m^3$) which is used in volume and capacity measurements

- The liter which equals 1000 $cm^3$. The liter is an acceptable unit of volume measure in SI because it can be defined in terms of an SI base unit.

### Units Derived from the Unit of Mass

Units of force, work (energy), power, and pressure are all derived from the base unit of mass.

*Force*. *Force* is measured in newtons (N). A *newton* is equal to (mass) · (acceleration) or kilogram · meter per second squared (N = kg · $m/s^2$).

> **A force of one newton applied to a one-kilogram mass produces an acceleration of one meter per second squared.**

*Work*. *Work* is measured in joules (J).

J = force · distance

= newtons · meters

= (N) · (m)

= N · m

> **One joule of work is produced when one newton is applied through a distance of one meter.**

*Power*. *Power* is measured in watts (W). A *watt* is equal to one joule per second or newtons (N) · meters per second (m/s). W = J/s = N · m/s.

> **One watt of power represents one joule of work completed in one second.**

*Pressure*. *Pressure* is measured in pascals (Pa). A *pascal* is equal to one newton per meter squared (Pa = $N/m^2$).

> **One pascal is equal to the force of one newton applied to an area of one square meter.**

### Units Derived from the Base Unit of Time (Second)

The hertz (Hz), velocity (v), and acceleration (a) are three units that are derived from the SI base unit of time, the second.

*Frequency*. The *hertz* (Hz) measures frequency or the number of cycles per second.

*Velocity*.  *Velocity* (v) refers to distance divided by time or meters per second.

*Acceleration*.  *Acceleration* (a) is the derived unit which expresses the rate of change in velocity. Acceleration is measured in meters per second per second or meters per second squared ($m/s^2$).

## Units Derived from the Base Electrical Unit (Ampere)

|  | Measurement | Derived Unit of Measurement | Formula |
|---|---|---|---|
| Base Electrical Unit: Ampere (A) | Electrical potential | volt (V) | $V = W/A$ |
|  | Electrical resistance | ohm ($\Omega$) | $\Omega = V/A$ |
|  | Electrical capacitance | farad (F) | $F = A \cdot (s/V)$ |
|  | Quantity of electricity | coulomb (C) | $C = A \cdot s$ |
|  | Electrical inductance | henry (H) | $H = Wb/A$ |
|  | Magnetic flux | weber (Wb) | $Wb = V \cdot s$ |

Table 8-1

## Units Derived from the Base Unit of Luminous Intensity

|  | Derived Units | Characteristics | Formula |
|---|---|---|---|
| Base Unit of Luminous Intensity: candela (cd) | Illumination, lux (lx) | Luminous intensity given by a luminous flux of one lumen per square meter. | $lx = lm/m^2$ |
|  | Luminous flux, lumen (lm) | Luminous flux emitted in a solid angle of one steradian by a point source with a uniform intensity of one. | $lm = cd \cdot sr$ |
| NOTE: 1 watt of electrical power = 17 lumens | | | |

Table 8-2

| Quantity | Base Units | Symbol |
|---|---|---|
| length (l) | meter | m |
| mass (m) | kilogram | kg |
| time (t) | second | s |
| electric current (I) | ampere | A |
| thermodynamic temperature (T) | kelvin | K |
| luminous intensity (I) | candela | cd |
| substance | mole | mol |
|  | **Supplementary Units** | **Symbol** |
| plane angle | radian | rad |
| solid angle | steradian | sr |

Table 8-3  Common SI metric units, symbols, and formulas

| Quantity | Derived Units (Selected) | Formula |
|---|---|---|
| acceleration (a) | meter per second squared | $m/s^2$ |
| area (A) | square meter | $m^2$ |
| density (D) | kilogram per cubic meter | $kg/m^3$ |
| electric capacitance (C) | farad (F) | $A \cdot s/V$ |
| electric charge (Q) (quantity of electricity) | coulomb (C) | $A \cdot s$ |
| Energy (E) | joule (J) | $N \cdot m$ |
| force (F) | newton (N) | $kg \cdot m/s^2$ |
| frequency (f) | hertz (Hz) | $(cycle)/s$ |
| magnetic flux density (B) | tesla (T) | $Wb/m^2$ |
| magnetomotive force | ampere (A) | |
| power (P) | watt (W) | $J/s$ |
| pressure (p) | newton per square meter | $N/m^2$ |
| thermal conductivity | watt per meter · kelvin | $W/m \cdot K$ |
| velocity (v) | meter per second | $m/s$ |
| work (W) | joule (J) | $N \cdot m$ |

Table 8-4  Common SI metric units, symbols, and formulas:  Derived Units

## CONVERSION FACTORS IN METRICATION

Tables of conversion factors are necessary and helpful when converting values between different measurement systems. Computations using these conversion factors will be required until all nations use a single acceptable system of measurement. A number of selected tables are included in this unit. These tables help to convert measurements from the customary system to SI metrics and vice versa.

The conversion factor tables contain up to eight place decimal values. As stated previously, the number of places used and the rounding off process for decimal values are determined by the degree of accuracy required.

The meter is usually given as the standard of length. Smaller and larger units of measure may be used by reviewing the relationships of values in the decimal system and the value assigned to the prefixes.

The reverse process of conversion may be performed by using the reciprocal of any multiplier as the divisor. For example, to convert from centimeters to inches, multiply the number of centimeters by 0.4 (rounded to one decimal place). Thus, 2.5 cm (approximately) = 1 in. If 2.5 is used as the reciprocal, then 1 cm = 1/2.5 in. or 0.4 in. (approximately).

### Conversion Processes and Conversion Factors

Conversions may be made from decimal multiple and submultiple values of SI units. The decimal point is moved according to the prefix of the unit. For example, for the base unit of length (the meter), submultiple decimal units such as milli (0.001) and centi (0.01), and multiple units such as kilo (1000), may be used to solve common problems. Thus, in place of the value 1 m = 39.37 in., the conversion factor in a table may be changed to 1 mm = 0.03937 in., 1 cm = 0.3937 in., or 1 km = 39 370 in. The way the conversion factor is expressed depends on the prefix used when a numerical value is stated.

Two simplified conversion tables, tables 8-5 and 8-6, illustrate the type of information that is available. Each table shows metric and customary units and values that are used in everyday problems involving measurements of length and mass. Additional conversion tables for other measurements such as area, capacity, or volume appear in the Appendix. Still other tables are available from professional engineering and scientific societies, handbooks, industrial literature, governmental agencies involved in weights and measures, and private technical publications.

| Metric Units | | |
|---|---|---|
| Conversion | | Conversion Factor (Multiply by) |
| From → To | | |
| meters (m) | inches | 39.370 08 |
| | feet | 3.280 840 |
| | yards | 1.093 613 |
| | miles | 0.000 621 37 |

| Customary Units | | |
|---|---|---|
| Conversion | | Conversion Factor (Multiply by) |
| From → To | | |
| miles | inches | 63 360 |
| | centimeters | 160 934.4 |
| | meters | 1 609.344 |
| yards | meters | 0.914 4 |
| | kilometers | 0.000 914 4 |
| feet | meters | 0.304 8 |
| | kilometers | 0.000 304 8 |
| inches | centimeters | 2.54 |

Table 8-5  Units of length

| Metric Units | | |
|---|---|---|
| Conversion | | Conversion Factor (Multiply by) |
| From → To | | |
| kilograms (kg) | grains | 15 432.36 |
| | avoirdupois drams | 564.383 4 |
| | avoirdupois ounces | 35.273 96 |
| | avoirdupois pounds | 2.204 623 |
| | Troy ounces | 32.150 75 |
| | Troy pounds | 2.679 229 |

| Customary Units | | |
|---|---|---|
| Conversion | | Conversion Factor (Multiply by) |
| From → To | | |
| avoirdupois pounds (lb) | kilograms | 0.453 592 37 |
| | metric tons | 0.000 453 59 |
| | grains | 7 000 |
| | avoirdupois drams | 256 |
| | avoirdupois ounces | 16 |
| | Troy ounces | 14.583 33 |
| | Troy pounds | 1.215 28 |

Table 8-6  Units of mass

Another form of table gives conversion factors for measurements arranged according to major categories. Table 8-7 is an abridged table which shows the multiplication factors by which a customary measurement may be either changed or converted to an equivalent metric unit.

## Metric Conversion Factors for Customary Measurement Units

| Category | Conversion From → To | | Multiply by |
|---|---|---|---|
| acceleration | ft./s$^2$ | m/s$^2$ | 0.304 8* |
| | in./s$^2$ | m/s$^2$ | 2.540 0 x 10$^{-2}$* |
| area | ft.$^2$ | m$^2$ | 9.290 3 x 10$^{-2}$* |
| | in.$^2$ | m$^2$ | 6.451 6 x 10$^{-4}$* |
| density | g/cm$^3$ | kg/m$^3$ | 1.000 0 x 10$^3$* |
| | lb. (mass)/ft.$^3$ | kg/m$^3$ | 16.018 5 |
| | lb. (mass)/in.$^3$ | kg/m$^3$ | 2.768 0 x 10$^4$ |
| energy | Btu (thermochemical) | J | 1.054 3 x 10$^3$ |
| | cal (thermochemical) | J | 4.184 0* |
| | eV | J | 1.602 1 x 10$^{-19}$ |
| | erg | J | 1.000 0 x 10$^{-7}$* |
| | ft. • lb. (force) | J | 1.355 8 |
| | kWh | J | 3.600 0 x 10$^6$* |
| | Wh | J | 3.600 0 x 10$^3$* |
| flow, liquid and solid | ft.$^3$/min. | m$^3$/s | 4.719 5 x 10$^{-4}$ |
| | ft.$^3$/s | m$^3$/s | 2.831 7 x 10$^{-2}$ |
| | in.$^3$/min. | m$^3$/s | 2.731 2 x 10$^{-7}$ |
| | lb. (mass)/s | kg/s | 0.453 6 |
| | lb. (mass)/min. | kg/s | 7.559 9 x 10$^{-3}$ |
| | tons (short, mass)/h | kg/s | 0.252 0 |
| force | dyne | N | 1.000 0 x 10$^{-5}$* |
| | kg (force) | N | 9.806 6 |
| | lb. (force) | N | 4.448 2 |
| heat | Btu (thermochemical)/ft.$^2$ | J/m$^2$ | 1.134 9 x 10$^4$ |
| | cal (thermochemical)/cm$^2$ | J/m$^2$ | 4.184 0 x 10$^4$* |
| | ft.$^2$/h | m$^2$/s | 2.580 6 x 10$^{-5}$ |
| length | A | m | 1.000 0 x 10$^{-10}$* |
| | ft. | m | 0.304 8* |
| | in. | m | 2.540 0 x 10$^{-2}$* |
| | $\mu$ (micron) | m | 1.000 0 x 10$^{-6}$* |
| | mil | m | 2.540 0 x 10$^{-5}$* |

*Exact

Fig. 8-7 (cont'd., page 73)

| Category | Conversion | | Multiply by |
|---|---|---|---|
| | From ⟶ | To | |
| mass | lb. (mass, avoirdupois) | kg | 0.453 6 |
| | oz. (mass, avoirdupois) | kg | $2.835\ 0 \times 10^{-2}$ |
| | ton, long = 2 240 lb. (mass) | kg | $1.016\ 0 \times 10^3$ |
| | ton, metric | kg | $1.000\ 0 \times 10^3$* |
| | ton, short = 2 000 lb. (mass) | kg | $0.907\ 2 \times 10^3$ |
| power | Btu (thermochemical)/min. | W | 17.572 5 |
| | cal (thermochemical)/min. | W | $6.973\ 3 \times 10^{-2}$ |
| | erg/s | W | $1.000\ 0 \times 10^{-7}$* |
| | ft. · lb. (force)/min. | W | $2.259\ 7 \times 10^{-2}$ |
| | h.p. (550 ft. · lb./s) | W | $7.457\ 0 \times 10^2$ |
| pressure (stress) | atm (760 torr) | $N/m^2$ | $1.013\ 2 \times 10^5$ |
| | dyne/cm$^2$ | $N/m^2$ | 0.100 0* |
| | g (force)/cm$^2$ | $N/m^2$ | 98.066 6* |
| | kg (force)/cm$^2$ | $N/m^2$ | $9.806\ 6 \times 10^4$ |
| | lb. (force)/in.$^2$ (or psi) | $N/m^2$ | $6.894\ 8 \times 10^3$ |
| | | kg (force)/mm$^2$ | $7.030\ 7 \times 10^{-4}$ |
| | torr (mm Hg at 0 degrees C) | $N/m^2$ | $1.333\ 2 \times 10^2$ |
| velocity | ft./min. | m/s | $5.080\ 0 \times 10^{-3}$* |
| | in./s | m/s | $2.540\ 0 \times 10^{-2}$* |
| | mph | m/s | 0.447 0 |
| | | km/h | 1.609 3 |
| volume | ft.$^3$ | m$^3$ | $2.831\ 7 \times 10^{-2}$ |
| | in.$^3$ | m$^3$ | $1.638\ 7 \times 10^{-5}$ |
| | liter | m$^3$ | $1.000\ 0 \times 10^{-3}$* |
| temperature | degree C | K | $t_K = t_C + 273.15$ |
| | degree F | K | $t_K = (t_F + 459.67)/1.8$ |

*Exact.

Table 8-7 Metric conversion factors for customary measurement units

## SUMMARY

- The International Bureau of Weights and Measures provides worldwide services in preserving metric standards, comparing national standards with metric standards, and conducting research on measurement.

- The seven base units of measurement currently recognized were developed cooperatively by the participants of the General Conference of Weights and Measures (CGPM).

  - The two fundamental units in the original metric system for the measurement of length, area, volume, capacity, and mass were the *meter* and *kilogram*.

    Historically, CGPM added the *second* in 1881 and formed the cgs system.

  - By 1900, the units of meters-kilograms-seconds provided the foundation for the MKS system of measurement.

  - The fourth base unit, the *ampere,* was accepted by CGPM in 1950.

  - The MKSA system of measurement was adopted at this time.

  - The units of *degree Kelvin* and *candela* were accepted in 1954.

  - The seventh base unit, the *mole,* was added during the 14th CGPM in 1971.

  - Throughout its 100-year history, CGPM has worked continuously to introduce, redefine, and rename more precisely SI metric units and the applications of these units to meet changing needs.

- SI metrics has the following advantages:

  - Only one unit is used for each physical quantity.

  - A standardized and defined set of symbols and abbreviations means that uniform communication and interpretation are possible.

  - It is a simplified system of computations involving powers of ten, multiples and submultiples of the base units, prefixes, and conversion factors.

  - It is a coherent system which yields accurate results.

  - Essential derived units make the system flexible to meet consumer and producer needs.

- The seven base units in SI metrics are as follows:

  - meter (m), the unit of length,

  - kilogram (kg), the unit of mass,

  - second (s), the unit of time,

  - ampere (A), the unit of electrical current,

  - degree Kelvin (K), the unit of thermodynamic temperature,

  - candela (cd), the unit of luminous intensity,

  - mole (mol), the unit of substance of a system.

- Plane and solid angles are represented in SI metrics by two supplementary units: the radian (rad) and the steradian (sr).

- Derived units represent other acceptable standards of measurement. These additional units are derived from the base and supplementary units.

  - For example, the derived units of force, work, power, and pressure relate to the base unit of mass; the derived units of potential, resistance, capacitance, inductance, quantity, and magnetic flux relate to the base electrical unit, the ampere. Derived units are applied daily in practical applications.

- SI metrics requires a working knowledge of the powers of ten (scientific notation system); prefixes and values; base, supplementary, and derived units of measurement; the expression of quantities and values in terms of technical standards for usage; and the application of conversion factors.

  — Tables of conversion factors list the numerical values to be used to convert a unit of measurement from a customary unit to SI metrics, and vice versa.

  — The number of decimal digits to be used in computations and for rounding off values is determined by the required degree of accuracy.

## ASSIGNMENT UNIT 8  INTERNATIONAL SYSTEM OF UNITS — SI METRICS

### HISTORICAL DEVELOPMENT OF SI METRICS

1. List six events or developments in the history of science that led to the evolution of SI metrics from the original fundamental units of the meter and gram.

### ADVANTAGES OF SI (SI METRICS)

1. List three distinct advantages of SI metrics over other metric systems or the British measurement system.

2. Briefly state one economic advantage and one scientific advantage of the process of metrication in the United States.

### BASE, SUPPLEMENTARY, AND DERIVED SI METRIC UNITS

1. Complete the following table for the seven base SI metric units.

   a. List the seven basic quantities of SI metrics.

   b. Identify the SI base unit for each quantity.

   c. Give the symbol for each SI base unit.

|   | Quantity (a) | Base Unit (b) | Symbol (c) |
|---|---|---|---|
| 1 |   |   |   |
| 2 |   |   |   |
| 3 |   |   |   |
| 4 |   |   |   |
| 5 |   |   |   |
| 6 |   |   |   |
| 7 |   |   |   |

2. State briefly the differences between:

   a. Base units and supplementary units in SI metrics.

   b. Derived units and base units in SI metrics.

3. a. Select and name one base unit.

   b. List three quantities that must be measured that relate to the base unit.

   c. State the derived unit of measurement for each of the three quantities.

   d. Give the formula for finding the value of each derived unit.

## CONVERSION FACTORS AND PROCESSES

Refer to a Table of Conversion Factors for both SI metric and customary units for the quantities of categories A through F in the following table. For problems 1 through 23, either the SI metric unit values or the customary unit values are given.

1. Indicate the conversion for each problem.

2. Compute the value of each measurement and round off each answer to the number of significant places indicated. Record and properly designate the unit in which the answer is given.

| Category | | | Measurement | | Conversion Factor (1) | Computed Measurement | |
|---|---|---|---|---|---|---|---|
| | | | Given Unit Value | Required Unit Value | | Significant Places | Rounded Off Value and Unit Designation (2) |
| A | Length | 1 | 5 in. | millimeters | | 2 | |
| | | 2 | 27 yd. | meters | | 3 | |
| | | 3 | 16 ft. 3 in. (16 1/4') | meters | | 1 | |
| | | 4 | 184 km | miles (statute) | | 2 | |
| | | 5 | 232.5 mm | inches | | 3 | |
| B | Mass | 6 | 100 kg | pounds/mass (avoir-dupois) | | 2 | |
| | | 7 | 100 kg | pounds/mass (apothe-cary-Troy) | | 3 | |
| | | 8 | 20 000 grains | kilograms | | 3 | |
| | | 9 | 62.4 lb. • m (Troy) | kilograms | | 2 | |
| | Mass Area | 10 | 17 400 kg/m$^2$ | ounce • mass/yard$^2$ | | 2 | |
| | Mass Time | 11 | 12.66 kg/s | pounds • mass/minute | | 2 | |
| | Mass Volume | 12 | 24 lb. • m/ft.$^3$ | kilogram/meter$^3$ | | 3 | |
| | | 13 | 2 156.9 kg/m$^3$ | pound • mass/gallon (US) | | 1 | |
| C | Pres-sure | 14 | 20 psi | pascal | | 1 | |
| | | 15 | 32 Pa | kilogram (force)/meter$^2$ | | 3 | |
| D | Power | 16 | 100 Btu/sec. | kilowatts | | 2 | |
| | | 17 | 20 000 watt | horsepower (metric) | | 1 | |
| | | 18 | 118 kc/min. | watts | | 2 | |
| E | Tem-pera-ture | 19 | 110 degrees C | degree Kelvin | | 2 | |
| | | 20 | 246.15 degrees K | degree Celsius | | 2 | |
| | | 21 | 104 degrees F | degree Celsius | | nearest whole number | |
| F | Veloc-ity | 22 | 80 km/hr. | miles per hour (mph) | | | |
| | | 23 | 80 mi./hr. | kilometer/hour | | | |

# Achievement Review
## SYSTEMS OF MEASUREMENT IN SCIENCE

### CUSTOMARY BRITISH AND METRIC MEASUREMENT SYSTEMS

Complete statements 1 to 10 by selecting the correct word(s) or value.

1. The (yardstick) (steel rule) (meter stick) is the most accurate device for taking linear measurements to an accuracy of 1/64 or 1/100 of an inch.

2. Measurements to accuracies of 0.0001 in. may be made with a (steel tape) (meter stick) (micrometer).

3. Precision laboratory measurements may be made within practical working limits of (0.1 in.) (0.000 002 in.) (0.000 000 002 in.).

4. The system of measurement commonly used by industry and technologists worldwide is the (English) (SI metric) system.

5. The square inch is (larger than) (smaller than) (the same size as) a square foot.

6. A cubic meter is (larger than) (smaller than) (the same size as) a cubic foot.

7. A liter is (larger than) (smaller than) (the same size as) a quart.

8. A cubic yard is (larger than) (smaller than) (the same size as) a cubic meter.

9. The volume of irregularly shaped objects is usually (computed) (located in tables) (found by the displacement method).

10. One thousand grams equal (one metric ton) (2.2 pounds, approximately) (20 centigrams).

11. Compute the cubical volume of cubes A and B, rectangular containers C and D, and cylinders E and F (English measure). Use $\pi = 22/7$.

|  | A | B | C | D |  | E | F |
|--------|------|--------|------|-----------|----------|------|--------|
| Length | 5″ | 1′ – 6″ | 10″ | 1 2/3 yd. | Diameter |  | 3′ – 6″ |
| Depth | 5″ | 1′ – 6″ | 5″ | 2′ – 4″ | Radius | 7″ |  |
| Height | 5″ | 1′ – 6″ | 2 1/2″ | 1′ | Length | 10″ | 36 yd. |

12. Compute the volumes (in the metric system) of rectangular solids A and B, and cylindrical solids C and D. Round off each answer to the nearest whole number.

|  | A | B |  | C | D |
|--------|-------|-------------------------|----------|-------|----------------------|
| Length | 2 m | 8 m | Diameter | 14 mm |  |
| Depth | 4.5 m | 10 dm 50 cm (10.5 dm) | Radius |  | 1 m 5 dm (1.5 m) |
| Height | 6.2 m | 1 m 20 cm (1.02 m) | Length | 25 mm | 2 m 6 dm 6 cm (2.66 m) |

13. Determine the volume in liters and the weight of water in kilograms required to fill containers A, B, and C to the height indicated. Use $\pi = 22/7$ and round off answers to the nearest whole number.

|  | Shape of Container | Inside Dimensions | Liquid Height |
|---|--------------------|----------------------|-----------------|
| A | Square | 10 cm | 5 cm |
| B | Rectangular | 10 m x 5 dm | 2 dm 5 cm (2.5 dm) |
| C | Round | 2 m 8 dm (2.8 m) diameter | 1 m |

14. Using a conversion table with BGS and MKSA measurement units, calculate equivalent values A through H in either the BGS or the MKSA systems, as required.

|  |  | Quantity | Measurement System | |
|---|---|---|---|---|
|  |  |  | BGS | MKSA |
| Length | A | 2 1/2 ft. |  | m |
|  |  |  |  | cm |
|  | B | 64 in. |  | m |
|  |  |  |  | cm |
|  | C | 16.4 miles | ft. | km |
| Volume | D | 169.8 liters | cu. ft. / gal. |  |
|  | E | 4.5 cu. m | ft.$^3$ | liters |
|  | F | 1 000 gal. | ft.$^3$ | m$^3$ |
| Mass | G | 1 500 kg | slugs |  |
|  | H | 29.2 slugs |  | kg |

## METRICATION AND THE EVOLVING INTERNATIONAL METRIC SYSTEM (SI)

1. State two advantages and two disadvantages to industries resulting from the conversion of products and processes from the British to the SI metric system of measurement.

2. For quantities A – D in the following table:

   a. State each quantity in terms of its base ten value.

   b. Compute the numerical value (expressed in terms of the quantity and the exponent).

   c. Restate the answer as a numerical value.

|  | | Solution | | |
|---|---|---|---|---|
| Quantity and Process | | Base Ten Value (a) | Quantities and Exponent (b) | Numerical Value (c) |
| A | 20 000 Hz · 17 900 Hz |  |  |  |
| B | 108 000 kV · 12 600 kV |  |  |  |
| C | $\dfrac{10\,800\ kg}{27\ kg}$ |  |  |  |
| D | $\dfrac{2\,160\ kg}{2.7(10^{-2})\ kg}$ |  |  |  |

## INTERNATIONAL SYSTEM OF UNITS (SI METRICS)

Secure a table of conversion factors for both SI metric units and customary units of measurement for length, mass, pressure, power, temperature, and velocity.

1. Insert the conversion factors required to change from the given value to the required unit value for problems 1 – 9 in the following table.

2. Compute the value of each measurement and round off the value to the stated number of significant places. Record the value and unit designation in the following table.

| Measurement Category | | | Measurement | | Computed Measurement | | |
|---|---|---|---|---|---|---|---|
| | | | Given Value | Required Unit Value | Conversion Factor | Required Significant Places | Rounded-off Value and Units |
| A | Length | 1 | 32 ft. 6 in. (32 1/2′) | meters | | 2 | |
| A | Length | 2 | 9 224 km | miles (statute) | | whole number | |
| B | Mass | 3 | 220 kg | pounds/mass (avoirdupois) | | 2 | |
| B | Mass | 4 | 486 lb. (mass) (Troy) | kilograms | | 2 | |
| B | Mass Time | 5 | 18.99 kg/s | lb. mass/min. | | 3 | |
| C | Pressure | 6 | 56 Pa | kg (force)/m² | | 2 | |
| D | Power | 7 | 177 kc/min. | watts | | 1 | |
| E | Temperature | 8 | 216.72°K | °Celsius | | 2 | |
| F | Velocity | 9 | 62 km/hr. | m/sec. | | 1 | |

# Section 3
# Mechanics, Machines, and Wave Motion

## Unit 9  Forces and Their Effects

A study of any machine or mechanism shows that each is made up of a number of movable parts. These parts transform a given motion to a desired motion. In other words, these machines perform work. *Work* is done when motion results from the application of force. Thus, a study of mechanics, machines, and wave motion deals with forces and the effects of forces on bodies.

A *force* is a push or pull. The effect of a force either changes the shape or motion of a body or prevents other forces from making such changes. Every force produces a stress in the part on which it is applied. Forces may be produced by an individual using muscular action or by machines with mechanical motion.

For example, when a person cuts a board, a downward and a forward push is applied. It is this muscular force which causes the saw teeth to cut. As another example, the jaws of a drill chuck may be tightened with a chuck wrench by applying a force by hand.

### KINDS OF FORCE:  TENSION, COMPRESSION, TORQUE, AND SHEAR

Forces are produced by physical or chemical change, gravity, or changes in motion. When a force is applied which tends to stretch an object, it is called a *tensile force*. A part experiencing a tensile force is said to be in *tension,* figure 9-1.

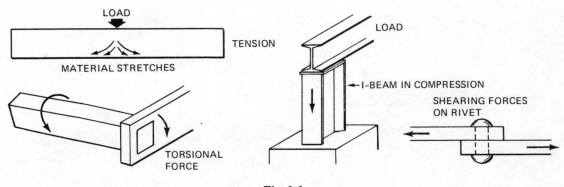

Fig. 9-1

A force can also be applied which tends to shorten or squeeze the object. Such a force is a *compressive force*. A part experiencing a compressive force is in *compression*, figure 9-1.

A third force is known as a *torsional force*, or a *torque* since it tends to twist an object. Still another kind of force, which seems to make the layers or molecules of a material slide or slip on one another, is a *shearing* force.

Each of these forces may act independently or in combination. For example, a downward force applied on a vertical steel beam tends to compress the beam. If this beam is placed in a horizontal position and a load is applied in the middle, the bottom of the beam tends to stretch and is in tension. At the same time, the top area is being pushed together in compression. If the compressive and tensile forces are great enough to make the layers of the material slide upon each other, a shearing force results.

The turning of a part in a metal lathe is another example of several forces in action, figure 9-2. As the work revolves and the cutting tool moves into the work, the wedging action of the cutting edge produces a shear force. This force causes the metal to seem to flow off the work in the form of chips. If this workpiece is held between the centers of the lathe, the centers exert a compressive force against the work.

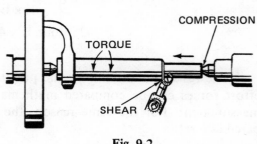

Fig. 9-2

The lathe dog which drives the work tends to produce a shearing force. The pressure of the cutting tool against the work produces tension and compression, as well as a shearing action.

## EFFORT FORCES AND RESISTANCE FORCES

An *effort force* refers to the force applied to a part. The resistance of the part to this applied force is the *resistance force*. For a 100% efficient machine, it can be shown that the effort force multiplied by the distance through which it acts is equal to the resistance force multiplied by its distance, at equilibrium, figure 9-3.

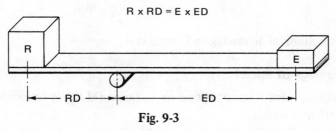

Fig. 9-3

This relationship is the basis for comparing effort (E), effort distance (ED), resistance (R), and resistance distance (RD). Such a comparison can be stated mathematically as an equation.

Effort (E) x Effort Distance (ED) =  Resistance (R) x Resistance Distance (RD)

$$\text{Thus, } (E) \times (ED) = (R) \times (RD)$$

$$\text{or, } \frac{E}{R} = \frac{RD}{ED}$$

The effort torque is E x ED and the resistance torque is R x RD. It can be seen that *torque* is the product of a force times the distance through which the force acts. When the effort is expressed in pounds of pull, and the distance through which it acts is given in inches, then the effort-torque is in pound-inches (the product of pounds x inches). Note that the effort and resistance forces are measured in like units (pounds); similarly, the distances are also measured in like units (inches). The distances may be given in feet as well.

The formula E x ED = R x RD can be used to find any one missing value when the three remaining values are given. For example, if a force of 25 pounds (E) is applied to an open-end wrench and the effort force operates through a distance of four inches (ED), the resistance offered by a square metal plug moving a distance of one inch is:

$$E \times ED = R \times RD$$
$$25 \text{ pounds} \times 4'' = R \times 1''$$
$$R = 100 \text{ pounds}$$

The resistance offered by the plug to the force is 100 pounds. The resistance and effort forces can be compared mathematically because they are in the same unit of measurement. For the same reason, the resistance and effort distances may be compared as a ratio.

## MECHANICAL ADVANTAGE OF FORCE AND SPEED

### Mechanical Advantage of Force

Whenever a small force is used to move a larger force, there is a gain which is called *mechanical advantage*. In the previous example, an effort force of 25 pounds moves a resistance force of 100 pounds. The mechanical advantage of force ($MA_f$) is the ratio of the resistance force to the effort force.

$$MA_f = \frac{R}{E}$$

Substituting values from the example yields:

$$MA_f = \frac{100}{25} = 4$$

Thus, the mechanical advantage of force is 4.

### Mechanical Advantage of Speed

The mechanical advantage of speed or distance ($MA_s$) is the ratio of the resistance distance to the effort distance.

$$MA_s = \frac{RD}{ED}$$

For the previous example, ED = 4″ and RD = 1″.

$$\text{Therefore, } \frac{RD}{ED} = \frac{1''}{4''} \text{ or } \frac{1}{4}$$

The mechanical advantage in speed is 1/4. It is obvious that the mechanical advantage in force is compensated by the loss in speed.

The principles underlying the nature and kinds of forces and the mechanical advantages of distance and force are basic to an understanding of the units which follow.

## WORK, POWER, AND EFFICIENCY

*Work* is defined as the product of a force and the distance through which the force acts. Work is expressed in terms of linear units and weight units, such as inch-pounds, foot-pounds, or millimeter-grams. If a wood plank weighing 60 pounds is lifted 4 feet from the floor, the work done equals (60 pounds) x (4 feet) or 240 foot-pounds, figure 9-4. Work is performed in lifting the plank because a motion results from the application of a force. However, if the plank weighs 600 pounds and an hour is spent trying to lift it without success, then, by definition no work is done.

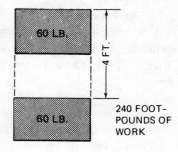

WORK = WEIGHT X DISTANCE

Fig. 9-4

Work is not a torque. The distance used to compute work is the distance the object moves in the same direction as the applied force. The distance used in torque calculations is perpendicular to the direction of the force.

### Power

The term *power* is applied to the rate at which work is done. In other words, a third unit of measurement (time) is applied to work. If a lift truck raises a 6600 pound casting 5 feet in one minute, the rate of doing work is 33 000 foot-pounds per minute; the amount of work done is 33 000 foot-pounds.

The rate at which work may be done is compared to a universally accepted standard. The *horsepower* (h.p.) is the standard unit of measurement used to express 33 000 foot-pounds of work per minute or 550 foot-pounds of work per second, figure 9-5. The power of motors and engines in horsepower units can be determined by dividing the known or computed rate in foot-pounds per minute by 33 000.

$$\text{h.p.} = \frac{\text{work (foot-pounds)}}{33\,000 \times \text{time (minutes)}}$$

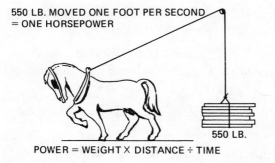

550 LB. MOVED ONE FOOT PER SECOND = ONE HORSEPOWER

POWER = WEIGHT X DISTANCE ÷ TIME

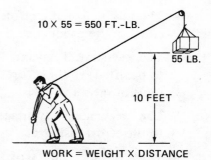

10 X 55 = 550 FT.-LB.

55 LB.

10 FEET

WORK = WEIGHT X DISTANCE

Fig. 9-5

### Efficiency

A machine transforms power from one form to another. There is some loss in this process as the usable power output is always less than the power input. The ratio of the output to the input power (expressed in the same units of measure) is known as the *efficiency* of the machine. Efficiency is given as a percent in many instances, and is always less than 100% (or 1.00). Friction and heat losses account for most of the losses in efficiency.

A crane lifts a 500-pound girder five feet. The work done is 2500 foot-pounds. To do this work, a cable moves through a distance of 20 feet and exerts a force of 150 pounds. The input is 20 x 150 or 3000 foot-pounds; the output is 5 x 500 or 2500 foot-pounds. The efficiency can be determined as follows:

$$\text{Efficiency} = \frac{\text{Output}}{\text{Input}} = \frac{2500}{3000} = 83\ 1/3\%$$

For this problem, the mechanical advantage of force is 500/150 or 3 1/3. The mechanical advantage of speed is 5/20 or 1/4. It is possible to lift heavy weights using lighter forces when there is a loss in the distance or speed.

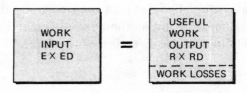

Fig. 9-6

---

### SUMMARY

- A force either changes the shape or motion of a body or prevents other forces from making such changes.

- A force may be produced by physical or chemical change, gravity, or change in motion.

- The four kinds of forces most used in industry are:  (1) tensile, (2) compressive, (3) torsional and (4) shearing.

- The product of the effort force and the distance through which it acts is equal to the product of the resistance force and its distance.

- The mechanical advantage of force is the ratio of the resistance (weight to be lifted) to the effort (force needed for the job), expressed in the same units of measure.

- The mechanical advantage of speed is the ratio of the resistance distance to the effort distance.

- Work is the product of a force and the distance through which it acts, expressed in linear and weight units.

- Power is the rate of doing work. The standard horsepower is the rate of doing 33 000 foot-pounds of work a minute.

- Efficiency is the ratio (percent) of output to input.

# ASSIGNMENT UNIT 9  FORCES AND THEIR EFFECTS

## PRACTICAL EXPERIMENTS WITH FORCES

### Equipment and Materials Required

| Experiment A<br>Kinds of Forces | Experiment B<br>Forces and Mechanical Advantage |
|---|---|
| Soft wood block, 3/4″ x 1″ x 2″<br>Bench vise<br>Narrow strip of plastic or wood, about 1/2″ square<br>Six or more thin wood strips, about 1/8″ thick x 1″<br>Two A-clamps<br>Wood blocks, 1″ x 1″ x 4″<br>Weights | 100-, 200-, and 400-gram weights<br>Yardstick<br>Wedge (triangular, wood) |

## EXPERIMENT A.  KINDS OF FORCES

### Procedure

1. Place a piece of soft wood in a vise and tighten the vise. Prepare a table such as the one shown for Recording Results. Indicate in this table the kind of force which the vise jaws are exerting on the wood.

2. Support both ends of a thin strip of wood, plastic, or other material. Apply a force near the center of the strip. Identify and record the kinds of force which act on the top and bottom surfaces of the strip.

3. Hold a small piece of wood, metal, or plastic (about 1/2″ square) between the fingers of both hands. Turn one end clockwise; turn the other end counterclockwise. Identify and record the kind of force exerted on the piece.

4. Place a number of thin wooden strips, one on top of another, to form a light beam.

5. Support the ends of this beam on two wood blocks.

6. Load the center of the beam until there is motion between the individual strips. Record what takes place.

7. Remove the weight and clamp the ends of the thin wood strips.

8. Weight the beam and record what happens to the top surface, bottom surface, and the separate strips.

### Recording Results

| | Condition | | Kinds of Forces |
|---|---|---|---|
| A | Wood in vise | | |
| B | Material across vise jaws | | |
| C | Material turned two directions | | |
| D | Weight applied on thin sections | | |
| E | Beam ends clamped | Top | |
| | | Bottom | |
| | | Sections | |

## Interpreting Results

1. Name the different kinds of forces that are demonstrated in this experiment.

2. Prepare a table as illustrated. List two parts or machines used in industry that either produce or are subjected to each kind of force indicated in the table.

| | Kind of Force | Examples |
|---|---|---|
| A | Tension | 1. |
| | | 2. |
| B | Compression | 3. |
| | | 4. |
| C | Torque | 5. |
| | | 6. |
| D | Shear | 7. |
| | | 8. |

NOTE: The effects of tension, compression, shear, and torsion may be shown by passing polarized light through a plastic material. The kind of force is readily determined by observing the lines of force as different kinds of force are applied.

## EXPERIMENT B. EFFORT AND RESISTANCE FORCES AND MECHANICAL ADVANTAGE

### Procedure

1. Place a 200-gram weight on one end of a yardstick.

2. Place a triangular wooden wedge at a point 12 in. from the center of gravity of the 200-gram weight.

3. Place a 400-gram weight near the opposite end of the yardstick and adjust it until the beam is balanced.

4. Prepare the table shown for Recording Results. Indicate the distance from the center of the wedge to the center of the 400-gram weight.

5. Now place a 200-gram weight 10 in. from the triangular wedge (fulcrum).

6. Place a 100-gram weight near the opposite end of the yardstick and adjust the weight until the beam is balanced. Determine and record the effort distance.

### Recording Results

| | Resistance Force | Effort Force | Resistance Distance | Effort Distance |
|---|---|---|---|---|
| A | 200 grams | 400 grams | 12 in. | |
| B | 200 grams | 100 grams | 10 in. | |

### Interpreting Results

1. Make a simple sketch of E, ED, R, and RD for both the 200-400 and the 200-100 gram combinations.

2. Compute ED for both combinations. Check these values with the results found in the experiment. Explain any differences.

3. Determine the mechanical advantage of force for each combination.

4. Check the mechanical advantage of speed of the two combinations.

## PRACTICAL PROBLEMS ON FORCES, WORK, POWER, EFFICIENCY

### Forces and Their Effects

1. State the kind of force caused by each of the following:

   a. filing on a lathe,

   b. cutting lips of a drill,

   c. jaws of a vise,

   d. plunger on hydraulic press.

2. The cutting edge of a lathe tool is positioned 30° toward the chuck. The end of the work is supported by a tailstock center. Make a simple sketch and identify with arrows the kind of force exerted by the lathe chuck, center, and cutting edge of the tool as the cut is taken.

### Effort Forces, Resistance Forces, and Mechanical Advantage

For statements 1 to 6, determine which are true (T) and which are false (F).

1. An effort force refers to the resistance a part offers.

2. Effort and resistance forces are given in the same units of measurement.

3. When E = 10 lb., ED = 10 in., and R = 20 lb., then RD = 15 in.

4. The mechanical advantage of force of 100 lb. moving a 200-lb. part is 2.

5. The mechanical advantage of speed in moving an object 2 meters with an effort distance of 4 meters is 1/2.

6. The mechanical advantage is equal to the resistance divided by the resistance arm.

### Work, Power, and Efficiency

Indicate the letter representing the words, phrases, or formulas which best complete statements 1 to 7.

1. Work is done when

   a. a force is exerted.
   b. motion is produced.
   c. effort is applied.

2. Work is equal to

   a. output ÷ input.
   b. resistance ÷ effort.
   c. force x distance.

3. Work input equals

   a. effort x resistance.
   b. resistance ÷ effort.
   c. effort x effort distance.

4. Efficiency equals

   a. output ÷ input.
   b. force x distance.
   c. input ÷ output.

5. Heat and friction losses cause

   a. a decrease in efficiency.
   b. an increase in efficiency.
   c. no change in efficiency.

6. Horsepower is a measure of

   a. foot-pounds of work.
   b. work ÷ time.
   c. efficiency x time.

7. One horsepower equals

   a. 550 foot-pounds.
   b. 33 000 foot-pounds.
   c. 550 foot-pounds per second.

8. Compute the horsepower required to deliver

   a. 66 000 foot-pounds per minute.

   b. 2750 foot-pounds per second.

   c. 38 775 foot-pounds per second.

   d. 8250 foot-pounds per minute.

9. Which letters indicate a situation where work is being performed?

   a. Measuring work length.
   b. Chuck jaws holding work.
   c. Lifting a saw horse.
   d. Tightening a motor nut.
   e. Bolt holding parts together.
   f. Computing wire size.

# Unit 10 Balance and Equilibrium–Parallel and Angular Forces

A mechanic must understand the conditions which affect the balance and stability of moving or stationary parts so that he will know when the work is properly anchored. If an object is to be balanced, it must be in *equilibrium*; that is, there must be no unbalanced forces tending to produce motion. An object balanced while at rest is said to be in *static balance;* an object balanced while in motion is in *dynamic balance.*

Materials which are not in static balance tend to turn or rotate in an effort to achieve balance. Forces which produce rotation are called *torque* (torsional) *forces* or *moment forces.* A moment is the product of the force and the distance through which it acts. This distance is the length of a perpendicular drawn from the axis of rotation to a line representing the direction in which the force is applied. The perpendicular passes through the point of application of the force. Since force is generally measured in pounds and distance is measured in feet or inches, the moment is expressed in pound-feet or pound-inches. Moments may produce clockwise or counterclockwise rotation. A body is in equilibrium when the clockwise and counterclockwise forces are equal and the body is free to rotate.

## DYNAMIC AND STATIC BALANCE AND EQUILIBRIUM

It is much easier to balance a part statically than it is to balance it dynamically. For example, a two-pound weight is placed at the right end of a short beam, figure 10-1. If the weight is placed three feet from the point at which the beam is held, the beam has a tendency to turn clockwise. The moment is equal to the force (two pounds) x the distance (three feet) or six pound-feet.

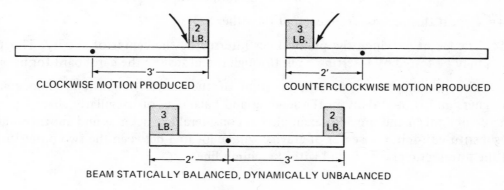

CLOCKWISE MOTION PRODUCED          COUNTERCLOCKWISE MOTION PRODUCED

BEAM STATICALLY BALANCED, DYNAMICALLY UNBALANCED

**Fig. 10-1**

If another weight of three pounds is placed two feet to the left of the point at which the beam is pivoted, the beam will turn in a counterclockwise direction. In this case, the moment is equal to the force (three pounds) x the distance (two feet) or six pound-feet.

This beam is in static balance since the moment which tends to turn the body in a clockwise direction (six pound-feet) is equal to the moment tending to rotate the body counterclockwise (6 pound-feet). However, there will be considerable vibration if the whole system turns. Because the system is out-of-balance dynamically, there will be undue wear on the moving parts, excessive vibration, and uneven motion. For these reasons, it is important that rotating parts be in dynamic balance. The location of the points where a part is out-of-balance may be determined by computation, by trial-and-error, or with special balancing equipment.

The degree to which a moving part is balanced depends on its use. A large turbine rotor turning at high speeds requires precision balancing to a far greater degree than does the wheel of an automobile. Dynamic balancing may be done either by removing material from the heavy side of a device or by attaching weights to the light side, as in the case of automobile wheels.

## Center of Gravity

The technician and engineer work with many irregularly shaped parts. It is important to determine the point where each of these parts is statically balanced, regardless of the position in which it is placed. This point is known as the *center of gravity*.

If it is necessary to find the center of gravity of an irregularly shaped piece of cardboard, the procedure is very simple, figure 10-2.

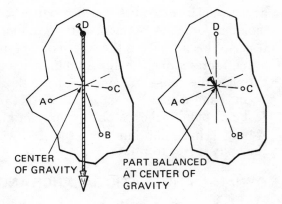

**Fig. 10-2**

1. Punch four or five holes near the outer edges of the cardboard.

2. Place the cardboard on a nail and drop a plumb line from this hole.

3. Mark the line of the plumb line on the cardboard.

4. Repeat this process from each of the other holes.

5. The point at which the plumb lines intersect is the center of gravity. This point may be tested by inserting a pin through it and testing the cardboard for balance.

The principle governing the location of the center of gravity is important to designers and is used daily in the loading and balancing of irregularly shaped parts. If the cardboard in the previous example has considerable thickness and another material is substituted for it, the center of gravity will be located between the two outer surfaces at the intersecting point found with the plumb line.

## Stability

An object may be in *stable* or *unstable equilibrium* depending upon the location of the center of gravity. Stability refers to the condition of balance of the object. When an object is tipped and it returns to its original position when released, it is in stable equilibrium. For any object in stable equilibrium, a plumb line dropped from the center of gravity will fall within the base of the object, figure 10-3.

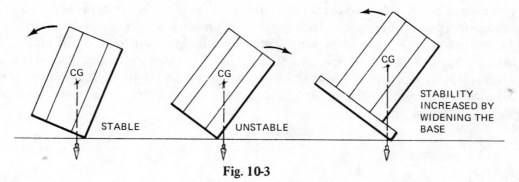

**Fig. 10-3**

The object is in unstable equilibrium when it is tipped so far that a plumb line from the center of gravity falls outside the base of the object. The center of gravity of an object seeks the lowest position possible. When a stable object is tipped, its center of gravity is raised. Thus, the object tends to return to its original position. If tipping the object lowers its center of gravity, it will not return to its original position. The center of gravity must be raised if the object is to return to its original position. Figure 10-3 shows when a body is stable and when it is unstable; in addition, it is shown how this stability may be increased by widening the base or lowering the center of gravity.

A third condition of equilibrium is known as neutral equilibrium. Consider objects such as ball bearings, cylinders, and cone-shaped parts. When these objects are moved on a level surface, they neither take on a new position nor return to the original one. When the tipping or rolling of an object neither raises nor lowers the center of gravity, the object is said to be in *neutral equilibrium.*

## BALANCING PARALLEL AND ANGULAR FORCES:
## ADDITION AND SUBTRACTION OF VECTORS

A force is a push or pull. The effect of a force is either to alter the shape of an object or to prevent other forces from making such a change. Forces may act singly or in combination with other forces. When forces act together, their combined force may be matched by a single force known as the *resultant.* Forces which have a known direction and known magnitude are called *vectors.* Vectors are lines which are drawn to scale to show the amount of force. Arrowheads are used to indicate the direction of force.

### Scalar and Vector Quantities

A *scalar quantity* represents a physical quantity that is completely specified by the magnitude of the scalar quantity (expressed by a number) and a unit. An object with a mass of 65 kilograms, an electrical energy rating of 10 amperes, a speed of 42 kilometers per hour, and a volume of 122 cm$^3$ are all examples of scalar quantities. In each instance, the quantities are expressed in *scalar statements.*

For other quantities, the direction of the quantity is as significant as the magnitude. A quantity having both magnitude and direction is called a *vector quantity.* There are three common vector quantities:

1. displacement
2. velocity
3. force

*Displacement* refers to a change in position. This change is expressed in a *vector statement*. If an automobile is traveling 250 kilometers east from a given point, its displacement at any point can be represented by a line (drawn to scale) to indicate the distance traveled, and an arrowhead at the end of the line to indicate the direction of travel. If other quantities and directions are involved in a problem, it is possible to solve for required values by graphing the given values on a vector diagram or by mathematical computation.

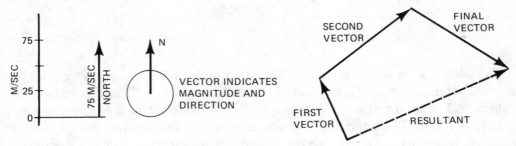

**Fig. 10-4 Representing a vector**

*Velocity* is represented by the speed of a body and the direction in which it travels. This information makes it possible to determine the path and location of the body at any point in time.

For the vector quantity of *force,* both the direction and magnitude of the force must be stated whenever the effect of the force is to be determined.

Vector problems may be simplified by making a simple drawing of each vector. A vector may be represented by a line that has magnitude and direction. For example, an upward force whose magnitude is 400 pounds may be represented on a scale of 1 cm equals 100 pounds. Thus, the vector representing this force is 4 cm long and the vertical direction of the force is indicated by a vertical arrowhead ( ↟ ) at the end of the vector.

The tail of each successive vector is placed at the point of the arrowhead of the previous vector. The original length and direction remain unchanged. The resultant is drawn as a vector which connects the tail of the first vector to the head (vertex of the arrowhead) of the last vector quantity.

## Forces Acting in the Same Direction (Parallel Forces)

One important principle concerning forces is that forces acting in the same direction are added to determine the resultant. If two people (A and B) pull the crate shown in figure 10-5 with forces of 60 pounds and 110 pounds, the combined force is 170 pounds. The resultant of the two forces acting in the same direction is their sum.

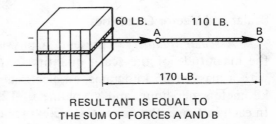

RESULTANT IS EQUAL TO
THE SUM OF FORCES A AND B

**Fig. 10-5**

The resultant of velocities can be found using the same method as that shown for forces. For example, an airplane traveling north at an air speed of 500 miles per hour has a south tailwind of 50 miles per hour. Since both the plane and the tailwind are traveling in the same direction, the speed of the plane is the sum of its speed (thrust of its engines) and the speed of the wind force. The resultant air speed is 550 miles per hour (see I in figure 10-6).

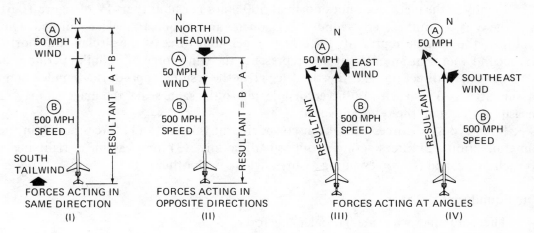

**Fig. 10-6**

## Forces Acting in Opposite Directions (Parallel Forces)

The resultant of forces acting in opposite directions is found by subtracting the forces. The same method applies to velocities. If this same airplane travels due north at the same air speed (500 miles per hour) and runs into a 50 mile an hour headwind from the north, then the airplane's speed is decreased. The resultant speed is the difference between the plane speed and the wind velocity; that is, 500 − 50 or 450 miles per hour (see II in figure 10-6).

## Forces Acting at an Angle to Each Other (Angular Forces)

All forces and velocities do not act in the same direction or in opposite directions. Refer to part III of figure 10-6. If an east wind exerts a wind force of 50 miles per hour, the plane will be blown off its course. If the vectors for the direction and speed of the airplane are drawn, they will form two legs of a right triangle. The resultant is the hypotenuse of a right triangle and may be determined mathematically.

## Vector Representation by Graphical Means

It is possible to lay out the triangle on graph paper using the same scale for both vectors (at right angles to each other). The resultant is the third leg of the triangle and can be measured to the same scale as the other two vectors. This is a practical method of finding the resultant in most situations. However, this method is not as accurate as finding the resultant mathematically.

The same resultant speed may be found using the parallelogram method of graphing the result, figure 10-7. In the parallelogram method, the vectors form two sides of a parallelogram. The two opposite and parallel sides of the parallelogram complete the figure. The diagonal of such a parallelogram is the

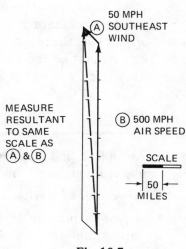

**Fig. 10-7**

resultant. If the vectors are laid out to the same scale and at the correct angles to each other, the resultant may be measured using the same scale.

Finally, if the plane continues north at 500 miles per hour (part IV of figure 10-6), and there is a southeasterly wind acting on it, an oblique triangle is formed. One vector of the triangle represents the 500 mile per hour force of the airplane due north, the second vector at an angle of 45° represents the wind force of 50 miles per hour in a northwestern direction, and the third leg gives the resultant speed. If a parallelogram of velocities is drawn with both vectors plotted to the same scale, the diagonal may be measured to obtain the resultant speed.

Four types of forces have been considered in this unit: (1) forces acting in the same direction, (2) forces acting in opposite directions, (3) forces acting at right angles to each other, and (4) forces acting at any angle to each other.

## The Equilibrant

There are times when a single balancing force is needed to prevent outside forces from moving an object. This balancing force which produces equilibrium is known as an *equilibrant*. The equilibrant is a force that is opposite and equal to the resultant and will balance the vector forces. A drawing which shows the magnitudes and directions of the vectors is known as a *vector diagram*. If a vector diagram consists of two forces (30 and 40 pounds) acting at right angles to each other, figure 10-8, the resultant force required to balance the two forces is 50 pounds. The equilibrant, therefore, is 50 pounds.

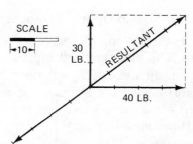

Fig. 10-8 Equilibrant equals a resultant but is opposite in direction.

The term *component* is often used to mean either of two forces which produce a given force as their resultant. Most force problems may be computed or plotted easily when a simple sketch is made to identify the direction and amount of each force. The amount of force is also called its *magnitude*. The point from which the forces act is called the *point of application*. The parallelogram method of determining the resultant is also called the *composition of forces* method.

The four combinations of forces considered in this unit, and the methods of computing or graphing the resultant force, are those that are commonly used to solve problems. The resultant or equilibrant force determined by one method should be checked for accuracy using a second method.

─────────────────────────── SUMMARY ───────────────────────────

- An object that is balanced while at rest is statically balanced; when the object is balanced in motion, it is dynamically balanced.

- The center of gravity is the point at which an object is statically balanced.

- A part or mechanism is in stable equilibrium when the center of gravity falls within its base. Unstable equilibrium results when the center of gravity of an object falls outside its base.

- A scalar quantity consists of a numerical value and a unit which completely specify the magnitude of a physical quantity.
- A vector quantity gives both direction and magnitude to completely specify a physical quantity.
- Vectors in a vector diagram originate from a point of application. A vector may be represented by a line whose length is proportional to its magnitude and which extends in a specific direction and terminates in an arrowhead.
- The resultant of two forces acting in the same direction is the sum of the forces.
- The resultant of two forces acting in opposite directions is the difference of the forces.
- The resultant of forces acting at angles to each other can be determined using the triangle of forces method or the parallelogram of forces method.
- The equilibrant is a single balancing force which is equal and opposite to the resultant; the equilibrant produces equilibrium.

### ASSIGNMENT UNIT 10 BALANCE, EQUILIBRIUM: PARALLEL AND ANGULAR FORCES

## PRACTICAL EXPERIMENTS WITH FORCES AND BALANCE

### Equipment and Materials Required

| Experiment A Dynamic and Static Balance | Experiment B Center of Gravity | Experiment C Composition of Forces |
|---|---|---|
| One-pound weight Three-pound weight Measuring rule 2″ x 2″ x 4′ wood beam Square block Dowel rod, 1/2-in. diameter Twine | Cardboard, 12″ x 18″ (approximately) Plumb bob and line Chalk | Two 200-gram spring balances 100- and 200-gram weights String Chalk |

## EXPERIMENT A. DYNAMIC AND STATIC BALANCE

### Procedure

1. Drill a 9/16-in. hole through the 2″ x 2″ wood beam, three feet from one end.
2. Place the wood beam on a block located approximately three feet from one end of the beam.
3. Place a one-pound weight on this end. Note the motion that takes place. Prepare a table for Recording Results and indicate the action.
4. Remove the one-pound weight and place a three-pound weight on the other end of the beam. Record what happens.

5. Remove the weight. Secure a 1/2-in. dowel in a vise and place the wood beam so that it can rotate around the dowel.

6. Strap the one-pound weight on the right end of the beam and the three-pound weight on the left end. Adjust the weights until the moments of force are equal. Note what happens, if anything.

7. Turn the beam and weights almost 90 degrees. Record what action, if any, takes place.

8. Rotate the beam gently and observe whether or not it is balanced dynamically.

## Recording Results

| | Condition | Action |
|---|---|---|
| A | 1-lb. weight on right end | |
| B | 3-lb. weight on left end | |
| C | Two weights on beam | |
| D | Rotating beam with weights | |

## Interpreting Results

1. What is the moment of force of the right side of the beam when it is loaded with the one-pound weight?

2. Determine the moment of force of the left end of the beam.

3. When the sums of the forces on both sides of the beam are equal, is the beam in neutral equilibrium, dynamic balance, or static balance?

4. From the results recorded in the actual test, determine if the beam is both statically and dynamically balanced.

NOTE: This same experiment may be performed with a bicycle wheel and lead weights placed on the spokes. The principles of balance in both cases are identical.

## EXPERIMENT B. CENTER OF GRAVITY

### Procedure

1. Using a piece of cardboard or sheet metal, cut an irregular shape.

2. Punch or drill four holes at four different places near the outer edge of the irregular shape.

3. Place the cardboard or metal plate over a nail so that it is free to move.

4. Suspend a plumb bob from the nail and snap a vertical chalk line. (This method is used by carpenters and masons to obtain straight lines between points.)

5. Repeat step 4 for the remaining three holes. Note that the lines intersect at a point.

### Interpreting Results

1. Identify the center of gravity by marking its location.

2. Describe a simple test for determining the center of gravity of the cardboard.

## EXPERIMENT C. COMPOSITION OF FORCES

### Procedure

1. Attach two 200-gram (or larger) spring balances to the frame of the laboratory blackboard about two feet apart.

2. Suspend a 100-gram weight from each spring balance so that the string is shorter on one side than the other.

3. Prepare a table for Recording Results and indicate the reading on each of the spring balances.

4. Chalk the strings on which each weight is suspended and snap lines on the board (or against a piece of paper that is placed behind the strings).

5. Select a scale, mark off the readings on the spring balances on the blackboard, and complete the parallelogram of forces.

6. Measure the resultant of the parallelogram and record this graphed reading.

7. Repeat steps 2-6 with 200-gram weights. Record the readings and check them by completing a parallelogram of forces.

### Recording Results

| | Condition | Reading | |
|---|---|---|---|
| | | Left spring balance | Right spring balance |
| **A** | 100-gram load | | |
| **B** | Resultant force (graphed) | | |
| | | Left spring balance | Right spring balance |
| **C** | 200-gram load | | |
| **D** | Resultant force (graphed) | | |

### Interpreting Results

1. Are the readings on both spring balances the same or are they different for the 100-gram and 200-gram weights? Explain briefly why this condition exists.

2. Is the resultant found by calculation the same as, or different from, the suspended weight? Should this resultant be equal to, greater than, or less than the suspended weight?

## PRACTICAL PROBLEMS WITH FORCES AND BALANCE

### Dynamic and Static Balance and Stability

For statements 1 to 8, determine which are true (T) and which are false (F).

1. A body that is balanced while at rest is in dynamic balance.

2. The forces that tend to produce rotation are called dynamic balance.

3. Equilibrium refers to a state of balance in which the clockwise and counterclockwise rotation are equal.

4. Uneven wear and excessive vibration of moving parts may be due to dynamic unbalance.

5. A bicycle wheel, automobile wheel, or pulley may be dynamically balanced by locating the points where weight may be added or removed, as needed.

6. A solid plate is stable when the center of gravity falls outside its base.

7. Car designers try to concentrate the weight and center of gravity of cars as low as possible for safety and stability.

8. A spherical ball bearing and objects that are cylindrical or cone-shaped are in neutral equilibrium because the center of gravity remains constant when they are rolled.

## Balancing Parallel and Angular Forces (Scalar and Vector Quantities)

Select the correct word or phrase to complete statements 2-6.

1. Define the following and give an example to illustrate each vector term.

    a. scalar quantity

    b. vector quantity

    c. displacement

2. Forces which are combined and act together may be matched by (a single force) (two or more forces).

3. The resultant of two forces pulling in the same direction is the (difference between) (product of) (sum of) the two forces.

4. The resultant of two forces pulling in opposite directions is the (product of) (sum of) (difference between) the two forces.

5. In a vector diagram of two forces acting at a right angle to each other, the hypotenuse of the right triangle formed is the (equilibrant) (resultant).

6. The equilibrant is a force that is (equal and opposite) (unequal and opposite) (has no relationship) to the resultant.

7. The airline distance between two points is 2040 miles. An airplane traveling at an air speed of 500 miles an hour due south runs into headwinds of 75 miles per hour due north. How many hours does it take the airplane to travel this distance?

8. Determine the resultant force in problems A, B, and C of figure 10-9. Check each problem using a second method. Show how each answer is determined.

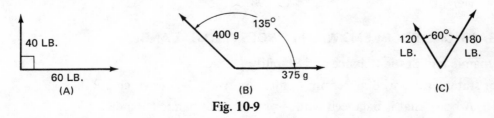

Fig. 10-9

9. A guy wire extends at an angle of $30°$ from the side of a building. The pull on this wire, which helps to support a sign, is 400 pounds. If the angle of the guy wire is changed to $45°$, will it require a greater or a lesser pull? How much?

10. Compute the force required to pull a 400-kilogram object up a $30°$ incline at a constant speed. Assume the forces are applied parallel to the plane. Neglect friction.

# Unit 11 Gravitation, Motion, and Mechanical Movements

Aristotle conducted many experiments with falling bodies and kept careful records of his findings. Although his work was later found to be in error, the importance of Aristotle's work lies in the fact that Galileo based his experiments on Aristotle's experiments. In the early 17th century, Galileo proved that the weight of a body has little to do with the speed at which a body falls. Two different bodies, such as a feather and a coin, fall at the same rate when dropped in a vacuum.

## FUNDAMENTAL LAWS OF GRAVITATION

### Galileo's Work with Falling Bodies

*Distance Covered by Objects Rolling Down Inclined Surfaces.* While experimenting with inclined surfaces, Galileo found that the distance covered by an object rolling down an incline is equal to the distance it travels in the first second multiplied by the square of the time traveled. If a one-inch diameter ball rolls down an inclined surface one foot in one second, it will travel four feet in two seconds, nine feet in three seconds, 16 feet in four seconds, and so on, figure 11-1.

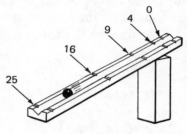

**Fig. 11-1**

The distance traveled can be expressed by the following formula:

$$D = d \times T^2$$

where  $D =$ total distance that the object travels, in feet
$d =$ distance covered for the first interval of time, in feet
$T =$ time, in seconds

By substituting values from the above example, $D = 1 \times 1^2$ or one foot the first second; $1 \times 2^2$ or four feet in two seconds; $1 \times 3^2$ or nine feet in three seconds; and $1 \times 4^2$ or 16 feet in four seconds.

*Distance Covered by Freely Falling Objects.* If an object falls straight down instead of rolling down an incline, the distance the object falls is again equal to the distance traveled in the first second multiplied by the square of the time. It is known, however, that freely falling bodies (neglecting any losses caused by the friction of the air) always fall 16 feet the first second.

Thus, the distance traveled by a freely falling body is given by the formula:

$$D = 16T^2$$

where  $D =$ total distance, in feet
$T =$ time, in seconds

In one second, then, an object falls 16 x 1² or 16 feet; in two seconds, 16 x 2² or 64 feet; in three seconds, 16 x 3² or 144 feet; in 10 seconds, 16 x 10² or 1600 feet; and in 30 seconds, 16 x 30² or 14 400 feet.

### Newton's Law of Gravitation

Many years after Galileo, Sir Isaac Newton continued experimenting with gravity and its effects on falling bodies. In simplified form, Newton's *law of gravitation* means that all objects in the universe attract all other objects and the closer two bodies are to each other, the greater is the attraction.

Newton also determined that the masses of the bodies influence the force of this attraction. All objects are attracted to the earth because the earth is larger than any object on it. The attractive force that pulls all things toward the center of the earth is called *gravity*. In addition, all falling objects gain speed at a uniform rate which is not affected by the weight of the object.

*Velocity of Freely Falling Objects.* A freely falling object starting at an initial velocity of zero reaches a speed of 32 ft./sec. at the end of the first second. For each succeeding second, the free-falling object increases its speed by 32 ft./sec. Therefore, the velocity of a free-falling object at any time (disregarding air resistance) is equal to 32 ft./sec. multiplied by the time in seconds.

Expressed as a formula, the velocity is:

$$V = 32T$$

where　$V$ = velocity, in feet/second
　　　　$T$ = time, in seconds.

Starting from rest, a freely falling object has a velocity at the end of one second of 32 x 1 or 32 ft./sec.; at the end of two seconds, 32 x 2 or 64 ft./sec.; at the end of three seconds, 32 x 3 or 96 ft./sec.; at the end of 10 seconds, 32 x 10 or 320 ft./sec., and so on.

### COMMON CONCEPTS ABOUT MOTION

A body may experience three kinds of motion. The body may move in straight or curved lines, it may rotate about its axis without moving from a fixed position, or it may vibrate. The term *motion* refers to a change of position. This motion may be either absolute or relative, figure 11-2. In *absolute motion,* the body moves away from a fixed point of reference (as shown at A in figure 11-2). On the other hand, if the body moves away from another point or object that is also moving, the motion, as shown in figure 11-2B, is *relative motion.*  A body is considered to be at *rest* when its position remains unchanged with respect to some given point or object.

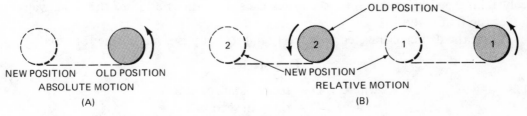

ABSOLUTE MOTION
(A)

RELATIVE MOTION
(B)

**Fig. 11-2**

When bodies are in motion, two terms are used to describe the motion: speed and velocity. *Speed* is defined as the distance traveled divided by the time of travel. *Velocity* includes both speed and direction. Both terms are used interchangeably in this and other units.

## Speed and Acceleration

The total distance covered by a moving object, divided by the time required to cover the distance, defines a quantity called the *average speed*. The term *constant speed* is applied to motion at unchanging speed; for example, a machine may produce wire or paper or plastic at a constant speed, such as 1000 feet per minute; this speed is also the average speed. If a car travels 60 miles in two hours, its average speed is 30 miles per hour. This speed probably was not a constant speed, but varied between 0 and 70 miles per hour. The rate at which speed changes is called *acceleration*. *Positive acceleration* implies a gain of speed; a loss of speed is *negative acceleration*. Numerically, acceleration equals the change in velocity divided by the time in which the change occurs. If a car increases its speed from 15 miles per hour to 25 miles per hour in two seconds, its acceleration is 5 miles per hour per second. If an object has *constant acceleration,* it is changing speed steadily, either speeding up or slowing down. A body that maintains a steady unchanging speed has *zero acceleration.*

## Average and Final Speeds and Accelerated Motion

*Average Speeds.* When a body accelerates so that it either gains or loses speed uniformly, the *average speed* is the sum of the initial and final speed divided by two. If an automobile travels at a speed of 30 miles per hour, figure 11-3, and gains speed at a constant rate until it reaches 60 miles per hour, then:

$$\text{Average speed} = (\text{initial speed} + \text{final speed})/2$$

Substituting values in this expression yields:

$$\frac{30 + 60}{2} = 45 \text{ miles per hour}$$

The same automobile climbing a steep hill may lose speed uniformly from 60 to 30 miles per hour. Again, the average speed is 45 mph as determined from the same formula.

*Final Speeds.* The *final speed* of a uniformly accelerating body is the speed at a given instant. Starting at rest, the final speed equals the elapsed time multiplied by the acceleration. If an object starting at rest gains speed down an incline at the rate of 10 feet per second each second, its final speed after half a minute is 30 seconds x 10 ft./sec./sec. = 300 ft./sec. The final speed of a body that loses speed uniformly is found in the same manner. The final speed ($S_f$) is equal to the product of the uniform acceleration (A) x time (T) plus the initial speed of the object ($S_i$), figure 11-4.

$$S_f = (A)(T) + S_1$$

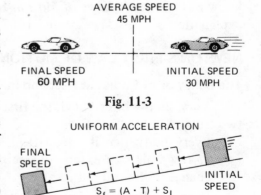

AVERAGE SPEED
45 MPH

FINAL SPEED
60 MPH

INITIAL SPEED
30 MPH

**Fig. 11-3**

UNIFORM ACCELERATION

FINAL SPEED

INITIAL SPEED

$S_f = (A \cdot T) + S_1$

**Fig. 11-4**

Substitute values from the preceding example in the expression to obtain the final speed:

$$S_f = \frac{10 \text{ ft./sec.}}{\text{sec.}} \times 30 \text{ sec.} + 0 = 300 \text{ ft./sec.}$$

Care must be taken to insure that the units of measure used in acceleration problems are practical and consistent. An aid to understanding and solving acceleration problems is the drawing of a simple sketch of the given conditions. Once the sketch is drawn, the values that must be found can be visualized and all values can be labeled.

*Uniform Motion.* *Uniform motion* is steady motion in a straight line at a constant speed. Uniform motion (distance) is equal to the product of the velocity (V) and time (T). A train moving at a uniform speed of 10 feet per second for 12 seconds travels a distance of:

$$D = (V)(T) = (10 \text{ ft./sec.})(12 \text{ sec.}) = 120 \text{ feet}$$

*Uniform acceleration* or *deceleration* indicates a constant increase or a constant decrease in speed, respectively. The velocity gained due to a constant acceleration is equal to the product of the acceleration and time. Conversely, the velocity lost by constant deceleration is the product of the rate of deceleration and time.

$$\mathbf{V_g = (A)(T)}$$
$$\mathbf{V_l = (D)(T)}$$

Where    $V_g$ = Velocity gained
          $V_l$ = Velocity lost
          A  = Acceleration
          D  = Deceleration
          T  = Time

Thus far, definitions have been given for the terms used to describe the kinds of motion at different speeds. These conditions are now studied in terms of the basic laws which affect motion. It is important to know these laws because they are applied daily in industry, on the farm, in business, and in the home. The three laws to be studied are known as *Newton's Laws of Motion* because of the work of Sir Isaac Newton and his explanation of these laws around 1687.

## NEWTON'S FIRST LAW OF MOTION

### The Behavior of Bodies at Rest and in Motion

Very simply interpreted, the first law states two truths.

1. Every body that is at rest tends to remain at rest unless an external force produces a change in the state of rest.

2. Every body in motion remains in motion unless an external force causes the body to change that motion.

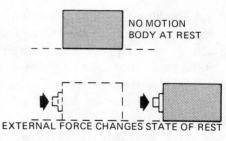

Fig. 11-5

Whenever bodies at rest or in motion are discussed, the term inertia is often used. *Inertia* describes the tendency of a body to remain at rest, or, if it is in motion, to continue its motion at the same speed.

The term *rotational inertia* is used with objects that spin around an axis and have a tendency to continue to rotate around the same axis. The principle of rotational inertia is used in the design of stabilizing devices such as the gyrocompass and the gyro-stabilizer in airplanes.

## NEWTON'S SECOND LAW OF MOTION

### Forces Affecting Motion

A body in motion has a certain quantity of motion which is called its *momentum*. Two factors control the momentum of a body at any given time: the body's *mass* and its *velocity*. Thus, momentum may be expressed as the product of the mass and the velocity of the body.

Newton's second law states that any change in the quantity of motion (momentum) is directly proportional to the force acting upon a body. This means that the less the time required to decelerate an object, the greater is the force acting on the object. Newton's second law is used in common applications such as the heavy flywheel of a gas or diesel engine.

A small force is used on engines to gradually build up the speed of the flywheel. Once the desired flywheel momentum is reached, little effort is required to keep the flywheel (or other heavy part) in motion. The energy in the momentum of a heavy flywheel can be recovered with a great increase in force by cutting the speed abruptly. This type of situation is found in a power press for stamping or forming materials. As the clutch mechanism transfers the rotary motion (momentum) of the flywheel to a reciprocating motion of the press ram, a tremendous force is exerted on the forming, cutting, or bending dies. These dies, in turn, cut and shape the material.

If a force is applied in the same direction as the motion of a body, the motion of the body increases. The direction in which the applied force acts and the magnitude of the applied force determine the amount of change in the momentum of the body.

Stated in another way, Newton's second law of motion says that the force required to accelerate an object is proportional to the mass of the object and to the acceleration produced, figure 11-6.

$$\text{Required Force} = \text{Mass} \times \text{Acceleration}$$

$$F = M \times A$$

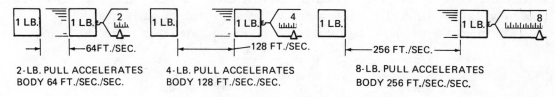

2-LB. PULL ACCELERATES BODY 64 FT./SEC./SEC.

4-LB. PULL ACCELERATES BODY 128 FT./SEC./SEC.

8-LB. PULL ACCELERATES BODY 256 FT./SEC./SEC.

Fig. 11-6  Acceleration is proportional to force.

## NEWTON'S THIRD LAW OF MOTION

### Action and Reaction

The third law of motion states that every action has an equal and opposite

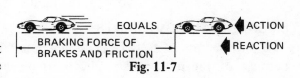

Fig. 11-7

reaction. That is, if object A pushes or pulls on object B, then, at the same time, B pushes or pulls on A with the same amount of force.

As an example of this law in action, consider that work must be fastened to machines with sufficient force to overcome the force exerted by a cutting tool on the work. A slab of steel is held in a milling machine vise to resist the forces of the cutter teeth when removing metal. The forward driving force of a jet plane is produced by the backward thrust of the jet engine.

## CENTRIFUGAL AND CENTRIPETAL FORCES

The inertia of a body tends to cause the body to travel in straight lines. The sparks (particles of material) thrown off grinding wheels fly from the wheel on paths which are tangent to the circumference of the wheel. A force known as *centrifugal force* tends to cause rotating objects to fly apart.

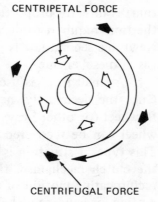

CENTRIPETAL FORCE

CENTRIFUGAL FORCE

**Fig. 11-8**

Considerable attention is given to the action of centrifugal force in grinding wheels. That is, the bonding agent that holds the abrasive particles on the wheel must be stronger than the forces which tend to make the revolving wheel fly apart at high speeds. For this reason, the speed of a grinding wheel should not exceed the safe surface speed limit specified by the manufacturer. Centrifugal force increases with speed.

The principles of centrifugal force are used in the design of centrifuge-type machines. Some centrifuges are used to separate chemicals; others are used to remove impurities in metals by centrifugal casting processes. Centrifugal force principles are also used in common appliances such as clothes dryers and in devices to control motor speeds and accelerate moving machines.

*Centripetal* force causes an object to travel in a circular path. This action is caused by the continuous application of forces which tend to pull the object to the center. In other words, the inward force which resists the centrifugal force is called the centripetal force. The centripetal force of objects spinning at a constant rate produces an acceleration toward the center which is equal and opposite to the centrifugal force.

The materials used in the construction of rapidly moving machine parts and mechanisms must be structurally sound and strong enough to provide the centripetal force required to hold the parts to a circular path. At the same time, the materials must be able to withstand the centrifugal force which tends to pull the parts apart.

Centripetal force and centrifugal force can be measured using the same formula:

$$C = \frac{W \times V^2}{g \times r}$$

where  $C$  = centrifugal or centripetal force
$W$  = weight of the object
$V$  = the speed or velocity of the object
$g$  = acceleration caused by gravity
$r$  = the radius of the path of the object

## MECHANICAL MOVEMENTS

Motion and the basic laws which affect motion are important considerations because of the numerous applications of these principles to produce work through mechanical devices. There are two primary mechanical motions: rotary and rectilinear. These terms suggest that *rotary motion* is a circular movement around a center line and *rectilinear motion* is a straight line motion. For either rotary or rectilinear motion, it is possible, with added mechanical devices, to produce other forms of motion such as intermittent motion and reciprocating motion.

## Transmitting Motion

*Rotary Motion.* The motion that is commonly transmitted is rotary motion. This type of motion may be produced with hand tools or power tools. Rotary motion is required to drill holes, turn parts in a lathe, grind tools, mill surfaces, cut through materials, or drive a generator or fan belt.

*Rectilinear Motion.* The feed of a tool on a lathe, the cutting of steel on a power saw, or the shaping of materials are all situations in which rectilinear or straight line motion produces work. In each of these situations a part or mechanism is used to change rotary motion to straight line motion. The screw of a micrometer, the threads in a nut, and the cross-feed screw on a lathe are still other applications where the direction of motion is changed from rotary to rectilinear.

*Harmonic and Intermittent Motion.* Any simple vibration, such as the regular back-and-forth movement of the end of a pendulum, is simple *harmonic motion.* However, many manufacturing processes require *intermittent* or irregular motion. For example, the fast return stroke of a power hacksaw or shaper ram is desirable because no cutting is done on the return stroke. Therefore, as more time is saved in returning the blade or cutting tool to the working position, the less expensive is the operation.

A device known as a *cam,* figure 11-9, is commonly used to change a uniform motion to an intermittent or irregular motion. The cam may rotate or oscillate. The cam has a surface or groove formed in such a way that it gives a desired motion to another part.

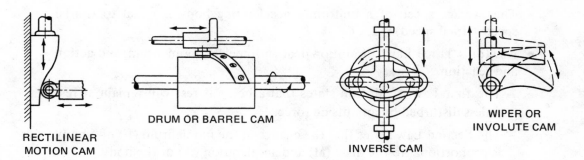

RECTILINEAR MOTION CAM    DRUM OR BARREL CAM    INVERSE CAM    WIPER OR INVOLUTE CAM

**Fig. 11-9**

The combinations of rotary and rectilinear motion obtainable are unlimited because of the large variety of parts such as gears, cams, pulleys, screws, links, and belts which can be combined in many arrangements.

## Computing Mechanisms

Mechanical movements and mechanisms are also incorporated in mathematical computation machines. Mechanical movements are used to make calculus computations, determine algebraic functions, and establish physical quantities involving angles and trigonometric functions. In each of these cases, the mechanism may be classed as an *analog computer*. The term *analog* means a relationship or similarity between two different things.

A specific length, distance, angle, or movement of a mechanism may represent time, distance, or a volume such as a liter or gallon per hour. Another mechanism may represent a specific force, work, or power output. The movement of one device is analogous to the information provided by another component of the mechanism.

Mechanical movements also are combined with electrical and electronic equipment to yield devices suitable for other types of computational applications. For example, a speedometer cable is rotated by the revolving of the steering wheel of an automobile. The cable, in turn, causes the rotation of a permanent magnet inside an electric coil. The amount of electric current generated (output) is proportional to the rate of change in the movement of the vehicle. Speed is measured by reading the position of the pointer with relation to the values calibrated (in miles per hour or kilometers per hour) on the face of the meter (speedometer).

## SUMMARY

- All bodies in the universe attract all other bodies.

- The mass of a body influences gravitation.

- All freely falling objects gain speed at a uniform rate of 32 feet per second per second regardless of the weight of the objects.

- Motion is the change of position of a body.

- Motion may be absolute or relative.

- A body has positive acceleration when it gains speed uniformly; a body with negative acceleration loses speed uniformly.

- The average speed of a uniformly accelerating body is equal to (initial speed + final speed) ÷ 2.

- Newton's Three Laws of Motion deal with inertia, momentum, and action and reaction.

  - The first Law of Motion states that a body at rest will remain at rest unless disturbed by an outside force.

  - The Second Law states that the change in the momentum (F) of a body is proportional to the mass (M) and acceleration (A) of the body.

$$F = M \times A$$

  - The Third Law states that for every force or action there is an equal and opposing force or action.

- Centrifugal force tends to cause a rotating object to fly apart; centripetal force tends to resist centrifugal force.

- Centrifugal or centripetal force is determined by the formula,

$$C = \frac{W \times V^2}{g \times r}$$

- Rotary motion is used in hand tools, machines, and mechanical devices to do work by moving parts or tools with a circular motion.

- Rectilinear motion is straight line motion which may be intermittent, uniform, or any combination of these.

## ASSIGNMENT 11 GRAVITATION, MOTION, AND MECHANICAL MOVEMENTS

### PRACTICAL EXPERIMENTS WITH GRAVITATION AND MOTION

**Equipment and Materials Required**

| Experiment A<br>Gravitation | Experiment B<br>Motion | Experiment C<br>Centrifugal Force |
|---|---|---|
| Steel ball bearing (1/2-in. diameter approx.)<br><br>Marble (1-in. diameter approx.)<br><br>Wooden or plastic ball (1 1/2-in. diameter approx.)<br><br>Grooved slide notched at 4 in., 20 in., 40 in., and 68 in. from a zero mark<br><br>Support for one end of slide<br><br>Stop watch | Movable cart, twine, and pulley<br><br>Weights:  two 2 lb.<br>           two 4 lb.<br>           two 1 lb.<br>           two 5 lb.<br>           misc. oz. weights | String<br><br>Chalk<br><br>Laboratory equipment for centrifugal force experiment |

## EXPERIMENT A. GRAVITATION

**Procedure**

1. Support one end of the grooved slide so that it is inclined from the zero notch.

2. Release a steel ball bearing at the zero mark and listen for the different sound or click as the ball passes over each notch.

3. Determine by the sound whether the interval of time between each click is the same or if it varies. Repeat step 2 a few times, then prepare a table for Recording Results and record if the sound intervals are the same or if they vary.

4. Repeat steps 2 and 3 with a marble and record whether the sound appears constant or varied.

5. Repeat steps 2 and 3 again with a wooden, plastic, or other lightweight ball. Record the sound interval.

6. Place the steel ball at the zero position. Release the ball and clock the time required for the ball to travel from the zero position to the farthest notched position. Record the time interval as measured by a stop watch. Recheck this interval a few times.

7. Repeat step 6 with the marble and the wooden or plastic balls. Record the results.

### Recording Results

| Object | Sound, Varied or Constant | Time, in Seconds |
|---|---|---|
| A  Steel ball bearing | | |
| B  Marble ball | | |
| C  Wooden or plastic ball | | |

### Interpreting Results

1. When the steel ball falls the distance between the 4-in., 9-in., 16-in., and 64-in. notches, is the sound interval for each the same, greater, or less? Explain.

2. Is the same time interval required (as clocked) for the steel marble and the plastic ball to drop from the zero to 64-in. positions? Explain.

## EXPERIMENT B. NEWTON'S SECOND LAW OF MOTION (F  = M x A)

### Procedure

1. Place a two-pound weight in the movable cart, figure 11-10.

2. Add weights as shown at (F) in the figure until the cart and load move slowly.

3. Prepare a table for Recording Results and note the weights and acceleration in each of the following steps.

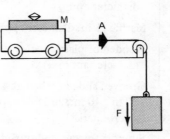

**Fig. 11-10**

4. Return the cart and load to the starting position. Double the force pulling the cart. Note the speed of the cart as compared with the speed noted in step 3. Record.

5. Increase the mass in the cart to four pounds. Add weights until the force again moves the mass. Record the force and the speed of acceleration produced.

6. Double the force on the cart. Note the change in speed (acceleration).

7. Repeat steps 5 and 6 with a six-pound mass in the cart. Record the findings.

### Recording Results

| | Mass Moved | Acceleration Produced | Force Needed |
|---|---|---|---|
| A | 2 lb. | | |
| B | 2 lb. | | |
| C | 4 lb. | | |
| D | 4 lb. | | |
| E | 6 lb. | | |
| F | 6 lb. | | |

### Interpreting Results

1. What effect does doubling the force have on the acceleration produced in each case?

2. Newton's Law states that Force = Mass Moved x Acceleration Produced. Check the formula in terms of the values recorded for this experiment.

3. Restate Newton's Law in another form that supports the results of the experiments.

## EXPERIMENT C.  CENTRIFUGAL AND CENTRIPETAL FORCE

### Procedure

1. Tie a small piece of chalk to one end of a string.

2. Hold the string two feet from the chalk and twirl the string from this point.

   NOTE:  The chalk must be tied securely; be sure no one is nearby as the string is twirled.

3. Repeat steps 1 and 2 using heavier materials. Twirl the chalk from the same point. Note the difference in the outward pull.

4. Increase the distance from the point where the string is held to the outside object. Twirl at about the same speed as before and note any differences due to the heavier objects and the greater distance.

   NOTE: This same experiment may be performed accurately with commercial laboratory equipment used to measure the centrifugal and centripetal forces.

### Interpreting Results

1. Make a simple sketch showing the centripetal and centrifugal forces acting on the string and the object as it revolves.

2. Does weight affect the centripetal and centrifugal forces? Explain.

3. Does the centrifugal force seem greater or smaller as the length of the string increases (the speed remains the same as for the shorter distance)?

## PRACTICAL PROBLEMS IN GRAVITATION, MOTION, AND MECHANICAL MOVEMENTS

### Early Experiments with Gravitation

Select the correct word or phrase to complete statements 1 to 6.

1. All objects in the universe (are) (are not) attracted to each other.

2. The (mass) (shape) of a body influences gravitation.

3. All freely falling objects gain speed at (an irregular) (a uniform) rate.

4. For an object rolling down an incline, the distance traveled is equal to $(d \times t^2)$ $(d \times t)$ $(d \times t^3)$, where d = distance and t = time.

5. As two bodies draw closer to each other, the attraction between them (decreases) (increases).

6. (Gravity) (Motion) is the attractive force that pulls all things toward the center of the earth.

7. Determine the total distance traveled by round parts as they move down an inclined, grooved conveyor at intervals of one second for each of five seconds. Use

speeds of 3 in., 4 in., 5 in., 6 in., and 8 in. per second, respectively, and the formula $D = (d)(t^2)$. Record the distances in a table similar to the one shown.

| Time Intervals (Sec.) | Speed (In./sec.) | | | | |
|---|---|---|---|---|---|
| | 3 | 4 | 5 | 6 | 8 |
| 1 | | | | | |
| 2 | | | | | |
| 3 | | | | | |
| 4 | | | | | |
| 5 | | | | | |

## Concepts of Motion

Complete statements 1 through 8.

1. _____ refers to a body that changes position.

2. Three common types of motion are: _____, _____, and _____.

3. The total distance covered divided by time is the _____.

4. A gain or loss in speed is referred to as _____.

5. The average speed is equal to _____.

6. The uniform acceleration of an object x time + the initial speed is the _____ _____.

7. In uniform motion, the distance covered is equal to _____ multiplied by _____.

8. The velocity an object loses by constant deceleration is equal to _____ multiplied by _____.

9. Determine the average speeds of a mechanism having the initial and final speeds shown in the table.

| | Initial Speed (Rpm) | Final Speed (Rpm) | Average Speed (Rpm) |
|---|---|---|---|
| A | 16 | 40 | |
| B | 225 | 675 | |
| C | 40 | 16 | |
| D | 675 | 225 | |

10. A series of motors accelerates uniformly in one second at the rate indicated in the table. Determine the final speed of each motor.

| Motor | Initial Speed (Revolutions) | Uniform Acceleration (Rev./sec.) | Time to Reach Final Speed (Sec.) | Final Speed (Revolutions) |
|---|---|---|---|---|
| A | At rest | 20 | 10 | |
| B | At rest | 35 | 8 | |
| C | 20 | 65 | 30 | |
| D | At rest | 420 | 4 1/2 | |

## Newton's Three Laws of Motion

Select the correct word or phrase to complete statements 1 to 5.

1. (Momentum) (Inertia) (Acceleration) is the tendency of a body to remain at rest, or, if in motion, to continue its motion at the same speed.

2. A body at rest (tends to remain at rest) (moves) when pulled upon by two equal and opposite forces.

3. The faster a body decelerates, the (smaller) (larger) the forces acting upon it.

4. The change in momentum of a body is determined by (the magnitude and direction of an applied force) (the inertia of the body) (the acceleration).

5. For every force, there is an (equal) (unequal) and (complementary) (opposing) force.

6. Give three examples from industry that show how every body remains at rest or continues in motion unless an external force produces a change.

7. List two industrial applications of Newton's second law of motion in which the force required to accelerate an object is proportional to the mass of the object and the acceleration produced.

8. List two industrial or commercial applications of Newton's third law of motion dealing with actions and reactions.

## Centrifugal and Centripetal Force

For statements 1 to 5, determine which are true (T) and which are false (F).

1. Centrifugal force decreases with speed and weight.

2. The centrifugal force of a revolving body causes the body to resist flying apart.

3. Centripetal force is an inward force tending to resist the centrifugal force.

4. At constant speed, the centripetal and centrifugal forces are equal.

5. The centripetal force decreases with speed and weight.

6. Explain briefly why a chipped or cracked abrasive wheel should not be used.

7. What safety precautions does the abrasive wheel manufacturer build into an abrasive wheel? Explain the answer in terms of the centripetal and centrifugal forces.

## Mechanical Movements

1. Name two mechanical devices which use rotary motion.

2. Describe what is meant by rectilinear motion.

3. List three industrial applications of rectilinear motion.

4. What is the difference between harmonic motion and intermittent motion?

5. Give an example of each type of motion: harmonic and intermittent.

6. A mechanical/electrical device is to be designed to compute the rate of change of one quantity with relation to another quantity.

   a. Identify a common application of the combination of a mechanical movement and an electrical device to compute (measure) changes of speed against time.

   b. Sketch the mechanism and label the major parts.

   c. State briefly how the mechanism works.

# Unit 12  Simple Machines: Levers

A machine is a mechanical device which uses given forces and directions to produce required work. The function of a machine is to change the magnitude or intensity of a force, the direction of a force, or the speed resulting from a force. Machines may be simple or they may be complex and involve many parts and mechanisms moving in one or more directions and at varying speeds. Examples of simple machines include a screwdriver, a wrench, and a hammer. Simple machines may be used alone or they may be combined, depending upon the kind of work and motion desired. The lever and the inclined plane are the two basic machines. The wheel and axle, the pulley, and the wedge and screws are adaptations of these basic machines.

This first unit on simple machines deals with the lever, its importance, and its application in industry. Attention is focused on the diagraming of levers, the principles of levers, the mechanical advantage of levers, and how a small force may be used to control or develop a larger force.

## CHARACTERISTICS OF LEVERS

The lever is one of the oldest and one of the simplest of all mechanical devices. A simple *lever* is nothing more than a rigid bar which turns around a fixed point. A straight steel bar used to tip up one end of a heavy crate is an example of a lever used as a simple machine. One end of the lever is placed under the crate. A block of wood located under the steel bar near the point of application is the fixed point around which the lever revolves as pressure is applied to the opposite end.

The fixed point (in this case, the block under the steel bar) is called the *fulcrum*, the crate is the *resistance* and the pressure is the *effort*. The fulcrum, resistance, and effort may be given or they must be computed for all lever problems. In addition, two other values are needed, the *effort distance* and the *resistance distance*, figure 12-1. Each value represents a distance between the fulcrum and the point of application of the effort and resistance, respectively.

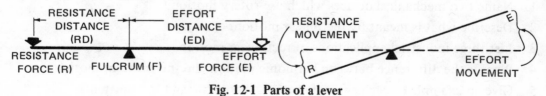

**Fig. 12-1  Parts of a lever**

Levers are usually diagramed to simplify the solution of problems and to help in visualizing the known and unknown values. The parts of a lever diagram are illustrated. The letters commonly used to indicate given values are shown for each of the three types of levers covered. Regardless of the type of lever, the terms used have the same meaning.

## TYPES OF LEVERS

### Levers of the First Class

Levers are classified according to the location of the effort and resistance forces with respect to the pivot point. For a *first class lever* the fulcrum or pivot point is located between the effort and resistance forces, figure 12-2. A first class lever increases the effort force and reverses the direction of motion. In other words, when a downward force is exerted on this type of lever, the force is applied upward against the resistance. A crowbar, a claw hammer and a pickaxe are all common examples of this class of lever. All of these tools use a smaller force to control or develop a larger force.

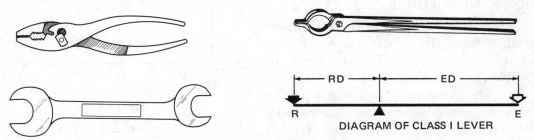

Fig. 12-2  Examples of Class 1 levers

### Levers of the Second Class

Levers of the *second class* are also used to gain force, but the fulcrum is located on one side of both the effort and resistance forces, and the direction of motion is the same. A wheelbarrow is one common example of a Class 2 lever, figure 12-3. Another example is a beam used to lift a heavy structure. In each case, the fulcrum is at one end of the lever and the load (R) is located between the fulcrum and the effort force (E).

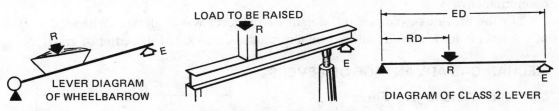

Fig. 12-3  Examples of Class 2 levers

### Levers of the Third Class

Levers of the *third class* increase speed or distance, but not force. The fulcrum of a Class 3 lever is at one side of the two forces just as it is for a Class 2 lever. The effort force, however, is applied between the pivot and the resistance, figure 12-4. Thus, the

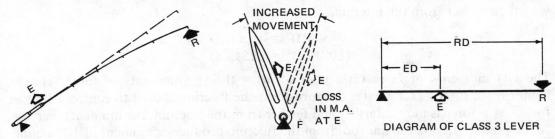

Fig. 12-4  Examples of Class 3 levers

distance between the pivot and applied force (effort distance) is less than the distance between the pivot and the load (resistance distance). The direction of movement is, of course, the same for both forces.

The fishing pole shown in figure 12-4 is an example of a Class 3 lever. The fulcrum is the point at which the pole is held against the body. The force applied by the fisherman's hands (E) increases the distance of movement at the tip of the rod (R).

Many indicating mechanisms make use of third class levers. For example, the pointer of a precision gauge shows a large movement on the dial face when a small force is applied in the working mechanism.

## MOMENTS OF FORCE

Both the effort force (E) and the resistance force (R) tend to turn the lever around the fulcrum. This tendency to revolve is called the *torque* or *moment of force*. The moment of force is equal to the force multiplied by the distance from

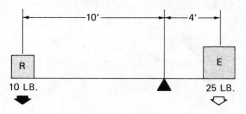

Fig. 12-5

the fulcrum. The (R) moments = R x RD and the (E) moments = E x ED. In figure 12-5, a resistance (R) of 10 pounds acts 10 feet from the fulcrum and tends to rotate the lever counterclockwise. The moment of force for (R) = (10 feet) x (10 pounds) = 100 pound-feet. Opposing the (R) moment of force is the effort (E) of 25 pounds acting four feet from the fulcrum. The (E) moment of force = (25 pounds) x (4 feet) = 100 pound-feet. When both the (E) and (R) moments are equal, the lever is balanced. This condition of balance indicates that the lever does not move. For a balanced lever, the *Law of Moments* states that the sum of the moments which tend to rotate a lever clockwise is equal to the sum of the moments which tend to turn the lever counterclockwise.

In the previous example, the weight of the lever was neglected. When the weight of the lever is to be considered, the weight is assumed to act at its center of gravity.

## MECHANICAL ADVANTAGE OF LEVERS

### Mechanical Advantage of Force

Levers are used principally to control larger and heavier forces. The control depends upon the distance of the forces from the fulcrum. As shown in figure 12-6, the 100-pound weight placed on a 10-foot beam may be raised by placing a 25-pound weight eight feet from the fulcrum.

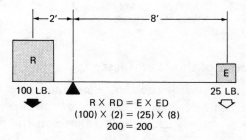

Fig. 12-6

$$R \times RD = E \times ED$$
$$(100)(2) = (25)(8)$$

The (R) moments of force (200 pound-feet) = the (E) moments of force. Thus, a lighter force placed at a greater distance from the fulcrum is used to control a heavier force. As a force is moved farther and farther from the fulcrum, the moments increase. A 100-pound force placed one foot from the fulcrum produces a moment of 100 pound-feet. At a distance of two feet from the fulcrum, the moment equals 200 pound-feet.

The moment depends on the force and the distance through which this force acts. Therefore, it is possible to use any combination of force and distance having the same product as the moment of force to be overcome. For the previous example, instead of a 25-pound effort placed at a distance of eight feet from the fulcrum, a 50-pound effort at four feet, or a 40-pound effort at five feet may be used.  Furthermore, a combination of two or more forces, equal to the same moment of force, may be used.

The mechanical advantage of force ($MA_f$) equals the resistance force (R) divided by the effort force (E):

$$MA_f = \frac{R}{E}$$

If a 50-pound effort overcomes a 100-pound resistance, the mechanical advantage of force is $100 \div 50 = 2$.

### Mechanical Advantage of Speed or Movement

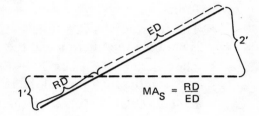

The distance that either the effort or the resistance moves and its speed depend on the position of the fulcrum. When a small force is used to move a larger force, the distance through which the (E) moment moves is greater than the distance of the (R) moment. Figure 12-7 shows that the effort arm (ED)

**Fig. 12-7**

is twice as long as the resistance arm (RD); therefore, the resistance moment moves one foot as compared with two feet for the (E) moment. In addition, the speed of the effort arm is twice that of the resistance arm because the effort arm moves twice as far in the same time. The mechanical advantage of speed ($MA_s$), however, is only one-half. As a result, the gain in force is balanced by the loss in having to move a greater distance to do the same work. The mechanical advantage of speed is expressed by the following formula.

$$MA_s = \frac{RD}{ED}$$

Levers may be compared by examining the ratio of resistance to effort (to find the mechanical advantage of force), or the ratio of resistance distance to effort distance (to find the mechanical advantage of speed).

### INDUSTRIAL APPLICATIONS OF LEVERS

One of the best known examples of a lever is the wrench. Tremendous force may be applied when tightening or loosening nuts with wrenches. This force is due to the fact that the effort force can be applied at a greater distance from the fulcrum.

Hammers of different types are also examples of the lever. Straps used for clamping work solidly to machine tables depend on leverage for increased forces. Vises also use the principles of the lever to apply force, figure 12-8.

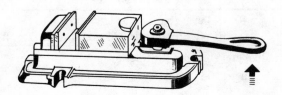
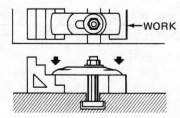

**Fig. 12-8**

The previous examples are but a few of the many applications of tools which depend on lever principles to control forces and their speeds. In later units the three types of levers are used in combination with other simple machines to produce and control complex and precise movements.

## SUMMARY

- A lever is a rigid bar which turns around a fixed point called a fulcrum.
- A lever diagram includes a simple line drawing of the lever; the known and required forces and distances are identified on the diagram.
- The fulcrum of a Class 1 lever is located between the effort and resistance forces.
- The fulcrum of a Class 2 lever is placed at one end and the resistance force is between the fulcrum and the effort force.
- The fulcrum of a Class 3 lever is placed at one end and the effort force is between the fulcrum and the resistance force.
- Regardless of the class of the lever, effort (E) x effort distance (ED) = resistance (R) x resistance distance (RD).
- The weight of the lever may be neglected in problems requiring only an approximate answer. When the lever weight must be computed, the weight is assumed to act at the center of gravity of the lever.
- The mechanical advantage of force of a lever is the ratio of the resistance to the effort force.

$$MA_f = \frac{R}{E}$$

- A lever produces an increase or decrease in the movement of either speed or force depending on the effort (E), resistance (R), effort distance (ED), and resistance distance (RD).
- A lever is balanced when the sum of the moments of force which tend to rotate the lever in one direction is equal to the sum of the moments of force which tend to rotate the lever in the opposite direction.

## ASSIGNMENT UNIT 12 SIMPLE MACHINES: LEVERS

### PRACTICAL EXPERIMENTS WITH LEVERS

**Equipment and Materials Required**

| Experiment A<br>Class 1 Levers | Experiment B<br>Class 2 Levers | Experiment C<br>Class 3 Levers |
|---|---|---|
| Graduated lever<br>Triangular wedge | Lever (band iron, 1/8″ x 1″ x 40″) | Lever (band iron, 1/8″ x 1″ x 40″)<br>(with holes drilled as shown in Exp. B) |
| Weights:<br>　two 1 lb.<br>　one 2 lb.<br>　one 4 lb. | 1/4-in. diameter steel pin for fulcrum<br>Spring balance (10-lb. capacity)<br>2-lb. weight | |

## EXPERIMENT A. CLASS ONE LEVERS

### Procedure

1. Balance a lever at its midpoint on a triangular wedge.

2. Place a one-pound weight at one end of the lever. Prepare a table for Recording Results as shown and note the direction of the lever movement.

3. Place a one-pound weight on the other end of the lever and note the direction of the movement.

4. Diagram the lever, giving dimensions and forces.

5. Move the fulcrum closer to one end of the lever. Adjust the other weight until the lever is balanced.

6. Diagram this second lever combination, indicating distances and weights.

7. Repeat the previous steps after the fulcrum is moved closer to one end. Use the two-pound weight as the resistance force and the four-pound weight as the effort force.

8. Diagram this third lever.

### Recording Results

| | Class One Lever Diagrams |
|---|---|
| **A** | |
| **B** | |
| **C** | |

### Interpreting Results

1. Compute the moments of force for each of the three levers.

2. Does the experiment prove or disprove the theory that the sums of the moments of force on both sides of the fulcrum of a balanced lever are equal? Explain briefly.

3. Give two advantages of the use of Class 1 levers.

## EXPERIMENT B. CLASS TWO LEVERS

### Procedure

1. Secure a 1/4-in. diameter pin in a vise and place one end of the lever over the pin so the lever is free to rotate.

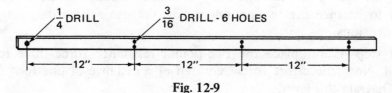

**Fig. 12-9**

2. Mount a two-pound weight at a point one foot from the fulcrum.

3. Place the hook of the spring balance one foot from the fulcrum. Note the effort force required to balance the two-pound resistance force.

4. Prepare a table for Recording Results and diagram this lever arrangement giving forces and dimensions.

5. Remove the spring balance and move it to the next hole two feet from the fulcrum. Note the effort required to balance the resistance force. Diagram this lever.

6. Repeat step 5 with the spring balance placed three feet from the fulcrum and the two-pound resistance moved until it is two feet from the fulcrum.

7. Observe the effort force required to balance the lever. Diagram the lever, giving all dimensions and forces.

### Recording Results

| | Class Two Lever Diagrams |
|---|---|
| **A** | |
| **B** | |
| **C** | |

### Interpreting Results

1. Compute the moments of force for each of the three Class 2 levers.

2. Are the computed moments of force equal in all cases? Explain.

3. Give two new examples of the use of Class 2 levers as simple machines.

   NOTE: This same experiment and the one which follows for Class 3 levers may be conducted using a saw horse or laboratory table, a 2″ x 4″ board about eight feet long as a lever, some heavy weights such as bricks or steel for the resistance force, and a spring balance of 50- or 100-pound capacity.

## EXPERIMENT C. CLASS THREE LEVERS

### Procedure

1. Use the same lever secured in a vise so that one end is free to rotate.

2. Prepare a table for Recording Results.

3. Place the two-pound weight at a point two feet from the fulcrum.

4. Attach the spring balance one foot from the fulcrum. Observe the effort force needed to balance the two-pound weight. Prepare a lever diagram giving forces and dimensions.

5. Repeat step 4, but place the two-pound resistance force three feet from the fulcrum. Note the effort force required at a distance of one foot from the fulcrum. Diagram this lever.

6. Move the spring balance to a point two feet from the fulcrum. Note the effort force needed to balance the resistance.

7. Diagram the lever and give all forces and dimensions.

## Recording Results

| Class Three Lever Diagrams | |
|---|---|
| A | |
| B | |
| C | |

## Interpreting Results

1. Compute the moments of force for each of the three Class 3 lever arrangements.

2. Give one advantage of using Class 3 levers.

3. Name two simple machines that are Class 3 levers.

## PRACTICAL PROBLEMS ON LEVERS

### Levers of all Classes

For statements 1 to 5, determine which are true (T) and which are false (F).

1. A machine may change the magnitude of a force, its direction, or its speed.

2. A lever diagram of a pair of pliers shows them to be a Class 2 lever.

3. The sum of the (E) moments of any lever is greater than the (R) moments.

4. The fulcrum is a fixed point around which a lever rotates.

5. A lever may be used to obtain the mechanical advantage of both force and speed at the same time.

Select the correct word or phrase to complete statements 6 through 10.

6. By increasing the length of a wrench handle, (less) (more) effort force is required to tighten a nut.

7. A wire-cutting pliers requires less effort to cut a wire when the wire is placed (closer to) (farther away from) the fulcrum.

8. A fluid control valve is regulated by a Class 2 lever. As the distance between the (E) force and the fulcrum is increased, the amount of pressure (R) needed to move the control lever (decreases) (increases).

9. The jaws of a pair of tongs are 3 in. long; the handles are 24 in. long. The mechanical advantage of force is (one-eighth) (eight).

10. Straps and bolts are used to hold parts securely against machine tables to resist the forces exerted by cutting tools. The work is held with (more) (less) force when (ED) is less than (RD).

## Class One Levers

1. Name a Class 1 lever which is different from the examples given in this unit.

    a. Diagram the lever.

    b. Identify all parts, forces, and distances.

2. Compute the missing values for A, B, and C in figure 12-10. Disregard the weight of each lever.

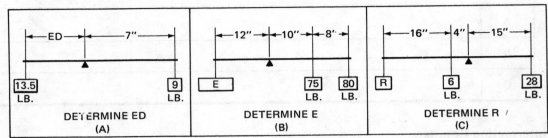

**Fig. 12-10**

## Class Two Levers

1. The center of gravity of a 150-pound load in a wheelbarrow is 16 in. from the fulcrum. What effort (E) is needed to lift the load if the handles are gripped 40 in. from the fulcrum?

2. Name two industrial applications of Class 2 levers as simple machines.

    a. Diagram each lever.

    b. Assign whole number values to the effort and resistance distances and the resistance force of one lever. Solve for the effort force (E).

    c. Assign different whole number values to the effort and resistance forces and the resistance distance. Determine (ED).

## Class Three Levers

1. A valve in an industrial production process is controlled by a Class 3 lever. An adjustable weight of two pounds is placed 14 inches from the fulcrum.

    a. Diagram the lever.

    b. Determine the value of the force which this lever arrangement controls (disregard the lever weight).

    c. Compute the (R) force that must be applied five inches from the fulcrum to balance the lever. Disregard the lever weight.

2. Name one other industrial Class 3 lever as a simple machine.

    a. Make a diagram of the lever.

    b. Assign values for (E), (ED), and (RD).

    c. Compute the value of (R).

# Unit 13 Simple Machines:
# Inclined Plane and Wedge

The second important simple machine is the inclined plane. The parts and principles of an inclined plane and some practical applications are presented in this unit.

A simple inclined plane is formed by a flat surface at an angle to another surface. The principal advantage of the inclined plane is that it makes it possible to move a load with less effort than is required to lift the same load a vertical distance.

## PRINCIPLES OF THE INCLINED PLANE

The parts, forces, and dimensions of an inclined plane include: (1) a load (resistance), (2) the force required to move the load (effort), (3) the distance the applied force moves (effort distance), and (4) the vertical height the load is lifted (resistance distance). Although these terms are the same as those used for levers, they are used in a slightly different manner.

An inclined plane is diagramed as a right triangle. The effort force (E) acts parallel to the inclined surface or *slope*. The length of this slope is the effort distance (ED). The object offering the resistance (R) is either pushed or pulled up the slope, figure 13-1.

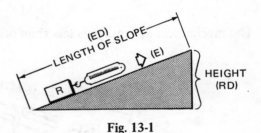

Fig. 13-1

Assume that a 500-pound machine is to be moved from the floor to a platform three feet high, figure 13-2. Using simple machines such as a lever, a set of rollers, and a reinforced inclined plane, this load can be raised and moved where needed.

Figure 13-2 shows a 150-pound effort force pulling the 500-pound weight on a path which is parallel to the 10-foot long slope. (Friction is neglected in this problem).

The effort force x the length of the slope is equal to the resistance force x the height the weight is raised. This statement can be expressed as a formula:

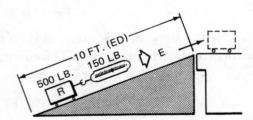

Fig. 13-2

$$(E) \times (ED) = (R) \times (RD)$$

## MECHANICAL ADVANTAGE OF AN INCLINED PLANE

### Mechanical Advantage of Force

The 150-pound effort force required to lift the 500-pound load indicates that there is some mechanical advantage to be gained by using the inclined plane. The mechanical advantage of force ($MA_f$) is the result of dividing the resistance force (R) by the effort force (E).

$$MA_f = \frac{R}{E}$$

Substituting the values from figure 13-2, the mechanical advantage of force is equal to:

$$MA_f = \frac{500 \text{ lb.}}{150 \text{ lb.}} = 3\frac{1}{3}$$

The mechanical advantage can also be found by dividing the distance of the slope (ED) by the vertical rise (RD), or

$$\frac{10 \text{ feet}}{3 \text{ feet}} = 3\frac{1}{3}$$

Thus, an inclined plane makes it possible to raise a large force using a small force.

### Mechanical Advantage of Distance

The gain in force is offset by a loss in distance; that is, the mechanical advantage of distance is equal to the height (RD) divided by the slope (ED).

$$MA_d = \frac{RD}{ED}$$

Substituting the values given in figure 13-2,

$$MA_d = \frac{3 \text{ feet}}{10 \text{ feet}} = \frac{3}{10}$$

The mechanical advantage is less than one.

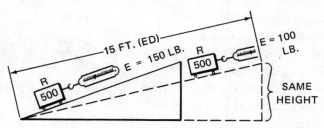

Fig. 13–3 Slope Angle Affects (E)

If the slope of this example is increased from 10 feet to 15 feet, figure 13-3, the effort required to move the 500-foot-pound load is:

$$(E) \times (15 \text{ ft.}) = (500 \text{ lb.}) \times (3 \text{ ft.})$$
$$(E) = \frac{1500 \text{ ft.-lb.}}{15 \text{ ft.}}$$
$$(E) = 100 \text{ lb.}$$

Figure 13-3 shows that the more gradual the slope, the less is the effort required to move a heavy load vertically.

## APPLICATION OF THE INCLINED PLANE TO A WEDGE

There are two basic differences between an inclined plane and a wedge. First, an inclined plane is fixed and does not move. In contrast, the wedge is movable. Second, the effort force moves parallel to the slope of an inclined plane. The effort force for a wedge is applied to the vertical height. The differences between an inclined plane and a wedge are shown in figure 13-4. Note that the resistance (R) moves along the slope of a wedge. The resistance distance (RD) is indicated by the height of the wedge under the resistance.

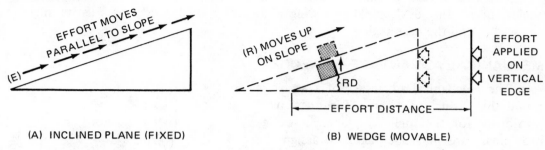

(A) INCLINED PLANE (FIXED)　　　(B) WEDGE (MOVABLE)

Fig. 13-4

### Wedges as Separating Devices

Wedges are used on tools and mechanical devices as separating or holding devices. Wedges are usually illustrated as triangles. However, there are many variations in the actual shapes of wedges. A chisel and an ax are two common applications of the wedge principle to cutting tools.

The cutting lips on a twist drill form a cutting angle. As the drill turns and moves into the work, the material to be drilled is separated in the form of chips. The teeth on a milling cutter, the teeth on a carpenter's saw, a plane blade, and a plumber's pipe cutting rolls all operate on the wedge principle as a separating device, figure 13-5.

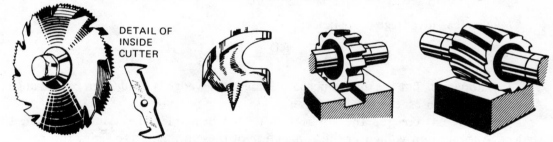

Fig. 13-5 Applications of the wedge as a separating device

### Wedges as Holding Devices

The wedge is also used for its holding power, figure 13-6. When a twist drill with a tapered shank is diagramed, it is represented as a wedge, figure 13-16A. Similarly, the

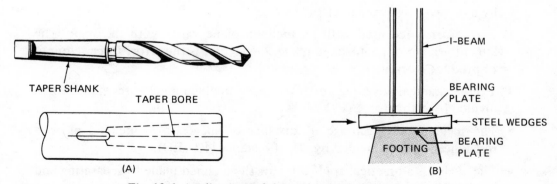

Fig. 13-6 Applications of the wedge as a holding device

bore of the drill press spindle in which the drill is placed is wedge-shaped. Another application of the wedge as a holding device is used in building construction as shown in figure 13-6B. A pair of wedges raises the I-beam and holds it securely in position.

Tool manufacturers have experimented for many years to determine the best angles for holding devices. The holding ability of a wedge decreases as its angle increases beyond the best angle.

## MECHANICAL ADVANTAGE OF WEDGES

The mechanical advantage of a wedge is found by using the same formulas used previously for the inclined plane. In wedge problems, care must be taken to assign the correct values for (E), (ED), (R), and (RD).

Cams are mechanical devices used on machines to impart motion. Cams operate on the wedge principle. A follower bearing on the wedge surface produces motion in a second mechanism.

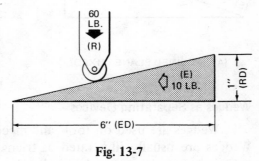

Fig. 13-7

A diagram of a cam, figure 13-7, shows that it has a mechanical advantage of force of six and a mechanical advantage of distance (or speed) of one-sixth.

Mechanical advantage of force:

$$MA_f = \frac{R}{E} = \frac{60 \text{ lb.}}{10 \text{ lb.}} = 6$$

Mechanical advantage of distance:

$$MA_d = \frac{RD}{ED} = \frac{1''}{6''} = \frac{1}{6}$$

It is apparent that the mechanical advantage of a wedge is influenced by the angle. As the angle increases, the $MA_f$ decreases.

This unit has covered the application of the principles of the inclined plane to wedges. Another unit will cover the application of these principles to screw threads.

─────────────────── SUMMARY ───────────────────

- An inclined plane is used when a large resistance is to be moved vertically by a comparatively small effort force.

- The effort force used with an inclined plane varies with the angle. The larger the angle of the inclined plane, the greater is the effort force required to produce work.

- The mechanical advantage of force of an inclined plane = resistance force divided by effort force (R/E).

- The mechanical advantage of distance or speed of an inclined plane = resistance distance divided by effort distance (RD/ED).

- The wedge is a practical application of the inclined plane to separating and holding devices.

- A wedge is diagramed as a right triangle with the base representing the effort distance (ED); the height is the effort (E); the resistance moving up the slope is (R); and the vertical distance the resistance moves is (RD).

- For any wedge, effort (E) x effort distance (ED) = (R) x (RD).

- The mechanical advantage of a wedge is computed using the same formulas as for the inclined plane.

## ASSIGNMENT UNIT 13  SIMPLE MACHINES:  INCLINED PLANE AND WEDGE

### PRACTICAL EXPERIMENTS WITH THE INCLINED PLANE AND WEDGE

#### Equipment and Materials Required

| Experiment A<br>Inclined Plane | Experiment B<br>Wedge |
|---|---|
| Adjustable inclined plane apparatus | Wedge apparatus with movable block |
| Inclined plane car | Wedges (10° and 20° angles) |
| Spring balance (10-lb. capacity) | Spring balance (10-lb. capacity) |
| Weights (1 lb. and 5 lb.) | Weights (1 lb. and 5 lb.) |

## EXPERIMENT A.  THE INCLINED PLANE

### Procedure

1. Set the slope of the inclined plane to a small angle (about 10°).

2. Prepare a table for Recording Results as shown. Measure the length and the height of the slope and record these values.

3. Weigh the inclined plane car with a 1-lb. weight attached to it. Record the combined weight as (R).

4. Attach a spring balance to the inclined plane car. Pull the car up the slope by exerting an effort force parallel to the slope.

5. Read the effort force (after the car is in motion) directly from the scale of the spring balance. Record this value in the table.

6. Repeat steps 3-5 using a 5-lb. weight in the car in place of the 1-lb. weight.

7. Increase the angle of slope of the inclined plane and replace the 5-lb. weight with the 1-lb. weight.

8. Measure (ED) and (RD).

9. Repeat the same steps to find (E). Record the value of (E) in the table.

10. Repeat the same procedure using a 5-lb. weight instead of the 1-lb. weight.

## Recording Results

| | | | Length of Slope (ED) | Height of Inclined Plane (RD) | Effort Force (E) | Resistance Force (R) |
|---|---|---|---|---|---|---|
| A | 1-lb. weight | Small slope angle (10°) | | | | |
| B | 5-lb. weight | | | | | |
| C | 1-lb. weight | Larger slope angle (20°) | | | | |
| D | 5-lb. weight | | | | | |

## Interpreting Results

1. Using the results of the experiment, prove or disprove the statement that a larger resistance force can be moved by a smaller effort force.

2. Explain briefly the effect of increasing the angle of the inclined plane.

3. Compute $MA_f$ for each case, using the data recorded in the table.

## EXPERIMENT B.  THE WEDGE

### Procedure

1. Weigh the resistance block with an added 1-lb. load. Prepare a table for Recording Results and note the combined weights.

2. Place a wedge with a small angle (10°) and the movable resistance in the vertical slide of the wedge apparatus.

3. Attach a spring balance to the end of the wedge. Pull the wedge so that it moves the resistance block. Read the effort force on the spring balance (after the wedge is moving) and record the value.

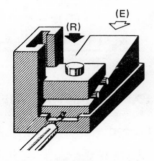

Fig. 13-8

4. Replace the 1-lb. weight with a 5-lb. weight. Repeat steps 1-3 and record the results.

5. Use a wedge with a 20° slope angle and repeat the steps 1-3 using first the 1-lb. weight and then the 5-lb. weight. Record (R) and (E) in each case.

## Recording Results

| | Load | Wedge Angle | Resistance Force (R) | Effort Force (E) |
|---|---|---|---|---|
| A | 1 lb. | 10° wedge | | |
| B | 5 lb. | | | |
| C | 1 lb. | 20° wedge | | |
| D | 5 lb. | | | |

## Interpreting Results

1. Indicate whether the wedge is used as a separating or a holding device in this experiment.

2. Is a larger or smaller (E) required to move (R)? Explain your answer briefly.

3. As the slope angle of the wedge is changed from 10° to 20° for the same (R), does the data in the table show that (E) is greater or smaller? Justify the conclusion given.

## PRACTICAL PROBLEMS WITH THE INCLINED PLANE AND WEDGE

### The Inclined Plane

Select the correct word or phrase to complete statements 1 to 6.

1. The vertical distance a load is moved up an inclined plane is the (effort distance) (resistance distance).

2. The effort force on an inclined plane acts parallel to the (slope) (base) (height).

3. Inclined planes are used because there is a (gain) (loss) in the mechanical advantage of force.

4. The $MA_f$ and $MA_d$ for an inclined plane are (the same) (different).

5. The $MA_f$ in each of the four diagrams in figure 13-9 (is equal) (varies).

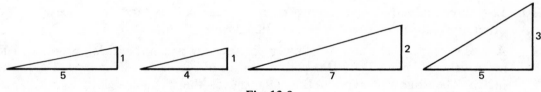

**Fig. 13-9**

6. As the (RD) on an inclined plane is increased, the $MA_f$ (increases) (decreases).

7. For parts A, B, C, D, and E in the following table, compute the forces or distances required to do the work specified using an inclined plane. (Disregard friction and round off numbers to the nearest whole number.)

| | Weight of Castings (R) | Height to be Raised (RD) | Effort Force (E) | Slope Length (ED) |
|---|---|---|---|---|
| A | | 4 ft. | 100 lb. | 8 ft. |
| B | 200 lb. | | 50 lb. | 16 ft. |
| C | 200 lb. | 4 ft. | | 4 ft. |
| D | | 4 ft. | 133 lb. | 6 ft. |
| E | 200 lb. | 4 ft. | 57 lb. | |

8. Make a graph of (ED) and (E) using the data in the table for Problem 7. Use graph paper with 1/4-in. squares. Represent (ED) by a horizontal scale of 1 in. = 2 ft., and (E) by a vertical scale of 1 in. = 50 lb. Indicate the scales on the graph.

9. Determine from the graph prepared for Problem 8 what (E) is required to move the 200-pound castings vertically four feet for slope lengths of 10 ft., 11 ft., and 12 ft.

10. Check the accuracy of the graph readings by computation. Show all mathematical processes.

**The Wedge**

Select the correct word or phrase to complete statements 1 to 5.

1.  A wedge differs from an inclined plane in that it is (fixed) (movable).

2.  The effort force on a wedge is applied along the (vertical edge) (slope) (base).

3.  The holding power of a wedge depends on the (angle of the wedge) (length of the slope).

4.  The (slope) (height) (base) represents the resistance distance of a wedge.

5.  The mechanical advantage of distance for a wedge is always (less than one) (more than one).

Add a word or phrase to complete statements 6 to 9.

6.  The wedge may be used as a _____ or a _____ device.

7.  The holding power of a wedge _____ as the inclined angle of the wedge increases beyond the best angle.

8.  The mechanical advantage of force of a wedge equals _____.

9.  The base of a wedge represents the _____.

10. List three industrial applications of the wedge as a separating device and two other applications of a wedge as a holding device.

11. A wedge-shaped cam and a follower are used in a manufacturing process. The input driving force (E) of 100 pounds moves at a constant distance (ED) of one inch. The follower, representing (R), moves the distances given in the table. Find (R) for cams A, B, C, and D.

|   | Wedge Angle of Cam | Driving Force (E) | Effort Distance (ED) | Resistance Force (R) | Follower Movement (RD) |
|---|---|---|---|---|---|
| A | 15° | 100 lb. | 1 in. |   | 1/3 in. |
| B | 30° | 100 lb. | 1 in. |   | 2/3 in. |
| C | 45° | 100 lb. | 1 in. |   | 1 in. |
| D | 60° | 100 lb. | 1 in. |   | 1 1/3 in. |

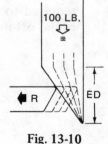

Fig. 13-10

12. Graph the relationship between the wedge angle of the cam and the resistance force (R) which can be overcome. Use a vertical scale of 1 in. = 15° and a horizontal scale of 1 in. = 50 pounds on graph paper with 1/4-in. squares. Indicate the scales and values on the graph.

13. What kind of application of the wedge is a cam and follower?

14. Interpret the graph prepared for Problem 12 in terms of the effect on (R) of increasing the wedge angle.

# Unit 14  Simple Machines: The Wheel and Axle

The wheel and axle is a simple machine consisting of a large handle or circular part (the wheel) which is rigidly secured to a smaller circular part (the axle). The wheel and axle serves one of two basic functions: (1) to transmit force or (2) to produce a change in speed.

## THE WHEEL AND AXLE DIAGRAM AND PARTS

A common example of the wheel and axle is a set of pulleys. The larger pulley is the wheel and the smaller pulley is the axle. Figure 14-1 shows that the radius of the axle is the resistance distance (RD) and the pull of the belt on the axle is the resistance (R).

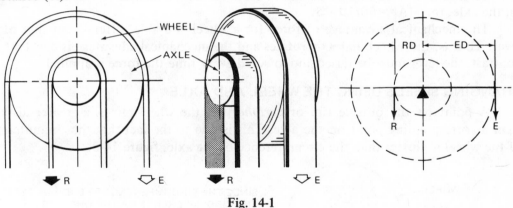

**Fig. 14-1**

The moments of force tend to revolve about the fulcrum (the fulcrum is at the center axis of the wheel and axle). The effort distance (ED) is the radius of the wheel and the effort force is represented by (E). Part of the diagram showing these forces and distances represents a Class I lever. Recall that the wheel and axle is a practical application of the lever. The symbols used on the diagram of the wheel and axle are the same as those applied to the lever.

## MOMENTS OF FORCE OF THE WHEEL AND AXLE

The moments of force of the wheel and axle are represented by the effort force and effort distance of the wheel and the resistance force and resistance distance of the axle. In the perfect wheel and axle, these moments of force are equal: (E) x (ED) = (R) x (RD).

Figure 14-2 shows an effort force of ten pounds applied to a wheel with a radius of five inches. The

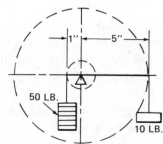

**Fig. 14-2**

resistance force acting on an axle with a radius of one inch (the force the effort force is to overcome) is found by the moment formula:  (E) x (ED) = (R) x (RD).  Substituting the values shown in figure 14-2, the result is:

$$(10 \text{ lb.}) \times (5 \text{ in.}) = (R) \times (1 \text{ in.})$$

$$\frac{50 \text{ pound-inches}}{1 \text{ inch}} = (R)$$

$$50 \text{ pounds} = (R)$$

Thus, a wheel and axle combination makes it possible to use a smaller force to produce or control a larger force.

## MECHANICAL ADVANTAGE OF FORCE

The amount of effort force needed to move a resistance depends upon the distance of both forces from the fulcrum. As the effort distance increases, the same effort force may be used to overcome a larger resistance force. In other words, the mechanical advantage of force is increased.

$$MA_f = \frac{R}{E}$$

If an effort force of 10 pounds on the wheel overcomes a resistance force of 50 pounds on the axle, the $MA_f = 50/10 = 5$.

The mechanical advantage of force for a wheel and axle differs from that of the lever. That is, the wheel and axle rotates and the mechanical advantage is constant. By contrast, the lever must be raised and lowered every time the force is applied.

## CHANGING SPEEDS USING THE WHEEL AND AXLE

A point on the outside rim or *periphery* of the wheel moves a greater distance than a corresponding point on the axle. This is due to the fact that the circumference of the wheel is greater than the circumference of the axle, figure 14-3.

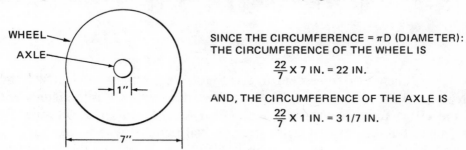

WHEEL
AXLE
1"
7"

SINCE THE CIRCUMFERENCE = $\pi$D (DIAMETER):
THE CIRCUMFERENCE OF THE WHEEL IS

$$\frac{22}{7} \times 7 \text{ IN.} = 22 \text{ IN.}$$

AND, THE CIRCUMFERENCE OF THE AXLE IS

$$\frac{22}{7} \times 1 \text{ IN.} = 3 \ 1/7 \text{ IN.}$$

Fig. 14-3

For the dimensions shown in figure 14-3, the circumference of the wheel is seven times greater than the circumference of the smaller pulley or axle. If the wheel is driving the axle, there is a loss in the mechanical advantage of speed. This loss is due to the fact that the driving belt travels seven times faster and farther than the driven belt on the axle.

Figures 14-4A and 14-4B show that any gain in the mechanical advantage of force is offset by a loss in the distance traveled or a loss of speed. Whenever the wheel and axle is used as a simple machine, only the force *or* the speed may be increased. The mechanical advantage of speed is equal to the resistance distance divided by the effort distance.

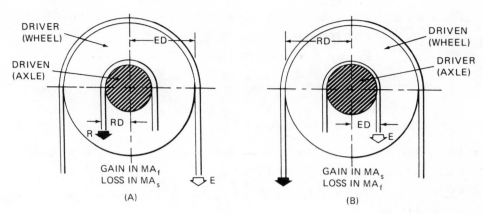

**Fig. 14-4**

## INDUSTRIAL APPLICATIONS OF THE WHEEL AND AXLE

The applications of the wheel and axle are designed to achieve either of two goals: (1) to obtain a gain in force, or (2) to control speed, figure 14-5. The first type of application includes cranks and handwheels having large radii which turn smaller axles to raise, lower, or adjust machine tables, columns, or other mechanical devices.

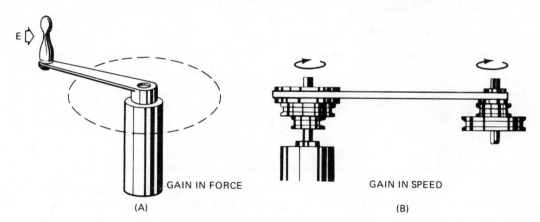

**Fig. 14-5 Wheel and axle applications**

Step-cone pulleys are an excellent example of the wheel and axle principle as it is used to control speed. Figure 14-5B shows the axle of a motor to be the driver with the largest pulley driving a smaller pulley. Such a combination produces a gain in speed on the machine spindle. Many other applications of the lever and wheel and axle principles are shown by simple gears, flywheels, and other rotating parts which gain either force or speed.

――――――――――――――――――― SUMMARY ―――――――――――――――――――

- In industry, the wheel and axle principle is used to transmit force or to change speed.

- The diagram of a wheel and axle is similar to a lever diagram and includes the same terms.

- For a wheel and axle, the effort moments equal the resistance moments.

Effort (E) x Effort Distance (ED) =
Resistance (R) x Resistance Distance (RD)

- The mechanical advantage of force = $\frac{R}{E}$.

- The mechanical advantage of speed = $\frac{RD}{ED}$

- When there is a gain in force, there is loss in speed. Conversely, a gain in speed is offset by a loss in force.

- On simple machines, it is impossible to gain both force and speed at the same time.

## ASSIGNMENT UNIT 14  SIMPLE MACHINES:  THE WHEEL AND AXLE

## PRACTICAL EXPERIMENTS WITH THE WHEEL AND AXLE

### Equipment and Materials Required

| Experiment A<br>Moments of Force | Experiment B<br>Control of Speed |
|---|---|
| Wheel and axle assembly with wheels of two different sizes | Wheel and axle assembly (same as for Experiment A) |
| Two pieces of stranded wire | Two pieces of stranded wire |
| Weights (2 lb. and 4 lb.) | Weights (1 lb., 2 lb., 5 lb., and miscellaneous sizes in ounces) |
| Spring balance | Measuring rule |
| Measuring rule | |

## EXPERIMENT A.  MOMENTS OF FORCE ON THE WHEEL AND AXLE

### Procedure

1. Measure the diameters of the large wheel and axle assembly. Prepare a table for Recording Results and record the radii.

2. Place one piece of stranded wire in the slot in the wheel. Wind the wire one and one-half turns clockwise on the wheel, figure 14-6.

3. Attach a spring balance to the end of the wire and pull gently until the wheel starts to move. Record the force required to overcome the starting friction.

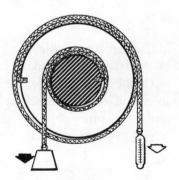

Fig. 14-6

4. Attach a two-pound weight to the second piece of stranded wire. This weight is (R).

5. Place the free end of the wire in the slot in the axle and wind the wire one and one-half turns counterclockwise.

6.  Pull on the spring balance until the wheel and axle assembly start to turn. Note the reading on the spring balance.

7.  Subtract the starting force from the final effort force and record this value as (E).

8.  Repeat steps 1-7 using a wheel twice the diameter of the first wheel. Record the force to overcome the starting friction, (RD), (ED), (R), and (E).

9.  Use the same combinations of the wheel and axle except reverse (E) so that it is applied to the axle to overcome the (R) on the wheel. Repeat steps 1-7 and record the results.

## Recording Results

|   | Radius of Wheel | Radius of Axle | Starting Friction | (R) | (E) |
|---|---|---|---|---|---|
| A |  |  |  | 2 lb. |  |
| B |  |  |  | 2 lb. |  |
| C |  |  |  | 2 lb. |  |
| D |  |  |  | 2 lb. |  |

## Interpreting Results

1.  Does the data prove or disprove the law of moments? Justify.

2.  Diagram each of the wheel and axle combinations. Omit the value of (E) and show how it can be computed if the starting friction is known.

3.  Compare the difference in effort force required to move (R) when (E) is applied to the axle and later to the wheel. Be brief and specific.

## EXPERIMENT B.  CONTROLLING SPEED WITH THE WHEEL AND AXLE

### Procedure

1.  Attach a two-pound weight to a piece of stranded wire and wind the wire clockwise around a wheel, figure 14-7.

2.  Secure another wire in the slotted axle and add weights until the (E) and (R) moments are equal and the weights are the same height from the laboratory table.

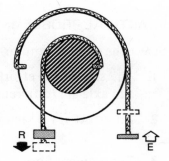

Fig. 14-7

3.  Prepare a table for Recording Results as suggested. Record the radius of both the wheel and the axle.

4.  Raise the (E) on the wheel one inch. Measure the distance traveled by the resistance force (R) on the axle.

5.  Raise (E) a total of two inches from the original position. Measure the total distance (R) moves. Record this value.

6.  Measure the radius of the second wheel on the wheel and axle assembly and record the value.

7. Change the wire and weights to the second wheel. Balance the (E) and (R) forces.

8. Lower the (E) force one inch. Measure the distance moved by the (R) force. Record this value.

9. Reverse the process by returning the wheel and axle to a balanced position. Now assume that (E) is acting from the axle and (R) is acting from the wheel.

10. Lower (E) three inches and note how far (R) moves up. Record this value.

## Recording Results

| | Radius of Wheel | Radius of Axle | Effort Force Movement | Resistance Force Movement |
|---|---|---|---|---|
| A | | | 1 in. up | |
| B | | | 2 in. up | |
| C | | | 1 in. down | |
| D | | | 3 in. down | |

## Interpreting Results

1. Compare the resulting speed when (E) is applied first to the axle and then to the wheel.

2. Prove from the data how a wheel and axle combination may be used to control speed.

3. Compute $MA_s$ for the four wheel and axle combinations in the two experiments.

## PRACTICAL PROBLEMS WITH THE WHEEL AND AXLE

Select the correct word or phrase to complete statements 1 to 6.

1. The sum of the effort moments and the sum of the resistance moments of a wheel and axle are (different) (equal).

2. An effort force, acting at a larger radius, will move a (heavier) (lighter) resistance force.

3. The moment formula for the wheel and axle is (E x R = ED x RD) (E x ED = R x RD).

4. The wheel and axle produce (both a gain in force and a gain in speed) (either a gain in force or a gain in speed).

5. The mechanical advantage of force is $\left(\frac{R}{E}\right) \left(\frac{RD}{ED}\right)$.

6. The mechanical advantage of speed is $\left(\frac{R}{E}\right) \left(\frac{RD}{ED}\right)$.

7. Compute either the effort force, the resistance force, or the resistance distance for the crank arm combinations indicated in the table at A, B, C, D, and E.

| Wheel and Axle Combinations | | | | |
|---|---|---|---|---|
| | Resistance | Effort | (RD) | (ED) |
| A | 75 lb. | 7.5 lb. | 1 in. | |
| B | 500 lb. | 125 lb. | | 12 in. |
| C | 1250 lb. | | 6 in. | 12 in. |
| D | | 960 lb. | 9 in. | 12 in. |
| E | 2250 lb. | 1800 lb. | 12 in. | |

8. Determine the mechanical advantage of force for the wheel and axle combinations given in the table in Question 7. Use the computed values for the missing values.

## Controlling Speed with the Wheel and Axle

Add a single word to complete statements 1 to 4.

1. A point on the rim of a wheel moves _____ than the corresponding point on the axis.

2. When the effort force is applied on the axle, the _____ force moves faster.

3. A gain in the speed of a wheel and axle combination is offset by the _____ in the force.

4. Speed is developed on a wheel and axle when the effort is applied to the _____ _____ .

5. Compute the $MA_s$ for the four pulley combinations given in the table for each of the three different sizes of pulleys shown.

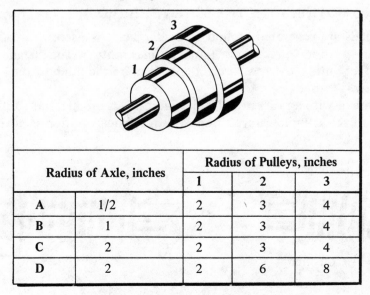

| Radius of Axle, inches | | Radius of Pulleys, inches | | |
|---|---|---|---|---|
| | | 1 | 2 | 3 |
| A | 1/2 | 2 | 3 | 4 |
| B | 1 | 2 | 3 | 4 |
| C | 2 | 2 | 3 | 4 |
| D | 2 | 2 | 6 | 8 |

# Unit 15  Simple Machines: The Screw Thread

A screw thread is one of the most practical applications of the principles governing the inclined plane, the wedge, and the wheel and axle. A *screw thread* is a ridge of uniform cross section formed around a cylinder or cone. There is a constant distance between each thread.

## SCREW THREAD USES AND TERMS

Screw threads are used (1) to transmit motion, (2) to apply tremendous forces with comparatively little effort, (3) to hold parts together, and (4) to obtain measurements.

Screw threads are specified according to their size, the number of threads per inch, and the form of the thread. Most screw threads are said to have a definite number of threads per inch. For example, in a single screw thread with ten threads per inch, the distance between two successive threads is one-tenth of an inch. This distance is called the *pitch,* figure 15-1. The pitch is important when computing the force a screw exerts, determining the mechanical advantage of a screw thread, and measuring.

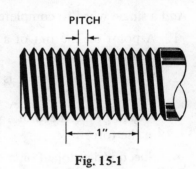

Fig. 15-1

$$\text{Pitch (P)} = \frac{1}{\text{number of threads per inch}}$$

## TRANSMITTING MOTION WITH THE SCREW THREAD

Screw threads are generated around a circular part. As a screw thread is turned, it transforms rotary motion to rectilinear (straight line) motion. The distance traveled by the part in contact with the screw thread depends upon the pitch of the screw and the amount the screw is turned.

A screw thread with a pitch of 1/10 in., figure 15-2, produces a rectilinear motion of 0.100 in. for each complete revolution. The rectilinear motion is governed by the

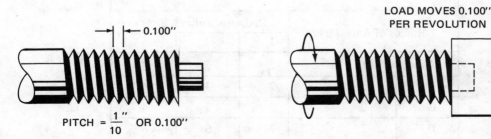

Fig. 15-2

circular motion of the screw thread. This characteristic of screw threads is applied to lead and feed screws. When such screws are turned, they move a table or column in a vertical or horizontal direction.

## SCREW THREADS USED IN MAKING ADJUSTMENTS AND MEASUREMENTS

One of the most common applications of the screw thread is to make fine adjustments such as in measuring devices. The legs on a drawing compass, a divider, a caliper, and the beam on a vernier caliper are all adjusted by a screw thread. The tables or elevating columns on precision machines are moved specified distances by attaching a graduated collar to a screw thread.

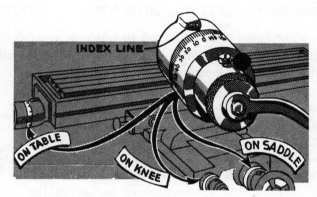

**Fig. 15-3  Machine applications of screw threads**

For example, if a lead screw on a machine has ten threads per inch, the pitch of the screw is one-tenth of an inch. As the screw advances one complete turn, the table to which the screw is attached advances this same distance (0.100 in.). If a circular collar having 100 graduations around its periphery is fastened to the lead screw, the distance the table moves from one graduation on the collar to the next is 1/100 of 0.100 in., or one-thousandth of an inch (0.001 in.).

This combination of the pitch of a screw thread and a graduated collar is commonly used on measuring instruments and mechanical devices. As a result, accurate measurements can be made easily.

At times a thread must be moved a required distance with fewer turns. When a screw thread moves twice the distance for each complete turn as it does for a single thread, it is known as a *double pitch* screw thread. Similarly, screws which advance three times as far in one turn have a *triple pitch* screw thread; a distance of four times as far indicates a *quadruple pitch* screw thread, and so forth.

Double, triple, and quadruple pitch threads are classed as *multiple threads*. The term *lead* is used with multiple threads to indicate how far the screw moves for each complete turn. Lead should not be confused with pitch. Pitch always means the distance between one thread and a corresponding point on the next thread. The *lead* of a multiple thread is the distance the screw moves with each complete turn.

## SCREW THREADS USED AS FASTENING DEVICES

Screw threads also act as wedges to give a holding action. Once screw threads lift a heavy load, they must be capable of holding the load in position. The screw on a vise has three functions: it is used to adjust the vise, it applies force to hold a workpiece, and the threads act as wedges to prevent the part from moving.

Machine and automotive bolts are common uses of the principle of the thread as a fastening device to hold parts together securely. In this type of application, the pitch of the screw threads is useful in determining the holding power.

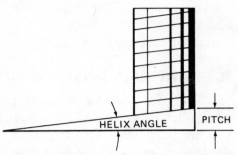

Fig. 15-4 Inclined plane applied to thread

## TRANSMITTING FORCE WITH SCREW THREADS

The screw thread is used to transmit great forces by either pushing or pulling. In this respect, the mechanical advantages of the inclined plane and the lever are combined to create a tremendous force.

One of the common applications of a screw thread is to a jack as shown in figure 15-5. There are several mechanical advantages of force which are provided by this device: (1) there is the mechanical advantage of the inclined plane which represents the *helix angle* of the thread, (2) there is the mechanical advantage of the lever which turns the screw, and (3) the combination of the screw thread and the lever result in a wheel and axle which provides another mechanical advantage.

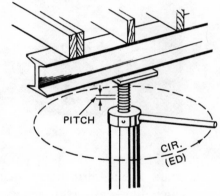

Fig. 15-5

The jack in figure 15-5 is raising an I-beam. This beam, in turn, supports another load. The screw has four threads per inch, or a pitch of 0.250 in. One complete turn of the 21-in. crank handle moves the I-beam 0.250 in. The effort force moves a distance equal to the circumference of a circle with a radius of 21 in. The circumference is equal to pi ($\pi$) x diameter. Thus, the circumference = 22/7 x 42 in. = 132 in.

The mechanical advantage is given by ED/RD.

$$MA = \frac{ED}{RD}$$

$$= 132 \div \frac{1}{4}$$

$$= 132 \times \frac{4}{1}$$

$$= 528$$

A mechanical advantage of 528 means that an effort force of 10 pounds is capable of moving a resistance force of 5280 pounds. (Friction losses are neglected in this example.) If the effort distance increases, the same effort force will lift an even greater load.

If the length of the lever handle is increased from 21 in. to 42 in. (by adding a length of pipe, for example), the ED = (pi) x (diameter) = 22/7 x 84 in. = 264 in. The mechanical advantage of this combination can be calculated by substituting (ED) = 264 and (RD) = 1/4 in the expression MA = ED/RD.

$$MA = 264 \div \frac{1}{4}$$
$$= \frac{264}{1} \times \frac{4}{1}$$
$$= 1056$$

In other words, a ten-pound effort force applied to the 42-in. handle will raise a load of (1056) x (10) or 10 560 pounds.

It is not possible to increase the effort distance indefinitely because of the physical limitations of materials. However, this simple example shows the way in which large resistance forces can be overcome by smaller effort forces. In actual practice, the mechanical advantage of the screw thread is somewhat less than the computed values because it is necessary to overcome friction losses.

The thread angle should be as small as possible and still have the required strength. The greater the thread angle, the greater is the force which tends to break the part into which the screw thread fits.

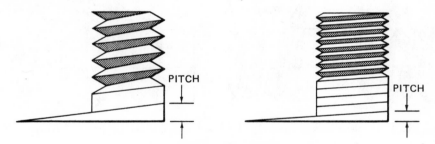

Fig. 15-6 Helix angle affects pitch

As the number of threads per inch increases, the helix angle decreases, figure 15-6. The *helix angle* corresponds to the angle of an inclined plane. The pitch of the thread is the same as the RD and the length of the slope is the ED. Since the greatest mechanical advantage is obtained when a small (E) force moves a large (R) force, a decrease in the helix angle means that the holding power of the thread is increased.

This holding power of the screw thread should not be confused with the force that a screw thread can exert. As the thread becomes finer (the pitch decreases), the thread cuts become shallower. As a result, the application of a great force to a screw with a fine pitch may strip the screw thread.

## CHARACTERISTICS OF COMMON THREAD FORMS

The most commonly used thread forms are the Unified and the American National forms. The only difference between these two forms is in the shape of the top (crest) and the bottom (root) of the thread. Both the crest and root of the American National thread form are flat. In contrast, the crest of the Unified thread form may be either flat or rounded and the root is always rounded. The characteristics of the two thread forms are shown in figure 15-7, page 140.

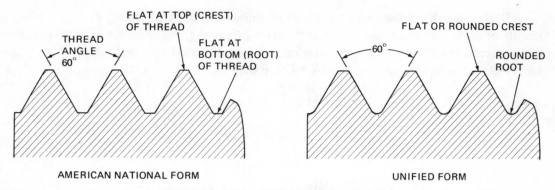

Fig. 15-7 Characteristics of thread forms

## Common Thread Forms

Figure 15-8 shows the pictorial representations of four common thread forms using the 60° included angle. The selection of the thread form to be used depends upon the application. For example, the 60° Stub thread is used where the Unified and American National forms may have too much depth for a particular part.

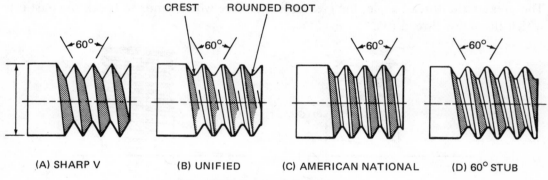

Fig. 15-8 Common thread forms

A square thread form or a modified form is used when great power or force is needed. The square form, figure 15-9A, is called a *square thread*. The more popular modified thread form has an included angle of 29° and is known as the *acme thread*, figure 15-9B. The *buttress thread*, figure 15-9C, is used where power is to be transmitted in one direction. The *knuckle thread*, figure 15-9D, is often used when a thread is molded (as in ceramic parts), or is rolled into thin metal parts (such as lamp sockets).

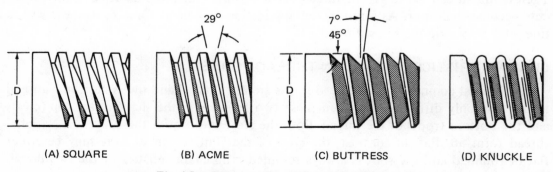

Fig. 15-9 Square and modified thread forms

## OTHER PRACTICAL APPLICATIONS OF SCREW THREADS

A nut on a saw arbor or the spindle of a grinder are typical applications of the screw thread as a holding device. The micrometer combines the principles of the screw thread and the graduated collar to make measurements to accuracies of (0.001 in.) and (0.0001 in.). The screw on a forming press has two functions in that it changes the rotary motion of a handwheel to a rectilinear motion and develops a large force against the part to be formed. The list of applications of screw threads is an endless one.

Screw threads of different forms are available with different types of heads. The square head machine screw, the hexagon head machine screw, and the Allen head setscrew are common types of screws used as fasteners.

--- SUMMARY ---

- A screw thread is a combination of two simple machines: an inclined plane which forms the helix angle and the wedge as applied to the teeth.
- Screw threads are used to:
  - change rotary motion to rectilinear or straight line motion,
  - increase force,
  - fasten parts together, and
  - make adjustments and take measurements.
- A third simple machine, the wheel and axle, is sometimes combined with a screw thread to produce a greater mechanical advantage of force.
- The pitch of a single thread screw is the distance from one point on one tooth to a corresponding point on the next tooth.
- The lead is the distance that a thread moves with each complete revolution.
- The pitch formula is:

$$P = \frac{1}{\text{number of threads per inch}}$$

- The helix angle of a screw thread determines the holding power of the thread if the physical properties of the material are not exceeded.

## ASSIGNMENT UNIT 15 SIMPLE MACHINES: THE SCREW THREAD

### PRACTICAL EXPERIMENTS WITH SCREW THREADS

#### Equipment and Materials Required

| Experiment A<br>Forming Threads | Experiment B<br>Transmitting Motion |
|---|---|
| Heavy paper<br>Rod, 1/2-in. diameter (wood or metal)<br>Rule | Screw thread plate, with coarse and fine<br>pitch screws<br>Rule |
| **Experiment C**<br>**Measuring** | **Experiment D**<br>**Transmitting Forces** |
| 0-1-in. Micrometer | Bench vise<br>Resistance block<br>Weights (1/2, 1, 2, 5, 10, and 20 lb.) |

## EXPERIMENT A. DEVELOPING A CONCEPT OF SCREW THREADS

### Procedure

1. Cut two strips of heavy paper with helix angles of 10° and 20° respectively.
2. Tape the starting point of the paper and wind the 10° inclined plane around the 1/2-in. diameter rod. Tape the second end.
3. Measure the distance between corresponding points on the inclined surface. Determine the number of such points in one inch.
4. Unwind the 10° inclined plane. Repeat step 2 using the 20° inclined plane.
5. Measure the distance between corresponding points on the inclined surface and determine the number of corresponding points per inch.

### Interpreting Results

1. State the two simple machines that can be combined to form a screw thread.
2. Is the pitch of a screw thread affected by the angle of the inclined plane? Justify the answer.

## EXPERIMENT B. SCREW THREADS USED TO TRANSMIT MOTION

### Procedure

1. Select the screw with the fine pitch and turn it into the screw plate until the identifying line on the head is vertical.
2. Measure the distance from the screw head to the plate.
3. Turn the screw one complete revolution. Measure the distance the screw moves. Prepare a table for Recording Results and note the distance the screw travels.
4. Repeat steps 1-3 with the coarse pitch screw thread. Record the distance moved.

### Recording Results

|   | Screw Pitch | Movement for One Turn | Pitch of Thread |
|---|-------------|----------------------|-----------------|
| A | Fine        |                      |                 |
| B | Coarse      |                      |                 |

### Interpreting Results

1. Determine the pitch of the fine and coarse screw threads.
2. Name the two kinds of motion transmitted by a screw.

## EXPERIMENT C. SCREW THREADS USED IN MEASURING

### Procedure

1. Turn the thimble of a 0-1-in. micrometer until the (10) graduation on the barrel is aligned with the (0) graduation on the thimble. At this position, the spindle and anvil faces are 1 in. apart.

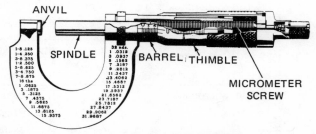

Fig. 15-10 Cutaway view of micrometer

2. Count the number of turns of the thimble required to bring the spindle and anvil together gently at the zero reading of the barrel and thimble.

3. Divide the total distance traveled (one inch) by the number of turns. Record this movement for one turn of the thimble.

4. Count the number of graduations around the thimble. Divide the pitch by this number to get the thread movement for each decimal fraction of a turn.

### Interpreting Results

1. State two scientific principles that are applied in a micrometer.

2. Briefly describe how fine measurements of 0.001 in. are obtained with a micrometer.

## EXPERIMENT D. SCREW THREADS USED TO TRANSMIT FORCES

### Procedure

1. Place a block of wood or metal in a bench vise. Turn the vise handle until the block can be moved with a slight pressure.

2. Apply an effort force of 0.5 pound on the vise handle. This can be accomplished by attaching a weight to the handle if it is in a horizontal position or by pulling with a 0.5-pound force on a spring balance tangent to the effort arm.

3. Attach weights to the block until it just begins to move down in the vise. Prepare a table for Recording Results and note both the effort force and the resistance force of the weights.

4. Add another 0.5-pound weight to the effort force exerted against the block and add more weights until the block moves. Record the effort force and the resistance force.

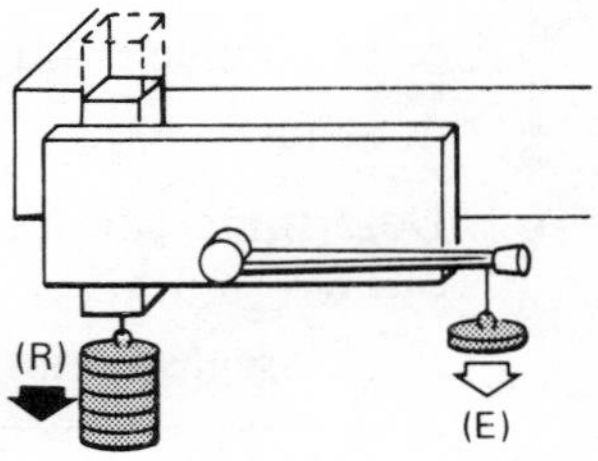

Fig. 15-11

5. Repeat the procedure with a two-pound effort force. Record both this force and the resistance force.

### Recording Results

|   | Effort Force | Pressure Exerted |
|---|---|---|
| A | | |
| B | | |
| C | | |

### Interpreting Results

1. Based on the effort forces exerted and the resistances recorded, do screw threads transmit a larger or smaller force? Justify this answer from the data obtained.

2. Determine the mechanical advantage of force due to the two-pound effort force.

## PRACTICAL PROBLEMS WITH FOUR MAJOR APPLICATIONS OF SCREW THREADS

For statements 1 to 7, determine which are true (T) and which are false (F).

1. The pitch of a thread refers to the shape of the thread.

2. For the same diameter, a screw with a coarser thread can exert a greater (E) than a screw with a fine pitch thread.

3. A screw thread is an application of the inclined plane and wedge.

4. A fine pitch screw thread travels a greater distance each revolution than does a coarse pitch screw thread.

5. The screws on an engine head serve as fasteners to hold the head securely to the engine block.

6. The threads on a jack used in building construction change a rotary motion of the screw to a straight line motion to move heavy loads.

7. The threads on a measuring instrument are used to transmit tremendous forces.

8. Prove or disprove by scientific reasoning the statement that the finer the pitch of a screw thread, the greater is the force that may be applied.

9. State three industrial applications (different from those given in this unit) of screws for each of the following:

   a. as measuring devices,

   b. to increase force,

   c. to change rotary motion to straight line motion, and

   d. as fasteners.

10. Determine the load which can be raised with the elevating screw combinations given in the table at A, B, C, D, E and F. (Disregard friction losses.)

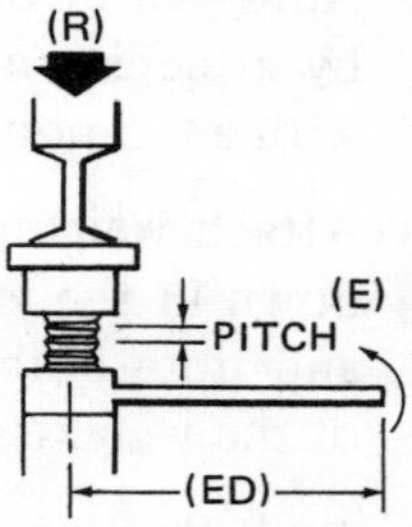

Fig. 15-12

| | A | B | C | D | E | F |
|---|---|---|---|---|---|---|
| Threads per inch | 10 | 8 | 4 | 2 | 4 | 2 |
| Overall radius of lever, inches (ED) | 28 | 28 | 28 | 28 | 42 | 42 |
| Effort force in lb. (E) | 40 | 40 | 40 | 40 | 90 | 180 |
| Load (R) | | | | | | |

11. Graph the approximate relationship between the pitch and the load for combinations A, B, C, and D (in problem 10) where the values of (ED) and (E) remain constant. Represent the threads per inch on a horizontal scale of 1 in. = 2 threads; the vertical scale is 1 in. = 12 000 pounds. Round off the computed load (R) to the nearest 1000 pounds.

12. Determine the approximate loads that screws with 3, 5, and 9 threads per inch can raise. Assume that (ED) is 28 in. and (E) is 40 pounds.

# Unit 16  Simple and Compound Gear Trains

Machines and other mechanical devices often depend upon gears and gear systems for their operation. Gears have four functions: (1) they transmit positive motion; (2) they are used to drive a second part or mechanism in either the same or the opposite direction to that of the motor or other driving force; (3) they are used to drive or control a second gear or part at the same speed, or at a slower or faster speed than the driver; and (4) they can be used to increase or decrease force. There are many types, shapes, and sizes of gears; however, the principles governing each of the four major gear functions apply to all gears.

## SIMPLE GEAR TRAINS

Motion can be transmitted from one smooth-surfaced cylinder to another by friction. If teeth are added around each cylinder, gears are formed. As one tooth in one gear drives against a tooth in a mating gear, there is positive contact and motion is transmitted.

When two gears are in mesh, the gears are known as a *pair of gears*.

As the number of gears is increased to three or more, the combination is called a *simple gear train,* figure 16-1. Simple gear trains can perform the same four basic functions listed previously. The following sections of this unit present each function as it is carried out by a simple gear train, and then, by a compound gear train.

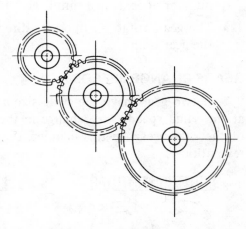

Fig. 16-1  Representation of a simple gear train

## GEARS TRANSMIT POSITIVE ROTARY MOTION

Motion can be transmitted by pulleys or shafts which either revolve against each other or are driven by belting. In these situations, there is a considerable loss in output due to slippage.

As the teeth added to circular parts are brought into contact, they will *mesh* if they have the same shape, depth, and thickness. The meshing action of one tooth on a driver gear against another tooth on a driven gear produces positive rotary motion without slippage. This feature is desirable when constant motion is required.

## GEARS CHANGE THE DIRECTION OF ROTATION

Gears in mesh with each other revolve in opposite directions. Note in figure 16-2 that as gear (A) revolves clockwise, it drives a second gear (B) counterclockwise. The gear that is rotated by the source of power is called the *driver* gear. The last gear on a train is known as the *driven* gear. If a third gear is added between the pair of gears, the driver and driven gears rotate in the same direction.

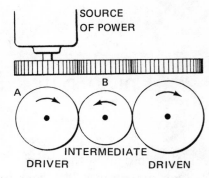

Fig. 16-2

The gears placed between the driver and driven gears are called *intermediate gears*. These gears control the direction of the driven gear. Intermediate gears also make it possible to compensate for changes in overall dimensions when gears of different sizes are used. The addition of a second intermediate gear causes the rotation of the driven gear to be in the direction opposite to that of the driver.

Regardless of the number of intermediate gears in a simple gear train, the ratio between the driver gear and the driven gear remains the same.

- The driver and driven gears rotate in the same direction when there is an odd number of gears in a simple gear train.

- The gears rotate in opposite directions when there is an even number of gears in the gear train.

## GEARS CHANGE SPEEDS

If two gears have the same size and number of teeth and are in mesh, they rotate at the same speed. Any change in the number of teeth in one gear affects the speed of the other gear. In this manner, the driven gear can be turned a desired number of revolutions.

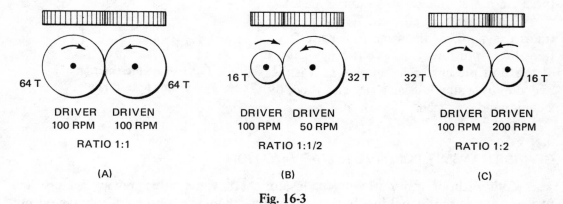

Fig. 16-3

Figure 16-3 shows three pairs of gears. For the gears at A, the ratio is 1:1 since both gears have the same number of teeth (64). Both the driver gear and the driven gear turn at 100 revolutions per minute (rpm). In figure 16-3B, a 16-tooth gear drives a 32-tooth gear. The driver gear must revolve twice to turn the driven gear one complete revolution. As a result, the speed of the driven gear is half the original value or 50 revolutions per minute. The gears in figure 16-3C consist of a 32-tooth driver gear and a 16-tooth driven gear. In one complete revolution of the driver, the driven gear makes two complete revolutions. Thus, a driver gear speed of 100 rpm results in a driven gear speed of 200 rpm.

### Ratio of Speed

The ratio of speed is the speed of the driver gear divided by the speed of the driven gear. For figure 16-3A, the speed ratio is 1:1. This ratio is obtained by dividing the driver gear speed (100 rpm) by the driven gear speed (100 rpm). For the gears in figure 16-3B, the speed ratio is:

$$\frac{100 \text{ rpm (driver)}}{50 \text{ rpm (driven)}} = \frac{2}{1} \text{ or } 2{:}1 \text{ (or } 1 : \frac{1}{2} \text{ as indicated)}$$

The ratio of the gears in figure 16-3C is:

$$\frac{100 \text{ rpm (driver)}}{200 \text{ rpm (driven)}} = \frac{1}{2} \text{ or } 1{:}2$$

In each of these examples, a comparison is made by finding the ratio of speed between the driver gear and the driven gear. As in other types of machines, it is impossible to gain speed and force at the same time.

### Ratio of Teeth for Gear Sets and Gear Trains

The number of teeth in a gear determines the speed. The ratio of teeth is found in the same manner as the ratio of speed. The ratio of teeth equals the number of teeth in the driver gear divided by the number of teeth in the driven gear. For the gear pairs in figure 16-3, the ratio of teeth for each set is determined as follows:

| (A) | (B) | (C) |
|---|---|---|
| **Ratio of First Set** | **Ratio of Second Set** | **Ratio of Third Set** |
| $\frac{\text{Teeth in Driver}}{\text{Teeth in Driven}} = \frac{64}{64}$ or 1:1 | $\frac{16}{32}$ or 1:2 | $\frac{32}{16}$ or 2:1 |

Assume that a certain machine has a gear train consisting of six gears. Three of the gears are drivers having 32, 40, and 48 teeth. These gears mesh with three driven gears having 24, 32, and 40 teeth, respectively. The ratio of teeth for the entire train is the product of the ratios for each set of gears.

$$\frac{\text{Teeth in First Driver}}{\text{Teeth in First Driven}} \text{ or } \frac{32}{24} \times \frac{40}{32} \times \frac{48}{40} \text{ or } 2{:}1$$

The addition of one, two, or more intermediate gears to any one of the three gear sets does not change the ratio of the driver gear to the driven gear.

## GEAR RATIOS AFFECT FORCE

Gears are an application of the wheel and axle. In addition, gear teeth will mesh only if they have the same size and shape. As a result, if one gear in a gear set has a smaller number of teeth, it must also be smaller in diameter. It follows from the previous statements that the effort force applied through the teeth from the smaller gear to the larger gear can overcome a larger resistance force in the larger gear.

For example, assume that a 16-tooth gear is driving a 32-tooth gear. The smaller gear revolves twice during each revolution of the large gear. The mechanical advantage of force of this gear combination is measured by the ratio of the teeth in the driven gear to the teeth in the driver gear. The $MA_f$ is 32/16 or 2. An effort force of 50 pounds on the 16-tooth gear can overcome a resistance force of 100 pounds on the 32-tooth gear. However, the gain in force is offset by a loss in speed. The speed decreases because the smaller gear must make two revolutions for each revolution of the driven gear.

### Compound Gear Trains

A *compound gear train* is formed when a driver gear and a driven gear are fixed to the same stud or shaft and cause another gear or train of gears to turn. A compound gear train provides a greater range of speed change. In addition the gears occupy less space than does a simple gear train. The advantage of this type of gear train is used in various automotive, electrical, and aviation industries where the speed of motors, engines, or other driving mechanisms must be changed, or where the force must be varied to produce a desired speed or force.

## ROTATION, SPEED, AND FORCE AFFECTED BY COMPOUND GEAR TRAINS

The gear set illustrated in figure 16-4 is one of the simplest examples of a compound gear train. Gear (A) is a driver turning gear (B). Gear (B) is securely fastened to a shaft so that it causes gear (C) to turn in the same direction and at the same speed as (B). Gear (C), in turn, is the driver for gear (D).

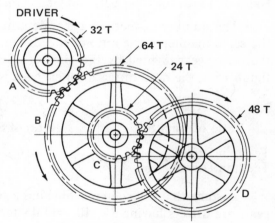

Fig. 16-4  Compound gear train

### Mechanical Advantage of Speed

In figure 16-4, note that gear (A) turns clockwise, gears (B) and (C) turn counterclockwise, and gear (D) turns clockwise. Gear (A) has 32 teeth and turns at 100 rpm. It drives gear (B) (64 teeth) at 50 rpm. Gear (C) also turns at 50 rpm. Since gear (C) has 24 teeth, it drives gear (D) (48 teeth) at 25 rpm. Thus, it is possible to reduce the speed by using a combination of gears in different sizes in a compound gear train.

The mechanical advantage of speed for any compound gear train is obtained by multiplying the $MA_s$ of each set of gears.

$$MA_s = \frac{\text{Teeth in First Driver}}{\text{Teeth in First Driven}} \times \frac{\text{Teeth in Second Driver}}{\text{Teeth in Second Driven}}$$

Substituting values from figure 16-4, the mechanical advantage of speed for this gear train is:

$$MA_s = \frac{32}{64} \times \frac{24}{48} = \frac{1}{4}$$

If the initial speed of the first driver gear of a compound gear train is 100 rpm and the $MA_s$ is 1/4, then the final speed is 25 rpm.

## PRACTICAL APPLICATIONS OF SIMPLE AND COMPOUND GEAR TRAINS

There are many machines for which unusual changes in speeds or feeds are best made by a simple gear train. Some of the older styles of lathes and milling machine dividing heads still use simple gear trains. In those situations where greater compactness is required and the gear ratios are fixed within certain limits, the use of compound gear trains is recommended.

One common example of a compound gear train is the transmission of an automobile, figure 16-5. The letters identify the gears which form the compound gear train.

Another use of the compound gear train is in the headstock of certain machines. By meshing gears with different numbers of teeth, a wide range of spindle speeds is possible. Compound gearing is used on quick change gear lathes to provide speed variation between the spindle and the lead and feed screws.

Gear reduction units use compound gear trains to gain tremendous reductions in speed with an equal gain in force that can be transmitted.

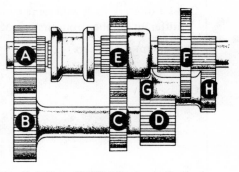

Fig. 16-5

## SUMMARY

- Gears are used to provide positive rotary motion.

- A simple gear train is a combination of three or more gears driving each other on three or more separate shafts.

- Simple gear trains are used to:
  - change the direction of rotation of shafts,
  - rotate shafts at the same or a different speed, and
  - produce changes in the mechanical advantage of force ($MA_f$) and the mechanical advantage of speed ($MA_s$).

- The mechanical advantage of speed ($MA_s$) of a simple gear train is the ratio of teeth in the driver divided by the number of teeth in the driven gear.

- Compound gear trains can produce the same speed changes as a simple gear train, but they occupy less space and use fewer gears.

- Compound gear trains can be used to change the direction of rotation of shafts as well as the speed of rotation.
- The mechanical advantage of speed ($MA_s$) of a compound gear train is the ratio of the speeds for each set of gears multiplied together.
- Whenever there is a gain in speed in either a simple or compound gear train, this gain is accompanied by a loss in force.

## ASSIGNMENT UNIT 16  SIMPLE AND COMPOUND GEAR TRAINS

## PRACTICAL EXPERIMENTS WITH SIMPLE AND COMPOUND GEAR TRAINS

### Equipment and Materials Required

| Experiment A<br>Simple Gear Trains | Experiment B<br>Compound Gear Trains |
| --- | --- |
| Standard gear demonstration board | Standard gear demonstration board |
| One set of gears, each with same number of teeth | 48-tooth gear (first driver)<br>24-tooth gear (first driven) |
| One set of gears, each with same number of teeth (different number from first set of gears) | 48-tooth gear (second driver)<br>24-tooth gear (second driven) |
| Two intermediate gears | 36-tooth gear (third driver)<br>18-tooth gear (third driven) |
| Steel tape | Steel tape |
| Chalk | Chalk |
|  | NOTE: Gears with different numbers of teeth may be used if the ratios are unchanged. |

## EXPERIMENT A.  TRANSMITTING POSITIVE ROTARY MOTION

### Procedure

1. On the demonstration board, mesh two gears having the same number of teeth.
2. Mark the gear tooth on the driver and the space on the driven gear with which it meshes.
3. Turn the driver one-half revolution and again mark the tooth and the corresponding space on the gear in which it meshes.
4. Count the number of teeth in the driver that were in mesh. Prepare a table for Recording Results and note this number.
5. Count and record the number of teeth in the driven gear that were in contact in this half revolution.
6. Turn the driver another half turn to complete the first revolution. Note the position of the chalk mark on both the driver gear and the driven gear.
7. Count the number of teeth that were in contact in both the driver and driven gears. Record this value.

8. Mark the direction of rotation on the driver. Turn the driver and note the direction of the driven gear.

9. Repeat steps 1-8 using two other gears with the same number of teeth in each gear.

## Recording Results

| | Revolutions | Number of Teeth in Contact | | Direction of Rotation | |
|---|---|---|---|---|---|
| | | Driver | Driven | Driver | Driven |
| A | 1/2 | | | | |
| B | 1 | | | | |
| C | 1/2 | | | | |
| D | 1 | | | | |

## Interpreting Results

1. What effect does the movement of one gear have on a mating gear?

2. Based on the data in the table, describe what positive motion means in terms of gear movements.

## EXPERIMENT B. CHANGING THE DIRECTION OF ROTATION AND SPEED

### Procedure

1. Add a third gear as an intermediate gear between the driver and driven gears used in Experiment A.

2. Turn the driver. Mark the directions of the driver, intermediate, and driven gears in the train.

3. Prepare a table for Recording Results and mark the direction of rotation of each gear.

4. Mark the two mating teeth on the driver, intermediate, and driven gears.

5. Turn the driver one complete revolution.

6. Count the number of teeth in the driver, and the number of teeth engaged in the intermediate and driven gears. Record these values.

7. Compute the ratio of speed between the driver gear and the driven gear.

8. Measure the overall length of the simple train (the length will be compared later to that of a compound gear train).

9. Add a fourth gear so the train has two intermediate gears.

10. Turn the driver gear and repeat steps 2-8 to find the direction of rotation of each gear and the number of teeth engaged for one complete revolution of the driver.

**Recording Results**

| | | Direction of Rotation | | | |
|---|---|---|---|---|---|
| | | Driver | Intermediate Gears | | Driven |
| | | | First | Second | |
| **A** | Direction of rotation | (⟲) | (•) | | (•) |
| | Number of teeth in mesh | | | | |
| | Overall length of train | | | | |
| **B** | Direction of Rotation | (⟲) | (•) | (•) | (•) |
| | Number of teeth in mesh | | | | |
| | Overall length of train | | | | |

**Interpreting Results**

1. Use a diagram to show the directions of rotation of (a) the driver gear and the driven gear in a gear set, (b) the gears in a simple gear train, and (c) a simple gear train with two intermediate gears.

2. Based on the data in the table, describe the effect that any number of intermediate gears has on the speed ratio between the driver gear and the driven gear.

## EXPERIMENT C. CHANGING THE DIRECTION OF ROTATION AND SPEED OF COMPOUND GEAR TRAINS

**Procedure**

1. Fasten a 48-tooth and a 24-tooth gear to the same intermediate shaft (the gears will turn together).

2. Mesh the 24-tooth intermediate gear with a 48-tooth driver gear.

3. Add the driven gear to a third shaft so that it meshes with the 48-tooth driver on the intermediate shaft.

4. Turn the driver shaft and mark the direction of rotation.

5. Prepare a table for Recording Results and show the directions of rotation of the driver gear, the intermediate gears, and the final driven gear.

6. Draw a chalk or pencil line across the gear tooth in each gear that is in mesh with a mating gear.

7. Turn the driver shaft one complete revolution. At the same time note the number of turns of the intermediate gears. Record these values.

8. Turn the driver gear slowly through another complete revolution. Note and record the number of turns of the second driver gear and the second driven gears.

9.  Determine the speed ratio of the driver gear to the driven gear.

10.  Measure and record the overall measurement of the gear train.

11.  Add a second set of intermediate gears with 48 and 24 teeth and form a compound gear train.

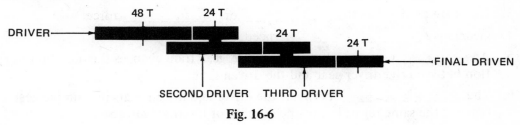

Fig. 16-6

12.  For this compound gear train, record the directions of rotation, the speed of each gear, and the overall measurement.

13.  Determine the speed ratio between the driver gear and the final driven gear.

## Recording Results

| Direction of Rotation and Overall Length | Number of Teeth | | | | | | Ratio of Final Dr to Dn | Overall Length |
|---|---|---|---|---|---|---|---|---|
| | Dr | Dn | Dr | Dn | Dr | Dn | | |
| | | | | | | | | |
| **Revolutions for One Rev. of Driver** | | | | | | | | |
| 1 rev. | | | | | | | | |
| | | | | | | | | |
| **Revolutions for One Rev. of Driver** | | | | | | | | |
| 1 rev. | | | | | | | | |

## Interpreting Results

1.  The direction of rotation of the final driven gear is affected in what manner by:

    a. one intermediate set of gears, and

    b. two sets of intermediate gears?

2.  Compare the overall length of the first compound gear train with a length measurement for a simple gear train such that the same speed ratio and direction of rotation are obtained.

## PRACTICAL PROBLEMS WITH SIMPLE AND COMPOUND GEAR TRAINS

Add a word or phrase to complete statements 1-10.

1.  Two gears in mesh are referred to as a _____ .

2.  Gears transmit _____ motion.

3.  Gears are used either to _____ or _____ force.

4.  Gears may _____ speed.

5.  An _____ gear in a simple gear train changes the direction of rotation between the driver gear and the driven gear.

6.  The _____ between the driver and driven gears in a simple gear train remains the same regardless of the number of intermediate gears.

7.  The ratio of the speed of a simple gear train is found by _____
    _____ .

8.  The relationship between the number of gear teeth in _____ and the number of gear teeth in _____ gear or intermediate gears affects force.

9.  One advantage of a gear train is that it is possible to get a _____ speed range in a smaller space.

10. The mechanical advantage of speed of a compound gear train is found by _____
    _____ the mechanical advantages (of speed) of each set of gears.

11. Name two industrial applications of (a) a simple gear train, and (b) a compound gear train.

12. Prepare a table similar to the one shown.

    a. Mark the directions of rotation of each gear combination A, B, C, and D.

    b. Compute the ratio of speed between the driver gear and the last driven gear in the simple gear trains (A and B).

    c. Find the ratio of speed between the driver gear and the last driven gear in the compound gear trains (C and D).

| | Direction of Rotation | Ratio of Speed First Driver to Final Driven Gear |
|---|---|---|
| A | 28 T <br> 64 T ( • ) ( • ) ( • ) 64 T | |
| B | 72 T ( • ) ( • ) ( • ) ( • ) 36 T <br> 36 T  48 T | |
| C | 96 T  48 T <br> 96 T  48 T | |
| D | 32 T  48 T <br> 24 T  64 T  48 T | |

13.  Which of the four gear trains in the table for question 12 can deliver the greatest force, if each gear train starts with the same force at the driver gear?

14.  Determine the speed in rpm of the driven gear shaft for each of the four combinations (A, B, C, and D) given in question 12, assuming that the first driver gear revolves at 200, 360, 175, and 480 rpm, respectively.

NOTE:  The rpm of the last driven gear = rpm of the first driven gear x the ratio of the gear train.

# Unit 17 Simple and Compound Machines

Simple machines have three basic functions: (1) they can change the direction of a force, (2) they can change the magnitude of a force, and (3) they can change the speed resulting from a force. Simple machines such as the lever and the inclined plane were studied in previous units as well as their applications to the wheel and axle, the wedge, the cam, the screw, and gears. Each of these machines can perform one or more of the basic functions of simple machines. Other machines such as pulleys and pulley systems and the worm and wheel are also commonly used to perform these functions. This unit covers these latter machines and then applies the principles of simple machines to compound machines.

## THE PULLEY AND PULLEY SYSTEMS

The pulley is used in construction work, power drives, on industrial and home workshop machines, and in any application where heavy loads must be lifted or moved with a large mechanical advantage. A pulley is another application of the principle of the lever. The word *sheave* identifies the number of grooves in a pulley. For example, *single sheave* indicates a pulley with a single groove, *double sheave* means a pulley with a double groove, and so on.

Pulleys may be fixed or movable. A *fixed pulley* stays in one position and does not rise or fall with any movement of the load. A stationary pulley fastened to a crane or some supporting beam is a fixed pulley. The fixed pulley operates on the principle that a resistance force can be moved by the same amount of effort force (neglecting friction). In addition, for the fixed pulley, both the effort and resistance distances are equal. Although fixed pulleys change the direction of a force, they do not increase the mechanical advantage.

Figure 17-1 shows that 50 pounds of effort moving downward one foot are required to raise a 50-pound resistance force one foot. However, if a movable pulley is used, the same resistance force of 50 pounds can be lifted by applying an upward effort force of 25 pounds. In this case, the effort distance is twice that of the resistance distance. Thus, the mechanical advantage of force is offset by the mechanical advantage of distance.

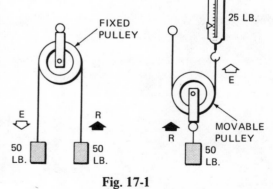

**Fig. 17-1**

## MECHANICAL ADVANTAGE OF THE PULLEY

When several pulleys are used in combination, the assembly is often called a *block and tackle*. This type of pulley combination increases the mechanical advantage of force so that tremendous loads can be lifted with little effort. The mechanical advantage of force is obtained when movable pulleys are used (recall that fixed pulleys change only the direction of the rope or cable from one pulley to the next). The mechanical advantage of force for a block and tackle depends on the number of ropes supporting the load on the movable pulley.

Figure 17-2 shows a 100-pound load being raised with two pulley combinations. Four ropes support the resistance in figure 17-2A. The effort force is 25 pounds.

In figure 17-2B, five ropes are supporting the same load. The 100-pound resistance in this case, can be moved with an effort force of 20 pounds. Thus, the load is lifted two feet when the effort force moves eight feet in figure 17-2A and ten feet in figure 17-2B. In each case, E x ED = R x RD. Any gain in effort is offset by a loss in distance.

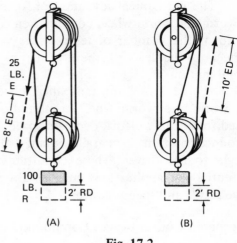

**Fig. 17-2**

## COMPOUND MACHINES

Simple machines such as the lever and the inclined plane, and the applications of these machines to other machines such as the wedge, the screw, gears, and the block and tackle, can be used alone or in combination with one another. The phrase *complex* or *compound machines* is used to describe two or more simple devices combined in such a way that they work as a single mechanism or device. The result of combining one simple machine with another is that the mechanical advantage of force or speed for the compound machine is the product of the mechanical advantages for each simple machine used.

### The Worm and Worm Wheel

The worm and worm wheel is an example of a compound machine which combines the principles of the inclined plane and the wheel and axle. This compound machine can (1) reduce speed, (2) increase force, (3) change the direction of motion, and (4) transmit positive motion.

As a worm turns, its threads engage the teeth on a worm wheel, figure 17-3. For each revolution of the worm, the worm wheel is moved a distance equal to just one tooth (for a single pitch screw worm). Figure 17-4 (page 158) shows a worm wheel with 64 teeth. If this gear is to make one complete revolution, the worm gear must make 64 complete revolutions. If the worm is part of a drive unit turning at 3200 rpm, the shaft to which the worm wheel is attached makes only 50 rpm. When ten pounds of effort are applied to turn the worm, the worm wheel can deliver a force of 640 pounds.

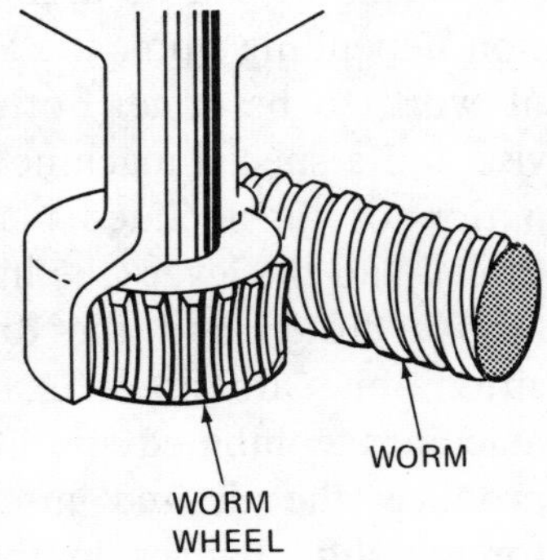

**Fig. 17-3**

The mechanical advantage of force for a worm and worm wheel combination is the ratio of the number of teeth in the worm wheel to the lead of the worm. For the combination in figure 17-4, the $MA_f$ is 64:1. This fact is important since this type of worm and worm wheel combination is used to multiply effort, decrease speed, and produce motion in parallel planes at an angle to the driver. These conditions are essential for applications such as automotive drives and speed reducers.

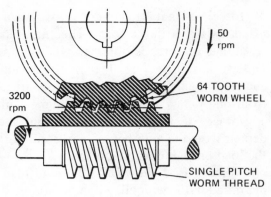

Fig. 17-4

## Mechanical Advantage of Compound Machines

The first step in determining the mechanical advantage of any compound machine is to identify the simple machines contained in it. The next step is to determine the mechanical advantage of each simple machine. The product of the mechanical advantages for these simple machines is the mechanical advantage of a compound machine. A common example of a compound machine is shown in figure 17-5. The student should be able to determine that this machine consists of an inclined plane and a pulley system. The mechanical advantage of force ($MA_f$) for the inclined plane is equal to the slope of the plane divided by the resistance distance, or 20 ft./4 ft. = $MA_f$ of 5. The $MA_f$ of the pulley system is equal to the number of ropes on the movable pulley, or three. The $MA_f$ of this compound machine is 5 x 3 or 15. As a result, a 300-pound effort force can move a resistance of 4500 pounds, neglecting friction.

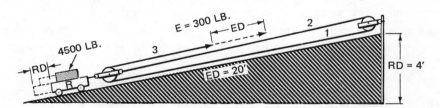

Fig. 17-5

Figure 17-6 shows just one example of a compound machine. The various simple machines can be used in any combination depending upon the amount and kind of work to be done. Both a woodworking vise and a special machine vise are a combination of the inclined plane of the screw thread and the lever (the handle). The automotive clutch and the differential on an automobile are other examples of simple machines combined in different forms to produce the desired motion, changes in forces, and changes in the magnitudes of forces.

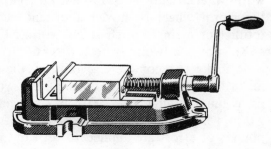

Fig. 17-6

─────────── SUMMARY ───────────

- A pulley is an application of the lever.

- The mechanical advantage of force of a pulley combination is determined by the number of ropes supporting the load on the movable pulley.

- A gain in the mechanical advantage of force ($MA_f$) of a pulley combination is offset by a loss in the mechanical advantage of speed ($MA_s$).

- A compound machine consists of two or more simple machines working in combination with each other.

- Compound machines produce large gains in either force or speed.

- The worm and worm wheel is a compound machine resulting from the combination of the inclined plane and the wheel and axle.

- The worm and worm wheel is used in industry to obtain tremendous reductions in speed and increases in force and to change the angle of rotation.

- The mechanical advantage of force ($MA_f$) of a compound machine is the product of the mechanical advantages of force ($MA_f$) of the simple machines combined to form the compound machine.

## ASSIGNMENT UNIT 17  SIMPLE AND COMPOUND MACHINES

### PRACTICAL EXPERIMENTS WITH PULLEY SYSTEMS AND COMPOUND MACHINES

**Equipment and Materials Required**

| Experiment A<br>The Pulley and Pulley Systems | Experiment B<br>Compound Machines |
|---|---|
| Two single sheave pulleys | Two single sheave pulleys |
| Weights (four at 20 lb. each) | Two double sheave pulleys |
| Spring balance (capacity 50 lb.) | Inclined plane equipment with car |
| Measuring stick or tape | Weights (two 50 lb., two 100 lb., and miscellaneous smaller weights totaling 100 lb. |
| Heavy cord or light steel stranded wire | Measuring stick or steel tape |
| | Spring balance |

### EXPERIMENT A.  THE PULLEY AND PULLEY SYSTEMS
**Procedure**

1. Suspend a single sheave pulley from a strong support.

2. Attach a 20-lb. weight as a resistance on one end of the rope and a similar weight as the effort on the other end.

3. Slowly lift the weight (E) 1 ft. Measure the distance that (R) moves down.

4. Prepare a table for Recording Results and in the steps that follow, record (E) (ED), and (RD), as required.

5. Weigh a single sheave pulley. Arrange the equipment as illustrated in pulley arrangement (B).

6. Pull the rope or wire taut and mark both the wire and the pulley. Continue to pull the wire until the resistance force is lifted 6 in. Measure (ED).

7. Arrange a pair of pulleys as shown at (C) and suspend a resistance force of 60 lb.

8. Mark the starting point on the rope and movable pulley.

9. Pull the free end of the wire until (R) is raised 1 ft. Note (E) and measure (ED).

10. Repeat steps 7-9 using the fixed and movable pulleys arranged as shown at (D). Add weights for (E).

## Recording Results

| | | Pulley Arrangement | | |
| | A | B | C | D |
|---|---|---|---|---|
| | | | | |
| Effort Force (E) | | * | * | * |
| Effort Distance (ED) | 1 ft. | | | |
| Resistance Force (R) | 20 lb. | 20 lb. | 60 lb. | 60 lb. |
| Resistance Distance (RD) | | 6 in. | 1 ft. | 1 ft. |
| *NOTE: Subtract the weight of the movable pulley to obtain (E). | | | | |

## Interpreting Results

1. For a single fixed pulley, are (E) and (R), and (ED) and (RD), equal or unequal? Prove your answer in terms of the data recorded for the experiment.

2. What is the mechanical advantage of a single sheave movable pulley?

3. From the data in the table for pulley arrangements (C) and (D), develop a rule for determining the mechanical advantage of fixed and movable pulley combinations.

## EXPERIMENT B. COMPOUND MACHINES

A single example is used to show experimentally how the mechanical advantage of each simple machine is used to obtain the combined mechanical advantage of a compound machine.

## COMBINATION ONE.  A PULLEY USED AS A SIMPLE MACHINE AND IN COMBINATION WITH AN INCLINED PLANE

### Procedure

1.  Set up a single sheave fixed pulley in Position I, figure 17-7. Determine (E) for the 50- and 100-lb. resistance forces (R), including the weight of the inclined plane car.

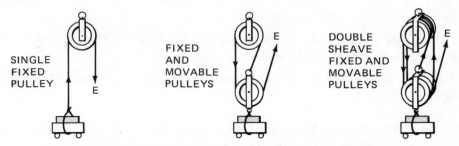

Fig. 17-7  Position I.  Load raised vertically by a simple machine.

2.  Prepare a table for Recording Results and record the values for each step.

3.  Refer to Position II in figure 17-8 and arrange the inclined plane with an (RD) of 1 ft. and the single sheave fixed pulley as shown.

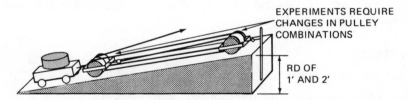

Fig. 17-8  Position II.  Load raised using a compound machine.

4.  Attach a spring balance to the free end of the rope. Pull slowly and as (R) moves, record the effort force for the 50- and 100-lb. loads, (A) and (C) in the table.

5.  Raise the inclined plane to an (RD) of 2 ft. and determine the effort force for loads (B) and (D).

### Recording Results

| | Load (R) | Inclined Plane | | Single Fixed Pulley (E) | | Single Fixed and Movable Pulleys | | Double Sheave Fixed-Movable Pulley Combination | |
|---|---|---|---|---|---|---|---|---|---|
| | | (RD) ft. | (ED) ft. | Pos. I | Pos. II | Pos. I | Pos. II | Pos. I | Pos. II |
| A | 50 lb. | 1 | 5 | | | | | | |
| B | 50 lb. | 2 | 5 | | | | | | |
| C | 100 lb. | 1 | 5 | | | | | | |
| D | 100 lb. | 2 | 5 | | | | | | |

## COMBINATION TWO. SINGLE SHEAVE FIXED AND MOVABLE PULLEY COMBINATIONS (SIMPLE AND COMPOUND MACHINES)

### Procedure

1. Set up the single sheave and movable pulley in Position I.

2. Determine (E) for the 50- and 100-lb. resistance forces (R).

3. Arrange the inclined plane so that the (RD) is 1 ft. Place the single sheave fixed and movable pulleys in Position II.

4. Attach a spring balance to the free end of the rope. Pull slowly and as (R) moves, determine and record (E) for loads (A) and (C).

5. Raise the inclined plane to an (RD) of 2 ft. and determine (E) for loads (B) and (D).

## COMBINATION THREE. DOUBLE SHEAVE FIXED AND MOVABLE PULLEYS (SIMPLE AND COMPOUND MACHINES)

### Procedure

1. Set up the double sheave fixed and movable pulleys in Position I.

2. Determine (E) for the 50- and 100-lb. loads (R).

3. Repeat steps 3-5 of Combination II, using the inclined plane and the double sheave pulleys in combination.

### Interpreting Results

Using the data recorded in the table at A, B, C, and D for Positions I and II, determine:

1. The $MA_f$ of the single sheave fixed pulley combinations,

2. The $MA_f$ of the single sheave fixed and movable pulleys, and

3. The $MA_f$ of the double sheave fixed and movable pulleys.

4. Based on the calculations for the $MA_f$, state three conclusions about simple and compound machines.

## PRACTICAL PROBLEMS WITH SIMPLE AND COMPOUND MACHINES

### Pulley and Pulley Systems

1. Match each item in Column I with the correct condition in Column II.

| Column I | Column II |
|---|---|
| a. Fixed pulley | 1. The pulley moves as the load moves. |
| b. Movable pulley | 2. Depends on the number of ropes supporting the load on the movable pulley. |
| c. Block and tackle | 3. Takes more effort to move a smaller resistance. |
| d. Mechanical advantage | 4. Several pulleys used in combination with each other. |
| | 5. The pulley remains in one position and does not rise or fall with the load. |

2. A 625-pound engine is to be raised one foot and placed on a skid. Two double sheave pulleys are available.

   a. Sketch the pulley arrangement that requires the least effort force to raise the load.

   b. Determine (E) and (ED).

## Compound Machines

Select the correct word or phrase to complete statements 1 to 7.

1. A worm and worm wheel is a (compound) (simple) machine.

2. A worm and worm wheel is a combination of these two simple machines: (inclined plane) (pulley) (wheel and axle).

3. A worm and worm wheel is used principally to (increase and decrease) (increase) (decrease) force.

4. A worm and worm wheel is used primarily to (increase) (decrease) speed.

5. A worm and worm wheel produces a $MA_f$ of (one) (more than one).

6. The (product) (sum) of each simple machine in a compound machine combination is the $MA_f$.

7. A compound machine consists of (a simple machine working alone) (two simple machines working independently of each other) (two or more simple machines combined).

8. From the dimensions and force given in figure 17-9, determine:

   a. the $MA_f$ of the gears,

   b. the $MA_f$ of the crank,

   c. the $MA_f$ of the combined machines, and

   d. the (R) that can be moved.

   NOTE: The radius of the small gear is 1 in.

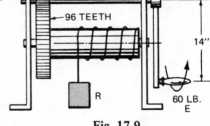

Fig. 17-9

9. A screw jack with two threads per inch and a lever arm with a radius of 35 in. are pulled with an effort force of 75 pounds.

   a. Name the simple machines combined in the jack.

   b. Determine the $MA_f$ of each simple machine.

   c. Determine the $MA_f$ of the combination.

   d. Give the weight of the load that may be lifted, disregarding friction.

10. Study a crane, a derrick, or a pneumatic power shovel on a construction project.

    a. Sketch the pulley combination and also the driving gear mechanism.

    b. Determine the $MA_f$ of the fixed and movable pulley combinations.

    c. Find the $MA_f$ between the driver and the driven gears on the drive mechanism.

    d. Compute the $MA_f$ of the entire compound machine.

# Unit 18 Mechanical Power Transmission, Friction, and Lubrication

*Power* is the term used to indicate the rate of work. Power is equal to the work done divided by the time required to do the work. It follows from this statement that a given amount of work may be done in either less or more time, depending on the power used.

## MECHANICAL POWER TRANSMISSION

Power in a usable form is transmitted from the source to the machine or mechanism which produces the desired motion or work. Common sources of power are gas engines, diesel engines, turbines, and electric motors (the most universal of all sources).

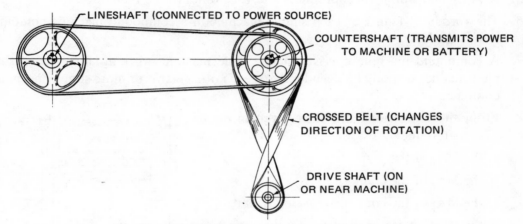

Fig. 18-1 Old "flat belt" system of transmitting power

Regardless of the source of power, it can be transmitted mechanically from the motor to a battery of machines by means of pulleys and belts. One problem with this method is that the great distances required for power transmission produce large power losses due to belt slippage. Another disadvantage arises because of the excessive weight and size of the entire lineshaft and pulley system.

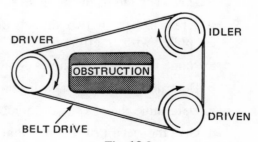

Fig. 18-2

In many situations, power systems must be designed for locations where there is an obstruction between the driver shaft and the driven shaft. One or more idler pulleys can be inserted in the pulley system to avoid the obstruction, figure 18-2. In theory, the use of idler pulleys does not affect the mechanical advantage of the pulley system.

## Modern Mechanical Power Transmission Devices

Direct power transmission in modern industrial equipment means that when a shaft or spindle revolves at a speed controlled within the motor itself or by an electrical device, the power can be delivered directly from the motor shaft. Polishing and grinding machines are examples of the application of direct power transmission.

Modern machine tools combine the mechanical advantage principles of compound machines and have a source of power integral with (included with or a part of) the machine. Such compact machines make it possible to reduce distances and part sizes. As a result there is less power loss than in older machine tools.

The gear is a device commonly used to provide positive power transmission without slippage, figure 18-3.

At higher speeds, a chain drive can be used to produce positive transmission. A chain drive does not require the same high initial tension necessary for either a V-belt or a flat belt.

Power can be transmitted from a device producing a rotary motion to a device requiring a straight line motion by the use of the screw principle. The forward or reverse feed movement of the cutting tool on a metal cutting lathe is controlled by a *feed screw*. This screw transmits rotary power through another part which converts the motion into the straight line motion of the carriage.

Power can be transmitted in the same direction as the motor shaft, or at an angle to the shaft, figure 18-4. A device known as a *coupling* is used to connect the source of power to a given shaft. The *rigid coupling* forms a direct connection or a direct drive. The *flexible coupling* provides a drive at an angle to the power source.

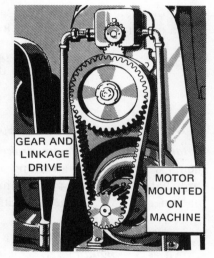

GEAR AND LINKAGE DRIVE

MOTOR MOUNTED ON MACHINE

Fig. 18-3

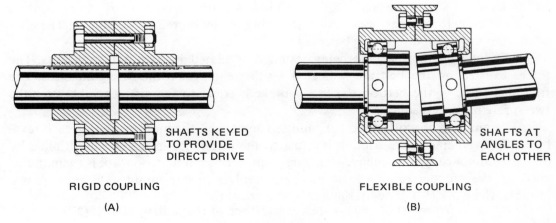

SHAFTS KEYED TO PROVIDE DIRECT DRIVE

SHAFTS AT ANGLES TO EACH OTHER

RIGID COUPLING

(A)

FLEXIBLE COUPLING

(B)

Fig. 18-4

The continuous transmission of power is not always required. Intermittent power use is possible with the use of mechanisms such as clutch arrangements which either engage or disengage the power source.

## Principles of Mechanical Power Measurement

*Reciprocating Machines.* The term *reciprocating machine* identifies a combination of parts (machine elements). These parts collectively produce a back and forth, and in and out reciprocating motion. A simple example is a piston which changes the rotary motion of the eccentric cam on a crankshaft into an up and down motion to produce work.

Power was defined as the rate of doing work. The standard measure of power is the horsepower. One horsepower (h.p.) is equal to 33 000 foot-pounds of work done in one minute. The horsepower required to do a given amount of work in a specified time is given by the following formula. Remember that force x distance = work.

$$\text{h.p.} = \frac{\text{Force (lb.) x Distance (ft.)}}{\text{33 000 ft.-lb./min. x Time (minutes)}}$$

For example, if a reciprocating compound machine exerts a force of 55 pounds over a distance of 100 feet, the horsepower (neglecting friction) is:

$$\text{h.p.} = \frac{\text{Force (lb.) x Distance (ft.)}}{\text{33 000 ft.-lb./min. x Time (min.)}} =$$

$$\frac{\text{55 lb. x 100 ft.}}{\text{33 000 ft.-lb./min. x 1 min.}} = \frac{5\ 500}{33\ 000} = \frac{1}{6}$$

Thus, a motor with a rating of 1/6 h.p. is required to do this work.

*Rotating Machines (Prony Brake).* *Rotating machines* consist of machine parts or elements which produce a turning or rotary motion. The flywheel of an engine, the armature shaft of an electric motor, and the spindle of a power tool are classified as rotating machinery. All of the machine elements just listed are actuated or controlled by other machine parts.

The mechanical power of a rotating machine can be measured with an instrument known as a *dynamometer.* This device absorbs the energy output of the machine, and converts it to heat or electrical energy which can be measured in watts. One type of dynamometer is the Prony brake, which measures the brake horsepower delivered to the flywheel shaft, or spindle, of a machine. A simple type of *Prony brake* consists of a band or belt that passes around a revolving pulley on the power source. The ends of the belt are secured to spring balances.

The load on the machine being tested is regulated by two adjusting nuts. The forces exerted on the belt are indicated by the readings on the spring balances. The frictional force of the belt on the Prony brake is equal to the difference between the two spring balance readings.

The amount of work done per minute (power) by the machine during each revolution to overcome brake friction is equal to the difference in readings multiplied by the circumference of the pulley. For one minute this amount of work is again multiplied by the number of times the machine revolves in one minute. Expressed as a formula, the work (W) done by the machine is equal to:

$$W = (F_1 - F_2) \text{ x (Circumference of the pulley, } \pi D) \text{ x (N)}$$

where N = revolutions per minute

$F_1, F_2$ = readings on the spring balances

D = diameter of the pulley

If the $F_1$ and $F_2$ readings are in pounds, and the circumference is given in feet, W is given as foot-pounds. The brake horsepower can be determined by dividing the amount of work done by the machine (in foot-pounds) by the 33 000 foot-pounds per minute.

$$\text{h.p.} = \frac{(F_1 - F_2) \times (\pi D) \times (N)}{33\,000 \text{ ft.-lb./min.}}$$

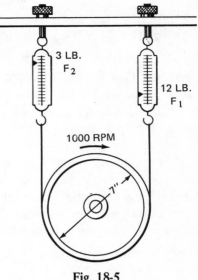

Fig. 18-5

Figure 18-5 shows a difference between $F_1$ and $F_2$ of 9 pounds acting on a 7-in. diameter pulley turning at 1000 rpm. Using the Prony brake formula and substituting values, the brake horsepower of the machine being tested can be obtained.

$$\text{h.p.} = \frac{(12 \text{ lb.} - 3 \text{ lb.}) \times \left(\frac{22/7 \times 7 \text{ in.}}{12 \text{ in.}}\right) \times (1000 \text{ rpm})}{33\,000 \text{ ft.-lb./min.}} = \frac{1}{2}$$

The Prony brake provides a fast and simple method of measuring the mechanical power of rotating machines.

## FRICTION: CAUSES AND TYPES

In many of the problems thus far, the student has been asked to disregard or neglect friction. Actually, friction is present to some degree whenever two parts are in contact and move on one another. The term *friction* refers to the resistance of two or more parts to movement.

### Valuable or Harmful Friction

Friction is harmful or valuable depending upon where it occurs. Friction is necessary for fastening devices such as screws and rivets which depend on friction to hold the fastener and the parts together. Belt drives, brakes, and tires are additional applications where friction is necessary.

The friction of moving parts in a machine is harmful because it reduces the mechanical advantage of the device. The increased effort force required to overcome friction between the turning parts causes a loss in power and a resulting increase in heat.

The heat produced by friction is lost energy because no work takes place. Also, greater power is required to overcome the increased friction. Heat is destructive in that it causes expansion. Expansion may cause a bearing or sliding surface to fit tighter. If a great enough pressure builds up because of expansion, the bearings may *seize* or freeze. In addition, as heat is developed, bearings made from low temperature materials may melt.

## Causes of Friction

Friction in solids results from any one or combination of three general causes: (1) surface finish, (2) the cohesion or adhesion of molecules, and (3) weight or pressure.

*Surface Finish.* A microscopic examination of two surfaces in contact with each other shows irregularities in both surfaces, figure 18-6. As the two surfaces move, the high and low points of each tend to interlock and prevent the surfaces from sliding freely over one another. As a result, a great force is needed to start or keep the parts moving. The surface irregularities cause a grinding or shearing action as the surfaces move over one another. This wear produces heat and subsequent power losses.

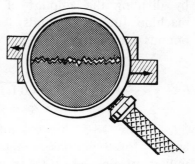

Fig. 18-6

*Intermolecular Attraction.* Friction is believed to be caused by *cohesion,* or the attraction of like molecules of the same material. Cohesion is greater as the surfaces of two like materials come closer in contact. Thus, for two like materials with highly polished surfaces, the following is true of the friction: (1) the friction caused by surface roughness is greatly reduced, and (2) the friction caused by cohesion is increased.

Friction is also caused by *adhesion,* or the attraction of unlike molecules in two different surfaces. The tendency of some soft metals to cling to rotating parts is due to adhesion. Similarly, the attraction between liquids and the walls of piping is the result of adhesion.

## Types of Friction

There are three types of friction which must be overcome in moving parts: (1) starting, (2) sliding, and (3) rolling. *Starting friction* is the friction between two solids that tend to resist movement. When two parts are at a state of rest, the surface irregularities of both parts tend to interlock and form a wedging action. To produce motion in these parts, the wedge-shaped peaks and valleys of the stationary surfaces must be made to slide out and over one another. The rougher the two surfaces, the greater is the starting friction resulting from their movement.

Since there is usually no fixed pattern between the peaks and valleys of two mating parts, the irregularities do not interlock once the parts are in motion but slide over one another. The friction of the two moving parts is known as *sliding friction.* As shown in figure 18-7, starting friction is always greater than sliding friction.

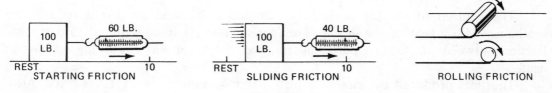

Fig. 18-7

*Rolling friction* occurs when roller devices are subjected to tremendous stresses which cause the parts to change shape or *deform.* Under these conditions, the material in front of a roller tends to pile up and forces the object to roll slightly uphill. This changing of shape, known as *deformation,* causes a movement of molecules. As a result, heat is produced from the added energy required to keep the parts turning and overcome friction.

## COEFFICIENT OF SLIDING FRICTION

The force required to overcome friction depends on the forces pressing the two surfaces together. These two forces are directly proportional. If a 100-pound force is required to slide a load, a 200-pound force is needed if the load is doubled. Similarly, a 300-pound force is needed if the load is tripled, and so on, figure 18-8.

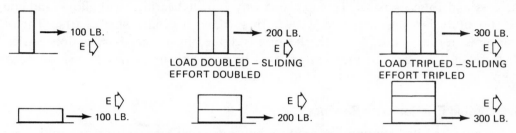

Fig. 18-8 Effort forces for same materials are proportional to load

The ratio of the friction force resisting motion and the perpendicular force pressing the two surfaces together is known as the *coefficient of friction*. When the object is on a horizontal surface and the force pressing the two surfaces together is perpendicular to the horizontal surface, then the coefficient of friction is expressed by the formula:

$$\text{Coefficient of Friction} = \frac{\text{Force (causes object to slide)}}{\text{Weight (of object)}}$$

For example, if 50 pounds are required to move a 100-pound block along a horizontal surface, the coefficient of friction is equal to force/weight: 50/100 = 0.5.

Since values of coefficients of friction are used so often, tables are available which give approximate values for pressures ranging between 14 and 20 pounds per square inch for various materials. Coefficients of friction must be increased for standing or starting friction and when higher pressures are used. A coefficient of friction table shows clearly those materials that may be used together when friction is needed. In addition, combinations of materials are shown which can be used when parts must slide freely over one another. When coefficient of friction tables are available and the weight of a sliding part is known, it is possible to determine the force pressing two surfaces together using the formula for the coefficient of friction.

Friction is considered independently of surface area. The friction between two materials remains the same regardless of the surface area (if the area is constant).

Figure 18-9 illustrates three objects of equal weight and material. Each object requires the same force to cause it to slide on the horizontal surface.

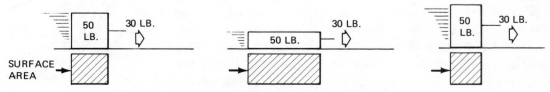

Fig. 18-9 Sliding friction of same materials is unaffected by surface area

## PRINCIPLES INVOLVED IN REDUCING FRICTION

The friction caused by the wedging action of surface irregularities can be overcome partly by the precision machining of the surfaces. However, even these smooth surfaces may require the use of a substance between them to reduce the friction still more. This substance is usually a lubricant which provides a fine, thin oil film. The film keeps the surfaces apart and prevents the cohesive forces of the surfaces from coming in close contact and producing heat.

Another way to reduce friction is to use different materials for the bearing surfaces and rotating parts. This explains why bronze bearings, soft alloys, and copper and tin oilite bearings are used with both soft and hardened steel shafts. The *oilite* bearing is porous. Thus, when the bearing is dipped in oil, capillary action carries the oil through the spaces of the bearing. This type of bearing carries its own lubricant to the points where the pressures are the greatest.

### Overcoming Friction with Bearings

Roller and ball bearings are used to reduce friction by substituting a rolling action for a sliding action. The roller or ball bearing moves over the surface irregularities and does not form a wedge with the irregularities as in the case of flat surfaces. There are many different uses for roller and ball bearings. If greater loads are to be supported, the roller bearing is preferred because the area of contact is greater than with the single point of contact obtained with a ball bearing, figure 18-10.

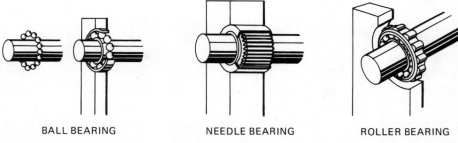

BALL BEARING          NEEDLE BEARING          ROLLER BEARING

Fig. 18-10

A needle bearing is also commonly used to overcome friction. The needle bearing is similar to the roller bearing except that the rollers are smaller and there are a greater number of them as compared to a roller bearing. A needle bearing has an advantage over the roller bearing in that for the same load-carrying capacity, the needle bearing is lighter and better adapted to machines and mechanisms where weight is important.

Ball and roller bearings are supported and spaced uniformly by a frame. The friction caused by the movement of the bearings in the frame can be overcome by lubrication. However, if the load on a bearing is increased to a point where the race in which the bearing rolls tends to become flat at the point of contact, the material piles up in front of the roller, figure 18-11. This effect creates molecular friction which in turn, produces heat. This condition cannot be corrected by lubrication. Either the shape or size of the bearings or the construction of the mechanism must be changed to overcome this condition.

Fig. 18-11

## PRINCIPLES OF LUBRICATION

Moving parts are lubricated to reduce friction, wear, and heat. The most commonly used lubricants are oils, greases, and graphite compounds. Each lubricant serves a different purpose. Essentially, a lubricant must meet three conditions:

(1) It must have sufficient body to be able to stand up under heavy loads or the operating conditions of the machine on which it is used. For example, since most mechanisms heat up during operation, an oil used as a lubricant must be able to withstand the heat of the unit and still provide the lubrication needed for continuous operation.

(2) The lubricant must flow freely so that a thin film is maintained between the moving parts. A lubricant that flows too slowly and does not easily reach all parts causes undue heat and wear.

(3) The composition of the lubricant should prevent chemical corrosion. Usually, a thin coating of oil is sufficient to prevent the formation of rust.

### Methods of Lubrication

The conditions under which two moving surfaces are to work determine the type of lubricant to be used and the system selected for distributing the lubricant. On slow moving parts with a minimum of pressure, an oil groove is usually sufficient to distribute the required quantity of lubricant to the surfaces moving on each other.

A second common method of lubrication is the splash or bath system in which parts moving in a reservoir of lubricant pick up sufficient oil which is then distributed to all moving parts during each cycle. This system is used in the crankcases of lawnmower engines to lubricate the crankshaft, connecting rod, and parts of the piston.

A lubrication system commonly used in industrial plants is the pressure system. In this system, a pump on a machine carries the lubricant to all of the bearing surfaces at a constant rate and quantity.

There are numerous other systems of lubrication and a considerable number of lubricants available for any given set of operating conditions. Modern industry pays greater attention to the use of the proper lubricants than at any other previous time because of the increased speeds, pressures, and operating demands placed on equipment and devices.

────────────────── SUMMARY ──────────────────

- Power is the rate of doing work. Power = Work ÷ Time:

$$\text{Power} = \frac{\text{Work (in foot-pounds)}}{33\ 000\ \text{ft.-lb./min.} \ \times \ \text{Time (in min.)}},$$

expressed in units of horsepower.

- Positive power transmission is provided by using gears, linkages and other mechanical devices, or by direct drive.

- Power losses are reduced when the source of power is brought as close as possible to the point of drive.

- Generally, friction is caused by surface finish, the cohesion and adhesion of the molecules of the moving parts, and the weight or pressure on the surfaces.

- Three common types of friction are: starting, sliding and rolling.

- The coefficient of friction is the ratio between the force required to move an object and the weight of the object.

- Friction due to surface irregularities can be reduced by precision machining, changing the materials in the parts that move, using ball and roller bearings, or by the use of lubrication.

- Lubricants are used to reduce heat, wear, and friction.

- Lubricants must have: sufficient viscosity to prevent moving surfaces from coming into contact; the ability to flow easily and to stand up under continued service without breaking down or becoming gummy; and freedom from rust-producing impurities.

## ASSIGNMENT UNIT 18 MECHANICAL POWER TRANSMISSION, FRICTION, LUBRICATION

## PRACTICAL EXPERIMENTS IN POWER MEASUREMENT, FRICTION, AND LUBRICATION

### Equipment and Materials Required

| Experiment A Power Measurement | Experiment B Friction | Experiment C Lubrication |
|---|---|---|
| Prony brake equipment | Spring balance | Grease (medium grade) |
| Speed indicator | Magnifying glass | Oil can (heavy oil) |
| Machine units to be tested, or electric motors (1/4, 1/2, 1 h.p.) | | Oil can (light oil) |
| | Rectangular blocks of aluminum, cast iron, and steel, each with one smooth side and one rough side | |
| | Flat plates of aluminum, cast iron, and steel, each with one smooth side and one rough side | |
| | Cast iron roller (undercut in center) | |

## EXPERIMENT A. MEASUREMENT OF MECHANICAL POWER WITH A PRONY BRAKE

### Procedure

1. Mount the smallest of the electric motors to be tested in the Prony brake frame.

2. Secure a small pulley to the motor shaft and adjust the band around the pulley until there is a slight pressure on both spring balances.

3. Prepare a table for Recording Results and record the forces, pulley diameters, and the rpm values for motors A, B, and C.

4. Start the motor and tighten the adjusting nuts until there is a reading on both spring balances.

5. Measure the speed of the motor with a speed indicator (tachometer).

6. Remove motor or machine (A) and replace with machine (B) (or the next largest motor).

7. Repeat steps 2–5.

8. Using motor (C), repeat steps 2–5.

## Recording Results

| | Force | | Pulley | |
|---|---|---|---|---|
| | $F_1$ | $F_2$ | Diameter | Rpm |
| Motor A | | | | |
| Motor B | | | | |
| Motor C | | | | |

## Interpreting Results

1. Determine the brake horsepower for each of the three motors or machine units from the data recorded in the table. Use the Prony brake formula to find the horsepower and round off the answers to the nearest 1/4 h.p.

2. What conclusions can be drawn about the relationship between the horsepower indicated on the motor plate and the calculated horsepower?

## EXPERIMENT B. DETERMINING STARTING, SLIDING, AND ROLLING FRICTION

### Procedure

1. Select three rectangular blocks, a single cast iron roller, and three plates on which these blocks and the roller can be moved.

2. Prepare a table for Recording Results. Record the friction readings for the indicated surfaces and conditions for each step that follows.

3. Place the rough surface of the aluminum block against the rough surface of the aluminum plate.

4. Pull the spring balance gently and note the greatest force required to start the block in motion (starting friction).

5. Continue to pull the aluminum block and record the force needed to slide the block on the rough aluminum plate surface. This is the value of the sliding friction.

6. Repeat steps 4 and 5 for the following conditions: the aluminum block placed on the rough cast iron and on the steel plate surfaces, and then on the smooth surfaces, of the aluminum, cast iron, and steel plates, respectively.

7. Using the smooth surface of the aluminum block, repeat steps 4–6 to determine the starting and sliding friction values when the smooth surfaces of the block and three plates are in contact.

8. Use the cast iron block next to obtain nine readings for starting friction and another nine readings for sliding friction.

9. Continue to test the steel block and the three plates.

10. Attach a spring balance to the undercut part of the cast iron roller and determine the effort required to roll the cast iron roller on the rough and smooth surfaces of the aluminum, cast iron, and steel plates.

## Recording Results

| | | Rough Block or Roller | | | | | | Smooth Block or Roller on Smooth Plate | | |
| | | on Rough Plate | | | on Smooth Plate | | | | | |
| | Plates → | Al | C.I. | Steel | Al | C.I. | Steel | Al | C.I. | Steel |
|---|---|---|---|---|---|---|---|---|---|---|
| | Type of Friction ↓ | Aluminum Block | | | | | | | | |
| A | Starting | | | | | | | | | |
| B | Sliding | | | | | | | | | |
| | | Cast Iron Block | | | | | | | | |
| A | Starting | | | | | | | | | |
| B | Sliding | | | | | | | | | |
| | | Steel Block | | | | | | | | |
| A | Starting | | | | | | | | | |
| B | Sliding | | | | | | | | | |
| | | Cast Iron Roller | | | | | | | | |
| C | Rolling | | | | | | | | | |

## Interpreting Results

1. Summarize the results of all of the experiments in terms of (a) the effect of the surface finish on starting, sliding, and rolling friction, and (b) the effect of different materials on the three types of friction.

2. Weigh each of the three blocks. Compute the coefficient of friction for each of the three conditions with each of the three different materials. Label each answer in the work space provided in the table.

3. Check the computed coefficients of friction with the table of coefficients in the Appendix. Justify any variation in the values.

## EXPERIMENT C. EFFECT OF LUBRICATION ON STARTING AND SLIDING FRICTION

### Procedure

1. Prepare three tables for Recording Results. The tables are to be identical with the one used in Experiment B. The tables are to be labeled as follows: (1) Recording

Results for Surfaces Lubricated with Grease; (2) Recording Results for Surfaces Lubricated with Oil (Heavyweight); and (3) Recording Results for Surfaces Lubricated with Oil (Lightweight).

2. Perform the same steps as for Experiment B. Use a grease as a lubricant in the first set of steps and record all of the forces required to start and keep the parts sliding.

3. Remove the grease and clean the surfaces thoroughly until dry. Then apply carefully a thin film of heavy oil and repeat the steps.

4. Replace the heavy oil with a lightweight oil and again repeat the same steps.

### Interpreting Results

1. Study the data recorded in the tables on friction and the use of lubricants and the table for which no lubricants were used.

2. Describe the effect of lubrication on the starting and sliding friction as compared with the same materials when dry.

3. Compare the difference in force between the starting and sliding friction of lubricated surfaces when different types or grades of lubricant are used. Make a general statement, based on the data in the tables, about any differences that may be noted.

## PRACTICAL PROBLEMS IN POWER MEASUREMENT, FRICTION, AND LUBRICATION

### Mechanical Power Transmission and Measurement

For statements 1 to 8, determine which are true (T) and which are false (F).

1. Power and work are the same.

2. Power refers to the number of foot-pounds of work done.

3. Power can be transmitted by direct drive using rigid or flexible couplings.

4. Gears and linkage systems provide less positive power transmission than flat belts.

5. One horsepower = 33 000 foot-pounds of work per minute.

6. One horsepower = 33 000 foot-pounds of work.

7. A clutch is preferred to a coupling in transmitting power continuously without interruption.

8. The Prony brake is a simple measuring instrument for determining the brake horsepower of rotating machines.

9. The results of Prony brake tests of three engines are given in the table. Compute the brake horsepower of each engine to the nearest 1/4 h.p.

| | Force (in lb.) | | Pulley | |
|---|---|---|---|---|
| | $F_1$ | $F_2$ | Diameter (inches) | Rpm |
| Engine A | 5 | 9 1/2 | 3 1/2 | 6000 |
| Engine B | 75 | 152 | 12 | 680 |
| Engine C | 10 | 259 | 14 | 903 |

## Friction

Select the correct word or phrase to complete statements 1 to 8.

1. Of the two types of friction, (sliding) (starting) friction is always greater.

2. As the weight of a moving part increases, the friction (remains the same) (decreases) (increases).

3. In general, two rough surfaces require a (larger) (smaller) force to move one over the other than two smooth surfaces.

4. If the weight of a part remains the same, the surface area (affects) (does not affect) friction.

5. The coefficient of friction is the ratio of the (effort force ÷ weight of the object) (weight of the object ÷ effort force).

6. Roller and ball bearings (decrease) (increase) friction.

7. Roller and ball bearings work on the principle of (sliding) (starting) (rolling) friction.

8. The (needle) (roller) (ball) bearing is best suited for heavy loads where light weight and the strength of the bearing and retainer are important.

9. List three common causes of friction.

10. List three common types of friction.

11. Study a car manufacturer's recommendations for breaking in a new car. Give five reasons for supporting these recommendations.

12. Name five parts or mechanisms in an automobile that depend on friction for safe operation.

13. List five parts or mechanisms in an automobile for which friction is undesirable.

14. Examine the surface finish of four different materials with a microscope. Make simple sketches of the surfaces and exaggerate the surface finish. Number the parts from 1 to 4 and range them in order from the coarsest to the smoothest, respectively.

    a. Which of the four surfaces provides the best friction?

    b. Which surface, in terms of surface finish, will wear the least and have the smallest starting friction?

15. Sketch the enlarged surfaces of two roughly machined parts that move on one another. Explain why starting friction is greater than sliding friction.

16. Wooden and metal rollers are used to move heavy equipment easily. Explain briefly why rollers are used.

17. The data from friction tests using different materials are given in table I. Determine the coefficient of sliding friction for combinations A, B, C, D, and E.

18. Arrange the materials in table I in order from 1 to 5 to indicate the lowest to the highest coefficients of friction, respectively. Use the letters to designate the materials.

| Materials | A Leather on Cast Iron | B Leather on Hardwood | C Bronze on Bronze | D Cast Iron on Hardwood | E Cast Iron on Cast Iron |
|---|---|---|---|---|---|
| Force (in lb.) | 140 | 111 | 215 | 70 | 2 |
| Weight of part or load (in lb.) | 250 | 335 | 1075 | 144 | 13 |
| Coeff. friction (sliding) | | | | | |

**Table I**

19. What combination of materials from the table should be used in parts which depend on friction for their operation?

20. What two materials from the table are best suited for parts that must slide continuously?

**Lubrication**

For statements 1 – 8, determine which are true (T) and which are false (F).

1. Lubricants are used to reduce starting and sliding friction.

2. In industry, careful attention is given to lubricants that have rust-prevention qualities.

3. The materials used to make moving parts have no influence on the type of lubricant needed.

4. The splash system of lubrication depends on the movement of parts in a reservoir of oil.

5. A slow moving lubricant always provides a sufficient flow to lubricate all parts.

6. The pressure or forced feed system of lubrication is the least expensive to install and is used on parts that move slowly under light loads.

7. Lubricants reduce cohesion and adhesion by providing a film of liquid between moving parts.

8. The lubrication of moving parts has no effect on the amount of heat produced.

9. Select the letters in the following list of those items that require lubrication.

    a. Flat leather belt         d. V-belt
    b. Flat bearing surface     e. Shaft bearing
    c. Machine spindle       f. Brake band

10. Locate two different types of roller or ball bearings and two other regular bearings. Name two parts or machines and two conditions where each of the four bearings can best be used.

11. Explain why it is easier to start a heavy drive shaft after a momentary stoppage than it is when the shaft remains stationary for a long time.

12. Why are lighter oils used in aircraft, automotive, and diesel engines in cold working temperatures?

13. Give three reasons why the used oil in the crankcase of an engine should be drained off periodically and replaced.

14. Give a brief explanation of why the use of a lubricant on two moving parts reduces wear, heat, and friction, and lengthens the life of a machine.

# Unit 19 Mechanics of Fluids at Rest and in Motion

A liquid has no definite form but takes the shape of its container. Similarly, a gas has neither form nor a definite volume. It, too, takes the shape of the vessel or container. In both cases, the word fluid can be applied to either the gas or the liquid because both substances can flow. The energy of the liquid or gas in a sealed container can be harnessed to do useful work.

This unit gives the basic principles and properties of fluids. Common applications of fluids to solve everyday needs are also provided. The unit begins with a review of the simple terms of pressure, density, buoyancy, and specific gravity as each of these properties apply to fluids.

## FLUID PRESSURE AND FORCE

*Pressure* refers to the force per unit area. The result is usually expressed as a pressure of so many pounds per square foot (lb./ft.$^2$), or as grams per square centimeter (g/cm$^2$), or as any other combination of weight and area measure. If five tons of concrete are poured over an area ten feet square, the pressure per square foot = 10 000 lb. ÷ 10 ft. x 10 ft., or 100 lb./ft.$^2$

There are times when the total force must be determined. At other times, the pressure for a unit area must be known. The force on the bottom of a tank is equal to the volume of the fluid x the density of the fluid.

For a rectangular or square tank, figure 19-1, the volume is the area of the base multiplied by the height of the fluid.

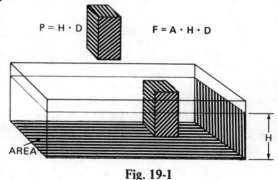

Fig. 19-1

**Force (F) = Area (A) x Height (H) x Density (D)**

In addition,

**Pressure (P) = Force (F) ÷ Area (A)**

If the value of (P) is substituted in the force equation, the areas (A) cancel out and (P) is equal to (HD).

$$F = AHD \text{ and } P = \frac{F}{A}$$

Therefore,

$$P = \frac{AHD}{A} = HD$$

The difference between pressure and force is that pressure depends on the area of contact and force depends on the volume area.

For example, compare the difference between the pressure and the force at the bottom of a square tank five feet on a side. The tank is filled with gasoline to a height of 10 feet (gasoline weighs 42 lb./ft.$^3$). The pressure at the bottom of the tank = HD = 10 ft. x 42 lb./ft.$^3$ = 420 lb./ft.$^2$.

For the same tank, the force (F) is equal to (AHD): F = 5 ft. x 5 ft. x 10 ft. x 42 lb./ft.$^3$ = 10 500 pounds. Fluid pressure is expressed as a weight per area such as 420 lb./ft.$^2$; fluid force is expressed by the total number of pounds, such as 10 500 lb.

Experimentation has shown that the pressure at a given depth in a fluid is the same in all directions. If the pressure is greater in any one direction, then the fluid will flow toward that area. This fact was first stated by Pascal and later became known as *Pascal's Law*. The law applies to all enclosed fluids at rest, both gaseous and liquid. Under these conditions the law states that any pressure applied to an enclosed fluid is transmitted equally in every direction without loss. In addition, the fluid acts with equal force on all surfaces.

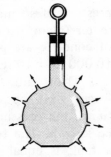

The action of this law is demonstrated by a special glass bottle as shown in figure 19-2. As pressure is applied to the liquid by the piston, the liquid is sprayed at the same pressure and in the same amounts from each of the small holes, regardless of the positions of the holes.

Assume that this same action is duplicated with a metal container having a number of low-pressure gas gauges attached. When air is pumped into the container, each gauge registers the same pressure. For both the glass bottle and the metal container, the fluids are at rest and enclosed in a container.

Fig. 19-2

## PRESSURES ON FLUIDS ARE UNAFFECTED BY SHAPE AND SIZE OF CONTAINER

The size and shape of a vessel have no effect on the gravity pressure of the fluid. *Gravity pressure* refers to the pressure exerted by the fluid itself in the absence of any additional external pressure. The fluid exerts pressure against any surface that it touches.

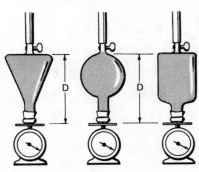

The fact that the size and shape of a container do not affect the pressure of a fluid can be demonstrated by three or four glass containers of different sizes, figure 19-3. Each container has the same neck size. If each container is filled to the same height with the same liquid, the neck of each is covered with a rubber diaphragm, and then each container is mounted in a test stand, the indicator readings for the pressures at the bottom of each vessel are identical. Note that the volume of liquid in each

Fig. 19-3

container and the container shapes are different. However, the height of the fluid in the container and the density of the liquid are constant. When these values are substituted in the formula P = HD, the value of P remains the same for each container.

This simple principle governing the action of fluids is used in engineering in the building of dams. Since pressure increases with depth, reservoir and power dams are built to withstand greater pressures at the bottom of the dam than at the top surface.

In large communities, water supply systems depend on these same principles to insure that water is delivered at the same pressure to all homes.

## APPLICATION OF FLUID PRESSURE TO MACHINES: HYDRAULIC PRESS

Practical applications of Pascal's Law are found wherever the energy of a fluid can be harnessed to produce useful work. Hydraulic jacks, air drills, automobile tires, and hydraulic presses are but a few of the applications. One of the widest uses of the principles of fluid pressure is in the hydraulic press. The operation of this device depends on the fact that pressure in a liquid at rest is transmitted unchanged (except for changes in level) in all directions. Hydraulic presses are capable of exerting enormous forces.

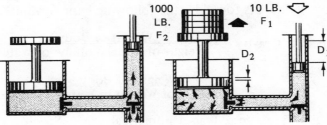

Fig. 19-4

A simple hydraulic press is shown graphically in figure 19-4. The press consists of a large piston and a small piston which each have a watertight fit in the press housing. The two pistons are connected by a small pipe. Two valves are provided so that additional liquid can be drawn into the small valve chamber on the upstroke while the valve in the large piston chamber is closed. As the small piston begins its downward stroke, the supply valve closes and the valve to the large piston chamber opens. In this manner, the pressure exerted by the small piston is transmitted to the large piston.

### Mechanical Advantage of the Hydraulic Press

If both pistons of the hydraulic press are the same size, the force transmitted by one piston is equal to the force of the other piston. In figure 19-4, one piston is 100 times larger than the small piston. Therefore, a ten-pound effort on the small piston produces a force of 1000 pounds on the large piston. The mechanical advantage of this combination (neglecting friction) is equal to R ÷ E = 1000 ÷ 10 = 100. In other words, the forces on both pistons are proportional to the areas of the pistons.

The increase in force from the small to the large piston must be compensated by a loss in distance. One of the fundamental laws for an ideal machine is that one force and the distance through which it acts is equal to another force and the distance through which this second force acts on the same machine.

$$Force_1 \text{ x } Distance_1 = Force_2 \text{ x } Distance_2$$

Thus, to compensate for the increased force on the large piston of the hydraulic press, the distance through which the small piston moves is increased. If the mechanical advantage of force ($MA_f$) = 100, the small piston travels 100 times farther than the large piston.

The hydraulic press is used in the automotive, aircraft, and structural industries to produce the enormous forces necessary to manufacture heavy forgings, car bodies, fenders, and structural shapes.

## SPECIFIC GRAVITY MEASUREMENTS OF FLUIDS

In one of the early units on the properties of liquids, Archimedes' principle of buoyancy was discussed. A fluid exerts an upward buoyant force upon a body that is wholly or partly submerged in it. Thus, the amount of the force is equal to the weight of the fluid displaced. The buoyancy is affected by the specific gravity of the fluid. Recall that the specific gravity of a body is the ratio of the weight of the body to the weight of a standard substance having the same volume as the body.

For liquids, the standard substance is water at 39.8 °F. At this temperature, water is at its maximum density. The specific gravity of a sinking (heavy) body, of a floating body, or of a liquid is found by the following basic formulas:

$$\text{Specific Gravity} = \frac{\text{Weight of Body}}{\text{Weight of Equal Volume of Water}}$$

$$= \frac{\text{Weight of Body}}{\text{Loss of Weight in Water}}$$

Specific gravity is expressed as a numerical value. Specific gravity values that are larger than one (1.00) indicate that the body is heavier than water; those values under one indicate that the body is lighter than water. A later unit of this text will cover the standard substance on which the specific gravity system of gases is based.

Specific gravity can be computed or it can be determined with an instrument known as a *hydrometer*. The hydrometer works on Archimedes' principle. Hydrometers usually consist of a weighted glass bulb. When the bulb is placed in a liquid, it sinks to the depth at which it is displacing exactly its own weight of the liquid. The specific gravity of the liquid determines how far the bulb will sink. The lower (smaller) the specific gravity of the liquid, the lower the bulb sinks in the liquid. A calibrated scale mounted on the bulb provides a simple method of reading the specific gravity of the liquid.

## FLUID FLOW, PRESSURE, AND SPEED: BERNOULLI'S PRINCIPLE

Pascal's law deals with fluids at rest in confined vessels where an external pressure is transmitted equally in all directions. As the fluid starts to move, the pressure is no longer the same in all parts of the vessel because a force is necessary to overcome friction. This fact is demonstrated by the bent copper tubing shown in figure 19-5. When the valve is closed and water fills the tubes, each gauge indicates the same pressure because the water is confined and the pressure is exerted equally in all directions. However, when the valve is opened, the gauges show a difference in pressure due to the friction within the tube. Thus, when liquids are to be piped for great distances, pumping stations must be installed at intervals so that a constant pressure can be maintained throughout the entire line.

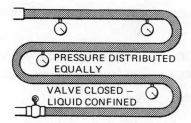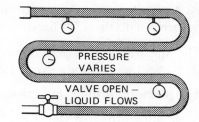

**Fig. 19-5**

Bernouilli's principle states that whenever the velocity of a fluid is increased at any point, its pressure is decreased. For example, if a stream of air is blown across the top of a glass tube dipped in a liquid, the liquid will rise in the tube. This action is due to the fact that the air flowing across the tube decreases the atmospheric pressure at the top of the tube. If the velocity of the air is great enough, the liquid is drawn completely up the tube and is blown with the passing air in a fine spray. Common applications of Bernoulli's principle include spray guns, atomizers, and carburetors. These devices act like suction pumps with no moving parts.

Another example of Bernoulli's principle is demonstrated when a pipe is reduced in cross-sectional area (*constricted*). In this case, the speed of the liquid through the pipe increases at the reduced point so that the same amount of liquid may pass through the constriction as through any other portion of the pipe. This increase in speed causes a decrease in pressure at this point, figure 19-6.

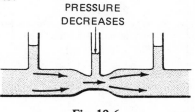

**Fig. 19-6**

The venturi gauge is based on this latter application of Bernoulli's principle. The venturi gauge is a tubelike device with a constricted portion. Fluid passing through the venturi increases in speed as it passes through the narrowed portion. At the same time, the pressure is reduced at the narrowed portion. This pressure reduction can be measured, and the rate of flow can be determined or controlled.

---

### SUMMARY

- The term *fluids* refers to both gases and liquids. The molecules of a fluid are free to move or to be pushed in all directions.

- Pressure is a force per unit area. Pressure is expressed in weight and area measure: P = HD.

- Force is a measure of area, height of fluid, and density: F = AHD. Force is expressed in weight measure.

- Pascal's law for confined fluids at rest states that any pressure applied to the fluid is transmitted in all directions without loss.

- Gravity pressure is the pressure exerted internally by the weight of the fluid itself and is not affected by the size and shape of the container.

- Machines such as the hydraulic press, hydraulic brakes, and pumps, are practical applications of the scientific principles of force, distance, and pressure combined to do productive work.

- At times, the specific gravity measurement of a fluid must be determined before calculations for pressure or force can be made.

- The specific gravity of sinking bodies, floating bodies, and liquids equals the weight of the body divided by the weight of an equal volume of water.

- The pressure of fluids in motion drops due to friction.

- Bernoulli's principle states that whenever the speed of a fluid is increased, the pressure decreases.

## ASSIGNMENT UNIT 19  MECHANICS OF FLUIDS AT REST AND IN MOTION

### PRACTICAL EXPERIMENTS WITH FLUIDS AND PRESSURES

### Equipment and Materials Required

| Experiment A<br>Pascal's Law | Experiment B<br>Fluid Pressure | Experiment C<br>Transmitting Fluid Pressure |
|---|---|---|
| Pressure apparatus similar to the one illustrated | Connected vessel apparatus<br><br>Test stand<br><br>Container of colored water<br><br>Container of second kind of fluid<br><br>Measuring stick or tape | Hydraulic press<br><br>Weights (1, 2, 4, 6, and 10 ounces and 1, 2, 3, 5, 10, 25, and 50 pounds)<br><br>Measuring stick or tape |

### EXPERIMENT A.  PRESSURE ON A CONFINED LIQUID

#### Procedure

1. Fill the pressure apparatus with water from a faucet. Close the intake valve, figure 19-7.

2. Prepare a table for Recording Results and record all gauge readings for the steps which follow.

3. Note the initial reading on gauges A, B, C, D, and E.

4. Place a ten-pound force (F) on the pump lever. Note and record the pressure on each of the gauges.

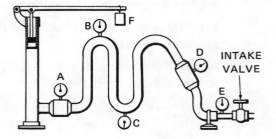

**Fig. 19-7**

5. Replace the ten-pound force with a 20-pound force and again note the pressure on each gauge.

6. Repeat step 5 with a 50-pound force.

## Recording Results

| | Force (F) on Lever (in lb.) | | Gauge Readings | | | | |
|---|---|---|---|---|---|---|---|
| | | A | B | C | D | E | |
| | Initial Reading | | | | | | |
| A | 10 | | | | | | |
| B | 20 | | | | | | |
| C | 50 | | | | | | |

## Interpreting Results

1. Do the first gauge readings indicate that water drawn from a faucet is under pressure? Justify.

2. When the ten-pound force is applied to the pump lever, are the pressure readings the same on all gauges? Why?

3. What effect does increasing the force to 20 and 50 pounds have on the pressure readings throughout the system?

4. How does the diameter or shape of a pipe affect the pressure? Explain.

## EXPERIMENT B. FLUID PRESSURE

### Procedure

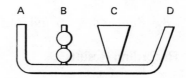
Fig. 19-8

1. Obtain or fuse four different shapes and sizes of vessels as illustrated in figure 19-8. Mount the apparatus vertically in a test stand.

2. Pour colored water into one of the tubes.

3. Wait until the water comes to rest in all of the vessels. Measure the height of the liquid in each vessel and record these values in the table prepared for Recording Results.

4. Repeat steps 2 and 3 by replacing the water with another fluid.

## Recording Results

| | Gauge Readings | | | |
|---|---|---|---|---|
| | A | B | C | D |
| Water | | | | |
| Other fluid | | | | |

## Interpreting Results

1. Compare the height of the water in vessels A, B, C, and D.

2. Does the size or shape of vessel or the fluid change the pressure at the bottom of each tube? Give the scientific reason which explains your observation.

## EXPERIMENT C. TRANSMITTING FLUID PRESSURE (HYDRAULIC PRESS)

### Procedure

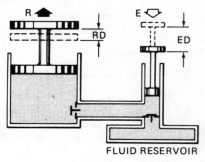

FLUID RESERVOIR

Fig. 19-9

1. Obtain a hydraulic press similar to the one shown in figure 19-9. Determine the sizes of the small and large pistons.

2. Prepare a table for Recording Results. Record the weights (E and R), the movement of the small piston (ED), and the movement of the large piston (RD).

3. Place a 1/2-pound weight on the small piston and note what happens to the large piston. Add weights to the large piston until the movement stops.

4. Measure the height of the small and large pistons. Add another small weight to the effort force (small piston). Let the small piston travel until the large piston moves one inch.

5. Measure the height that the small piston moves to raise the large piston one inch.

6. Place a one-pound weight on the small piston. As the small piston moves up, keep adding weights to the large piston until the movement of the small piston stops.

7. Measure the movement of the small piston needed to raise the large piston one inch.

8. Repeat steps 6 and 7 using a five-pound and a ten-pound weight, respectively.

### Recording Results

|   | Piston Movement | | Force Applied | |
|---|---|---|---|---|
|   | Small (ED) | Large (RD) | Small Piston (E) | Large Piston (R) |
| A |   | 1 in. | 1/2 lb. |   |
| B |   | 1 in. | 1 lb. |   |
| C |   |   | 5 lb. |   |
| D |   |   | 10 lb. |   |

### Interpreting Results

1. Determine the ratio of the effort force (E) to the resistance force (R) for combinations A, B, C, and D.

2. Find the ratio of small piston movement to large piston movement.

3. Determine the mechanical advantage of force and the mechanical advantage of speed for each of the four combinations given in the table.

4. Justify the data in the table comparing the ratio of force to the ratio of distance. Are the mechanical advantages of speed and force for a hydraulic press equal or different? Justify this answer.

## PRACTICAL PROBLEMS WITH FLUIDS AND PRESSURES

### Fluid Pressure and Force

Select the correct word or phrase to complete statements 1 to 6.

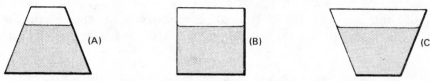

VESSELS A, B, AND C ARE FILLED TO THE SAME HEIGHT WITH WATER.

Fig. 19-10

1. The force on the bottom of A is (greater than) (less than) (equal to) the weight of the water.

2. The force on the bottom of B is (greater than) (less than) (equal to) the weight of the water.

3. The force on the bottom of C is (greater than) (less than) (equal to) the weight of the water.

4. The pressures on the bottom of A, B, and C are (not comparable) (the same) (different).

5. Pressure depends on (area and density) (area, height, and density) (area and height).

6. The force on a container depends on (area and height) (area and density) (area, height, and density).

Add the word or words needed to complete statements 7 to 10.

7. A liquid with a specific gravity of 0.70 _____ in water.

8. Carbon tetrachloride with a specific gravity of 1.5 _____ in water.

9. A substance with a density of 5.4 lb./ft.$^3$ _____ in a liquid with a density of 62.4 lb./ft.$^3$

10. A liquid with a density of 49.0 lb./ft.$^3$ _____ in a liquid with a density of 36.8 lb./ft.$^3$

11. What is the pressure at the bottom of a water column 50 ft. high?

12. Explain why the flow of water in a faucet in the home decreases when another faucet is opened.

13. How is the pressure of a water system affected by (a) increasing the diameter of the standpipe to a water tower, and (b) by increasing the length or height of the standpipe?

14. State the advantages of applying Pascal's Law to a system of hydraulic brakes.

15. Find the weights of each of the following liquids, using an assumed weight for water of 8 lb./gal.

    a. 5 gallons of cutting oil (specific gravity 0.70)
    b. 20 gallons of gasoline (specific gravity 0.80)
    c. 1 quart of carbon tetrachloride (specific gravity 1.50)
    d. 1 pint of sulphuric acid (specific gravity 1.82)

16. A tank 21 feet in diameter is filled with gasoline to a depth of ten feet. The gasoline weighs 0.78 times the weight of the water.

   a. Find the pressure on the bottom of the tank.
   b. Compute the force on the bottom.

17. Compute the force on the bottom of containers A through F and the pressure in each container per square foot. Use 62.4 lb./ft.³ as the weight of water.

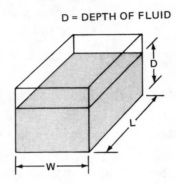

D = DEPTH OF FLUID

| | Dimensions of Fluid | | | Specific Gravity | Force | Pressure |
|---|---|---|---|---|---|---|
| | L | W | D | | | |
| A | 1 ft. | 1 ft. | 1 ft. | 1 | | |
| B | 2 ft. | 2 ft. | 1 ft. | 1 | | |
| C | 2 ft. | 2 ft. | 6 in. | 1 | | |
| D | 2 ft. | 2 ft. | 2 ft. | 1 | | |
| E | 10 ft. | 10 ft. | 10 ft. | 0.8 | | |
| F | 10 ft. | 10 ft. | 5 ft. | 0.8 | | |

## Application of Fluid Pressure to Hydraulic Machines

1. The pistons of a hydraulic lift have an area ratio of 1 to 600. A force of 50 pounds is exerted on the small piston.

   a. Determine the force on the large piston.
   b. Determine the force distance through which the large piston moves when the small piston makes 25 four-inch strokes.

2. The small piston of a hydraulic jack has an area of 6 in.² ; the large piston has an area of 12 in.². What force on the small piston is needed to raise a 3200-pound load?

3. The diameters of the pistons on a heavy hydraulic press are 1 1/2 in. and 24 in. A force of 75 tons is needed to stamp and form a metal part. What effort must be applied to the small piston?

4. The piston of a hydraulic lift for automobiles is 14 in. in diameter. The device is operated by water from a city water system. What water pressure is necessary to lift a car if the total load is 3456 pounds?

5. Determine the missing values for problems A, B, C, and D from the data given in the table for a simple hydraulic press. NOTE: The computed values for (ED) at C and D are the total distances that the small pistons move, and not the distance moved for each stroke.

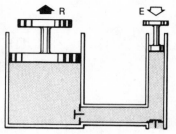

| | E | Area of Small Piston | ED | R | Area of Large Piston | RD |
|---|---|---|---|---|---|---|
| A | 10 lb. | 0.5 in.$^2$ | 5 in. | | 5 in.$^2$ | |
| B | 50 lb. | 2 in.$^2$ | 5 in. | 500 lb. | | |
| C | 100 lb. | 1/2 in.$^2$ | | 50 tons | 500 in.$^2$ | 10 in. |
| D | 75 lb. | 5 in.$^2$ | | | 125 in.$^2$ | 4 in. |

6. A hydraulic lift actuated by a lever is shown in figure 19-11. Compute the missing dimensions in the table for conditions A, B, C, and D.

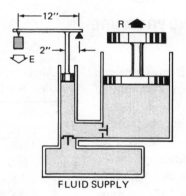

**Fig. 19-11**

| | Effort Force (E) | Diameter Small Piston | Diameter Large Piston | Resistance Force (R) |
|---|---|---|---|---|
| A | 10 lb. | 1 in. | 5 in. | |
| B | 25 lb. | | 5 in. | 750 lb. |
| C | 100 lb. | 1/2 in. | | 6 tons |
| D | | 2 in. | 10 in. | 1 1/2 tons |

7. The hydraulic lift in problem 6 is 80% efficient for conditions A and B. Determine the $MA_f$ for each condition.

8. Increase the effort distance for conditions C and D in problem 6 to 20 in. Using an efficiency of 75%, determine
   a. the mechanical advantage of force and
   b. the mechanical advantage of speed.

# Unit 20 Atmospheric Pressure: Principles and Applications

When the term *atmospheric pressure* is used, it refers to the pressure caused by an air ocean surrounding the earth's surface. Air has weight. Thus, air causes pressure on the surface of the earth because the force of gravity of the earth exerts a force on the atmosphere through a height of several miles. A cubic foot of air weighs about 1.2 ounces as compared to the same volume of water which weighs nearly 62.5 pounds. In other words, water weighs approximately 800 times as much as air. The total amount of air pressure exerted by this ocean of air will support a column of water at a maximum height of 34 feet. The air pressure required to support this column of water is equal to height x density: 34 ft. x 62.5 lb./ft.$^3$ = 2125 lb./ft.$^2$. This value can be reduced to a pressure of 14.7 pounds per square inch or 1033.6 grams per square centimeter.

## MEASURING ATMOSPHERIC PRESSURE

### The Barometer

While fluctuations in air pressure can be measured with a 34-foot column of water, such a measuring device is impractical. Evangelista Torricelli, an assistant to Galileo, perfected a simple device for measuring air pressure. In place of water, Torricelli used mercury which has a specific gravity of 13.6. The rise of mercury in this device always reached about 30 inches (at sea level). As a result, a 30-inch column of mercury indicated the same atmospheric pressure as a 34-foot column of water. The mercury-filled measuring instrument became known as a Torricelli *barometer*. Torricelli's experiments showed that if, for any reason, the air pressure is lowered, the 30-inch height of mercury cannot be supported.

| Altitude, Ft. Above Sea Level | Pressure | |
|---|---|---|
| | In. Hg | Lb./In.$^2$ |
| 000 (Sea Level) | 29.9 | 14.7 |
| 1,000 | 28.9 | 14.1 |
| 2,000 | 27.8 | 13.6 |
| 3,000 | 26.8 | 13.2 |
| 4,000 | 25.8 | 12.6 |
| 5,000 | 24.9 | 12.2 |
| 6,000 | 24.0 | 11.8 |
| 7,000 | 23.1 | 11.3 |
| 8,000 | 22.2 | 10.9 |
| 9,000 | 21.4 | 10.5 |
| 10,000 | 20.5 | 10.1 |
| 15,000 | 16.9 | 8.3 |
| 20,000 | 13.8 | 6.8 |

**Table 20-1  Atmospheric pressure versus altitude**

Pascal discovered that as a barometer is carried to higher altitudes, the mercury drops almost one inch for every 1000 feet of elevation, table 20-1. When a barometer is used for measuring altitude, it is usually called an *altimeter*.

Since air pressure varies with distance above the surface of the earth, it was necessary to establish a standard to provide a constant value from which to work. The standard pressure was selected to be a column of mercury 29.92 in. or 76 centimeters high at sea level. While pressure should be given in units of weight and square measure, it is common practice to refer to air pressure in terms of 29.52 in., or 62 centimeters, or 752 millimeters. Each of these values refers to the height of a column of mercury which the atmospheric pressure will support.

In summary, the atmospheric pressure at sea level is equal to:

14.7 lb./in.$^2$
29.92 in. Hg
76 cm Hg
34 ft. of water

In metric units, atmospheric pressure at sea level is

$$1.013 \times 10^5 \text{ newtons/m}^2.$$

**Fig. 20-1**

### The Aneroid Barometer

It is impractical to use a 30-inch column of mercury in many applications. A smaller and more compact device for measuring changes in air pressure is called the aneroid barometer.

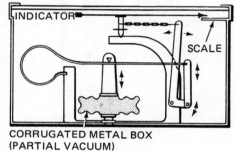

The *aneroid barometer,* figure 20-2, consists of a corrugated metal expansion box from which the air is removed. Variations in air pressure cause the metal box to be compressed or returned to its original shape. A system of levers magnifies the movement of the corrugated metal box and transmits this movement to a pointer. The reading is indicated directly on a scale calibrated in inches of mercury.

CORRUGATED METAL BOX
(PARTIAL VACUUM)

**Fig. 20-2**

If a continuous reading of pressure is required, a recording pen and chart and a timing mechanism are added to the barometer. In this manner, a permanent record is formed of changes in air pressure (or the pressure of other gases) for specific periods of time. This unit has described the simple mercury barometer and the mechanical aneroid barometer. However, there are numerous other types of barometers available for measuring, recording, and controlling the pressure and volume of fluids.

## ATMOSPHERIC PRESSURE AT WORK

Air pressure is important to the degree that it can be harnessed to do productive work. Before applications of air pressure can be made, there are several basic principles governing the action of air pressure that must be understood. Robert Boyle, an English scientist, found in 1662 that a high pressure is created by reducing the volume of a gas while maintaining a constant temperature. In addition, it was found that the pressure of an enclosed gas is inversely proportional to its volume, figure 20-3.

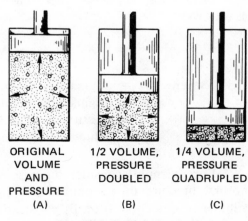

ORIGINAL VOLUME AND PRESSURE

(A)

1/2 VOLUME, PRESSURE DOUBLED

(B)

1/4 VOLUME, PRESSURE QUADRUPLED

(C)

**Fig. 20-3 Boyle's Gas Law**

### Boyle's Gas Laws

Figure 20-3 (page 191) illustrates *Boyle's Law.* Starting at rest in (A), the volume of the gas in the cylinder is decreased to one-half at (B) and one-fourth the original volume at (C). For these same cylinders, the pressure is doubled at (B) and quadrupled at (C) (if the effects of heating due to compression are overcome).

Boyle's Law can be explained on the basis of the principle which states that all matter is made up of tiny particles known as molecules. It is the bombardment of these molecules against the walls of a container that produces pressure. Thus, in figure 20-3A, the gas molecules exert a normal pressure against the walls of the container. When the volume of gas is compressed to half the original volume, the number of molecules per cubic inch doubles, figure 20-3B. In addition, the bombardment of molecules against this same piston area results in a pressure that is twice that of the original pressure. As the molecules are again compressed into one-fourth the original volume, the number of molecules and the pressure of these molecules per square inch is multiplied four times, figure 20-3C.

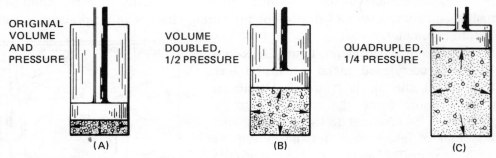

ORIGINAL VOLUME AND PRESSURE (A)

VOLUME DOUBLED, 1/2 PRESSURE (B)

QUADRUPLED, 1/4 PRESSURE (C)

**Fig. 20-4**

To this point, the unit has been concerned with the compression of gases. Figure 20-4 illustrates what takes place as the gas is expanded. If the original volume at (A) is increased to twice the original volume, as shown at (B), the molecules occupy twice as much space. The bombardment of molecules against the walls of the container is reduced to half the number of the original. As a result, the pressure is also reduced to one-half the pressure of (A). When the volume is expanded to four times the original volume, as shown at (C), the pressure is decreased to one-fourth the pressure at (A).

Boyle's Law applies equally to low or high pressures and is expressed by a formula as follows:

$$P_1 \times V_1 = P_2 \times V_2$$

where  $P_1$ = original pressure of a gas
$V_1$ = original volume of a gas
$P_2$ = pressure of a gas under a second set of conditions
$V_2$ = volume of a gas under a second set of conditions

For example, consider the case in which four cubic feet of a certain gas at a normal pressure of 15 pounds per square inch are compressed into a tank with a volume of one cubic foot. By substituting the given values in the formula, it is possible to determine the pressure that the tank must be able to withstand.

$$P_1 \times V_1 = P_2 \times V_2$$
$$15 \text{ lb./in.}^2 \times 4 \text{ ft.}^3 = P_2 \times 1 \text{ ft.}^3$$
$$P_2 = 60 \text{ lb./in.}^2$$

Thus, the pressure increases to compensate for the reduction in volume.

### Changing the Amount of Air

The simple air pump, figure 20-5, is the basis for more complex compressors which squeeze or force air under heavy pressures into closed vessels. The air pump consists of a piston moving up and down in a cylinder which also contains intake and outlet openings controlled by valves.

The upstroke of the piston causes the volume of air in the cylinder to increase and the pressure of the air to decrease. Normal atmospheric pressure forces air through the intake opening.

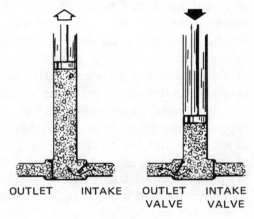

OUTLET   INTAKE   OUTLET   INTAKE
                  VALVE    VALVE

Fig. 20-5

On the downstroke, the piston compresses the air in the cylinder. The intake valve is forced to close and the outlet valve opens. The air pump exhausts the air from one container and compresses it into another.

### The Siphon

The action of a siphon is the result of air pressure. A *siphon* is a U-shaped tube with arms of unequal lengths, figure 20-6. This device is used to move liquids from a higher elevation to a lower point. The siphon is not self-starting, and must be filled with liquid to initiate the flow.

In figure 20-6, assume that water is to be siphoned from the container on the left. The atmospheric pressure on the surface of the water in each container is the same

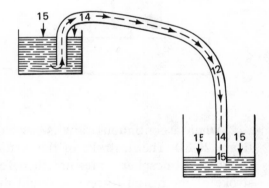

Fig. 20-6

(approximately 15 lb./in.$^2$). The pressure within the tube varies. Atmospheric pressure on the surface of the liquid maintains the liquid column in the left side of the tube. At a higher level, there is less pressure on the liquid than at a lower level. Therefore, the pressure at the first (left) bend in the tube may be 14 lb./in.$^2$. At the right (outlet) end, the pressure is atmospheric pressure (15 lb./in.$^2$). Again, the pressure on the fluid decreases with the height above the outlet. The pressure difference from left to right (from 14 lb./in.$^2$ to 12 lb./in.$^2$) along the horizontal section of the tube causes the fluid to flow to the right (the fluid tends to flow toward a low-pressure area). Once the fluid reaches the area of the lowest pressure, 12 lb./in.$^2$, the weight of the water causes it to fall into the lower container.

## OTHER APPLICATIONS OF MECHANICAL DEVICES USING FLUIDS

### Older Types of Lift and Force Pumps

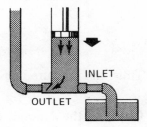

The principles of air pressure can be combined with a pump to produce useful work to move fluids, figure 20-7. If the inlet side of a pump is connected to a liquid source, a pressure will force this liquid through the pump outlet. As the piston moves up, the volume of air in the cylinder increases and the pressure is lowered. Atmospheric pressure on the liquid forces it into the pump cylinder. On the downstroke of the piston, the pressure exerted on the liquid by the piston closes the inlet valve, opens the outlet valve, and forces the liquid through the valve.

**Fig. 20-7**

Water and other liquids can be raised by a simple lift pump, figure 20-8A. This type of pump depends on atmospheric pressure to start the upward movement of the liquid. The first few strokes of the piston usually reduce the pressure in the pipe until atmospheric pressure can force the water into the cylinder. The flow of liquid from a lift pump is intermittent and depends on the piston action. The lift pump cannot lift water from a well more than 34 feet deep. The lift pump works by reducing air pressure and it has been shown that air pressure will not lift water beyond a height of 34 feet.

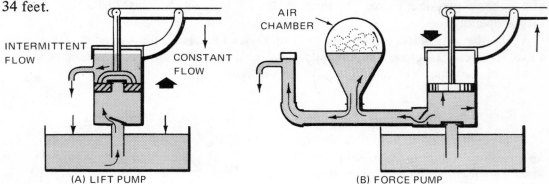

**Fig. 20-8**

When a continuous flow is needed, a force pump with an air chamber is used, figure 20-8B. The upstroke of the piston causes a reduction of the pressure inside the cylinder. Atmospheric pressure then forces the liquid into the cylinder. The downstroke of the piston forces the liquid through the outlet valve into an outlet pipe. The pressure of the downward stroke also compresses the air in the air chamber. On the upward stroke of the piston, the outlet valve closes and the compressed air continues to force the liquid through the outlet pipe to insure a continuous flow.

Figure 20-9 illustrates several types of commonly used fluid pumps. Figure 20-9A shows that an electric fan can create low- and high-pressure areas with air. The centrifugal pump, figure 20-9B, can pump large volumes of liquid under pressure. The rotary vane pump, figure 29-9C, is used when low pressures must be developed. The movement of the eccentric reduces the fluid pressure at the inlet side and forces the gas through the outlet side. Such a pump makes it possible to remove gas from a container until the pressure remaining is reduced to one-millionth part of one atmosphere. Another common type of fluid pump is the gear pump, figure 20-9D. This type of pump produces a flow at a constant rate and pressure.

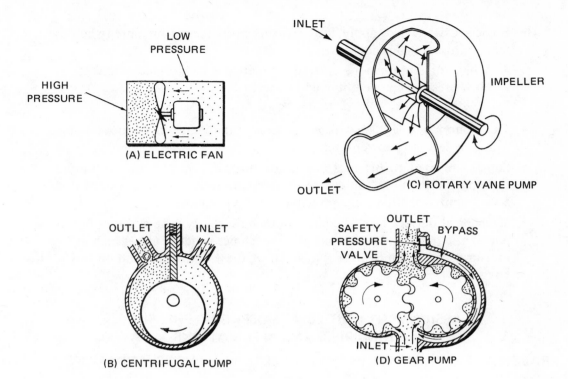

**Fig. 20-9**

## USING FLUID PRESSURES IN INDUSTRY

Several of the numerous uses of air and liquid pressures were illustrated in this unit. The operation of many mechanical devices depends upon the use of air at greater than atmospheric pressure. As one example, energy can be transmitted great distances by piping air pressure which has been built up by a compressor. Air pressure is used in underwater construction such as tunnels and bridge piers to force water out of huge reinforced steel chambers called caissons.

All compressors, compressed air and liquid devices, and internal combustion and steam engines depend on Boyle's Law. Other industrial processes using compressed air include welding and cutting, air conditioning, and spraying.

### ———————————— SUMMARY ————————————

- The atmosphere exerts pressure equally in all directions.
- Atmospheric pressure at sea level can support a column of water 34 feet high or a column of mercury 30 inches high.
- The normal atmospheric pressure at sea level is 14.7 lb./in.$^2$, or 29.92 in. Hg, or 76 cm Hg.
- Atmospheric pressure decreases almost one inch for each rise of 1000 feet.
- The Torricelli barometer is an instrument which measures the height to which atmospheric pressure, under varying conditions, supports a column of mercury.
- The aneroid barometer is a practical measuring device for determining and recording air pressure.

- Boyle's Laws, which apply to decreasing or increasing pressure (when the temperature remains constant), state:
  - A high pressure can be created by reducing the volume of a gas; a low pressure can be obtained by increasing the volume of a gas.

$$P_1 \times V_1 = P_2 \times V_2$$

- The siphon is a U-shaped tube, or similar device, used to move liquids from one level to a lower level.

- Pumps are used to lift or force fluids through a mechanism or a piping system. The pump action can result in an intermittent or a constant flow. Such pumps operate on the principles of Boyle's Laws.

- Mechanical devices such as compressors, fans, rotary vane and gear pumps, pneumatic tools, and numerous other machines produce either high or low pressures and volumes with fluids. All of these devices depend on Boyle's Laws.

## ASSIGNMENT UNIT 20  ATMOSPHERIC PRESSURE: PRINCIPLES AND APPLICATIONS

### PRACTICAL EXPERIMENTS WITH BAROMETERS, SIPHONS, AND PUMPS

### Equipment and Materials Required

| Experiment A<br>Torricelli Barometer | Experiment B<br>The Siphon | Experiment C<br>Air Pump |
| --- | --- | --- |
| Glass tube, 36 in. long, 1/4-in. inside diameter<br>Rubber tube (10 in. long) to fit glass tube<br>Thistle tube<br>Mercury, to fill glass tube and half the length of the rubber tube | Glass tube<br>Rubber tube<br>Two beakers with colored water<br>Test stand | Simple air pump<br>Two rubber balloons<br>String |
| | **Experiment D**<br>**Boyle's Law** | |
| Beaker<br>Measuring stick<br>Connection to vacuum pump | Graduated vessel or glass graduate<br>J-shaped glass tube (fine) (Boyle's Law apparatus, if preferred) | Mercury supply<br>Rule |

## EXPERIMENT A.  THE TORICELLI BAROMETER

### Procedure

1. Assemble the glass tube, rubber tube, and thistle tube as shown in figure 20-10.
2. Pour mercury into the thistle tube until the glass tube is filled.
3. Form a U-shaped tube with the open end of the thistle tube up.
4. Secure the mercury tube and thistle tube to a test stand.

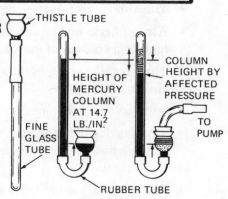

Fig. 20-10

5. Note that a vacuum is created in the top of the tube. Note also, that while some of the mercury flows back into the thistle tube, the column of mercury seems to be at a constant height.

6. Measure the height of the mercury column.

7. Connect the open end of the thistle tube to a vacuum pump and slowly reduce the pressure on the mercury in the thistle tube.

8. Note the rise of mercury in the thistle tube and how the column in the glass tube is lowered. Measure the height of the mercury above the height in the thistle tube.

9. Reverse the process by raising the pressure with the pump. Watch the effect on the column of mercury and the height in the thistle tube.

### Interpreting Results

1. Does the experiment confirm or reject Torricelli's findings about the height of a column of mercury that can be supported by atmospheric pressure?

2. Compute the atmospheric pressure per square inch at sea level.

3. What will the approximate barometric reading be at altitudes of 10 000, 20 000, and 50 000 feet above sea level?

## EXPERIMENT B. THE SIPHON

### Procedure

1. Fill a U-shaped glass tube from the long end with water.

2. Cover both ends of the tube. Insert the short end in a beaker of colored water.

3. Prepare a table for recording the results of the following steps.

4. Place the long end of the glass tube in a second beaker or jar at a lower level.

5. Uncover both ends of the tube. Trace the path and action of the water in:
   a. the long end
   b. neck
   c. the short end

6. Replace the glass tube with a rubber tube and repeat steps 1, 2, 4, and 5.

7. Raise the longer arm of the rubber tube until the end is at the same level as the level of water in the first beaker. Record what action, if any, takes place.

8. Continue to raise the tube and jar and again note and record what happens to the flow of water.

### Recording Results

| | Condition | | Action |
|---|---|---|---|
| **A** | Long end | | |
| | Neck | | |
| | Short End | | |
| **B** | Liquid level in both beakers is the same | | |
| **C** | Liquid level raised above original level | | |

### Interpreting Results

1. Make a simple sketch of the siphon (with a long arm and a short arm) and the two beakers or jars.

2. On the sketch, indicate the approximate atmospheric and air pressures by adding whole numbers in the tube and on the surface of the liquid.

3. What takes place when the water in the long end of the tube and the first beaker reach the same level?

4. Explain what happens in the tube and the beakers when the beaker at the lower level is raised so that the water level in the beaker and the long arm of the siphon is higher than it was at the start of the experiment.

## EXPERIMENT C.  SIMPLE AIR PUMP

### Procedure

1. Attach an inflated rubber balloon to the intake side of a hand-operated air pump.

2. Attach a deflated balloon to the outlet side of the pump.

3. Start pumping slowly. Note what happens to both balloons.

### Interpreting Results

1. Make a simple sketch of a hand-operated air pump. Label the parts of the pump.

2. Explain briefly how the pump works.

3. Convert the hand-operated air pump to a liquid pump. Sketch this new pump.

4. Letter the drawing to show how atmospheric pressure is applied in its operation.

## EXPERIMENT D.  VOLUME AND PRESSURE RELATIONSHIPS (BOYLE'S LAW)

### Procedure

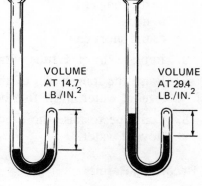

1. Mount a fine diameter J-shaped glass tube in a test stand.

2. Pour mercury from a graduated vessel slowly into the open end of the tube until the level in both sides is the same. Note the amount of mercury used.

3. Prepare a table for Recording Results and record the volume and pressure for each of the following steps.

4. Measure the height of the air column in the closed end of the tube. This height represents the volume of air under normal atmospheric pressure (14.7 lb./in.$^2$).

5. Add a volume of mercury equal to the volume already in the tube.

6. Note what happens to the air column. Measure and record this height (which represents the volume of air under a pressure equal to 29.4 lb./in.$^2$).

7. Repeat step 6 with a third equal volume of mercury (if the end of the tube is long enough).

## Recording Results

|   | Pressure (Lb./In.$^2$) | Volume (Ratio of Height) |
|---|---|---|
| A | 14.7 | |
| B | 29.4 | |
| C | 44.1 | |

## Interpreting Results

1. Compare the volume of the original column of air under normal atmospheric pressure with the volume when the pressure is doubled or tripled.

2. Use the Boyle's Law formula ($V_1$ x $P_1$ = $V_2$ x $P_2$) to check the volumes under pressures of 29.4 lb./in.$^2$ and 44.1 lb./in.$^2$

3. Do the values found in this experiment prove or disprove the formula expressing Boyle's Law?

4. Explain in outline form the important parts of Boyle's Laws in terms of volume and pressure.

## PRACTICAL PROBLEMS IN APPLYING ATMOSPHERIC PRESSURE

### Measuring Atmospheric Pressure

Select the correct word or phrase to complete statements 1 to 7.

1. The weight of a cubic foot of air is (62.5 lb./ft.$^3$) (14.7 lb.) (1.2 oz.).

2. The standard atmospheric pressure at sea level is (14.7 lb./in.$^2$) (14.7 lb.) (14.7 lb./ft.$^2$).

3. Atmospheric pressure at sea level supports a column of mercury (30 in.) (34 ft.) (62.5 in.) high.

4. Atmospheric pressure at sea level supports a column of water (30 in.) (34 ft.) (62.5 in.) high.

5. A change of 1000 feet in elevation causes (a change of 2 in.) (a change of 1 in.) (no change) in the height of a column of mercury.

6. The (Torricelli) (aneroid) barometer is a compact device for measuring and recording changes in pressure.

7. Boyle's Laws (are) (are not) affected by a raising or lowering of temperature.

8. State the relationship between the density and the pressure of gases.

9. As a weather or other balloon rises, does the pressure (a) in the balloon increase, and (b) against the balloon increase? Explain.

10. Compare (a) the air pressure at both ends of a siphon, and (b) the pressure on the liquid in both containers.

11. List two commercial applications of the siphon and two household uses.

12. Compare the advantages of the aneroid barometer to those of the mercury barometer in terms of (a) sensitivity, (b) accuracy, and (c) convenience in handling, reading, and recording.

13. From the data given in the table determine (a) the column height of fluids A and B which will be supported by atmospheric pressure at sea level, and (b) the specific gravity of fluids C and D.

|   | Specific Gravity | Column Height |
|---|---|---|
| A | 6.8 | |
| B | 0.82 | |
| C | | 68 ft. |
| D | | 23 ft. 8 in. |

**Atmospheric Pressure at Work**

Add a word or phrase to complete statements 1 to 12.

1. The pressure of gases in a confined space is caused by the bombardment of _____ against the walls of the container.

2. When the original volume of a gas is compressed, the _____ increases.

3. Any change in pressure on a confined gas produces a change in _____.

4. When the temperature of a gas is constant, doubling the pressure _____ the volume.

5. Without any change in temperature, decreasing the pressure 50% _____ the volume _____.

6. If the temperature remains constant, a reduction of the volume of a gas produces a corresponding _____ in pressure.

7. Boyle's Law, stated mathematically, is: _____.

8. A _____ is a simple tube device for moving liquids from one level to a lower level.

9. The siphoning action _____ when the liquid level in one container is the same height as the level of liquid in the long arm of the siphon.

10. For a siphon, the maximum height to which water can be raised by atmospheric pressure is _____ .

11. The siphon works on the principle of a difference in _____ between the pressure inside the tube and the pressure on the surface of the fluid.

12. In addition to the diameter of the tube, the rate of flow by siphoning depends on the _____ between the two liquid levels.

13. Explain how an air pump works. Include a simple sketch with the important parts labeled.

14. On what principle does a siphon operate? Explain and include a sketch.

15. Using a simple sketch, explain how a vacuum cleaner works.

16. Compute the height to which fluids (A) and (B) may be raised for the values given in the table. Assume an atmospheric pressure of 15 lb./in.$^2$.

| Pressure (Lb./In.$^2$) | Height – Fluid A (Sp. G. 0.8) | Height – Fluid B (Sp. G. 1.5) |
|---|---|---|
| 5 | | |
| 10 | | |
| 15 | | |
| 25 | | |
| 40 | | |

17. Graph the values determined in question 16 for volume (height) and pressure. Use a horizontal scale for the volume of 1/2 in. = 10 ft. (height of column) and a vertical scale of 1/2 in. = 5-lb. pressure.

18. Refer to the graph prepared in question 17 and (a) tell what happens when the pressure is increased, and (b) determine the volume of fluid (A) and fluid (B) for pressures of 20, 30, and 35 lb./in.$^2$.

## Applications with Mechanical Devices

1. Match each type of pump or fan in Column I with the appropriate use given in Column II.

| Column I | Column II |
|---|---|
| a. Centrifugal fluid pump | 1. Exhausts gases from tubes to 1/1 000 000 th part of an atmosphere. |
| b. Electric fan | |
| c. Rotary vane pump | 2. Increases or decreases atmospheric pressure by the action of straight-line motion pistons. |
| d. Gear pump | 3. Pumps large volumes of fluids at a fairly constant rate. |
| | 4. Creates high or low pressure areas with air. |
| | 5. Pumps fluids at a constant rate and pressure. |

Select the correct word or phrase to complete statements 2 to 5.

2. The simple lift pump used to raise fluids produces (a constant) (an intermittent) flow.

3. The force pump (which uses the force of a piston and an air chamber) produces (a constant) (an intermittent) flow.

4. (Air pressure) (Water pressure) is used to prevent the flow of water into an underwater caisson.

5. Acetylene gas is supplied at a controlled volume and pressure from cylinders. To do this, the original volume (is reduced at atmospheric pressure) (is reduced under greatly increased pressure) and stored in a cylinder.

6. The 3 1/2-inch diameter piston of a single cylinder engine compresses a fuel mixture to one-sixth of its original volume of 22 cubic inches. Determine the final pressure on the piston at the end of the compression stroke.

7. The table furnishes data on the volume and pressure for gases A, B, C, and D. The volume consumed for each hour of operation is also given. Assuming the atmospheric pressure is 15 lb./in.$^2$, determine the volume of gas that can be used by reducing the pressure to atmospheric pressure. Determine the number of hours the fuel can be used as the pressure is decreased to atmospheric pressure.

| | Volume (In.$^3$) | Pressure (Lb./In.$^2$) | Vol. when P is Reduced to 15 Lb./In.$^2$ | Hourly Consumed Vol. (In.$^3$) | Hours Supply of Gas |
|---|---|---|---|---|---|
| A | 1500 | 45 | | 1000 | |
| B | 1500 | 165 | | 1000 | |
| C | 3000 | 165 | | 750 | |
| D | 3000 | 457.5 | | 500 | |

# Unit 21 Fluid Power: Principles and Applications

Fluids under pressure are widely used in systems that transmit and control power. When the fluid is air or any other gas, the system is identified as a *pneumatic* system. A system that uses oil or another type of liquid is known as a *hydraulic* system. These two types of systems are sometimes combined into a common system called *fluid power*.

## PRESSURE, FORCE, AND VOLUME RELATIONSHIPS

Fluid power applications depend on the principles of pressure to exert a force or torque and the principle of *flow*. The output of the system in terms of the speed and amount of motion produced, is related directly to flow.

When a pressure is applied to a gas, the molecules of the gas are pushed closer together and are compressed into a smaller space. The greater the force pushing the molecules together, the greater is the effort of the molecules to move apart. As the pressure on a gas in a closed container increases, the volume decreases, figure 21-1. Conversely, as the volume of a gas is increased, the pressure decreases.

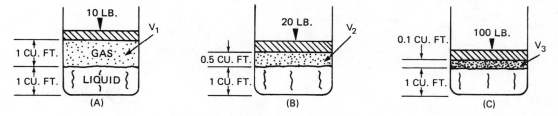

**Fig. 21-1 Compressibility of fluids**

Figure 21-1 reviews Boyle's Law. Part (A) shows a cubic foot of gas under ten pounds of pressure. The volume is decreased at (B) to one-half cubic foot under 20 pounds of pressure. At (C) the volume is reduced further to one-tenth cubic foot under 100 pounds of pressure. Using the formula $P_1 \times V_1 = P_2 \times V_2$, the volume of the gas ($V_2$) under 20 lb. of pressure is 0.5 cu. ft.: $(10) \times (1) = 20 (V_2)$. The same formula can be applied to (C) with the result that, $V_3 = 0.1$ cu. ft.

In contrast to gases, a liquid resists efforts to be compressed and reduced in volume. The volume of the liquid shown in figure 21-1, for practical purposes, is the same at (A), (B), and (C). However, the same force can be transmitted by either the liquid or the compressed air or gas.

The effect of heat on fluids is another important consideration in the control and transmission of power. The effects of temperature and heat energy are covered in later units. However, at this point, it should be noted that fluid power systems must be designed to compensate for the expansion of fluids due to changes in temperature resulting from pressure and flow factors.

## Transmitting Power

Power can be transmitted efficiently through a fluid system. The pressure and direction of the force may be changed by the design and length of the transmission line or pipe through which the fluid flows. The change of direction does not affect the transfer of force, figure 21-2.

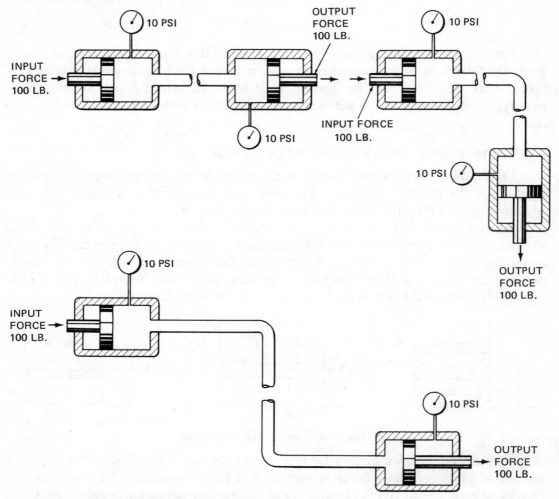

Fig. 21-2 Transfer of force is not affected by a change of direction.

## Transmission of Force as a Pressure

Hydraulic systems have great flexibility. Forces can be applied over long distances by the use of fluid power. In situations where locations are inaccessible to standard mechanical linkages, the force can be transmitted by the addition of piping and liquid. All power systems are governed by the principle that a gain in force results in a loss of distance. The output force of a hydraulic system can be changed by changing the area of the surface upon which the force acts.

A pneumatic system normally uses a compressor to develop the pressure in a gas. A storage tank holds the gas at this pressure. The pressure is then transmitted through the system to the point of application, figure 21-3. The 100-psi pressure in the system can deliver a force of 250 pounds on a piston area of 2 1/2 square inches.

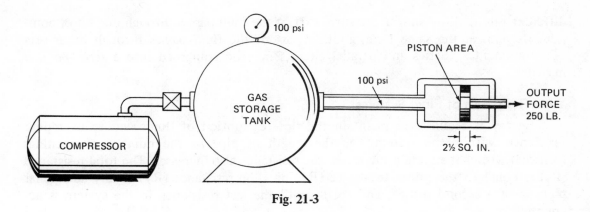

Fig. 21-3

## OPPOSITION (RESISTANCE) TO FLOW

*Opposition* is present whenever a force acts to change a motion or flow and another force is present to oppose the original force and limit the change of motion. In electrical circuits, the opposition is called *resistance;* in magnetic circuits, it is known as *reluctance;* in mechanical systems, *friction;* and in fluids, *fluid resistance.*

Fluid resistance limits the change of motion or flow and transforms some of the energy into heat (thermal) energy. All fluids have varying degrees of viscosity. Viscosity accounts for the internal friction (drag) of one part of a fluid on a neighboring part of the fluid in motion. The fluid also exhibits different types of motion. The flow may be *laminar* (steady), or it may be classified as *turbulent flow,* figure 21-4. Turbulent flow has whirls and eddies due to the rapid motion or to obstructions in the flow path. Friction losses are minimized in laminar flow where the layers of liquid move smoothly over one another.

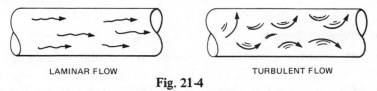

LAMINAR FLOW      TURBULENT FLOW

Fig. 21-4

There are two basic types of fluid circuits. In a *series circuit,* figure 21-5(A), all of the fluid flows through each component in the system. For a *parallel circuit,* however, the fluid flows through branches which are designed to provide the same or a

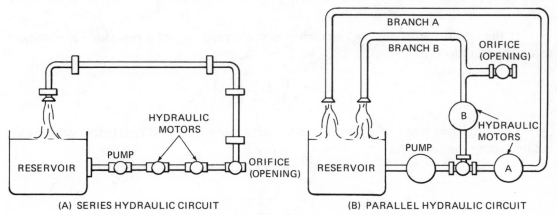

(A) SERIES HYDRAULIC CIRCUIT      (B) PARALLEL HYDRAULIC CIRCUIT

Fig. 21-5

different rate of flow and/or pressure. Part of the fluid passes through one set of components and, at the same time, another part of the fluid passes through other sets of components. Series and parallel circuits may be combined into a *series-parallel network*.

## Total Resistance of Fluid Systems

Energy is required to maintain the flow or motion of fluids and to overcome resistance within the system. As the length of pipe or the number of openings (orifices) within the system increases, the resistance also increases. The total resistance ($R_t$) is equal to the initial resistance ($R_1$) plus the resistance ($R_2$) of any additional pipe or of a second orifice, and so on until the last resistance in the system is accounted for.

$$R_t = R_1 + R_2 + \ldots + R_n$$

where $R_n$ is the last resistance in the system.

## Resistance of a Series Circuit

As the pressure is increased in a series circuit, the quantity of fluid per unit time which passes through the circuit is increased. This quantity of fluid per unit time is the *fluid flow rate*. Any opening (orifice) placed in the circuit will decrease the flow rate for the same pressure.

The resistance (R) of a series circuit is equal to the pressure difference ($\Delta P$) across the circuit divided by the quantity of fluid (Q) transported per unit of time in the system.

$$R = \frac{\Delta P}{Q}$$

## Resistance of a Parallel Circuit

In a parallel circuit, the total resistance can be calculated using one of two formulas. When the resistance in each branch of a parallel circuit is equal, the total resistance ($R_t$) is equal to the resistance (R) divided by the number of branches (N).

$$R_t = \frac{R}{N}$$

If the resistance of the two branches is not equal, the total resistance is obtained using the equation:

$$R_t = \frac{(R_1) \times (R_2)}{R_1 + R_2}$$

The total resistance of a parallel circuit with more than two resistances can be calculated using the equation:

$$R_t = \frac{1}{\dfrac{1}{R_1} + \dfrac{1}{R_2} + \dfrac{1}{R_3} + \ldots + \dfrac{1}{R_n}}$$

### Series — Parallel Fluid Flow

Some fluid flow systems require a combination of any number of series and parallel circuits. Pressure differences are necessary to maintain the fluid in motion. To compute the resistance of a combination series-parallel system, an easy method is to consider the system as a simple series circuit. The total resistance of the system then is the sum of the resistances of the equivalent series component parts. The total resistance is expressed in psi-min./gal. The flow rate may be expressed in *gpm* (gallons per minute) or *cu. ft./min.* (cubic feet per minute).

## STATIC STORAGE — POTENTIAL AND KINETIC ENERGY

A fluid system requires a storage tank and reservoir. This reservoir serves three prime functions in that it (1) maintains a constant pressure head, (2) allows heat to be removed, and (3) stores energy. The stored energy is *potential energy* when the fluid has the ability to do work because of its position with respect to other bodies. The potential energy (PE) of a body (in this case the fluid) is equal to the weight of the body (W) multiplied by the height (H) the body is lifted.

$$PE = (W) \times (H)$$

Since weight (W) = mass (M) times the acceleration due to gravity (g),

$$PE = (M \times g) \times (H)$$

When the mass is released, the stored-up potential energy becomes the energy of motion or *kinetic energy*. The velocity of the mass when the body reaches the lowest position is the same as the speed of a body falling from rest from height (H). Torricelli developed a theorem to express this condition.

$$V = \sqrt{2gH}$$

This expression says that the speed at which the liquid is discharged is the same as the speed of a body falling from rest from height H.

The kinetic energy (KE) is equal to one-half the mass (M) multiplied by the square of the velocity (V).

$$KE = 1/2M(V^2)$$

The flow rate at which the liquid leaves the valve or orifice in the storage tank is given by the following equation:

$$Q = (V)(A) = A\sqrt{2gH}$$

where  Q = flow rate, in cu.ft./sec.
          V = velocity, in ft./sec.
          A = effective area of the opening, in ft.$^2$
          g = 32 ft./sec.$^2$
          H = height from top of liquid to orifice, in ft.

There are several other factors that are important in any consideration of fluid systems. For example, thermal resistance may be present as a result of temperature differences which cause heat flow from one body to another. There is a time factor to consider since time is required to flow a charge or mass through the system. The units for thermal flow are expressed in British thermal units (Btu) per second or minute.

The resistance of the system also depends on the condition of the components. Corroded pipes and valves require an increase in pressure difference to maintain a given flow.

## Time Constants

The term *time constant* is used to indicate the period of time required to complete the fluid movement in a system when the transfer continues at its initial rate. Any change from one pressure level to another requires time. If a pneumatic system is subjected to a sudden change of input, there is a time lapse before the system responds completely to the change. A system action of this type is known as a *transient response* condition. A *steady-state* condition means that the system has reached equilibrium and is constant.

## Widespread Application of Fluid Power in Industry

In industry, pneumatic systems are classified as power systems and control systems. Pneumatic power systems are used for light loads and in situations where speed is important. Hydraulic systems are used with heavy loads where speed is of secondary importance. Pneumatic lifts, machine clamping devices, and pneumatic punches are typical applications of pneumatic power systems.

Pneumatic controls are widely used because of the advantages they have over electrical controls and other types of control systems. Air valves are fireproof and can be used in explosive areas, do not overheat and burn out, and operate over long periods of time without malfunction.

Logic control methods are another application of pneumatic controls. Pneumatic logic control takes place without moving parts by using fluids in motion. This field is known as *fluidics. Fluidic control devices* have the capacity to determine an alternate course of action when the directed course proves to be undesirable. Some advantages of fluidic control devices include: compact size, low cost, and long operational life. However, the response time for fluidic controls is usually slower than that of a sophisticated electronic control system.

Hydraulic systems are used to operate forming and cutting presses, machine cutters, mills, and for other heavy-duty operations. The principles of fluids at rest and in motion are applied to motor vehicle components such as the brake system. This system transmits motion, transmits pressure to develop friction between the brake shoes and drums, and multiplies the force applied at the brake pedal to the pistons which actuate the brake shoes.

Transmission units, hydraulic jacks, hoists, and landing gear on aircraft are other typical examples of hydraulics at work. The automobile carburetor is an excellent example of fluids in motion and the application of Bernoulli's principle.

## SUMMARY

- The term fluid power is applied to pneumatic (air/gas) systems and hydraulic (liquid) systems.

- Fluid power systems are used to transmit and control power, to apply force in places that are inaccessible to mechanical linkages, and in fluidic devices that regulate courses of action for automated mechanisms.

- The output force of a fluid power system can be changed by changing the surface area over which the force acts.

- Opposition (resistance) to flow results from forces that act to change the motion or flow.  Fluid viscosity, the nature of flow, and the type of system are some factors which provide the opposition to flow.

- The total resistance of a fluid power system is equal to the initial resistance plus the resistances of additional pipes, orifices, valves, and other components.

$$R_t = R_1 + R_2 + \ldots + R_n$$

- The total resistance of a series circuit is equal to the pressure difference across the circuit divided by the quantity of fluid transported per unit of time.

$$R_t = \frac{\Delta P}{Q}$$

- The total resistance of a parallel circuit can be determined as follows:

  - For two equal branches, total resistance equals the resistance divided by the number of branches:

$$R_t = \frac{R}{N}$$

  - For two unequal branches, the total resistance is:

$$R_t = \frac{(R_1) \times (R_2)}{(R_1) + (R_2)}$$

- The total resistance, expressed in psi-min./gal., is the sum of the equivalent series component parts.

- The potential energy of a fluid body equals the weight of the body times the height the body is lifted. $PE = (W)(H)$.

- The kinetic energy that a body possesses due to motion equals $1/2M(V)^2$.

- The velocity of the mass at the lowest position equals the square root of $2(gH)$: $V = \sqrt{2gH}$.

- The flow rate of the liquid at the orifice in a storage tank equals the effective area of the opening in ft.$^2$ multiplied by the square root of 2 times g (32 ft./sec.$^2$), times the height (H) from the top of the liquid to the orifice, in feet.

$$Q = A\sqrt{2gH}$$

- A time constant in a fluid system refers to the time required to complete the fluid movement process at the initial rate.

- Pneumatic and hydraulic systems are widely used in industry, agriculture, and commerce. In addition, these systems control the operation of many mechanisms that serve individuals such as the automobile, other forms of transportation, and computer devices.

## ASSIGNMENT UNIT 21  FLUID POWER:  PRINCIPLES AND APPLICATIONS

### PRACTICAL EXPERIMENTS WITH OPPOSITION AND ENERGY

### Equipment and Materials Required

| Experiment A<br>Opposition (Resistance) | Experiment B<br>Fluid Energy Storage |
|---|---|
| Hydraulic pump with relief valve, reservoir and return line on a hydraulic bench | Storage tank (16-gallon capacity) with bottom outlet and valve |
| Hydraulic hoses (1/4-inch diameter, 20-foot and 1-foot lengths) | Stop watch |
| Quick connecting coupler | Gallon container |
| Relief valve | Measuring stick |
| Check valve | |
| Pressure Gauges (2) | |
| Flowmeters (2) | |

## EXPERIMENT A.  OPPOSITION (RESISTANCE) IN FLUID SYSTEMS

### Procedure

1. Connect the hydraulic circuit components as illustrated.
2. Start the system and adjust the relief valve at 90 psi.
3. Prepare a table for Recording Results and record the rate of liquid flow (flow rate).
4. Replace the 20-ft. line with a one-foot length of hose.  Read and record the line pressure and the rate of flow.

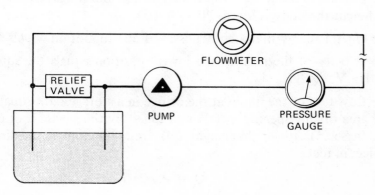

5. Replace the one-foot line with the 20-ft. line. Insert a quick connection coupler between the pressure gauge and the flowmeter. Record the line pressure and the rate of flow.

6. Repeat the test by replacing the quick connector with (a) a relief valve, (b) a check valve, (c) a pressure gauge and (d) a flowmeter. Read and record the line pressure and the rate of flow.

7. Calculate the resistance values of each component. Record in the table.

## Recording Results

|   | System Component | Pressure (Psi) | Flow Rate (Gal./Min.) | Resistance (Psi-min./gal.) | Resistance Values |
|---|------------------|----------------|----------------------|---------------------------|-------------------|
| A | 20-ft. line (1/4-in. diameter) | | | | |
| B | 1-ft. line | | | | |
| C | Quick connector | | | | |
| D | Relief valve | | | | |
| E | Pressure gauge | | | | |
| F | Flowmeter | | | | |

## Interpreting Results

1. Study the resistances that were computed for each component in the circuit. State what happens to the flow rate in series and parallel fluid circuits when:

   a. the length of pipe is increased,

   b. a quick connection coupler is added to the circuit, and

   c. different components are inserted in the circuit and their effect is measured.

2. State and prove the effect on fluid flow if all of the components that were inserted singly into the experiment were an integral part of the circuit.

## EXPERIMENT B. FLUID ENERGY STORAGE

### Procedure

1. Measure the tank dimensions.

2. Calculate the tank capacity. Prepare a table for Recording Results and record the tank capacity.

3. Determine the mass of the water; use the equation $M = \frac{gV}{32}$.

4. Fill the tank with water.

5. Draw eight separate one-gallon samples of water. In each case, record (a) the time in seconds required to draw each gallon, and (b) the height of the water remaining in the storage tank after each gallon is removed.

6. Calculate the flow rate in gpm. Record this value.

7. Find the potential energy of the water drained from the tank. Record.

$$PE = (M_1 \times g \times H_1) - (M_2 \times g \times H_2)$$

8. Calculate the kinetic energy of the water leaving the tank. Record this value.

$$KE = 1/2(M_1 V_1)^2 - 1/2(M_2 V_2)^2$$

9. Plot a graph of the flow rate (in gpm) and the water height (in inches).

## Recording Results

| Sample | Tank Capacity (cu. ft.) | Mass of Water $\dfrac{lb.-sec.^2}{ft.}$ | Outlet Diameter = 0.5 in. |
|---|---|---|---|
|  | Time (seconds) | Water Level (inches) | Flow Rate (gpm) |
| 1 |  |  |  |
| 2 |  |  |  |
| 3 |  |  |  |
| 4 |  |  |  |
| 5 |  |  |  |
| 6 |  |  |  |
| 7 |  |  |  |
| 8 |  |  |  |
| Potential Energy (ft:lb.): | | Kinetic Energy (ft:lb.): | |

## Interpreting Results

1. Describe the effect on the flow rate that results when the water level is reduced from 30 inches to each successive level as each gallon is withdrawn.

2. State how the flow rate in gpm, as measured in the experiment and plotted on the graph, can be verified.

3. Identify three factors that may affect the accuracy of the data.

## PRACTICAL PROBLEMS WITH FLUID POWER CIRCUITS

### Opposition (Resistance) in a Fluid Power Circuit

1. Calculate the resistance values and pressure drop across a series fluid circuit that has a rate of flow of six gallons per minute and components A, B, and C.

   Note: a. Record the resistance values for components A, B, and C as determined in Experiment A in the following table. Compute the resistance for the number of components in the circuit and the total resistance ($R_t$) of all components.

b. Determine the pressure drop resulting from each set of components and the total pressure drop across the circuit.

| Component | | Resistance | | Pressure Drop (Psi-min./Gal.) |
|---|---|---|---|---|
| | | *Unit Value | Total | |
| A | 50-ft. hydraulic hose, 1/4-in. diam. | | | |
| B | Quick disconnect couplers (3) | | | |
| C | Flowmeters (2) | | | |
| * Established in Experiment A | | | $R_t$: | Total P: |

2. Record the unit resistance values for components A, B, and C (as determined in Experiment A). The components are part of a series fluid system which maintains a pressure of 240 psi.

a. Compute the resistance for each set of components and the total of all components.

b. Calculate the rate of flow for the system.

| Component | | Resistance | |
|---|---|---|---|
| | | Unit Value | Total for Components |
| A | 40-ft. hydraulic hose, 1/4-in. diam. | | |
| B | Pressure gauges (4) | | |
| C | Quick disconnect couplers (8) | | |
| | Pressure drop: | Rate of Flow: | |

3. Determine the resistance of a fluid power system which has a flow rate of 42 gal./min. under a pressure of 516 psi.

## Fluid Energy

1. A fluid flows from a four-inch outlet of a storage tank at a speed of 18 ft./sec. Determine the flow rate of the fluid.

2. A fluid power system has an orifice which passes fluid at the rate of 9.6 ft./sec. and has a fluid flow rate of 300 gallons per minute. Compute the diameter of the outlet opening.

3. Determine the kinetic energy of 400 ft.$^3$ of water in a storage tank when the fluid passes an outlet valve at 12.2 ft./sec.

# Unit 22 Wave Motion: Transfer of Energy

Energy can be transferred from one place to another by mechanical and other physical means of moving material. Energy can also be transferred by nonphysical transport. This unit deals with wave motion principles and phenomena which are developed further in later units through applications to heat, sound and light energy, and to selected electromagnetic waves.

## MECHANICAL WAVES

*Wave motion* refers to the transfer and transport of energy by means of a disturbance in a medium. A *mechanical wave* must be generated by a mechanical source and must have a material medium through which it can move. In other words, a mechanical wave is considered to be a physical disturbance in an elastic medium.

As an example of a mechanical wave, consider the wave that is created when an object is dropped into a body of water. The disturbance spreads in the water as a series of concentric circles. That is, successive water particles transfer the energy from the point of contact in the fluid to an object that may be floating in the fluid. Thus, the waves transmit energy from one place to another through the motion created by a change in the medium, figure 22-1.

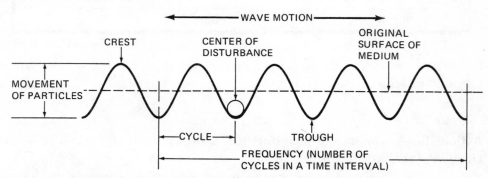

Fig. 22-1 Wave motion produced by a disturbance of the medium

The student must recognize that disturbances (energy) can be transmitted physically on mechanical waves or without a physical medium. For example, energy is transmitted in a nonphysical medium when it is propagated by electromagnetic waves caused by electrical and magnetic disturbances.

## CHARACTERISTICS OF WAVES

### Types of Waves

There are two major types of waves: transverse and longitudinal. In a *transverse wave* the vibration of the individual particles of the medium is perpendicular to the direction of wave propagation. In the example of an object dropped into water, the resulting waves travel away from the point of impact and the particles of water move up and down at right angles to the line of travel of the waves, figure 22-2.

Thus, a single disturbance, called a *pulse,* is sent across the medium. As the individual particles of water move up and down, the disturbance moves to the left or right with a velocity (v).

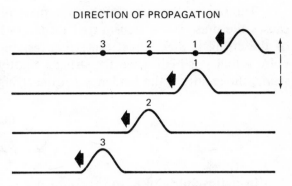

DIRECTION OF PROPAGATION

**Fig. 22-2 Characteristics of a transverse wave**

In a *longitudinal wave,* the vibration of the individual particles moves in a determined direction which is parallel to the direction of wave propagation. A coiled spring can be used to demonstrate the action of a longitudinal wave. The spring windings at the left of figure 22-3(A) are close together (condensed). When part of a wave is compressed in this manner, it is said that *condensation* is taking place. When the force causing the condensation is removed, a *condensation pulse* is created and propagated through the length of the spring, as shown in figures 22-3(B) and (C). The wave is a longitudinal wave because the material particles (the coils of the spring) are displaced in the direction of the disturbance.

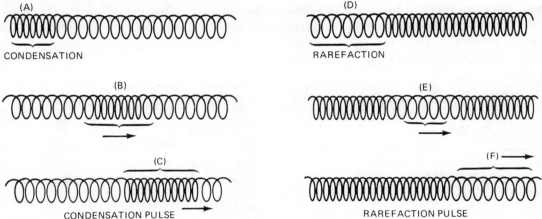

**Fig. 22-3 Longitudinal motion**

If the coils in the spring are forced apart, figure 22-3(D), a *rarefaction* is formed. Again, when the disturbance force is removed, a longitudinal *rarefaction pulse* is propagated along the spring, as shown at (E) and (F) of figure 22-3.

A longitudinal wave generally consists of a series of condensations and rarefactions which move in a specified direction.

### Pulses

A string can be used to demonstrate the simplest kind of wave phenomenon. When a completely flexible and uniform string is used, the pulse keeps the same shape as it moves along the stretched string with a constant velocity (v). The velocity depends upon the size of the string and the amount of tension caused by the stretching of the string.

The size of the string means the mass per unit length. A string with a large mass causes the pulse being transmitted to have a low velocity. The inertia of each segment of the string is high; thus, the string responds slowly to the forces acting on it. The pulse velocity is high when the string is tightly stretched. The higher the tension of the string, the greater is the tendency for the string to straighten.

$$\text{velocity} = \sqrt{\frac{\text{tension}}{\text{mass/length}}}$$

$$v = \sqrt{\frac{T}{m/l}}$$

In a traveling pulse, the forward portion of the pulse moves upward and the back portion of the pulse moves downward, figure 22-4. If the end of the string is securely fastened, the arrival of an upward pulse at the attached end of the string exerts an upward force on the support, figure 22-5.

According to Newton's Third Law of Motion, the support exerts an equal and opposite reaction force on the fixed end of the string. As a result of this reaction force, an inverted pulse is propagated in the opposite direction, figure 22-5.

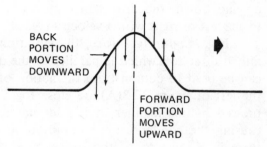

**Fig. 22-4**

The inverted pulse has the same shape as the original pulse and moves back along the string to the free end. In summary, when an upward pulse is reflected at the fixed end, it becomes a downward pulse; similarly, a downward pulse becomes an upward pulse.

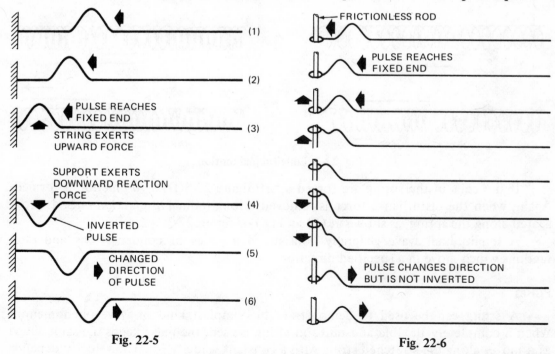

**Fig. 22-5**

**Fig. 22-6**

If the end of the string is free to move, figure 22-6, the pulse first moves upward on the string and then down again. While the direction of the pulse is reversed upon reflection, the reflected pulse has the same shape as the original pulse. If the tension on the string is between that of a completely taut string and a completely free string, the pulse disappears and is not reflected when it reaches the opposite end.

### Pulses Transmitted Between Media of Low and High Mass Per Unit Length

It has been shown that pulses can travel in a material of uniform mass per unit length. Pulses can also travel through materials having different masses per unit length. In figure 22-7, a light string of low mass per unit length is attached to a heavier string that is rigidly fastened. A pulse, generated by the up-and-down motion of the free end of the string, passes from the light string to the heavy string at their junction. The pulse is transmitted in the same direction to the heavy string. The greater inertia of the heavy string causes a reactive force in the opposite direction. This reactive force produces an *inverted reflected pulse.*

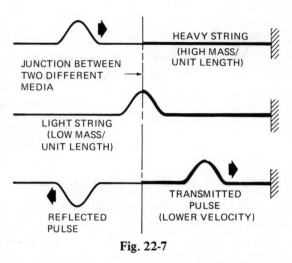

Fig. 22-7

When the pulse is transmitted from the heavy string to the light string, the inertia of the light string allows it to follow the pulse movement of the heavy string. However, in this case, the pulse velocity in the light string is greater and the length of the pulse is longer. The reflected pulse remains right side up. The preceding demonstration of the reflection and transmission of a pulse at a junction between different media also applies to other types of waves.

### PRINCIPLE OF SUPERPOSITION

The previous paragraphs considered a pulse as a single train (a single pulse transmitted from a single source along a medium). When two or more pulses move along a medium toward each other, there is a larger pulse at the instant the pulses come together, figure 22-8. The separate pulses then reappear unchanged and continue in the original directions of motion. The pulses are unaffected by their crossing. The principle explaining this type of behavior is called *superposition.* This principle states that when two or more wave trains (pulses) exist simultaneously in the same medium, the displacement at that point where the pulses meet is equal to the algebraic sum of the instantaneous displacements of each wave. In other words. each wave travels through the medium as if the other wave were not present.

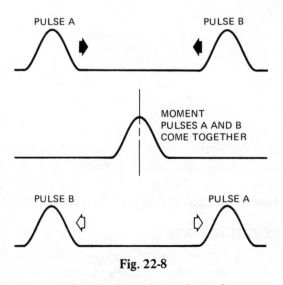

Fig. 22-8

When two identical pulses exist in the same medium, but one pulse is inverted with respect to the other pulse, the displacements cancel out, figure 22-9. However, after complete cancellation, the pulses reappear unchanged and continue in their original directions of motion.

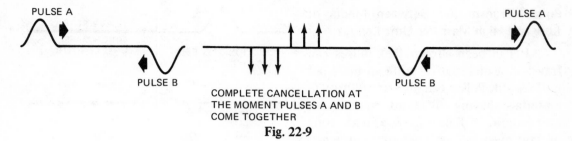

Fig. 22-9

## Constructive and Destructive Interference

When two waves of the same wavelength come together (coincide) so that the crests and troughs of the waves meet, the waves are said to *interfere constructively*, figure 22-10A. The amplitude of the resulting wave is greater than the amplitude of either of the original waves. According to the superposition principle, the amplitude is equal to the algebraic sum of the instantaneous amplitudes of the individual waves.

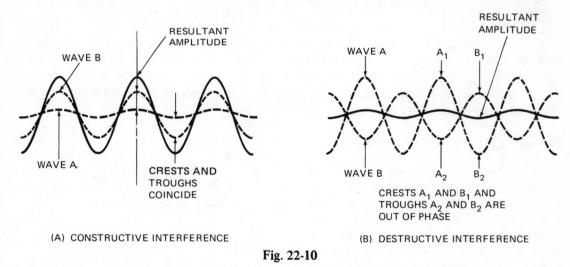

(A) CONSTRUCTIVE INTERFERENCE     (B) DESTRUCTIVE INTERFERENCE

Fig. 22-10

Similarly, when two waves of equal wavelength come together so that a crest meets (coincides with) a trough and the trough meets a crest, the waves *interfere destructively*, figure 22-10B. The amplitude of the resultant composite wave is less than the amplitude of the larger of the two original waves.

## PERIODIC WAVES

A *periodic wave* may describe sound, light, water, and other types of waves where one pulse follows another in regular succession. The shape of the respective waves (the *waveform*) is repeated at regular intervals. A wave train of periodic waves consists of a number of crests and troughs which move at a constant rate of speed.

The distance between two adjacent crests or troughs in the wave train is called the *wavelength*. The symbol for wavelength is the Greek letter lambda ($\lambda$). In a periodic wave train the wavelength ($\lambda$) is the distance between two particles that are in phase. Two particles are *in phase* when they are moving in the same direction and have the same displacement at the same time.

### Description of Periodic Waves

Periodic waves are described by three related quantities: wave velocity (v), wavelength ($\lambda$), and frequency (f). The distance a wave moves per second (*wave velocity*) equals the number of waves passing a given point per second (*frequency*) times the distance between adjacent crests or troughs of the wave (*wavelength*).

$$\text{Wave velocity (v)} = \text{frequency (f)} \times \text{wavelength } (\lambda)$$

$$v = f \times \lambda$$

The *period* of a wave is equal to the time (T) required for one wavelength to pass a given point.

$$\text{Time (seconds)} = \frac{1}{\text{frequency}}$$

$$T = \frac{1}{f}$$

$$\text{Wave velocity (v)} = \text{frequency (f)} \times \text{wavelength } (\lambda)$$

$$= \frac{\text{frequency (f)}}{\text{period of wave (T)}}$$

Frequency is measured in units of waves per second, cycles per second, or hertz (Hz). Wavelength is expressed by a linear unit and velocity is expressed in linear units per second, figure 22-11.

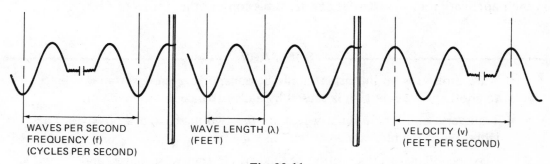

WAVES PER SECOND
FREQUENCY (f)
(CYCLES PER SECOND)

WAVE LENGTH ($\lambda$)
(FEET)

VELOCITY (v)
(FEET PER SECOND)

**Fig. 22-11**

Frequency can be expressed as follows:

1 kilocycle per second (kc/s) = 1 kilohertz (kHz) = $10^3$ cycles/sec.

1 megacycle per second (Mc/s) = 1 megahertz (MHz) = $10^6$ cycles/sec.

For example, 40 MHz = 40 x $10^6$ Hz = 4 x $10^7$ cycles/sec.

The following example illustrates the procedure for determining the frequency, velocity, and wavelength of periodic waves.

The distance between wave crests is 80 feet. In a one minute interval, ten waves pass a given point. Determine the frequency and velocity of the waves.

$$f = \frac{1}{T \text{ (sec.)}} = \frac{1}{6} = 0.17 \text{ Hz}$$

$$v = \frac{\lambda}{T} = \frac{80}{1/6} = 480 \text{ ft./sec.}$$

### Amplitude

A wave can also be described in terms of its amplitude. As shown in figure 22-12, *amplitude* is the height of the wave crest or the depth of the wave trough as measured from the original position of the particle. Thus, the amplitude (A) of a wave is its maximum displacement from the normal position of the particle as it oscillates (moves back and forth) while the wave moves.

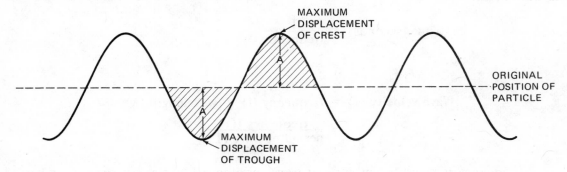

Fig. 22-12 Amplitude (A) of a wave

This unit stresses the importance of understanding the principles of wave motion in their application to the transfer and transmission of other forms of energy.

---

### SUMMARY

- Wave motion is an important mode of transporting energy from one place to another. Wave motion is caused by a disturbance in the medium.

- Transverse and longitudinal waves are the two major types of waves. The vibration of the individual particles in a medium is:

  - perpendicular to the motion of the wave in transverse waves, and

  - for longitudinal waves, parallel to the direction of motion of the waves in the medium.

- Removal of a force causing condensation results in the propagation of a condensation pulse. A rarefaction pulse is propagated by removing the force producing rarefaction.

- The velocity of a pulse $= \sqrt{\dfrac{\text{tension}}{\text{mass/length}}} = \sqrt{\dfrac{T}{m/l}}$.

- When a pulse reaches a fixed end, the pulse is subjected to an equal and opposite reaction force. This force produces an inverted pulse moving in the opposite direction.

- When two or more waves of the same type travel past a given point at the same time, the resultant amplitude is equal to the sum of the amplitude of each wave.

- Interference is the interaction of different wave trains.

  − Constructive interference occurs when the amplitude of the composite wave is greater than the amplitudes of the original waves.

  − Destructive interference occurs when the amplitude of the composite wave is less than the amplitudes of the original waves.

- Periodic waves are described by the related quantities wave velocity (v), wavelength ($\lambda$), and frequency (f).

$$\text{velocity} = \text{frequency} \times \text{wavelength}$$

$$v = (f)(\lambda)$$

- The period of a wave is measured by the time (T) in seconds required for one wavelength to pass a given point.

$$\text{Time (sec.)} = \frac{1}{\text{frequency (f)}}$$

$$T = \frac{1}{f}$$

- Frequency is measured in waves or cycles per second or hertz.

$$\text{Frequency} = \frac{1}{\text{time (sec.)}}$$

$$f = \frac{1}{T}$$

- Wavelength is expressed in a linear measurement; velocity is expressed in feet per second.

$$\text{Velocity} = \frac{\text{wavelength}}{\text{time}}$$

$$v = \frac{\lambda}{T} = (f)(\lambda)$$

- The amplitude of a wave is the maximum height or depth that a particle of the medium is displaced on either side of the normal position of the medium as the wave passes.

## ASSIGNMENT UNIT 22  WAVE MOTION:  TRANSFER OF ENERGY

## PRACTICAL EXPERIMENTS WITH WAVE MOTION AND WAVE PHENOMENA
### Equipment and Materials Required

| Experiment A<br>Wave Motion | Experiment B<br>Wave Phenomena |
|---|---|
| Fluid container<br><br>Block (float)<br><br>Stone | String (two pieces heavy, two pieces light). One piece of light string and one of heavy string are each to be 3 ft. long. The remaining pieces are to be 6 ft. long.<br>Rigid support<br>Frictionless rod<br>Ring (to provide a free moving body along the rod) |

## EXPERIMENT A.  WAVE MOTION:  DISTURBANCE AND ENERGY TRANSFER

### Procedure

1. Place a wooden block or float into a container that is partly filled with water. The block should be at rest.

2. Drop a stone into the water.

3. Observe (a) the nature of the disturbance in the medium and (b) the effect of the disturbance on the floating block.

### Interpreting Results

1. Illustrate graphically the disturbance caused by dropping the stone in the medium.

2. Explain the effect of the disturbance on the floating block.

## EXPERIMENT B.  WAVE PHENOMENA

## PART I.  CHARACTERISTICS OF A PULSE:  FIXED END
### Procedure

1. Tie one end of the 3-ft. long light string (light in terms of mass per unit length) securely to a rigid support.

2. Pull the string until it is stretched.

3. Give the string a quick up-and-down motion to produce a pulse in the string.

4. Prepare a table for Recording Results and note the direction of the pulse (a) at the beginning of the string and (b) after it reaches the rigid support.

5. Observe and record the nature of the pulse from the beginning to the rigid support.

   NOTE:  This experiment may be performed more accurately by passing each string in succession over a pulley and attaching a weight to the loose end. This procedure insures that the distance and tension are equal in all cases.

6. Repeat steps 1 through 5 using the longer length of light string and then the two lengths of heavy string. Record the results.

## Recording Results

| | String Characteristics (Mass/Unit Length) | Pulse Direction | | | Nature of Pulse |
|---|---|---|---|---|---|
| One End Secured in a Fixed Position | | Start | After Reaching Fixed End | Start | At Fixed End of String |
| One End Secured in a Fixed Position | Light (3 ft.) | | | | |
| One End Secured in a Fixed Position | Light (6 ft.) | | | | |
| One End Secured in a Fixed Position | Heavy (3 ft.) | | | | |
| One End Secured in a Fixed Position | Heavy (6 ft.) | | | | |
| Free End to String | Light (3 ft.) | | | | |
| Free End to String | Light (6 ft.) | | | | |
| Free End to String | Heavy (3 ft.) | | | | |
| Free End to String | Heavy (6 ft.) | | | | |

## PART II. CHARACTERISTICS OF A PULSE: FREE END

### Procedure

1. Tie one end of the 3-ft. long light string to a ring that moves freely up and down a frictionless rod.

2. Hold the string taut. Give the string a quick up-and-down motion to produce a pulse.

3. Record the direction of the pulse (a) at the beginning of the string and (b) as it travels and reaches the free end.

4. Repeat steps 1, 2, and 3 using the longer length of light string and then the two lengths of heavy string. Maintain the same degree of tension with each string.

### Interpreting Results

1. Make a simple illustration to show a pulse (a) as it moves to the free end of the string and (b) when the pulse is reflected.

2. Explain the effect of the reaction force at the free end of the string.

3. Compare the pulse velocity of the (a) 3-ft. and 6-ft. pieces of light string, (b) the two lengths of heavy string, and (c) the same lengths of light and heavy string (3-ft. lengths and the 6-ft. lengths respectively).

4. Illustrate the shape and direction of a pulse which travels to and returns from the free end of all the strings.

## PRACTICAL PROBLEMS WITH WAVE MOTION AND PHENOMENA

1. Define each of the following terms or quantities:

   a. Mechanical wave     d. Longitudinal wave     g. Constructive interference
   b. Wave motion     e. Condensation pulse     h. Frequency
   c. Transverse wave     f. Rarefaction pulse     i. Amplitude

Of the choices given for statements 2–9, determine which one(s) correctly complete(s) each statement.

2. The individual particles of a transverse wave move in a direction that is _____ _____ to the direction of travel of the wave.

   a. parallel                    c. circular
   b. perpendicular               d. opposite

3. In longitudinal waves, the individual particles of a medium vibrate back and forth in the _____ direction as that in which the waves travel.

   a. same                        c. perpendicular
   b. reverse                     d. elliptical

4. The instantaneous amplitudes of two waves that meet at a specific time are A and B. The combined amplitude is

   a. $\dfrac{A + B}{2}$.          c. A – B.

   d. A + B.                      d. $\dfrac{A - B}{2}$.

5. The period of a wave is the time required for

   a. one pulse.
   b. a number of waves to pass a point per second.
   c. each wave to move a specific distance per second.
   d. one complete wave to pass a given point.

6. Three related quantities that are used to describe periodic waves are:

   a. distance                    d. crests
   b. frequency                   e. wavelength
   c. constructive interference   f. wave velocity

7. At the junctions between different mediums, all types of waves exhibit

   a. constructive interference.  c. instantaneous amplitude.
   b. destructive interference.   d. reflection and transmission.

8. The maximum displacement from the normal position of the moving particles as a wave passes is identified as the _____ of a wave.

   a. frequency.                  c. velocity.
   b. wavelength.                 d. amplitude.

9. Wave velocity equals

   a. frequency x wavelength.     c. frequency x cycles.
   b. the sum of the displacement. d. frequency x amplitude.

10. Make generalized statements to explain each of these phenomena:

   a. In a uniform metal wire under tension, a pulse reaching a fixed end is inverted upon reflection.
   b. When a pulse reaches a free end of a metal wire, it is not inverted upon reflection.
   c. In two strings of different diameters but under the same tension, a pulse travels slower in the heavier string.

11. State the principle of superposition. Illustrate the principle by a simple line drawing.

12. Compute the speed of longitudinal waves A, B, C, and D. The frequency and wavelength for each wave is given in the table. Round off values C and D to the nearest whole number.

| Wave | Frequency (f) (Cycles/Sec.) | Wavelength (Feet) | Wave Velocity (v) (Ft./Sec.) |
|------|------|------|------|
| A | 100 | 6 | |
| B | $20 \times 10^3$ | 10 | |
| C | 365 | 9.5 | |
| D | 29.2 | 6.8 | |

# Achievement Review
## MECHANICS, MACHINES, AND WAVE MOTION

### FORCES AND THEIR EFFECTS

Indicate the letter corresponding to the item which correctly completes statements 1–7.

1. A number of forces acting on a body act (a) one at a time (b) in combination.

2. Effort force (E) x effort distance (ED) = (a) E x R (b) R x ED (c) R x RD (d) RD x ED.

3. The mechanical advantage of force where an effort force of 20 pounds moves a 100-pound load is (a) 50 (b) 5 (c) 1/5 (d) 2000.

4. The mechanical advantage of speed for a machine whose resistance moves through a distance of 36 feet to move an effort force nine feet is (a) 4 (b) 1/4 (c) 32.

5. Work output equals (a) E x ED (b) E x R (c) R x RD.

6. Efficiency equals (a) output x input (b) E x ED (c) output ÷ input.

7. Horsepower, the measure of the rate of doing work, equals (a) 550 x work x time (b) $\dfrac{\text{Work (Foot-pounds) x Time (Min.)}}{33\ 000}$ (c) $\dfrac{33\ 000}{\text{Work x Time}}$.

8. Compute the horsepower needed to do the work indicated in the table at A, B, C, and D.

|   | Rate of Work | Horsepower |
|---|---|---|
| **A** | 99 000 foot-pounds/min. | |
| **B** | 1100 foot-pounds/sec. | |
| **C** | 28 825 foot-pounds/min. | |
| **D** | 137.5 foot-pounds/sec. | |

### BALANCE, EQUILIBRIUM – PARALLEL AND ANGULAR FORCES

For statements 1 to 5, determine which are true (T) and which are false (F).

1. A dynamically balanced part is not balanced while in motion.

2. The vectors of a vector diagram indicate both the direction and magnitude of one or more forces.

3. An object is stable when its center of gravity falls within its base.

4. The resultant of two forces acting in directly opposite directions is the sum of the forces.

5. A single force that is equal and opposite to a resultant force is the equilibrant.

6. Select a scale to represent the values of the vectors in figure 1 at A, B, and C.
   a. Lay out the vector diagrams accurately.
   b. Determine the value of the resultant in each case to the nearest whole number.

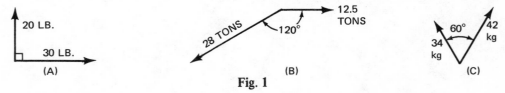

Fig. 1

7. Compute the resultant mathematically for the problems represented in figure 1.

a. Check the computed values (to the nearest whole number) against the scaled values.

b. Rework any value that is different.

## GRAVITATION, MOTION, AND MECHANICAL MOVEMENTS

Complete statements 1 to 3 correctly by indicating the missing word, phrase, or value.

1. Regardless of the weight of a body, a freely falling body gains speed uniformly at the rate of _____ .

2. Mass x _____ equals change in momentum.

3. Rectilinear motion may be (a) _____ , (b) _____ , or any combination of these.

4. Determine the final speed and the average speed for each of the three engines in the table, according to the conditions given.

| Engine | Starting Speed (Rev.) | Uniform Acceleration (Rev./Sec.) | Time to Reach Final Speed (Sec.) | Final Speed (Rev.) | Average Speed (Rev.) |
|--------|------|------|------|------|------|
| A | Rest | 100 | 10 | | |
| B | 1000 | 100 | 10 | | |
| C | Rest | 900 | 3.5 | | |

5. Identify two rotating parts, machines, or mechanism where the manufacturer must dynamically balance the moving parts to overcome any centrifugal force.

## SIMPLE MACHINES: LEVERS

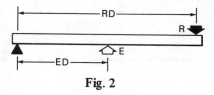

Fig. 2

Select the correct word or phrase to complete statements 1, 2, and 3.

1. A (an) (wedge) (inclined plane) (lever) is a rigid bar which turns around a fulcrum.

2. Of the three major classes of levers, (Class 1) (Class 2) (Class 3) has the resistance force between the fulcrum at one end and the effort force.

3. Figure 2 represents a (Class 1) (Class 2) (Class 3) lever.

4. Find either E, ED, or RD as required for levers A, B, and C. Neglect the weight of the lever and prove each answer.

| Lever | Class | E | ED | R | RD |
|-------|-------|------|------|------|------|
| A | 1 | 62.5 lb. | 8 ft. | 40 lb. | |
| B | 2 | | 22.5 ft. | 125 tons | 12 ft. |
| C | 3 | 648 lb. | | 540 lb. | 9 ft. |

5. a. For a mechanical or electrical device, name two parts that serve as levers.

   b. Identify each lever as to class by making a simple diagram and labeling the important parts.

   c. Describe briefly the function of each of the two levers.

## SIMPLE MACHINES: INCLINED PLANE AND WEDGE

For statements 1 to 4, determine which are true (T) and which are false (F).

1. The larger the angle of the inclined plane, the greater is the effort force needed to produce work.

2. Tapered holding devices are held securely by a wedging action.

3. The effort force on a wedge moves parallel to the slope.

4. The mechanical advantage of force of an inclined plane is equal to RD/ED.

5. Compute the forces or distances missing in the table for inclined planes A, B, and C. Neglect friction. Round off answers to the nearest whole number.

| | Effort Force (E) | Slope Length (ED) | Load (R) | Height (RD) |
|---|---|---|---|---|
| A | | 14 ft. | 6 tons | 4 ft. |
| B | 320 lb. | | 1275 lb. | 5 ft. |
| C | | 15 1/2 ft. | 5 1/4 tons | 4 1/2 ft. |

6. Compute the resistance force for the four wedge angles given in the table.

| Wedge Angle | Drive Force (E) | Effort Movement (ED) | Resistance Force (R) | Follower Distance (RD) |
|---|---|---|---|---|
| 20° | 200 lb. | 1 in. | | 1/2 in. |
| 40° | 200 lb. | 1 in. | | 1 in. |
| 60° | 200 lb. | 1 in. | | 1 1/2 in. |
| 80° | 200 lb. | 1 in. | | 2 in. |

7. Plot a graph of the resistance force (R) of a wedge-shaped cam from the information in the table given in question 6. Use a vertical scale of 1/4″ = 5°, and a horizontal scale of 1/4″ = 20 pounds.

8. Refer to the resistance force and wedge angle graph plotted in question 7.

   a. Determine the value of the resistance force for wedge angles of 30°, 45°, and 75°.

   b. Explain the effect on the resistance force of increasing the wedge angle from 20° to 80°.

## SIMPLE MACHINES: THE WHEEL AND AXLE

1. The following table gives distances and forces for four wheel and axle combinations.

   a. Make a wheel and axle diagram, labeling (R), (E), (ED), and (RD).

   b. Compute the missing values.

| Wheel and Axle Combinations | | | | |
|---|---|---|---|---|
| | E | ED | R | RD |
| A | 25 kg | 2.4 m | 60 kg | |
| B | 350 lb. | 12 in. | | 15.5 in. |
| C | | 24 in. | 2.6 tons | 15.5 in. |
| D | 3500 lb. | | 4200 lb. | 18 1/2 in. |

2. a. Determine the mechanical advantage of speed of the four wheel and axle combinations given in problem 1.
   b. Determine the mechanical advantage of force.

## SIMPLE MACHINES: THE SCREW THREAD

1. Screw threads are used: (a) to increase force, (b) as measuring devices, (c) to change the direction of motion, and (d) as fasteners. Give two new applications of each of the four uses of screw threads.

2. Compute the load which may be raised by the construction jacks A, B, C, and D for the dimensions and forces shown in the table. Disregard friction losses.

| Jack | Threads Per In. | Lever Radius (ED) | Effort Force (E) | Load (R) |
|---|---|---|---|---|
| A | 4 | 35 in. | 80 lb. | |
| B | 3 | 35 in. | 80 lb. | |
| C | 2 | 35 in. | 80 lb. | |
| D | 10 | 35 in. | 80 lb. | |

3. Prepare a graph showing the relationship between threads per inch and load (R) as given in the table of problem 2.
   a. Use a horizontal scale of 1/2″ = one thread per inch.
   b. Use a vertical scale of 1/4″ = 12 500 lb.

4. Determine from the graph in problem 3 what loads may be raised by changing the threads per inch to 6, 8, and 12.

## SIMPLE AND COMPOUND GEAR TRAINS

1. Determine the direction of rotation of each of the simple and compound gear trains shown in the table at the top of page 230.

2. Select the gears from the given sizes which will produce the required speed ratios for trains A, B, C, and D.

3. Compute the speed in rpm of the final driven gear using initial driver speeds of 40 rpm, 200 rpm, 320 rpm, and 400 rpm respectively.

| | Gear Train (Dr = Driver) Available Gears: Two Each of 24, 30, 32, 36, 48, 56, 64, and 96 Teeth | Speed Ratio Driver to Final Driven Gear |
|---|---|---|
| A | | 1 to 1 |
| B | | 4 to 1 |
| C | | 4 to 1 |
| D | | 6 to 1 |

## SIMPLE AND COMPOUND MACHINES

Match the numbered items on the left with the lettered items on the right.

1. $MA_f$ of a compound machine
2. $MA_f$ of pulleys
3. A pulley
4. The worm and worm wheel

a. A compound machine consisting of the wheel and axle and the inclined plane.

b. The product of the $MA_f$ of each separate machine.

c. The number of grooves in a pulley.

d. The number of ropes supporting the load on the movable pulley.

e. A change in the direction and magnitude of a force.

f. An application of the lever in a circular shape.

5. A single pitch worm gear revolving at 1440 rpm requires an effort force of 12 lb. to drive a 96 T gear. Mounted on the same solid shaft is a 32 T gear which drives a 128 T gear.

   a. Determine the resistance which this gear combination can move.
   b. Compute the mechanical advantages of force and of speed.

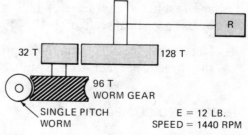

Fig. 3

## MECHANICAL POWER TRANSMISSION, FRICTION, AND LUBRICATION

Indicate the correct word or phrase to complete statements 1, 2, and 3.

1. When the source of power is brought close to the driven unit, some _____ losses are eliminated.

2. Friction is caused by (a) _____ , (b) _____ , and (c) _____ .

3. Friction may be reduced by (a) _____ , (b) _____ , and (c) _____ .

4. Select a common machine and name three parts which depend on friction for their operation.

5. The results of a prony brake test for three motors are shown in the table.

   a. Compute the amount of work done by each motor in one minute.
   b. Convert the work value to the nearest fractional horsepower for each motor.

| Motor | Force (Lb.) | | Diam. | Rpm | Work (1 min.) | H.p. |
|---|---|---|---|---|---|---|
| | $F_1$ | $F_2$ | | | | |
| A | 12 | 6 | 7 in. | 1000 | | |
| B | 12 | 6 | 3 1/2 in. | 2000 | | |
| C | 26.4 | 12.6 | 2.5 in. | 5600 | | |

## MECHANICS OF FLUIDS AT REST AND IN MOTION

Select the word or words which correctly complete statements 1 to 4.

1. Gravity pressure is (an externally applied pressure) (an internal pressure exerted by the weight of the fluid).

2. Hydraulic machines require the combined application of (force, distance, and pressure) (force and pressure) (pressure and distance).

3. The pressure of a fluid (does not change) (increases) (decreases) whenever the speed of a liquid is increased as it passes a constriction.

4. The (specific gravity) (gravity pressure) (force) of sinking and floating bodies equals the weight of the body divided by the weight of an equal volume of water.

5. Compute the force on the bottom of tanks A, B, C, and D. Use the weight of water as 62.4 lb./ft.$^3$.

| | Volume of Fluid | | | Specific Gravity | F (Lb.) | P (Lb./Ft.$^2$) |
|---|---|---|---|---|---|---|
| | L | W | D | | | |
| A | 1 ft. | 1 ft. | 1 ft. | 0.5 | | |
| B | 1 ft. | 1 ft. | 10 ft. | 0.5 | | |
| C | Diameter | | | | | |
| | 3 1/2 ft. | | 5 ft. | 1.0 | | |
| D | Radius | | | | | |
| | 14 ft. | | 9 in. | 0.8 | | |

6. Determine the pressure on tanks A, B, C, and D, using the values from the table.

7. Refer to the drawing of the lever-actuated hydraulic lift. Use the given dimensions and the data in the table to find the missing values.

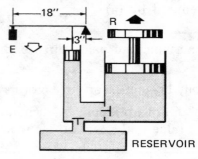

Fig. 4

| | Resistance Force (R) | Diameter Large Piston | Diameter Small Piston | Effort Force |
|---|---|---|---|---|
| A | | 4 in. | 2 in. | 20 lb. |
| B | 1200 lb. | 10 in. | 2 in. | |
| C | 12 000 lb. | | 1 in. | 100 lb. |
| D | 120 tons | 20 in. | 1 in. | |

## ATMOSPHERIC PRESSURE: PRINCIPLES AND APPLICATIONS

1. Determine the heights to which the columns of Fluids I, II, and III can be raised for the pressures given at A, B, and C. Use the weight of water (62.4 lb./cu. ft.) as the standard. Round off answers to one decimal place.

| | Pressure | Fluid I | Fluid II | Fluid III |
|---|---|---|---|---|
| | | | Specific Gravity | |
| | | Water | 0.6 | 2.5 |
| A | 14.7 lb./sq.in. | | | |
| B | 100 lb./sq.ft. | | | |
| C | 212 lb./sq.ft. | | | |

2. Compute the volume of gases A, B, C, and D when the pressures are reduced as indicated in the table. Round off to the nearest whole number.

| | Pressure (P) | Volume 1 | Volume 2 When P is Reduced to | Hourly Consumption | Gas Supply |
|---|---|---|---|---|---|
| | Lb./Sq.In. | Cu. In. | 15 Lb./Sq.In. | Cu. In. | Hours |
| A | 30 | 1000 | | 250 | |
| B | 60 | 1000 | | 500 | |
| | Lb./Sq.Ft. | Cu. Ft. | 2160 Lb./Sq.Ft. | Cu. Ft. | Hours |
| C | 4320 | 1000 | | 40 | |
| | Tons/Sq.Ft. | | | | |
| D | 12.6 | 1000 | | 33 | |

3. Use the computed volume (V) and the hourly consumption of each gas and determine the length of time each gas can be used. Round off each answer to the nearest hour.

## FLUID POWER: PRINCIPLES AND APPLICATIONS

1. Describe the differences between hydraulic, pneumatic, and fluid power systems.
2. Identify three design factors, conditions, or phenomena which affect fluid flow.
3. Illustrate a simple series fluid power circuit and a simple parallel circuit. Name the major components.
4. Name and describe two types of motion that are characteristic of fluid flow.
5. A parallel fluid system has four branches. The resistances in each branch are given in the table. Determine the total resistance of the system.

| Branch | Resistance (Psi-min./Gal.) |
|--------|----------------------------|
| A | 16 |
| B | 24 |
| C | 32 |
| D | 48 |

6. Three separate series circuits have the pressures and flow rates shown in the table. Determine the resistance of each circuit.

| Circuit | Pressure (Psi) | Flow Rate (Gpm) | Resistance (Psi-min./Gal.) |
|---------|----------------|-----------------|----------------------------|
| A | 396 | 8.4 | |
| B | 438 | 9.6 | |
| C | 724 | 10.2 | |

7. Use the resistance values given in the table for the components of a series fluid circuit having a flow of eight gallons per minute.
   a. Compute the total resistance for each set of components.
   b. Find the total resistance for all components.
   c. Determine the pressure drop for each set of components.
   d. Determine the pressure drop for the whole system.

| | Component | Resistance Unit Value | Resistance Total for all Components | Pressure Drop (Psi-min./Gal.) |
|---|-----------|------------|-----------------------|-------------------------------|
| A | 60 ft. of hydraulic hose, 1/4-in. diam. | 1.8 | | |
| B | Pressure gauges (3) | 5.4 | | |
| C | Flowmeters (2) | 0.9 | | |
| D | Quick disconnect couplers (8) | 5.2 | | |
| | Total for System | | | |

8. Determine the (a) potential energy and (b) the kinetic energy of 608 cubic feet of water in a storage tank. The fluid passes through an outlet at the rate of six feet per second at the beginning of the process. At the end of the process, the water passes through the outlet at the rate of four feet per second.

$H_1$ = 40 in.

$M_1 = \dfrac{4.23 \text{ lb.-sec.}^2}{\text{ft.}}$

$H_2$ = 20 in.

$M_2 = \dfrac{2.16 \text{ lb.-sec.}^2}{\text{ft.}}$

$g$ = 32 ft./sec.$^2$

## WAVE MOTION: TRANSFER OF ENERGY

1. Draw two waves having the same wavelength and different amplitudes. The crests and troughs of the waves are to coincide. Describe the magnitude of the composite wave and the conditions that are created.

2. Calculate the equivalent values in cycles/sec. for A, B, C, and D, according to the values given in the table.

| Medium | Frequency (f) | Equivalent Value in Cycles/Sec. |
|---|---|---|
| A | 60 megahertz (MHz) | |
| B | 30 kilocycles (kc/sec.) | |
| C | 25 megacycles (Mc/sec.) | |
| D | 2.5 kilohertz (kHz) | |

3. The distance between adjacent wave crests ($\lambda$) and the time interval of waves A, B, C, and D are given in the table. Determine both the frequency and the velocity of each wave.

| Wave | Wavelength ($\lambda$) | Time (s) per Passing Event | Frequency (f) | Velocity (v) |
|---|---|---|---|---|
| A | 80 ft. | 2.25 | | |
| B | 160 ft. | 2.25 | | |
| C | 80 ft. | 4.5 | | |
| D | 160 ft. | 4.5 | | |

# Section 4
# Magnetism and Electrical Energy

## Unit 23 Static Electricity and the Electron Theory

The scientific phenomenon known as static electricity is not new. More than 2 000 years ago a Greek philosopher named Thales observed that a piece of amber rubbed with another material attracted other small objects. The word electricity is derived from the Greek word *elektron* meaning amber. *Static electricity* means electricity that is still or stationary as compared to electricity which is said to flow in a wire.

### THE ELECTRON THEORY

All studies of electricity and electronics are based on the existence of minute particles called *electrons*. Electrons can either accumulate (build up) in one place or move from place to place. *Electricity* results when electrons move from one point to another, or when there is an excess or lack of electrons in a body.

To describe electrons, consider the composition of an ordinary drop of water. If a single drop is divided into two smaller drops and these smaller drops are then further divided, a microscopic examination reveals that each drop has the same characteristics. Each drop exhibits the properties of water. The process of dividing is continued until the smallest possible droplet having the chemical characteristics of water is produced. This smallest particle is called a *molecule*.

When this molecule of water is examined under the microscope at extremely high magnification, its structure is found to consist of three bodies, two tiny bodies and a larger body. Each of these bodies is called an *atom*. The two small bodies are the same and are the atoms of hydrogen. The larger body differs from the small bodies and is an atom of oxygen. Thus, one molecule of water is composed of two atoms of hydrogen and one atom of oxygen, figure 23-1. While water is composed of two different kinds of atoms, there are over 100 different kinds of atoms which combine in various ways to form the molecules of all other materials. Each different type of atom is known as an *element*.

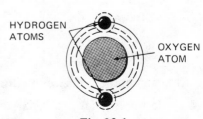

Fig. 23-1

If the hydrogen atom is examined more closely, it looks like a
sun with one planet circling it. The sun is the *nucleus* of the atom
and the planet is the electron, figure 23-2. For all atoms, the nucleus
contains a positive charge of electricity and the electron has a negative charge of electricity. The positive charge is called a *proton*.

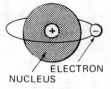

ELECTRON
NUCLEUS

**Fig. 23-2**

It has been found that for any atom, the number of positive
charges in the nucleus is exactly equal to the number of negative
charges (electrons) outside of the nucleus, figure 23-3. In addition to the positively
charged protons, the nucleus contains electrically neutral particles called *neutrons*.
These particles are shown in the figure as an electron and a proton bonded together
( ⊕⊖ ). Electrons are in a constant state of motion about the nucleus and the protons
and neutrons are stationary within the nucleus.

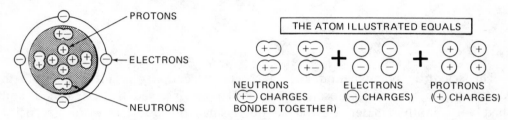

**Fig. 23-3**

An electric current is the result of the movement of free electrons. *Free electrons*
are those farthest from the nucleus. These electrons can be forced from their orbits
fairly easily. In contrast, *bound electrons* closer to the nucleus cannot be forced easily
from their orbits.

In summary then, electricity is the effect of too many or too few electrons or
is due to the movement of electrons from place to place.

## POSITIVE AND NEGATIVE CHARGES

A material is said to have a *negative charge* when it has an excess of electrons
( ⊖ charges), figure 23-4. A material that has a normal number of positive charges
in the nucleus, but lacks electrons, is said to have a *positive charge*. When there is
neither an excess nor a lack of electrons, the body is *uncharged*. All of these conditions are caused by the flow of electrons (with the positive charges stationary within
the nucleus).

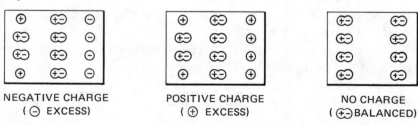

**Fig. 23-4**

### Attraction and Repulsion of Charges

Materials that are charged with static electricity either attract or repel each
other. Attraction takes place between unlike charges because the excess electrons
of a negative charge seek out the charge having a lack of electrons. Unlike charges

( ⊕ and ⊖ ) attract and like charges ( ⊖ and ⊖ or ⊕ and ⊕ ) repel each other, figure 23-5.

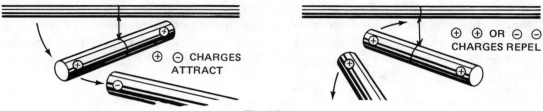

⊕ ⊖ CHARGES ATTRACT

⊕ ⊕ OR ⊖ ⊖ CHARGES REPEL

Fig. 23-5

## PRODUCING STATIC ELECTRICITY

Electricity may be produced by any one of six basic sources of energy: friction, heat, pressure, light, magnetism, or chemical action. Each of these forms of energy is discussed in a separate unit of this text. This unit covers the production of electricity by friction, the simplest basic source.

Friction is caused by the movement of two or more parts or materials on one another. During this contact, some electrons move from one material to the other. This transfer of electrons produces the effect that is known as *static charge* or *static electricity*.

Materials differ in their ability to build up and transfer static electricity. Depending on which material gives up the electrons, the static charge may be either positive or negative.

Silk, rayon, nylon, flannel, and hard rubber build up a static charge. When a glass rod is rubbed with silk, it becomes positively charged because electrons are transferred to the silk, figure 23-6; when a hard rubber rod is rubbed with fur, electrons are transferred from the fur and the rubber rod becomes negatively charged.

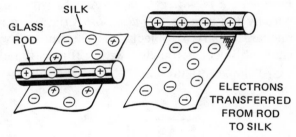

SILK

GLASS ROD

ELECTRONS TRANSFERRED FROM ROD TO SILK

Fig. 23-6

### Transferring Static Electricity by Contact

*Producing a Positive Charge.* Static charges can also be produced by transferring a a charge. One of the ways of transferring a charge is by contact. If a rod having a positive charge (a lack of electrons) is brought close to an uncharged part, the electrons ( ⊖ ) are attracted to the rod, figure 23-7. When the positively charged rod touches the uncharged part, some of the free electrons move to the rod. Thus, the rod, which was positively charged originally, becomes less positively charged and the part that originally was uncharged becomes more positively charged.

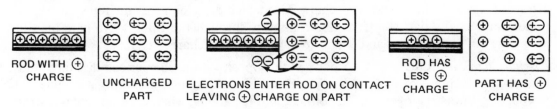

ROD WITH ⊕ CHARGE

UNCHARGED PART

ELECTRONS ENTER ROD ON CONTACT LEAVING ⊕ CHARGE ON PART

ROD HAS LESS ⊕ CHARGE

PART HAS ⊕ CHARGE

Fig. 23-7

*Producing a Negative Charge.* If a rod with a negative charge (having an excess of electrons ⊖ ) is touched to an uncharged part, a negative charge is produced on the part. When a negatively charged rod is brought near an uncharged part, the electrons in the part are repelled and move to the opposite side as shown in figure 23-8. The positively charged end of the part is then near the electrons of the rod. When the rod touches the part, the electrons move from the rod to the part to neutralize the positive charge, figure 23-8.

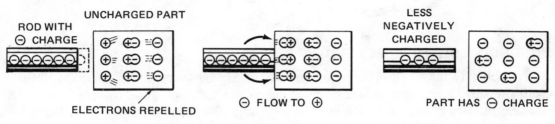

**Fig. 23-8**

The result of this action is that the rod loses some of its free electrons and becomes less negatively charged. This process continues until both the rod and part have the same amount of charge. As the rod is moved away, the originally uncharged part now has an excess of electrons and a negative charge.

## Transferring Static Electricity Through Induction

*Producing a Negative Charge.* A negative charge can also be produced in an uncharged part by the process of *induction.* Consider a rod with a positive charge placed near an uncharged part or object. The electrons flow toward the positive charge. If the uncharged object is touched with a finger, the electrons in the finger are attracted to the ( ⊕ ) charges, figure 23-9.

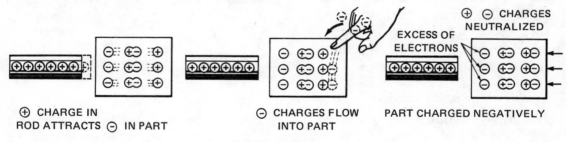

**Fig. 23-9**

When the finger is removed, the ( ⊕ ) charges and the ( ⊖ ) charges neutralize each other leaving an excess of ( ⊖ ) charges. The bar thus has a ( ⊖ ) charge.

*Producing a Positive Charge.* To produce a positive charge in a part, start with a rod having an excess of electrons ( ⊖ ). In this case, the positive charges in the part are attracted toward the negatively charged rod and the electrons are repelled. As the finger touches the part, the free electrons flow through the finger. The part then has an excess of ( ⊕ ) charges and is said to have a positive charge.

*Discharging Static Charges.* Two parts with opposite static charges can discharge these charges in a number of different ways. The three most common methods include: (1) connecting the parts together, (2) contact, and (3) arcing, when each body has a strong enough charge, figure 23-10.

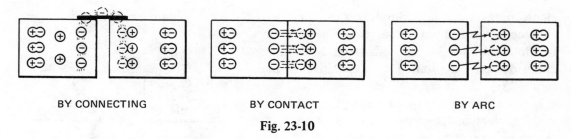

BY CONNECTING      BY CONTACT      BY ARC

**Fig. 23-10**

Regardless of the method of static discharge used, the ( ⊖ ) charges flow toward the ( ⊕ ) charges until the charges are neutralized.

## ELECTRONS IN MOTION

This explanation of electron theory serves as the background for further study of electromagnetism, magnetism, and electricity. The continuing movement of atoms and molecules in a material depends on the material itself, the temperature, and the pressure.

### Causes of Current

Those electrons which are farthest away from the nucleus and are free to flow are attracted out of their orbits by protons. An atom that loses an electron from one of its outer orbits becomes positively charged and attracts another free electron. It should be noted that the movement of free electrons is equal in all directions. However, when most of the movement is in the same general direction, the electrons flow in one direction to produce what is known as a current.

In other words, a *current* is the movement of free electrons as they are attracted to atoms that are positively charged because they have lost electrons in their outer orbits. The current, then, is caused by the loss and gain of electrons among the atoms. The movement of electrons is started by a change in pressure, the characteristics of the material itself, or a change in temperature.

### Properties of Materials; Conductors and Insulators

The ability of a material to conduct electricity depends on the degree of attraction between the positive nucleus of an atom and the electrons in the outer orbit. The atoms of some materials have a strong attraction between the positive charge of the nucleus and the negative charges (electrons) in the outer orbit. Materials such as wood, rubber, porcelain, and other clay products are examples of materials having the strong attraction between the nucleus and the electrons of the atoms of the material. These materials are not good conductors of electricity because they release few free electrons. Thus, these materials insulate against the flow of electrons.

Good conductors are those materials in which there is less attraction between the positive charge of the nucleus and the free electrons in the outer orbit. Copper, silver, iron, and other metals are examples of good conductors of electricity.

### Direction of Current

Electron flow takes place at any point in a wire at the same moment because any movement of an electron exerts a force (attracting or repelling) on the next electron. As a result, a simultaneous chain reaction occurs throughout the part. The direction of the electron flow proceeds from a negatively charged body or source to a positive charge, figure 23-11.

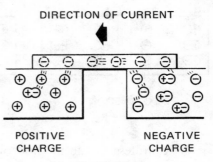

**Fig. 23-11**

This phenomenon of current is explained by the electron theory which states that free electrons are attracted by positive charges. As a result of this attraction, there is a movement from a negative charge to a positive charge.

### Importance of the Electron Theory

To appreciate the importance of the electron theory, the student should realize that electrical instruments, appliances and devices, electronic communication and control equipment, parts such as tubes and bulbs, the simple battery, and complex generating and distributing equipment all depend on current to produce work.

## DETECTING AND USING ELECTRICAL CHARGES

### The Electroscope

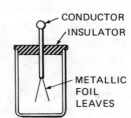

**Fig. 23-12**
**An electroscope**

It is possible to perform some simple experiments to show visually how static electricity may be produced, stored, and used. These experiments also demonstrate the nature of the electric charges. The experiments given in the assignment portion of this unit make use of a laboratory device known as an *electroscope*. This device is simply a jar and an insulated cap. Secured to the cap is a metal rod with a ball on one end and gold or aluminum foil leaves on the other end.

The leaves of the electroscope can be charged as described earlier in the unit, by contact or by induction. The charge received by the electroscope is indicated by the action of the leaves. If an excess of electrons or protons remain in the electroscope, the leaves repel each other. When the charges are neutralized, the leaves fall together. If the charges remain in the electroscope for any time, the leaves start to come together, indicating there is a minute release of charges to the air. This is due to an ionizing effect which is also seen when heat is brought close to the knob of the electroscope.

### The Leyden Jar

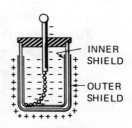

**Fig. 23-13**
**Leyden jar**

One of the oldest types of laboratory devices used to store and build up small charges of electricity is the Leyden jar. The *Leyden jar* acts as a capacitor and stores an electrical charge so that desired work can be done. In any capacitor, the plates are usually separated from each other so that the two plates have opposite charges. The jar portion of the Leyden jar serves as an insulator to separate a metallic covering on the outside of the jar from a similar covering on the inside.

Charges are introduced to the Leyden jar, either by induction or contact, through a metal rod with a ball on one end and a chain extending to the inside metallic shield on the other end. The outer shield is positively charged and attracts electrons to the inner shield, figure 23-13. When the unlike charges of the inside and outside shields are connected, there is a discharge of current.

Because a Leyden jar is bulky and has a limited capacity, commercial applications require the use of capacitors to control both the quantity of electrical flow and the intensity. There are two general types of capacitors: the *fixed capacitor* where only a definite amount of electricity is stored and used at a constant rate, and the *variable capacitor*. A radio capacitor is a good example of a variable capacitor. The variation is produced by turning a series of movable metal plates between another series of fixed plates. Thus, the capacity is increased or decreased depending on the areas of the metal surfaces that are brought close together.

## SUMMARY

- The electron theory is the basis for the study and explanation of the effect of electricity on the design and operation of all electrical and electronic devices, instruments, and appliances.

- The atom is composed of a heavy nucleus which does not move and one or more electrons which travel in orbits around the nucleus.

- The nucleus has a heavy positive charge (protons) and heavy neutrons which have no electrical charge.

- The electron is an almost weightless negative charge which is located in an orbit about the nucleus of the atom.
  - Free electrons are those which can be easily forced out of their orbits.
  - Bound electrons are those closer to the nucleus; these electrons cannot be forced out of their orbits very readily.

- A part is negatively charged when it has an excess of electrons. A part is positively charged when there is a lack of electrons.

- Like electrical charges repel each other; unlike charges attract.

- Either positive or negative electrical charges can be transferred by contact, induction, or arcing.

- Current is the movement of free electrons in one direction. The movement is controlled by pressure, the material, and the temperature.

- According to the electron theory, electrons move from a negative to a positive charge.

- Conductors are materials in which the free electrons in the atom are given up easily.

- Insulators are materials that resist the flow of electrons.

- Fixed and variable capacitors are used to store or distribute charges at either a constant rate or a variable rate.

## ASSIGNMENT UNIT 23 STATIC ELECTRICITY AND THE ELECTRON THEORY

### PRACTICAL EXPERIMENTS WITH ELECTRICAL CHARGES

#### Equipment and Materials Required

| Experiment A<br>Producing Electrical Charges | Experiment B<br>Removing and Reducing Charges |
|---|---|
| Two hard rubber rods<br>Fur or wool cloth<br>Two glass rods<br>Silk cloth<br>Small pieces of paper | Simple pith ball electroscope with<br> one and two pith balls<br>Rubber rod<br>Wool cloth<br>Sponge with small quantity of water |

### EXPERIMENT A. PRODUCING AND DETECTING ELECTRICAL CHARGES

#### Procedure

1.  Suspend a hard rubber rod by a string.
2.  Rub the rod with a piece of fur or woolen cloth until it is charged.
3.  Rub a second rubber rod with the same fur or wool cloth.
4.  Bring one end of the rod near the suspended rod. Prepare a table for Recording Results and note the action.
5.  Rub the glass rod with the silk cloth to produce a charge. Bring the glass rod near the suspended rubber rod. Note the action.
6.  Suspend the glass rod and recharge it. Rub the second glass rod with the silk cloth. Then bring it close to the suspended rod. Note the action.
7.  Recharge one glass rod and touch it to some small pieces of paper. Note the action.
8.  Repeat step 7 with the rubber rod.

#### Recording Results

| | Conditions | Action Observed |
|---|---|---|
| A | Two rubber rods rubbed with woolen cloth | |
| B | Rubber rod and glass rod | |
| C | Two glass rods | |
| D | Glass rod and paper | |
| E | Rubber rod and paper | |

#### Interpreting Results

1.  Is the charge in the rubber rod the same as that of the glass rod? Explain.
2.  Do like charges attract or repel each other? Prove the answer in terms of the results of this experiment.
3.  Using the results of this experiment, prove or disprove the statement that unlike charges attract each other.

## EXPERIMENT B.  REMOVING AND REDUCING STATIC CHARGES

### Procedure

1. Suspend two pith balls by a silk thread in a simple electroscope.
2. Rub a rubber rod with a woolen cloth. Then touch each pith ball with the rubber rod. Prepare a table for Recording Results and note what happens to the pith balls.
3. Touch one pith ball with a finger. Record the action.
4. Dampen one of the pith balls. Recharge the rubber rod and touch the dampened pith ball. Note if the two pith balls move apart as quickly as they did when they were dry.

### Recording Results

|   | Condition | Action |
|---|-----------|--------|
| A | Two pith balls — same charge |  |
| B | One pith ball — grounded |  |
| C | Same charge — one pith ball dampened |  |

### Interpreting Results

1. From the actions recorded in these experiments, state at least two methods for reducing or removing static electricity.
2. What method of reducing static electricity is recommended for an electric motor? For a moving belt?

## PRACTICAL PROBLEMS ON THE ELECTRON THEORY AND STATIC ELECTRICITY

For statements 1 to 10, determine which are true (T) and which are false (F).

1. Static electricity is a new scientific phenomenon.
2. The molecule is the smallest particle of a material having the same properties as the original material.
3. Atoms are made up of molecules.
4. An atom contains a nucleus and electrons in one or more orbits.
5. An electron is a positive charge.
6. A neutron is a negative charge.
7. A neutron is an electrically neutral charge.
8. A proton is a negative charge.
9. A part that is negatively charged has an excess of protons.
10. A part that is positively charged has an excess of protons.

Select the correct word or phrase to complete statements 11 to 15.

11. An uncharged body has (an excess of) (a lack of) (neither an excess nor a lack of) electrons.

12. Unlike charges attract because the excess electrons in the negatively charged body seek out a positively charged body which has a (lack) (excess) of electrons.

13. Like charges (repel) (attract) each other.

14. Static electrical charges may be transferred by (contact, induction, and arcing) (light, heat, and chemical action).

15. Electricity may be produced by (contact, induction, and arcing) (light, heat, magnetism, and chemical action).

16. Explain with sketches and a brief outline how a static charge may be transferred by induction.

17. Define current flow and use a sketch to simplify the explanation.

18. Why are some materials good conductors of electricity?

    a. Explain the answer in terms of the electron theory.
    b. Give two *new* examples of materials that are good insulators.

19. Explain why some materials are good insulators. Give two commercial examples of insulators.

20. Why are rubber pads used on factory floors around dangerous machines or where there are many moving belts?

21. Explain briefly how lightning rods protect property when lightning strikes.

22. A moving belt produces considerable quantities of static electricity in an industrial plant. This belt is a safety hazard. Draw a sketch of a simple device which removes the static electricity.

23. Explain briefly why gasoline trucks and trucks carrying combustible materials have a chain extending from the body to the ground.

24. Identify two devices used on electrical or electronic devices to store electrical charges.

    a. Make a rough sketch of one of these devices.
    b. Explain briefly how the device as sketched works.

# Unit 24  Magnetism

Much of the electrical energy that is produced and controlled today depends upon magnetism for its origin. The generation, transmission, distribution, and application of electricity are best understood after the principles of magnetism and the action of magnets are known. The terms used in magnetism and some of the special properties of magnets are discussed in this unit.

The name *magnet* is given to metals that have the property of attracting pieces of iron or steel. This property is found to a lesser degree in nickel, cobalt, chromium, and magnesium. This natural type of magnet is called a permanent magnet. Historically, magnets have been known and used for thousands of years. Almost 4 000 years ago, in the city of Magnesia in Asia Minor, a black rock (magnetite) was found to have the special property of attracting iron.

These pieces of black rock were called magnets — a name derived from the city near where they were found, Magnesia. The first practical use for a magnet was made in the twelfth century. It was learned that a magnet which was free to rotate would swing until it came to rest in a north-south direction, figure 24-1. Following this discovery, early explorers built compasses to aid in navigation and exploration to open unexplored territories, new trade routes, and new markets. The magnets in these first compasses were made by stroking a piece of metal with a natural magnet.

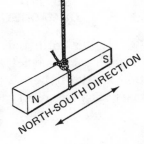

**Fig. 24-1**

## MAGNETIC AND NONMAGNETIC SUBSTANCES (DIAMAGNETIC AND PARAMAGNETIC)

Magnetic and nonmagnetic materials are classified into several categories. Materials such as iron and steel are said to be *ferromagnetic*. The properties of ferromagnetic materials are important in the design of devices, systems, and mechanisms that depend for their operation on magnetic phenomena. Artificial materials such as the alloys *permalloy* and *alnico* are used to make magnets that possess exceptional magnetic properties.

Another classification of materials is nonmagnetic substances. When certain materials are placed in a strong magnetic field, they are repelled; other materials under the same conditions are slightly attracted. Gold, zinc, mercury, and sodium chloride are common examples of substances that are slightly repelled. These substances possess the property of *diamagnetism* and are said to be *diamagnetic*.

Aluminum, wood, platinum, oxygen, and copper sulphate are slightly attracted by a strong magnet. These substances possess the magnetic property of *paramagnetism* and are identified as paramagnetic materials.

## POLES OF A MAGNET

If a magnet is placed near iron filings, most of the particles of iron are drawn about both ends of the magnet. If the magnet is free to rotate, it aligns itself in a north-south direction.

If the end of another magnet is brought close to the suspended magnet, the two magnets either attract each other or move away from each other.

FIRST LAW
OF MAGNETISM
| **Unlike Magnetic Poles Attract Each Other** |

The earth itself is a magnet; the south magnetic pole of the earth is located about a thousand miles away from the true geographic north pole. The end of the magnet that swings to a position of rest pointing to the south magnetic pole is attracted to it. This end is known as the north pole of the magnet; the other end of the magnet is the south pole (or the north-seeking and the south-seeking poles, respectively).

Regardless of the position in which the magnet is turned, once it is free to rotate it always rests in the north-south direction. In summary, each magnet has two poles: a north pole and a south pole. When two magnets are placed near each other, figure 24-2A, the unlike poles attract.

SECOND LAW
OF MAGNETISM
| **Like Magnetic Poles Repel Each Other** |

If the north pole of one magnet is brought near the north pole of a suspended magnet, each magnet will tend to push or move away from the other. The same condition is true if both south poles are brought close together, figure 24-2B. Thus, like magnetic poles repel each other.

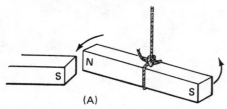 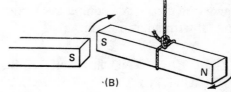

Fig. 24-2

## DOMAIN THEORY OF MAGNETISM

In the years following the published reports of William Gilbert's experiments with natural magnets (1600), later discoveries by Coulomb, Oersted, and Ampere added to the knowledge of magnetic behavior. Present scientific knowledge relates the magnetic properties of magnets to the movements of electrons within the atoms of substances.

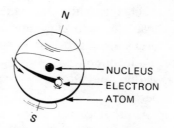

Fig. 24-3 Concept of single electron spinning on its own axis

Since electrons are considered to be charged particles, the modern concept of magnetism suggests that *magnetism is a property of the change of state of electron motion.* Two kinds of electron motion are important in this concept:

• An electron revolving about the nucleus of an atom gives a magnetic property to the atomic structure.

- The electron spins on its own axis. Each spinning electron acts as a minute permanent magnet.

Magnetic properties are associated with both kinds of electron motion.

As shown in figure 24-4, electrons can spin in opposite directions; these directions of spin are designated either as a positive (+) spin or a negative (–) spin. Electrons spinning in opposite directions become *paired electrons*. When electrons are paired, their magnetic properties are neutralized. When there is an imbalance between the orbits and the spins of the electrons, the atoms possess permanent magnetic properties.

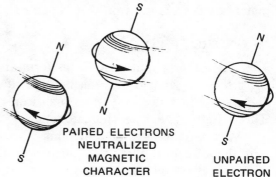

PAIRED ELECTRONS
NEUTRALIZED
MAGNETIC
CHARACTER

UNPAIRED
ELECTRON

**Fig. 24-4**

The atoms, depending on the interactions among the electron spins, are grouped in microscope magnetic regions known as *domains*. Within each substance there are crystal axes. The atoms in each domain are polarized magnetically to be parallel to a crystal axis. Since crystal axes are oriented in every possible direction within the substance, the polarities of the magnetized atoms in each domain tend to cancel one another. As a result, the net magnetism of each domain, and hence of the entire material, is almost zero. However, when the material is placed in an external magnetic field, some of the domains are reoriented in the direction of the magnetic field and the material becomes magnetized, figure 24-5. The degree to which the material is magnetized depends to a large extend on the strength of the magnetic field.

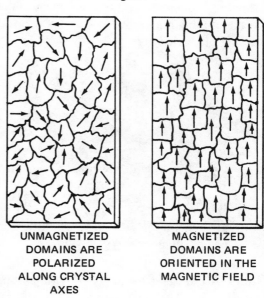

UNMAGNETIZED
DOMAINS ARE
POLARIZED
ALONG CRYSTAL
AXES

MAGNETIZED
DOMAINS ARE
ORIENTED IN THE
MAGNETIC FIELD

**Fig. 24-5**

According to the current theory of magnetism, when a ferromagnetic material is placed in an external magnetic field and becomes magnetized, the domains that are oriented in the direction of the magnetic field may increase in size. Adjacent domains decrease in size. Still other domains are influenced and become favorably oriented to the external magnetic field. The material becomes permanently magnetized when the domain boundaries remain extended after the magnetizing force is removed.

Another phenomenon is that the domains in iron resist being turned or magnetically aligned. However, the greater the strength of the magnetic field, the greater is its ability to rotate the domains further. The greater the distance the domains are rotated, the longer they remain in the newly aligned positions after the magnetic field is removed.

Physicists are able to examine magnetic materials under an electron microscope and photograph the microscopic domains or regions. The domains possess the north and south pole characteristics of a magnet. The magnetic properties of ferromagnetic materials are affected by temperatures approaching the melting points of the substances. The critical point at which the domain regions disappear and the material becomes paramagnetic is known as the *Curie point*. The Curie points for some common metals are as follows: iron, 770°C; cobalt, 1131°C; and nickel, 358°C.

## MAGNETIC FIELD

The pattern that iron filings form around the magnet can be seen clearly by placing a sheet of glass over a magnet and then sprinkling iron filings on the glass. When the glass is tapped gently, the small iron particles form a definite pattern, figure 24-6. This pattern has certain characteristics.

The broken lines of iron filings show the actual lines of force in a magnetic field and the direction they take, figure 24-8. When the iron filings are brought into the magnetic field zone of the magnet, they become magnetized. Thus, each of these iron particles becomes a small magnet. The combination of the two forces (that of the large magnet and that of each small iron particle magnet) produces the lines of force in the magnetic field. To produce action, there must be two magnetic fields acting on each other.

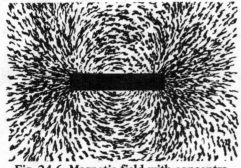

Fig. 24-6 Magnetic field with concentration of lines of force at poles

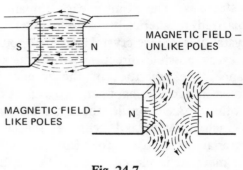

MAGNETIC FIELD – UNLIKE POLES

MAGNETIC FIELD – LIKE POLES

Fig. 24-7

## Lines of Flux and Intensity of a Magnetic Field (Magnetic Flux Density)

A *line of flux* represents the path of an independent north pole in a magnetic field. The path is so drawn that a tangent to it at any point indicates the direction of the magnetic field. The lines of force which can be used to map the magnetic field and show the intensity of the field are called the *magnetic flux*.

The magnetic flux has both direction and force. The direction of the magnetic flux is shown in figure 24-8. The lines flow through the magnet from south to north. Outside the magnet, the direction of the flux is from north to south.

The pattern formed by the iron filings shows a concentration at both ends of the magnet. Thus, the intensity of the magnetic field is greater at the poles and decreases as

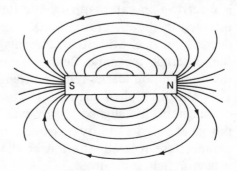

Fig. 24-8 Direction and pattern of magnetic lines of force

the distance from the magnet increases. A magnet is most effective when work is done as close to the poles as possible.

The *magnetic flux density* is defined as the number of lines of flux per unit area that permeate the magnetic field. Flux density is denoted by the unit webers per square meter (Wb/m$^2$); magnetic field intensity is represented by the unit newtons per weber (N/Wb).

## Force Between Magnetic Poles

The force between two magnets is expressed by Coulomb's law for magnetism: *The force between two magnetic poles is directly proportional to the products of the strengths of the poles and is inversely proportional to the square of the distance between the poles.* Although a single isolated pole is physically impossible because every magnet has two well-defined poles, the concept of a *unit pole* must be assumed. A *unit pole* is viewed as a pole which repels an exactly similar pole with a force of one dyne at a distance of one centimeter (1 dyne = $10^{-5}$ newton).

## Permeability and Residual and Induced Magnetism

*Permeability* (represented by the symbol $\mu$) is that property of a material which allows it to change its flux density in a magnetic field as compared to its value in air (unity). In other words, permeability is the degree to which lines of force are formed about a magnet with relation to the intensity of the magnetic field. The intensity of the magnetic field also depends on the material used in the magnet and the ease with which the lines of force pass through or around the magnet.

Diamagnetic substances have permeability values that are less than unity. Paramagnetic substances have permeability values that are slightly greater than unity.

When a magnetic field produces magnetism in a ferromagnetic substance, it is called *induced magnetism.* *Residual magnetism* is the magnetic property retained by a body after being magnetized by induction and then removed from the magnetic source.

## SHIELDING

### Nonmagnetic Materials

Not all materials can be magnetized. For example, if powdered glass or sawdust is sprinkled on the glass plate over a magnet instead of the iron filings, no pattern is formed as the magnetic field of the magnet has no effect on the glass or sawdust. As stated previously, the iron filings become magnetized and the magnetic fields of the filings and the magnet itself interact to produce the force needed to twist and move the small particles into the pattern.

### Paths of Least Resistance

A magnet that is free to rotate can be either pushed away from or pulled toward another bar magnet. The result will be the same if nonmagnetic materials are placed between the two magnets, figure 24-9, page 250.

However, when a magnetic material is placed between the two magnets, the fields fail to interact. Rather, the magnetic material offers less resistance than does the air to the flow of magnetic lines of force. As a result, the magnetic lines of force take the *path of least resistance* through the magnetic shield. This action is called *shielding,* and has important applications for magnetic materials that must be shielded from outside magnetic influence. A casing of iron or steel is an effective shield for magnetic materials.

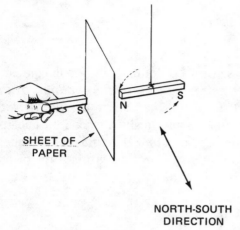

SHEET OF PAPER

NORTH-SOUTH DIRECTION

**Fig. 24-9 Nonmagnetic material permits lines of magnetic force to pass through.**

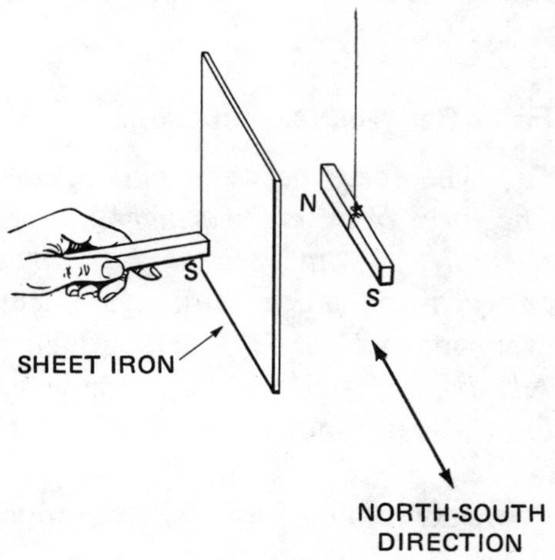

SHEET IRON

NORTH-SOUTH DIRECTION

**Fig. 24-10 Magnetic material diverts magnetic lines of force.**

--------- SUMMARY ---------

- Each magnet has two poles (north and south).
- Like poles repel each other.
- Unlike poles attract each other.
- A magnetic field surrounds a magnet.
- The intensity of a magnetic field is greatest at the poles.
- Nonmagnetic materials do not affect the lines of force.
- Magnetic lines of force follow the path of least resistance.
- Diamagnetic materials are those which are minutely repelled in a magnetic field; paramagnetic materials are slightly attracted when placed in a strong magnetic field.
- Magnetic flux has both direction and force. These properties may be illustrated by the lines of force.
  - Magnetic flux density = $Wb/m^2$.
  - Magnetic field intensity = N/Wb.
- The force between two magnetic poles is directly proportional to the product of the pole strengths and is inversely proportional to the square of the distance at the poles.
- Permeability is the degree to which lines of force are formed about a magnet in relation to the intensity of the magnetic field.
- Shielding occurs when a magnetic material is placed between two magnetic sources to prevent outside magnetic influence.

- The domain theory of magnetism describes magnetism as a property of the change of state of electron motion. According to the domain theory:
  - An electron revolving around the nucleus of an atom imparts a magnetic property to the atom structure, and
  - Each electron spins on its own axis and serves as a minute permanent magnet.

- Atoms are grouped in minute magnetic regions called domains; these domains are polarized magnetically parallel to the crystal axes of the material.

- The force exerted by an external magnetic field orients the domain boundaries by rotating the domains. The greater the rotation, the longer the domains remain in a magnetically aligned position.

- The Curie point is the critical temperature at which the domain regions disappear and the material becomes paramagnetic.

## ASSIGNMENT UNIT 24 MAGNETISM

### PRACTICAL EXPERIMENTS WITH MAGNETISM

### Equipment and Materials Required

| Experiment A Polarity | Experiment B Magnetic Fields | Experiment C Shielding |
|---|---|---|
| Two bar magnets | Same as for A | Same as for A |
| Stand | Glass plate | Sheet of paper |
| String | Adhesive tape | Sheet of iron |
| | Iron filings in shaker | |

## EXPERIMENT A. POLARITY

### Procedure

1. Suspend a bar magnet so that it rotates freely.

2. Bring the north pole of a second magnet close to the north pole of the suspended magnet.

3. Prepare a table for Recording Results. Record the action that occurs between the two magnets.

4. Bring the south poles of the two magnets close together. Record what happens in the table.

5. Now bring the north pole of the magnet to the south pole of the suspended magnet. Record the action.

6. Bring the south pole of the one magnet to the north pole of the suspended magnet. Record the action.

## Recording Results

|   | Condition | Action |
|---|---|---|
| **A** | North pole to north pole | |
| **B** | South pole to south pole | |
| **C** | North pole to south pole | |
| **D** | South pole to north pole | |

## Interpreting Results

1. State two simple principles of magnetism which explain the attraction and repulsion action of the two magnets.

2. Test four different materials with a bar magnet. Give the name of each material and indicate whether it is magnetic or nonmagnetic.

## EXPERIMENT B. MAGNETIC FIELDS

### Procedure

1. Secure a magnet in the center of a glass plate with adhesive tape.

2. Sprinkle iron filings from a shaker over the top of the glass. Observe the filings to see the pattern that they take.

3. Tap the glass carefully. Note that the filings assume a definite pattern.

4. Make a simple sketch of the magnet and the pattern which the iron filings form.

   NOTE: An accurate drawing can be made by placing an unexposed piece of blueprint paper over the glass, and then sprinkling iron filings on the paper. When the paper is exposed to light, the shape and pattern of the filings appear as white lines against the blue background.

### Interpreting Results

1. Indicate on the sketch of the magnet:

   a. The poles of the magnet (N and S).
   b. The places where the greatest number of lines of force are concentrated.
   c. The places where the smallest number of lines of force appear.

2. State two other simple principles of magnetism relating to a magnetic field and the lines of force.

## EXPERIMENT C. SHIELDING

### Procedure

1. Suspend a bar magnet so that it rotates freely. Place a sheet of paper between the north pole of the suspended magnet and the south pole of a second magnet. Prepare a table for Recording Results and record what happens.

2. Repeat step 1 and shield the north poles of both magnets with the sheet of paper. Note and record what happens.

3. Place a thin sheet of iron between the north pole of the suspended magnet and the south pole of the second magnet. Record whether or not there is any movement between the two magnets.

4. Repeat step 3 by inserting a sheet of iron between the two like poles of the magnets. Record any action that takes place.

### Recording Results

| | Condition | Action |
|---|---|---|
| A | N to S pole shielded with nonmagnetic material | |
| B | N to N pole shielded with nonmagnetic material | |
| C | N to S pole shielded with magnetic material | |
| D | N to N pole shielded with magnetic material | |

### Interpreting Results

1. Do nonmagnetic materials shield lines of force? Explain.

2. Will two magnets attract or repel one another when they are separated by a magnetic iron sheet?

3. State two principles of magnetism which explain the shielding action (if any) of (a) nonmagnetic materials, and (b) magnetic materials.

## PRACTICAL PROBLEMS IN MAGNETISM AND THE DOMAIN THEORY

### Polarity, Magnetic Fields, and Shielding

1. State one of the early practical uses of magnets.

2. a. Name the poles of a magnet.
   b. State two basic laws of magnets.

3. Briefly state why the magnetic properties of materials are important to industry.

For problems 4–11, indicate the letter of the answer which correctly completes each statement.

4. The south magnetic pole is (A) at the true geographic north pole (B) within 1000 miles of the true geographic north pole (C) at the true geographic south pole (D) at the equator.

5. The north pole of a magnet swings to a position of rest which points to the (A) north magnetic pole (B) south magnetic pole (C) true geographic north pole (within 1000 miles) (D) true geographic south pole (within 1000 miles).

6. A magnet that is free to rotate without external influences from another magnet will come to rest in a (an) (A) east-west (B) southwest by northeast (C) southeast by northwest (D) north-south direction.

7. The south pole of one magnet will (A) attract (B) repel (C) not be influenced by the north pole of another magnet.

8. The north pole of one magnet will (A) attract (B) repel (C) not be influenced by the north pole of another magnet.

9. A line of flux represents (A) the path or direction of the magnetic field (B) the force between two magnets (C) magnetic shielding.

10. Those nonmagnetic materials which are placed in a strong magnetic field and are minutely repelled are classified as (A) permeable (B) excellent shields (C) diamagnetic (D) paramagnetic.

11. The substance that offers least resistance to magnetic lines of force is (A) air (B) water (C) wood (D) copper (E) iron.

12. Prepare a simple table with two vertical columns marked magnetic and nonmagnetic materials. Add three horizontal columns and identify three magnetic and three nonmagnetic materials used in automobiles.

13. Briefly explain why a watch with a steel case is less likely to become magnetized when near rotating electrical machinery than is one with a gold case.

## Domain Theory of Magnetism

Identify the condition or phenomenon from Column II which correctly describes each term in Column I.

| Column I | Column II |
|---|---|
| 1. Electrons | a. Action of an electron revolving around the nucleus of an atom. |
| 2. Magnetic properties | |
| 3. Minute permanent magnet | b. Spinning in an opposite direction, the magnetic properties are neutralized. |
| 4. Magnetic property imparted to the atom structure | c. Electrically charged particles. |
| 5. Paired electrons | d. Atoms polarized magnetically to be parallel to the axes of crystals in a material. |
| 6. Domains | e. A change of motion; movements of electrons within the atoms of substances. |
| | f. An electron spinning on its own axis. |
| | g. The cancelling of magnetized atoms. |
| | h. The Curie point at which a material becomes paramagnetic. |

# Unit 25  Electromagnetism

Less than 200 years ago, electricity was discovered accidentally. An Italian professor, working on an anatomy experiment with a dead frog, had tied an iron wire around the frog's legs. When the iron wire was touched to a brass wire, the dead frog's legs moved violently. Another scientist analyzed this phenomenon a few years later and determined that the iron wire (tied to the frog's legs) was connected to the brass wire by the fluid within the frog's body.

Further experimentation with different solutions and different materials led to the development of an *electric cell* which was first used in 1800 for experiments. This electric cell consisted of alternately spaced copper and zinc plates immersed in a solution of salt water.

## MAGNETIC FIELD AROUND A CURRENT-CARRYING WIRE

Using the electric cell as a source of electricity, other scientific experiments were conducted. One of the first discoveries was that an electric current in a wire created a magnetic field around the wire, figure 25-1. The magnetism produced by an electric current is called *electromagnetism*.

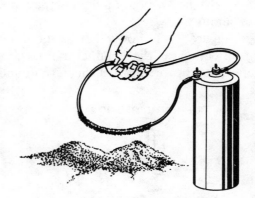

Fig. 25-1  Producing a magnetic field

### Characteristics of an Electromagnetic Field

A magnetic compass needle moves as it is brought near a current-carrying wire to indicate the presence of a magnetic field. The characteristics of this magnetic field can be studied by using a dry cell battery, a piece of wire, cardboard, and iron filings. The wire is passed through the cardboard as shown in figure 25-2. When the wire is connected to the source of current, a magnetic field is produced. This magnetic field can be seen by sprinkling iron filings on the cardboard.

When the cardboard is tapped sharply, the iron filings arrange themselves in continuous circles around the wire. In other

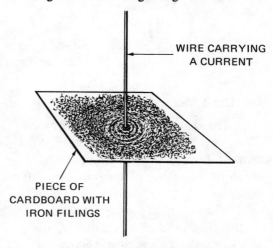

WIRE CARRYING A CURRENT

PIECE OF CARDBOARD WITH IRON FILINGS

Fig. 25-2  Magnetic field

words, the iron particles have become magnetized and thus indicate the pattern of the magnetic field about the wire.

**A magnetic field formed around a current-carrying wire is known as electromagnetism.**

The concentric circles of iron filings are closer together near the current-carrying wire than they are as the distance from the wire increases. As the current is increased, more and more iron particles line up in the concentric rings. Again, the greatest concentration of particles is near the wire. The current-carrying wire is also called a *current-carrying conductor.*

**The intensity of electromagnetism is greatest near the current-carrying conductor. The magnetic field decreases as the distance from the conductor increases.**

## DIRECTION OF CURRENT AND FLUX

The lines of force around a current-carrying conductor (wire) are referred to as the *flux*. This flux has a magnetic direction which can be shown by placing several small magnetic compasses around the wire conductor at a number of points, figure 25-3. The north pole of the compass points in the direction of the magnetic field or flux about the wire.

Figure 25-3 shows that as the direction of current in the conductor is changed, the direction of the conductor flux is reversed. Two common methods are used to determine the direction of the current and the resulting flux direction.

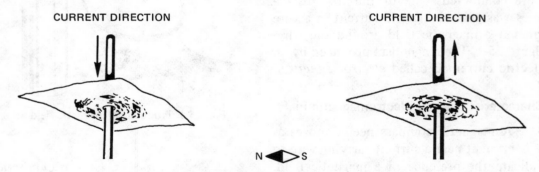

Fig. 25-3  Current direction determines flux direction.

### Dot-Cross Method

The dot-cross method as shown in figure 25-4 uses a *dot* ⊙ to indicate the flow of current toward the observer, and a *cross* ⊗ to show the flow of current away from the observer.

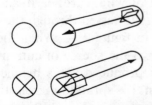

Fig. 25-4  Dot-cross method of indicating current direction

When a magnetic compass is placed in the flux around a conductor, the compass shows the flux direction, figure 25-5. The figure also shows (a) how the flux around a current-carrying conductor is distributed, (b) the density of the flux (greatest near the wire), and (c) the fact that the concentric lines of flux are closed rings.

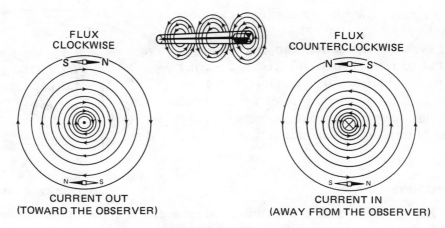

**Fig. 25-5 Density and direction of flux**

Note the clockwise direction of the flux when the current is toward the observer, and the counterclockwise direction of the flux when the current is away from the observer.

### Left-Hand Rule

The relationship between the flow of electrons in a conductor and the flux direction can also be determined by the *left-hand rule*. This rule can be used when either the current or flux direction is known.

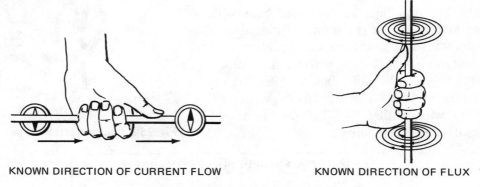

**Fig. 25-6**

Assuming that the direction of the current is given, the conductor is grasped in the left hand with the thumb pointing in the direction of the current. The fingers curl around the wire and point in the direction of the magnetic field around the wire. When the direction of flux is known and the fingers of the left hand point in this direction, then the thumb indicates the current direction.

It is important that both methods of determining current and flux directions be understood because practical applications of current-carrying wires and coils are based on a knowledge of these directions.

## ELECTROMAGNETIC COILS AND ELECTROMAGNETS

Thus far, the current-carrying conductors have been magnets without poles. The poles are those points where the flux lines either leave or enter the magnet.

Thus, a straight current-carrying conductor does not have poles.

However, if the current-carrying conductor is wound to form a loop, it becomes a magnet with poles, figure 25-7. As more loops are formed and the conductor takes the shape of a coil, the magnetic strength increases as the number of turns in the coil increases.

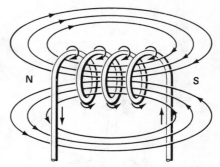

Fig. 25-7  Direction of magnetic flux and polarity of a coil

## Ampere Turns

When magnetic circuits are considered in electrical applications, the term used to describe the magnetic strength of the coils is *ampere turns*. This term indicates that the magnetic strength of a coil varies with the value of the current and the number of loops in the coil.

Figure 25-7 shows that the magnetic lines pass through the coil. Each line makes a complete circuit by returning along a path outside the coil. The end of the coil at which the flux emerges is the north pole; the other end where the lines of flux reenter the coil, is the south pole.

## Polarity of a Coil

The magnetic polarity of a coil can be found by applying the left-hand rule with a slight adaptation. If the thumb is placed on the conductor so it points in the direction of the current, the fingers then will point toward the north pole of the coil, figure 25-8. The same rule can be used to determine the direction of current when either the flux direction or the poles of the coil are known.

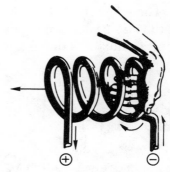

Fig. 25-8 Applying left-hand rule to a coil

> **The magnetic strength of coils depends on:**
>
> - **The number of turns**
> - **The amount of current**
> - **The material used as a core (inside the coil)**

## PRACTICAL APPLICATIONS OF ELECTROMAGNETISM

The principles of electromagnetism are applied in many important electrical devices and machines.

### The Electromagnet

It is known that air offers greater resistance to the flow of magnetic lines of force than does soft iron. Thus, a current-carrying coil wound around a soft iron core becomes a powerful magnet. This device is commonly referred to as an *electromagnet*.

## Magnetic Saturation

The strength of an electromagnet cannot be increased indefinitely. The iron core reaches a point of *magnetic saturation* when all of the molecules in the iron are lined up as magnets. Beyond this point, additional ampere turns do not appreciably increase the magnetic strength.

## Application of Electromagnets

Electromagnets are used daily in the home, in business and industry, and on the farm for numerous devices such as motors, chimes, and door bells. Greater use is made of electromagnets in industry for heavy lifting magnets, large motors, transformers, relays, and a host of other electrical and electromechanical machines, control devices, and instruments.

--- SUMMARY ---

- An electric current in a wire produces a magnetic field about the wire. This resulting field is known as electromagnetism.

- A reversal of the direction of the current reverses the direction of the magnetic flux.

- The dot-cross method and left-hand rule are two means of determining the direction of current and the magnetic flux.

- A current-carrying wire becomes a magnet with poles when it is wound to form a coil.

- The magnetic strength of a current-carrying coil increases (until a saturation point is reached) depending on the number of turns in the coil, the amount of current, and the material used inside the coil.

- An electromagnet results when the current-carrying coil is wound around an iron core.

- Magnetic strength is increased in electromagnets (using an iron core) because there is less resistance to the flow of magnetic lines of force through an iron core than there is through air.

- Electromagnets are used universally in electrical control devices, machines, instruments and other similar devices.

## ASSIGNMENT UNIT 25 ELECTROMAGNETISM

### PRACTICAL EXPERIMENTS IN ELECTROMAGNETISM

#### Equipment and Materials Required

| Experiment A Magnetic Fields | Experiment B Current and Flux | Experiment C Electromagnets |
|---|---|---|
| Copper wire | Same as A | Same as A and B |
| Iron filings | Cardboard sheet | Magnetic compass |
| Dry cell battery | | |

## EXPERIMENT A.  MAGNETIC FIELD AROUND A CURRENT-CARRYING WIRE

### Procedure

1. Connect one end of a copper wire to one terminal of a battery.
2. Place a small pile of iron filings near the battery.
3. Prepare a table for Recording Results as shown.
4. Bring the wire near the iron filings. Record the resulting action, if any.
5. Attach the second end of the wire to the battery.
6. Bring the iron filings near the current-carrying conductor. Note and record the resulting action.
7. Disconnect one end of the wire. Note and record what happens to the iron filings.

### Recording Results

|   | Condition | Action |
|---|-----------|--------|
| A | Wire placed in filings, without current | |
| B | Current-carrying conductor moved in iron filings | |
| C | Electric current removed | |

### Interpreting Results

1. Explain briefly what occurs when an electric current is passed through a wire.
2. Describe the action that occurs when the source of electric current is removed.

## EXPERIMENT B.  DIRECTION OF CURRENT AND FLUX

### Procedure

1. Place a small cardboard sheet at right angles to a wire.
2. Suspend the single wire vertically so that it is held tightly and is insulated from the stand.
3. Connect the two ends of the wire to the two battery terminals.
4. Sprinkle iron filings on the cardboard. Observe what takes place when the card is tapped sharply.
5. Prepare a table for Recording Results as shown. Record all observations.
6. Disconnect one end of the wire so no electrons flow through the conductor. Tap the cardboard. Note and record what happens to the filings.
7. Reverse the direction of the current and tap the cardboard again. Record the action.

### Recording Results

|   | Condition | Action |
|---|-----------|--------|
| A | Current-carrying wire, current "on" | |
| B | Source of current removed | |
| C | Source of current reversed | |

## Interpreting Results

1.  Describe briefly what happens to the iron filings when:

    a. the wire is a current-carrying conductor,
    b. the source of current is removed, and
    c. the source of the electric current is reversed.

2.  Explain what observations are made about the pattern formed by the iron filings

    a. around a current-carrying conductor,
    b. when the source of current is removed, and
    c. when the source of current is reversed.

# EXPERIMENT C. ELECTROMAGNETIC COILS AND ELECTROMAGNETS

## PART I. CURRENT-CARRYING CONDUCTORS

### Procedure

1.  Suspend a single wire vertically and pass an electric current through it.
2.  Make two simple cross-sectional drawings as shown in figure 25-9A and B. Make the drawings twice the size of those shown.

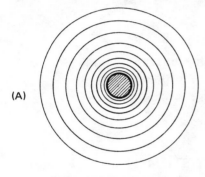 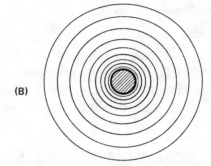

(A)  Electrons Flowing in One Direction      (B)  Electrons Flowing in Opposite Direction

**Fig. 25-9**

3.  Place a magnetic compass on a cardboard near the current-carrying conductor. Draw the position of the magnetic compass needle at the top of the cross-sectional drawing at (A) in figure 25-9.

    NOTE: Use the symbol (◁▶) for the compass needle with the black portion indicating the north pole.

4.  Move the magnetic compass clockwise through an arc of 90°. Note the changed direction of the needle. Again, draw the symbol in cross section (A) to indicate the direction of the magnetic flux.

5.  Repeat step 4 at approximately 180° and 270° intervals from the first position. Draw the position of the compass needle at each interval.

6.  Reverse the direction of current in the wire.

7.  Take compass readings at four points around the current-carrying conductor (90° apart).

8.  Illustrate the position of the compass needle at each of four places on the second cross-sectional drawing (B).

**Interpreting Results**

1. Indicate the flux direction by inserting arrows on the concentric rings of figure 25-9A and B.

2. Place a dot or cross in the center circle at (A) and (B) to indicate the direction of the current.

## PART II. ELECTROMAGNETIC COILS

**Procedure**

1. Form a number of large loops in a current-carrying conductor.

2. Bring the magnetic compass near one end of the coil. Note the position of the magnetic needle.

3. Repeat step 2 at the opposite end of the coil. Note the position of the magnetic needle.

4. Move the magnetic compass in the magnetic field around each loop. Observe in each case the position of the magnetic needle.

**Interpreting Results**

1. Make a simple sketch of the current-carrying coil.

   a. Use arrows to indicate the direction of the current.
   b. Determine the flux direction by using the left-hand rule.
   c. Label the poles on the sketch.
   d. Show the flux direction, using arrows.

## PRACTICAL PROBLEMS ON ELECTROMAGNETISM

**Magnetic Field Around a Current-Carrying Wire**

For statements 1 to 5, indicate which are true (T) and which are false (F).

1. A current-carrying wire does not have magnetic properties.

2. A straight current-carrying wire has a north pole and a south pole.

3. Iron filings arrange themselves in concentric circles around a current-carrying wire.

4. A magnetic field increases in intensity as the distance from the current-carrying wire increases.

5. Magnetic lines of force are concentrated near the current-carrying wire.

**Direction of Current and Flux**

Select the correct word or phrase to complete statements 1 to 5.

1. When the direction of current in the conductor is reversed, the direction of the flux (remains the same) (is reversed).

2. In the dot-cross method, the (dot) (cross) shows that the electrons are flowing away from the observer.

3.  When electron current is directed away from the observer, the flux travels in a (clockwise) (counterclockwise) direction.

4.  Flux density is (greatest) (least) near the current-carrying conductor.

5.  If the thumb of the left hand points in the direction of current, the fingers point in the direction of the (magnetic field) (polarity).

## Electromagnetic Coils and Electromagnets

For statements 1 to 4, determine which are true (T) and which are false (F).

1.  A current-carrying coil has a magnetic polarity.

2.  The magnetic lines of force make a complete path through a coil and around the outside of the coil.

3.  An air core offers less resistance to magnetic flux than does a soft iron core.

4.  When magnetic saturation is reached, additional ampere turns increase the strength of an electromagnet to a large degree.

5.  List six applications of electromagnets:  three in the home and three in industry or business.

6.  Sketch an electromagnet.

    a. Letter the poles.
    b. Show by arrows and broken lines the flux pattern and the direction of the flux.

# Unit 26  Electricity in Motion (Direct Current)

Current was defined in an earlier unit as the movement of electrons in one direction. If a measurement is to be made of how many electrons flow through a material in a given time, it is necessary to know those terms and measurements that relate to electron flow.

## THE MEANING OF CURRENT AND ITS MEASUREMENT

The smallest quantity of electricity is the electron. Regardless of whether the electrons are moving or still, 6 300 000 000 000 000 000 ($6.3 \times 10^{18}$) electrons are required for one coulomb of charge. The *coulomb* is a basic unit of measurement for an electric charge. As the electrons move through a material, the flow rate is expressed by another unit of measurement known as the ampere. An *ampere* describes both the amount of an electrical charge and the time. In other words, the ampere refers to the quantity of electrons moving and the rate at which they move. One ampere = one coulomb per second.

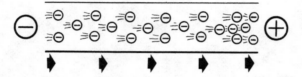

**Fig. 26-1  Movement of free electrons (electron flow)**

Measurements of current can range from 1000 amperes to less than a millionth part of one ampere. Because of this range of measurements, it is impractical to use the ampere alone as a unit of measurement. The metric system prefix *milli* is used to denote values of one-thousandth part of an ampere. The prefix *micro* denotes one-millionth part of an ampere. One ampere, therefore, is equal to 1000 milliamperes (1000 mA) or 1 000 000 microamperes (1 000 000 $\mu$A).

Measurements of current are made in amperes, milliamperes, and microamperes. When working with these values, it is often necessary to change from one unit of measurement to another. For example, to change from amperes to milliamperes, the decimal point must be moved three places to the right and three zeros added; to change from amperes to microamperes, the decimal point is moved six places to the

| Changing Larger to Smaller Units | | Changing Smaller to Larger Units | |
|---|---|---|---|
| | **Move Decimal Point to Right and Add Zeros** | | **Move Decimal Point to Left** |
| 10 Amperes = | 10 000 milliamperes 10 000 000 microamperes | 10 000 milliamperes 10 000 000 microamperes | = 10 Amperes |

**Fig. 26-2**

right and six zeros are added. The process is reversed to change from a smaller unit to a larger unit; for example, from microamperes to milliamperes or amperes.

Whenever current is to be measured, an instrument called an *ammeter* is used. Ammeters are available in three basic ranges: (1) the standard ammeter range measures amperes of current, (2) the milliammeter range is between one ampere and one milliampere, and (3) the microammeter range is for small current measurements between one microampere and 1000 microamperes. In all cases, the measurements are taken from a calibrated scale.

## ELECTROMOTIVE FORCE AND ITS MEASUREMENT

The electrons in a material remain stationary until a force causes them to move. When the electrons move, an electrical current is produced.

The electrical potential energy of a coulomb of electrons is measured in a unit called the *volt*. Since the potential energy can cause electrons to move, this energy also gives rise to an electron-moving force or *electromotive force* (emf). Electromotive force is expressed in volts. The amount of work done to create a coulomb of static charge is measured in volts. A stored electric charge is actually a value of potential energy.

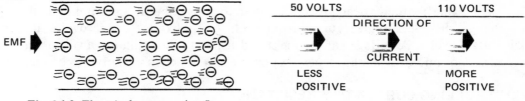

Fig. 26-3  Electrical pressure (emf)
causes current

Fig. 26-4

Electrical charges do not move until there is a difference between the potential energy of one charge as compared to another charge. This difference in energy or electromotive force is expressed in volts. The difference in potential between two unequal charges is said to be the *voltage*. Voltage is a comparison of the electromotive forces between two unequal charges. Figure 26-4 shows that there is a voltage of 60 volts representing the difference in potential between the two charges.

### Maintaining the Electromotive Force

Electrons tend to flow continually when a difference in potential exists. The continuous flow of electrons in batteries is produced by the chemical action within the battery. Since the major source of electrical energy is magnetism, it is the continuous movement of generator wires through magnetic fields that provides a constant unequal charge at both terminals of a generator. As various devices such as motors, lights, and other electrical equipment are connected to the source of the electrical energy, the current causes bulbs to light, motors to turn, and other electrical devices to operate.

*Voltage and Current.* Every electrical device requires a fixed amount of voltage and current for its operation. A light bulb operates on 117 volts and small motors usually operate on 117 or 234 volts. These voltages produce the amount of current needed for efficient operation. If the voltage is less than the amount required, the devices operate less efficiently: the bulb is dimmer and the motor cannot do the work that it is rated to do. A greater voltage than is required produces an excessive current which causes the bulb, motor, or other device to burn out. Protection is provided against such overloads by a *fuse* installed either in the power line or on the device, or both.

### Measuring the Electromotive Force

The difference in potential between two unequal charges can range from a fractional part of a single volt to thousands of volts. Again, the metric prefixes milli and micro are used to express one-thousandth part of a volt and one-millionth part of one volt, respectively. In other words, 1000 millivolts or 1 000 000 microvolts = one volt. For larger units of measure, the prefix kilo indicates 1000; thus, 1000 volts = one kilovolt (kV).

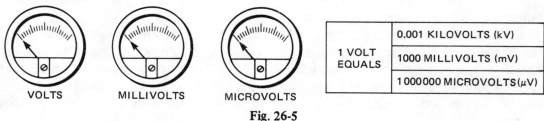

VOLTS    MILLIVOLTS    MICROVOLTS

| 1 VOLT EQUALS | 0.001 KILOVOLTS (kV) |
|---|---|
| | 1000 MILLIVOLTS (mV) |
| | 1 000 000 MICROVOLTS (μV) |

**Fig. 26-5**

The same mathematical process is used to move the decimal point to the right to change the measurement units from volts to millivolts or microvolts and from kilovolts to volts. A voltage can be measured using a kilovoltmeter, a voltmeter, millivoltmeter, or microvoltmeter, depending on the difference in potential to be measured.

## CONTROLLING CURRENT BY RESISTANCE

### Kind of Material Affects Resistance

Current is the movement of free electrons. The amount of this movement is controlled by the opposition to the movement of the electrons through a material. This opposition is called *resistance*.

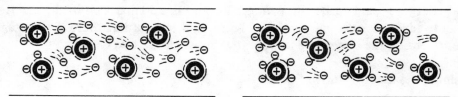

**Fig. 26-6 Different materials offer different resistance**

Resistance is present in varying degrees in different materials, figure 26-6. Those materials which are good conductors of electricity offer a minimum of resistance to current. In contrast, the best insulators are those materials which offer greater opposition to the flow of free electrons through the material. Thus, the material selected is one factor in controlling the resistance. Metals such as silver, copper, aluminum, and steel each offer a different resistance to electron flow. The resistance depends on how easily the electrons in the outer orbit of the atom move.

### Length of Material Affects Resistance

The longer the path of an electric current, the greater is the resistance offered by the material to the current flow. A nichrome wire 20 feet long offers a greater resistance than does a one-foot piece of the same size nichrome wire, figure 26-7. As a matter of fact, the resistance is 20 times greater because the electrons are held by 20 times as many atoms.

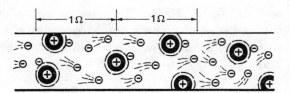
Fig. 27-7  Resistance increases with length

### Cross-sectional Area Controls Resistance

The cross-sectional area of a material also affects the resistance offered to a current. The greater the cross-sectional area of a conductor, the less is the value of the resistance offered, figure 26-8. If the cross-sectional area is doubled, twice as many electrons are available for movement and the current is doubled.

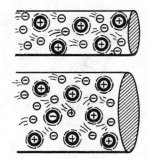
Fig. 26-8

### Temperature Controls Resistance

A fourth condition controlling resistance is the temperature of the conductor. Temperature affects the ease with which the electrons flow until the maximum heat of the conductor is reached. The greater the conductor temperature, the greater is the resistance to electron flow due to atom vibration, figure 26-9.

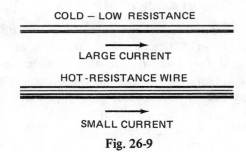
Fig. 26-9

Conversely, the colder a conductor is, the greater is the ease with which the electrons flow. The variation in resistance due to temperature differences in the material depends on the material. The effect of temperature changes is slight on conductors such as copper and aluminum.

### Measuring Resistance

The resistances of different materials are compared using a unit of measurement called an ohm. One *ohm* is the resistance offered to the flow of one ampere under a pressure of one volt. Again, the prefix micro is used to indicate a resistance of one-millionth part of an ohm; kilo is used to represent a thousand ohms; and the term megohm is equivalent to 1 000 000 ohms.

The instrument used to measure resistance is called an ohmmeter. *Ohmmeters* are available in various ranges. The ohmmeter range selected is determined by the range of the resistances to be measured.

### Application of Resistors

To control the flow of electrons (current) in a motor or in other electrical equipment, a resistance is added. This resistance is added in the form of a device known as a *resistor,* figure 26-10. The resistor can be as simple as a piece of resistance wire or it can be a small component having either a fixed or a variable resistance value. The *variable resistor* allows an adjustment to be made to control the amount of current under varying conditions.

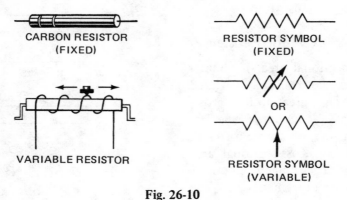

Fig. 26-10

## DETERMINING VALUES FOR CURRENT, VOLTAGE, AND RESISTANCE (OHM'S LAW)

The unit has thus far covered the meaning and the measurement of current, voltage, and resistance. To summarize, current refers to the movement of negatively charged free electrons in the direction of a more positive charge. The voltage is the pressure-energy (electromotive force) applied to cause electrons to flow through a resistance. Resistance is the opposition offered to the flow of free electrons. The amount of current (in amperes), the differential in pressure (voltage), and the resistance (in ohms) are interrelated.

### Application of Ohm's Law

As an illustration of this relationship, assume that the resistance remains the same as the voltage decreases. Then, the amount of current also decreases. Using a constant voltage, a material with a low electrical resistance means that a larger current is possible than if a higher resistance is used. The relationship of current, voltage, and resistance was formulated by George S. Ohm, a mathematician. In 1827, Ohm stated this relationship mathematically according to the formula:

$$\text{Current (I)} = \frac{\text{Voltage (E)}}{\text{Resistance (R)}}$$

or

$$\text{Amperes} = \frac{\text{Volts}}{\text{Ohms}}$$

This formula is known as *Ohm's Law* and is one of the most widely used laws of electricity. The importance of this law lies in its practical use in finding any one value when the other two values are known. Ohm's Law may be written in three forms which are shown in figure 26-11. When any one of these formulas is used, the electrical units must be expressed in amperes, volts, and ohms.

Three simple electrical diagrams are given in figure 26-12 to show how each value can be determined mathematically by applying Ohm's Law.

| | | |
|---|---|---|
| Current (I, Amperes) = $\dfrac{\text{Voltage (E, Volts)}}{\text{Resistance (R, Ohms)}}$ | | $I = \dfrac{E}{R}$ |
| $\dfrac{\text{Voltage (E)}}{\text{Current (I)}}$ = Resistance (R) | | $R = \dfrac{E}{I}$ |
| Current (I) x Resistance (R) = Voltage (E) | | $E = IR$ |

**Fig. 26-11**

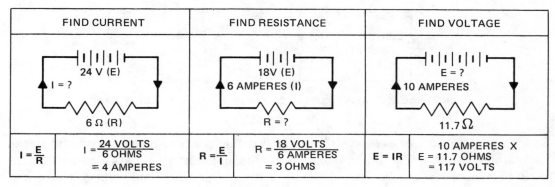

| FIND CURRENT | | FIND RESISTANCE | | FIND VOLTAGE | |
|---|---|---|---|---|---|
| 24 V (E), I = ?, 6 Ω (R) | | 18V (E), 6 AMPERES (I), R = ? | | E = ?, 10 AMPERES, 11.7 Ω | |
| $I = \dfrac{E}{R}$ | $I = \dfrac{24 \text{ VOLTS}}{6 \text{ OHMS}}$ = 4 AMPERES | $R = \dfrac{E}{I}$ | $R = \dfrac{18 \text{ VOLTS}}{6 \text{ AMPERES}}$ = 3 OHMS | $E = IR$ | 10 AMPERES X E = 11.7 OHMS = 117 VOLTS |

**Fig. 26-12**

## ELECTRICAL ENERGY AT WORK (DIRECT-CURRENT ELECTRICITY)

Current, voltage and resistance are especially important when electrical energy is put to work. The current used in this unit is a type which is known as *direct-current electricity*. This means that the electrons flow in one direction only. The major sources of direct current include dry cell batteries, wet cell batteries, and direct-current generators.

### Types of Circuits

*Open and Closed Electrical Circuits.* Electrons flow between two unequal charges when a path is provided through a conductor between the charges. When there is a complete electrical pathway from the negative terminal of a battery to a lamp and back to the positive terminal of the battery, an *electrical circuit* is formed. When this pathway is complete, a *closed circuit* results. If the circuit is broken, the term *open circuit* is used, figure 26-13. Every complete circuit has a current, a voltage, and a resistance.

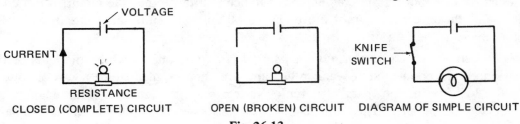

CLOSED (COMPLETE) CIRCUIT    OPEN (BROKEN) CIRCUIT    DIAGRAM OF SIMPLE CIRCUIT

**Fig. 26-13**

A convenient device used to open and close a circuit safely is a *switch*. Switches are available in a variety of sizes and shapes depending on the nature and quantity of the current and the circuits to be controlled.

*Short Circuits.* A circuit can be opened or closed at will when switches are added to the circuit. The opposite condition prevails when a circuit is shorted. A *short* may be caused by two bare wires touching each other. The circuit is also shorted when the terminals of a resistance are connected together or when other improper wiring exists. A short has the effect of cutting out or reducing the total resistance in the circuit. As a result, a larger value of current exists in the circuit. Excessive current produces heat. If a circuit is not protected properly with fuses or other electrical devices to open an overloaded circuit, the components in the circuit will be damaged by the excessive heat.

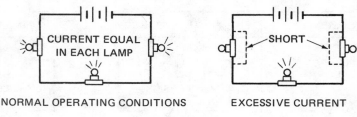

NORMAL OPERATING CONDITIONS          EXCESSIVE CURRENT

Fig. 26-14

*Series Circuits and Connections.* There are two fundamental types of circuits: series and parallel. As stated before, a complete electrical circuit must have a current, resistance, and voltage. A circuit containing only current, resistance, and voltage is known as a *simple circuit.*

When the resistances of a circuit are connected end to end in series fashion so that there is only one path in which the electrons can flow, the circuit is known as a *series circuit.* A simple series circuit containing four lamps connected to a battery is shown in figure 26-16. Note that the lamps are connected end to end and there is only one path for the current.

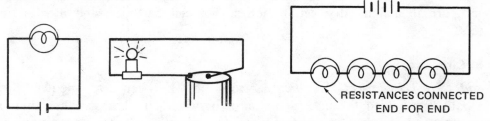

RESISTANCES CONNECTED
END FOR END

Fig. 26-15  Simple circuit

Fig. 26-16  Current same through all lamps:
series circuit

In this series circuit, the current is the same in all parts of the circuit. Thus, if one lamp burns out or the conductor resistance is interrupted, the circuit is not complete and the electrons cannot move. The total voltage of the series circuit shown in figure 26-16 is the sum of the separate voltages across each resistance.

*Parallel Circuits and Connections.* A *parallel circuit* is one in which the ends of the resistances are connected to each other and are then connected to a power source, figure 26-17. In the parallel circuit, shown in the figure, the three lamps are connected in parallel. The voltage of each resistance in a parallel circuit remains constant and is equal to the voltage of the power source. However, the current through each resistance (lamp) depends on the amount of resistance. In the home, for example, devices such as radios, televisions, lamps, refrigerators, and other lighting fixtures and appliances depend for normal operation on a 117-volt power line. Where heavier loads are connected to the line, the voltage supplied must be either 117 volts or 234 volts. Regardless of the voltage supplied, which is the same for each of the electrical devices in a parallel circuit, the current through the path of least resistance is greater.

In a parallel circuit, the resistances connected in parallel provide different paths for the current. The total resistance offered by a parallel circuit to electron flow is less than the smallest single resistance in the circuit.

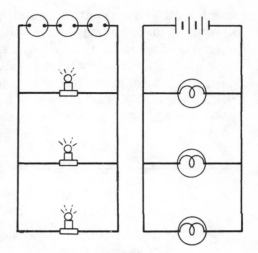

Fig. 26-17  Resistances connected in parallel

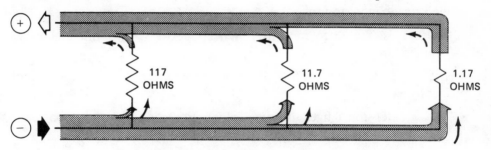

Fig. 26-18  Current divides according to resistance.

Figure 26-18 shows graphically how the current from a 117-volt dc power line divides for a parallel circuit. As the resistance decreases and the 117-volt source remains constant, the amount of current (amperes) in the various loads connected to the line increases. Using Ohm's Law, the current in the first electrical appliance having 117-ohms resistance is one ampere; in the second load of 11.7 ohms, the current is 10 amperes; the third load of 1.17 ohms has a current of 100 amperes. Each of these current values can be found by using the Ohm's Law formula, $I = E \div R$ (where E = 117 volts and R equals the resistance of each load in turn).

### Applications of Direct-current Electricity

All of the laws and terms used in this unit to describe and determine values for current, resistance, and voltage become important as they are applied to electrical circuits and electrical equipment such as instruments and devices. Later units of this text deal with electrochemical sources of electrical energy, and the generation, transmission, and distribution of electricity and its application to numerous types of equipment.

While alternating current is the most widely used type of electricity, there are applications for which direct current is more practical and efficient. One advantage of direct current over alternating current is that it may be used directly to charge storage batteries. Another application of direct current is its use in plating and refining operations. Direct-current motors have superior operating characteristics for variable speed control. Such control is essential in certain fabricating mills and for certain types of production. Holding devices for securing ferrous metal parts during machining operations depend on direct current as a power source for the electromagnets.

When alternating current is supplied, motor-generator sets and other electronic devices are used to convert the power supply from alternating current to direct current. The principles of motors and the application of alternating-current electricity are covered in later units of this text.

─────────────────── **SUMMARY** ───────────────────

- Current is the movement of electrons in one direction, from a negative polarity to a positive polarity.

- The ampere is the unit of measurement for current which is the quantity rate of electron flow.

- The electromotive force is the electrical pressure-energy which causes an electrical charge to move.

- Voltage is the difference in potential between two unequal electrical charges.

- Different materials offer different resistances to the movement of electrons.

- The greater the length through which electrons flow in a material, the greater is the resistance encountered by the electrons.

- The greater the cross-sectional area of a conductor, the less is the resistance offered to electron flow.

- The temperature of a conductor affects the resistance.

- The ohm is the unit of measure for the resistance. One ohm is the resistance offered to the flow of one ampere of current at a pressure of one volt.

- Ohm's Law is one of the most universally used electrical laws for determining current, voltage, or resistance.

$$I = \frac{E}{R}$$

- Electrical circuits are open when the circuit is broken; when the circuit is complete it is said to be closed.

- The two basic types of circuits are the series circuit and the parallel circuit.
  - In a series circuit the resistances are connected end to end so that there is one path through which the electrons may flow.
  - In a parallel circuit, the resistance ends are connected side by side to each other and then are connected to a power source.

- The resistances in a parallel circuit provide different paths for the current. The amount of current in each path depends on the resistance of each load.

- The basic principles of current, pressure, and resistance govern the operation of all electrical equipment and devices for generating, transmitting, distributing, and converting electrical energy to enable it to do useful work.

## ASSIGNMENT UNIT 26 ELECTRICITY IN MOTION (DIRECT CURRENT)

### EXPERIMENTS WITH CURRENT, PRESSURE, AND RESISTANCE

### Equipment and Materials Required

| Experiment A<br>Battery Voltages | Experiment B<br>Voltage and Current | Experiment C<br>Controlling Resistance |
|---|---|---|
| Six 1 1/2-volt dry cells | | Dry cell |
| Insulated lead wires for connections | | Knife switch |
| Dc voltmeter, 10-volt range | | Ammeter |
| Three small lamp sockets with bases | | Nichrome wire (10-in. and |
| Three 6-volt lamps | | 20-in. lengths) |

## EXPERIMENT A. MEASURING BATTERY VOLTAGES

### Procedure

1. Connect the (+) terminal of one dry cell to the (+) terminal of a voltmeter as shown in figure 26-19. Then connect the (–) terminals.

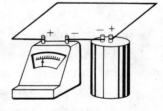

Fig. 26-19

2. Prepare a table for Recording Results. Read the voltage of the single cell on the voltmeter and record this value.

3. Test the voltage of each of the five remaining cells in the same manner. Record the voltage in each case.

4. Connect two cells in series by connecting the (–) terminal of one cell to the (+) terminal of the next cell.

5. Place a voltmeter in the circuit and connect the (+) terminal of the second cell to the (+) terminal of the voltmeter. Then connect the two negative terminals in the same manner. Read and record the voltmeter reading in the second column of the table.

6. Repeat step 5 with three cells connected in series, and then four, five, and six cells connected in series. Note and record the voltage in each case.

## Recording Results

| Voltages Of | | | |
|---|---|---|---|
| Single Cells | | Cells in Series | |
| A | | A | |
| B | | B | |
| C | | C | |
| D | | D | |
| E | | E | |
| F | | | |

## Interpreting Results

1. Compare the sum of the voltage of the first two cells with the voltmeter reading of the two cells in series.

2. Compare the sum of the voltages of the first three, four, five, and then six cells respectively, with the total voltage for each series combination of cells.

3. Briefly name and describe two electrical principles or facts that are observed and proved by these simple experiments.

## EXPERIMENT B. MEASURING VOLTAGE AND CURRENT

### Procedure

1. Connect six dry cells in series so that the (–) terminal of the first cell is connected to a lamp socket and to the (–) terminal of a voltmeter, figure 26-20.

2. Connect the (+) terminal of the voltmeter to the second terminal of the lamp socket. To this same terminal connect a lead wire so that the opposite (bare) end of the lead can be touched to the (+) terminal of any of the battery combinations.

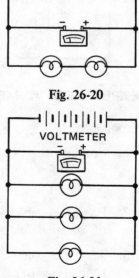

Fig. 26-20

Fig. 26-21

3. Touch the bare end of the wire to the (+) terminal of the first three cells in series.

4. Record the voltage shown on the voltmeter and note whether the lamp lights.

5. Now connect the bare end of the lead to four cells in series, then five, and finally six cells in series. In each case, record the voltage. Observe the brilliance of the lamp as the voltage is increased.

6. Add a second lamp socket and lamp in series with the first lamp and repeat steps 3-5.

7. Disconnect the last two lamps. Reconnect the three lamps so that they are wired in parallel with a voltmeter in the circuit.

8. Connect the first three cells in series, then four, then five, and finally six cells in series.

9. In each case, record the voltage. Note the brilliance of the lamp.

## Recording Results

| Number of Cells Connected in Series | Voltages | | | | | |
|---|---|---|---|---|---|---|
| | Lamps in Series | | | Lamps in Parallel | | |
| | One | Two | Three | One | Two | Three |
| 3 | | | | | | |
| 4 | | | | | | |
| 5 | | | | | | |
| 6 | | | | | | |

## Interpreting Results

1. As the voltage on the single lamp is increased, does the current increase or decrease? Briefly explain the reason for the action.

2. What effect on voltage do the two, and later, three lamps in series have? What is the effect on the current? Prove these statements in terms of the recorded and observed results.

3. What effect on voltage is there due to the two and three lamps in parallel? What effect is there on the current? The answer must be based on the observations made in this experiment.

## EXPERIMENT C. FACTORS CONTROLLING RESISTANCE

### Procedure

1. Connect the (–) terminal of a dry cell to the (–) terminal of an ammeter.

2. Connect the (+) terminal of the cell to one terminal of a knife switch.

3. Connect the 20-in. length of nichrome wire (or other resistance wire) from the (+) terminal of the ammeter to the second terminal of the switch.

4. Use the switch to close the circuit and note the value of the current. Prepare a table for Recording Results and record the current for the nichrome wire in the first current (amperes) column.

5. Allow the current to flow for a minute. As the wire becomes hot, note and record in the (amperes) column the reading that seems to remain constant.

6. Open the circuit, allow the nichrome wire to cool, and then remove the wire. Fold the wire in two and twist it so that it is half the length of the original wire and is twice the original cross-sectional area.

7. Insert this resistance in the circuit. Close the switch to complete the circuit and record the current reading.

8. Allow the wire to heat as in step 5 and record the current that remains constant.

9. Again open the switch, remove this wire, and replace it with a 10-in. length. Close the switch and note the current value when the wire is cold and when it is heated. Record the current values.

10. Repeat steps 3 through 9 for the 20-in. and 10-in. lengths of copper wire.

## Recording Results

| Conditions Affecting Resistance | | | | |
|---|---|---|---|---|
| Material | Length | Diameter | Amperes (Cold Temp.) | Amperes (Hot Temp.) |
| Nichrome | 20 in. | 1 D | | |
| | 10 in. | 2 D | | |
| | 10 in. | 1 D | | |
| Copper | 20 in. | | | |
| | 10 in. | | | |
| | 10 in. | | | |

## Interpreting Results

1. Based on these experiments, state the effect that different materials have on the resistance offered to current.

2. By doubling the cross-sectional area of the resistance wire (for both copper and nichrome), is the resistance the same, halved, or doubled? Prove the answer given.

3. What effect does temperature have on resistance?

## PRACTICAL PROBLEMS WITH DIRECT-CURRENT ELECTRICITY IN MOTION

### Current, Pressure, and Resistance and Their Measurement

For statements 1 to 8, determine which are true (T) and which are false (F).

1. A coulomb is a larger unit of measurement of electrical charge than an electron.

2. Electromotive force can be compared with a pressure applied to water to cause it to move.

3. A voltage is a comparison between two unequal electrical charges.

4. Different materials offer the same resistances to current.

5. Insulators offer a minimum resistance to current flow.

6. The material in a conductor influences its resistance.

7. One ohm is the resistance to the flow of one ampere of current under a pressure of one volt.

8. Resistances are measured with ammeters.

Select the correct word or phrase to complete each statement.

9. An ampere means (the amount) (the flow rate, in quantity per second) (the flow rate, as the speed of electron movement) of an electrical charge.

10. One ampere equals (1000 milliamperes) (1000 microamperes) (1 000 000 milliamperes).

11. The (ammeter) (milliammeter) (microammeter) measures larger amounts of current than either of the other two meters.

12. The electrical term (ampere) (volt) (ohm) is used in measuring the pressure exerted to create an electrical charge.

13. When less than the required voltage is applied to a lamp, the brilliance of the lamp is (increased) (decreased) (not affected).

14. One kilovolt equals (1000 volts) (1000 microvolts) (1000 millivolts).

15. The (longer) (shorter) the length of a conductor, the less is the resistance.

16. The (smaller) (greater) the cross-sectional area of a wire, the less is the resistance.

17. The resistance (is) (is not) affected by the temperature of the conductor.

18. Ohm's law states that Current (I) equals (E ÷ R) (E x R).

19. Match each term in Column I with the correct description in Column II.

| Column I | Column II |
|---|---|
| (a) Resistance | (1) The movement of electrons in one direction. |
| (b) Electromotive force | (2) The rate of flow of electrons, quantity per second. |
| (c) Resistor | (3) The electrical energy which causes a charge to move, thus causing current. |
| (d) Ampere | (4) A resistance placed in a circuit to control the amount of current. |
| | (5) The opposition with which a material opposes the movement of electrons. |

20. Determine the missing values for resistance, voltage, and current in the accompanying table. Use Ohm's Law.

| | Resistance (Ohms) | Voltage (Volts) | Current (Amperes) |
|---|---|---|---|
| A | 15 | 30 | |
| B | 30 | 30 | |
| C | 15 | | 3 |
| D | 30 | | 3 |
| E | | 117 | 117 |
| F | | 234 | 117 |

## Circuits and Electrical Diagrams

Select the correct word or phrase to complete statements 1 to 6.

1. An electrical circuit is (open) (closed) when there is a complete path from the current source through the parts in the circuit and back to the power source.

2. An electrical circuit is usually opened or closed with a (switch) (fuse).

3. Short circuits are caused by (a smaller load than the capacity) (reducing the total resistance) of an electrical circuit.

4. A simple circuit has (a current and a resistance) (a current and any number of resistances) (a current, resistance, and voltage).

5. A (series) (parallel) circuit has the resistances connected side by side to each other with the ends connected to the power source.

6. (Switches) (Fuses) break or open electrical circuits automatically.

7. Match each term in Column I with the correct description in Column II.

| Column I | Column II |
|---|---|
| a. Series circuit | 1. Resistances are connected end to end to each other and to the power source. |
| b. Parallel circuit | 2. The voltage of the power source. |
| c. Total voltage of a series circuit | 3. The current in any resistance. |
| d. Voltage of a parallel circuit | 4. The separate voltages of each resistance. |
| | 5. Resistances are connected side by side with ends connected to the power source. |

8. Sketch (using electrical symbols) a simple closed circuit having a dry cell, a small lamp, and a knife switch. Indicate the direction of current and letter the three parts of any circuit.

9. Sketch an open series circuit connecting two lamps and a motor to a direct-current power source.

10. Show a parallel circuit in which two lamps and a motor are connected to a 117-volt direct-current source.

# Unit 27 Electrochemical Sources of Electrical Energy

Energy in any one of its six basic forms can be used to produce electricity. The six energy forms include friction, magnetism, chemical action, pressure, heat, and light. Friction and magnetism were covered in earlier units of this text. In addition, some applications of batteries were given. However, detailed information was not given previously about the battery as an electrochemical source of electricity.

The battery is comparatively simple in principle and operation. In some cases, a battery is the most practical way of supplying electricity. Batteries serve as a portable source of electricity. Numerous applications of batteries are made in the automotive, aeronautical, marine, and other industrial areas. Batteries are used to provide a standby reserve source of power in the event of an electrical power failure. All batteries operate on the principle of converting chemical action and energy into electrical energy.

## THE PRIMARY CELL

### Principle of Operation of a Cell

Chemical energy is changed into electrical energy by a device known as a *cell*. When two or more of these cells are connected in series or parallel they form a *battery*. The term *primary cell* is used to indicate a type of cell which, once it is completely discharged, can only be recharged using new materials.

In its simplest form, there are three parts to a primary cell: the two cell plates and an electrolyte. The cell plates are made of two different metals which are separated from each other and immersed in a liquid solution called an *electrolyte*. The electrolyte helps to transport electrons from one plate to the other plate. This movement of electrons causes one plate to build up a negative charge as the other plate becomes more positively charged. In other words, one plate becomes the negative ( – ) terminal and the other plate becomes the positive ( + ) terminal, figure 27-1. In this process, the electrolyte reacts chemically with the plates to produce electrical energy. When as many electrons accumulate on the negative terminal as possible, no further movement of electrons takes place until the two plates are connected. The electrons then move from the negative plate to the positive plate and the cycle continues.

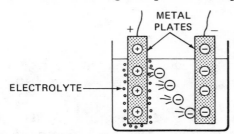

Fig. 27-1  A simple cell

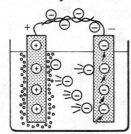

Fig. 27-2

A close examination of the cell during this process shows that gas bubbles appear on the positive plate as the negative plate is dissolved. When the negative plate ( – ) is worn away, the cell is dead and the plate must be replaced.

### The Dry Cell

One of the most common types of cells is the dry cell. There are many uses for dry cells including emergency lighting, portable lamps, portable radios, instruments, and numerous household and industrial applications. Dry cells range in size from a pencil type to cells which are very heavy and have large capacities. Regardless of the size or shape of the cell, the construction, operation, and care in use are the same for all dry cells.

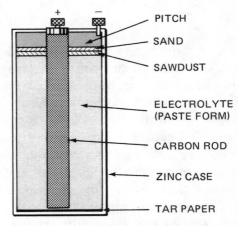

Fig. 27-3  Cross-section of a dry cell

The ordinary *dry cell* includes a zinc case (the negative plate), a carbon rod suspended in the center (the positive plate), and an electrolyte, figure 27-3. The electrolyte is usually ammonium chloride (a salt in paste form). The carbon rod is separated from the zinc case by tar paper.  At the top of the cell, the pitch, sand, and sawdust packing prevent leakage of the electrolyte. The movement of the electrons between the positive plate and negative plate as a result of the chemical action of the electrolyte produces electricity. This cell operates until either the zinc case or the electrolyte is consumed. The chemical reaction then stops and the cell is dead. The voltage of a single cell is not very large. To increase the voltage, the cells are connected in series to form a *dry cell battery*.

Dry cells last longer when they are used for intermittent operation. When a dry cell is in continuous use, the bubbles (hydrogen) which are liberated around the positive carbon rod tend to form a shield around the carbon terminal so that, in effect, it becomes a hydrogen terminal. The reaction between the hydrogen and the zinc plate reduces the voltage of the cell. Normal activity within the cell resumes when the circuit is broken long enough for the hydrogen bubbles to disperse from the carbon plate. The formation of the hydrogen bubbles around the carbon plate is called *polarization*.

## THE SECONDARY CELL

### Storage Batteries

The storage battery is used where large amounts of current are required for comparatively short periods of time. It is possible to recharge a storage battery for subsequent use.

The storage battery, figure 27-4, consists of a series of secondary cells. *Secondary cells* differ from primary cells in that once they are discharged they can be recharged without necessarily replacing any of the parts.  Each secondary cell in the battery consists of two groups of coated lead plates known as *electrodes*. One group of plates forms the *positive* electrode; the other group of plates forms the *negative* electrode. The plates are separated from each other by wood, glass, or rubber separators. The electrolyte is usually a weak solution of sulphuric acid. The electrolyte and plates are housed in a hard rubber container.

One set of plates is connected to a single ( − ) terminal; the other set is connected to a ( + ) terminal. Vent caps are provided on the battery to permit the escape of gases during the charging operation and to make it possible to add distilled water.

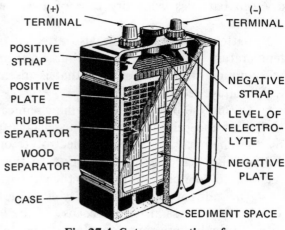

Fig. 27-4  Cut-away section of a
lead-acid secondary cell

***Battery Ratings.*** The conventional storage battery consists of six secondary cells connected in series. In general, each cell of a storage battery is rated at 2.0 volts. Thus, the six-cell battery has a 12.0-volt rating. The voltage may be increased by adding more cells to a storage battery.

***Battery Capacities.*** The current output of a battery is governed by the condition of the electrolyte, and the size, condition, and number of plates. The total charge that a battery can deliver is rated at a capacity given in *ampere-hours*. In other words, a 100 ampere-hour car battery is capable of delivering 10 amperes continuously over a ten-hour period. The ampere-hour rating depends on the size and number of plates. (Note that the 2.0-volt rating of a cell does not depend on these factors.)

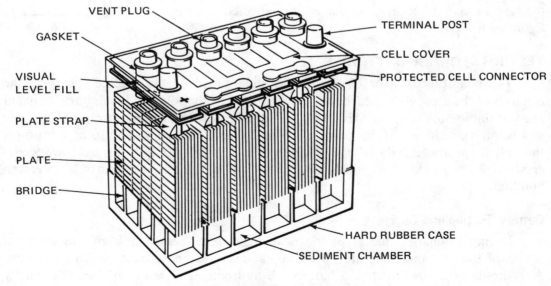

Fig. 27-5  Parts of a storage battery

The storage battery is constructed so that it can withstand short heavy loads. The average 12-volt car battery with a capacity of 100 ampere-hours does the same amount of work in starting an automobile engine that can be supplied by 50 four-cell dry cell batteries using dry cells of 1 1/2 volts and 2 ampere-hour capacities for a short time.

Battery performance is also affected by the rate of discharge and the battery temperature. When current is withdrawn from a battery, the voltage drops. The greater the amount of electric current that is discharged (higher discharge rate), the greater is the voltage reduction.

During the discharge cycle, the chemical reaction of the electrolyte decreases the ability of the cells to maintain an electric pressure or voltage. Voltage is produced by the chemical action of the electrolyte on the lead compounds in the plates. The voltage also decreases in proportion to the amount of time needed for the chemical reaction to take place to produce the electrical energy.

The rate of chemical activity is reduced as the battery temperature decreases. The battery is less efficient because of the high discharge rate at the colder tempera-ture. As the battery runs down, an increasingly higher charging rate is required to force current through it. At a rapid discharge rate and a lower level of chemical activity, there is insufficient time to charge the battery to produce enough voltage to start the automobile. Under these conditions, a dead battery may result.

*Formula for Ampere-hour Capacity.* A simple formula can be used to determine how long a fully charged battery will last. The ampere-hour ($A_h$) capacity is equal to the load current (I) times the time in hours (T):

$$A_h = I \times T$$

In the example in the previous paragraphs, $A_h$ = 10 amperes x 10 hours = 100 ampere-hour capacity. By sub-stituting two known values in the formula, the third one can be computed easily.

A 12-volt storage battery is represented on a circuit diagram by a symbol showing six cells connected in series, figure 27-6.

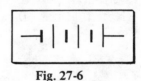

Fig. 27-6

## TESTING STORAGE BATTERIES

The condition of a battery is usually determined by the specific gravity of the electrolyte. In a fully charged battery, nearly 40% of the solution is sulphuric acid and the remaining 60% is water. Although the specific gravities of equal volumes of water and sulphuric acid are in the ratio of 1.8 to 1 (for a 40% to 60% ratio of solutions), the actual specific gravity of the electrolyte is 1.300 for a full charge. The specific gravity of a normally discharged cell is 1.100. This value is also expressed as eleven hundred.

### Battery Testing and Charging

During discharge, the sulphuric acid electrolyte (resulting from the chemical action of the + and − plates) breaks up and is replaced by water which also forms. A hydrometer is used to test a battery. A hydrometer measures the specific gravity of a sample of liquid drawn from the cell into the barrel of the hydrometer. The specific gravity is read on a scale at the level of the liquid. Batteries can also be checked with an ammeter load circuit. A specific gravity of 1.100 indicates that the battery must be recharged. A specific gravity of 1.300 indicates a fully charged battery.

The storage battery with its secondary cells can be recharged. During discharge, the chemical activity in a battery decreases. However, when electrical energy is pumped back through the storage battery, the chemical energy is renewed and the voltage increases to the normal value.

For example, in an automobile there is a great drain on the battery when the engine is started. Once the engine is running, the alternator connected to the engine charges the battery until the normal voltage is reached. The battery then does not receive a charge and the electrical energy for the lights, radio, and other accessories is furnished by the alternator.

The storage battery can be discharged by disuse or from normal discharge. In either case, completely discharged batteries should be recharged immediately because the chemicals forming on both electrodes harden and cause a reduction in both the battery capacity and life.

*Charging Batteries.* If in good operating condition, a discharged battery having sufficient electrolyte can be charged with direct current, or it can be charged by changing alternating current into direct current before it is applied to the battery. The direction in which the current enters the battery for charging must be opposite to the direction of discharge. In automobiles, aircraft, marine engines, and other applications, the alternating current supplied by the alternator is rectified by diodes and reaches the battery as direct current for charging purposes. As battery current is used to start an engine, the alternator immediately supplies the current necessary to build up the battery to its normal operating capacity.

## BATTERY CARE AND MAINTENANCE

There are two basic systems of recharging batteries: slow charge and fast charge. In the slow charge system, a relatively small amount of current is supplied over a long period of time. The fast charge system refers to the large amounts of current delivered for a short period of time. Of the two methods, the slow charge is preferred because it prevents the electrolyte from becoming too hot. If the electrolyte temperature should exceed 100°F, gases develop in the cells and destroy the operation of the battery.

Eight simple rules can be followed to insure the maximum output and life of the battery.

- A completely discharged battery should be charged immediately.

- The level of the electrolyte in each cell should be checked regularly to insure that the liquid covers the separators by at least 3/8 in.

- Distilled water should be added before the battery is recharged because the charging process causes hydrogen and oxygen gases to be produced from the chemical reaction with the water in the electrolyte.

- In cold weather the storage battery should be tested at more frequent intervals to prevent a discharged battery from freezing.

- Fast charging rates should be avoided. However, it is good practice to check the temperature of the electrolyte carefully to insure that it does not exceed 110°F. The presence of gases coming freely from any cell indicates that the charging rate is too high.

- Storage batteries should be checked regularly and recharged when not in use or when partly discharged.

- Any electrolyte spilled during testing should be replaced. Care must be taken when adding the acid to the water slowly.

- The greatest strain on a battery results from a heavy discharge of current over long periods of time. Heavy discharges of current for short periods of time result in longer battery life.

––––––––––––––––––––––––––––– SUMMARY –––––––––––––––––––––––––––––

- A primary cell is a device in which chemical energy is converted to electrical energy. When a primary cell is discharged, new materials must be added before the cell can be recharged.

- There are three parts to a cell: plates of two different metals and an electrolyte.

- Electrical charges are built up on the plates of a cell by the movement of electrons due to chemical action.

- The dry cell battery is capable of changing chemical energy into electrical energy in small amounts for short periods of time.

- Dry cells provide better service when used for intermittent rather than continuous operation.

- The storage battery consists of a number of secondary cells connected in series; the battery produces large amounts of current for short periods of time.

- The essential parts of a storage battery include two groups of lead plates separated from each other and immersed in an electrolyte.

- The ampere-hour capacity of a storage battery depends on the number and size of the plates. The voltage is influenced only by the number of cells.

- The ampere-hour capacity may be computed by the formula: $A_h = I \times T$.

- The condition of a storage battery can be determined by the specific gravity of the electrolyte.

- Discharged batteries can be recharged using direct current as a source of electrical energy. This energy is converted to chemical energy by reversing the direction of current during the charging process.

- The storage cell is used where heavy demands for electricity are made for short periods of time and at places where other sources of energy are not available.

## ASSIGNMENT UNIT 27 ELECTROCHEMICAL SOURCES OF ELECTRICAL ENERGY

### PRACTICAL EXPERIMENTS WITH PRIMARY AND SECONDARY CELLS

### Equipment and Materials Required

| Experiment A<br>Primary Cells | Experiment B<br>Secondary Cells |
| --- | --- |
| Two metal plates (zinc and copper, or other combinations) | Three two-volt secondary cells |
| Glass container | Voltmeter |
| Dilute solution of sulphuric acid (2:5) | Ammeter |
| Insulated copper wires | Battery hydrometer |
| Lamp socket and small lamp | Insulated copper wires |
| Voltmeter | |

### EXPERIMENT A. PRIMARY CELLS

### Procedure

1. Immerse two small plates of different metals in a dilute solution of sulphuric acid (two parts sulphuric acid to five parts distilled water).

   CAUTION: *Avoid touching the acid or getting any on clothing to prevent burns and damage to cloth.*

2. Connect one metal plate to one terminal of a voltmeter and connect the other plate to the second terminal. Wait for a short period of time; if there is no reading, reverse the wire connections. Note what takes place in the solution around the plates.

3. Disconnect the voltmeter and insert a small lamp in the circuit.

4. Break the circuit and permit the chemical action to continue for a short time.

5. Complete the circuit again and note any change in the brilliance of the lamp.

6. Keep the circuit closed for a few minutes. Determine the effect of such action on the lamp and plates.

### Interpreting Results

1. Explain in terms of the results observed in the experiment how chemical energy is converted into electrical energy.

2. Before the plates build up any charge, is the charging rate greater or less than the discharge rate when the lamp circuit is completed? Explain.

3. After the plates receive a maximum charge and the lamp circuit is complete, is the charging rate equal to, less than, or more than, the rate of discharge (approximately)? Base the answer to this question on the results of the experiment.

## EXPERIMENT B. SECONDARY CELLS

### Procedure

### Determining Voltage

1. Connect a single two-volt secondary cell to a voltmeter. Note the voltage. Prepare a table for Recording Results and record the voltage.
2. Repeat step 1 with the second and third two-volt cells. Record the voltage in each instance.
3. Connect the three cells in series. Add a voltmeter and note the total voltage.
4. Disconnect the voltmeter and add one automobile headlight to the storage battery circuit. Note the brilliance of the headlight.
5. Add a second headlight and determine if both lights have the same brilliance as the single light.

### Determining Current Output

1. Start again with the first two-volt capacity secondary cell. Put a small lamp in a circuit with the cell and add an ammeter. Record the current in the table.
2. Repeat step 1 with the second and third cells. Then repeat step 1 with the three cells in series. Record the ammeter reading in each case.

### Testing the Electrolyte for Condition and Charge

1. Use a battery hydrometer and enough electrolyte from the first cell to float the hydrometer bulb.
2. Read the scale bulb on the float of the hydrometer to determine the specific gravity of the electrolyte and the condition of the cell charge.
3. Record this reading in the table. Repeat steps 1 and 2 with the second and third two-volt cells. Record the specific gravity in both cases.

### Recording Results

|   | Condition | Voltage | Current | Reading of Electrolyte |
|---|-----------|---------|---------|------------------------|
| A | First two-volt cell | | | |
| B | Second two-volt cell | | | |
| C | Third two-volt cell | | | |
| D | Three two-volt cells in series | | | |

### Interpreting Results

1. From the results recorded, determine if there is a difference in voltage (a) between each cell and (b) when the three cells are connected in series to form a storage battery.
2. Compare the current in amperes for (a) each of the three separate cells and (b) the storage battery (the three cells in series).

3. Compare the specific gravity readings of the three cells. If there was any variation in current reading of any cell, explain the relationship between current and the condition of the electrolyte.

## PRACTICAL PROBLEMS ON ELECTROCHEMICAL SOURCES OF ELECTRICAL ENERGY

### Primary Cells

Add a word or phrase to complete statements 1 to 8.

1. The smallest number of parts to a primary cell is _____ .

2. A primary cell converts _____ energy to _____ energy.

3. The _____ between two dissimilar metal plates permits electrons to flow.

4. The series of plates that build up a negative charge form the _____ terminal; the plates with the positive charge form the _____ terminal.

5. In the ordinary dry cell, the chemical action takes place between a (an) _____ case, a (an) _____ core, and a paste form of the _____ .

6. When hydrogen bubbles form around the carbon plate (or other plate in a cell), the charging rate is _____ .

7. The capacity of a dry cell is increased when it is used for _____ service.

8. A primary cell is that type of cell which, once the cell is completely discharged, can be recharged only by using _____ .

9. Referring to a cutaway model of a dry cell, make a simple sketch of the cell and (a) label six of the most important parts and (b) describe one important function of each of the six parts.

10. Make a closed circuit diagram of a flashlight, including (a) labeling the parts and (b) explaining how chemical energy causes the bulb to light.

### Secondary Cells

For statements 1 through 8, determine which are true (T) and which are false (F).

1. A cell is a group of positive and negative plates suspended in an electrolyte with separators between plates.

2. Regardless of the number of plates, each cell of a standard battery supplies two volts.

3. The active materials in a battery are always active regardless of whether an electrolyte is present.

4. Secondary cells are connected in parallel to overcome any weaker cell that may cause a stronger cell to discharge.

5. A discharged battery may produce as much electricity as a fully charged battery.

6. The electrolyte in a fully charged battery is heavier than in a discharged battery.

7. Alternating current from a controlled source may be used directly for charging batteries.

8. Excessive gas forming at any cell during charging is an indication that the charging rate is just right.

Select the correct word or phrase to complete statements 9 to 18.

9. One advantage that a secondary cell has over a primary cell is that it (is) (is not) possible to recharge the secondary cell for a subsequent use.

10. The electrolyte in a storage battery is usually a weak solution of (tannic) (nitric) (sulphuric) acid.

11. The two sets of coated lead plates are separated from each other by some (current-conducting) (nonconducting) material.

12. Any increase in the number of plates in a cell (increases) (decreases) (does not affect) the voltage of a cell.

13. Increasing the size and/or number of plates in a cell (increases) (decreases) (does not affect) the ability to produce current.

14. The construction of a storage battery means that it can be used for (short heavy loads) (loads making a continuous heavy demand).

15. The ampere-hour rating of a battery depends on the (voltage of one cell) (the size and number of plates) (number of plates) (the voltage of all cells).

16. Secondary cells are connected in (series) (parallel) to prevent a weaker cell from causing the other cells to discharge.

17. A specific gravity reading of 1.10 indicates that the battery is (fully charged) (discharged).

18. A storage battery that is in operating condition but is discharged, can be recharged by converting (electrical) (chemical) energy to (electrical) (chemical) energy.

19. Match the condition in Column II with the correct item in Column I.

| Column I | Column II |
|---|---|
| a. Battery recharging | 1. A series of dry cells that may not be recharged unless the materials are replaced. |
| b. Fast charge | |
| c. Storage battery | 2. Direct current is fed into the battery in a direction that is opposite to the direction of discharge. |
| d. Ampere-hour capacity | |
| e. Fully charged battery | 3. The specific gravity of the electrolyte at 90°F is between 1.300 and 1.280. |
| | 4. A series of secondary cells connected in series. |
| | 5. Delivering large amounts of direct current to charge a battery. |
| | 6. A battery rating found by multiplying the current output by time (in hours). |

20. Name eleven main parts of the storage battery illustrated. Describe the function of each part.

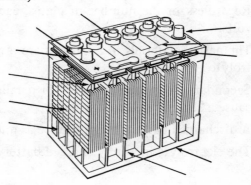

# Unit 28 Generation, Transmission, and Distribution of Electricity

In a previous unit, direct current was said to be electron flow in one direction. In the unit on electrochemical sources of electrical energy, it was shown that the current from storage batteries and dry cell batteries is direct current and moves in one direction.

*Alternating current* is the rapid back and forth motion of electrons and is another widely used form of electricity. Alternating current first must be generated, then is transmitted from the source to the place of need, and then is distributed at the voltage required for the specific purposes.

## CONVERTING DIFFERENT FORMS OF ENERGY TO MECHANICAL ENERGY

A generating station is used to change the stored energy of coal, oil, falling water, or the nuclei of atoms. Of these four sources of energy, almost 65% of the electricity generated comes from fuel-burning stations. One of the earliest generating stations was constructed in New York City by Thomas A. Edison in 1882. Steam engines were used at this station to drive generators. In turn, these generators produced electrical energy. Since that date, more efficient steam turbines have replaced the steam engine.

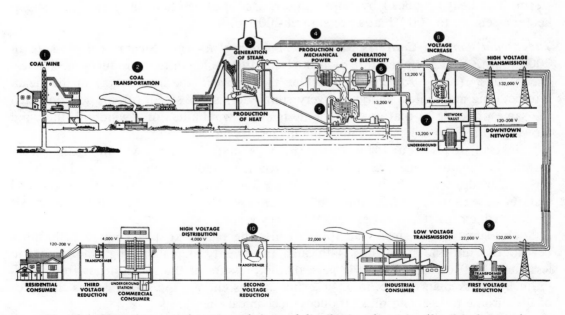

Fig. 28-1  Stages in generating, transmitting and distributing electricity (fossil fuel system)

Another 20% of the installed generating capacity is produced by hydroelectric plants which depend on free-falling water to drive hydroturbines. Five percent of the generated electricity is produced by diesel-driven generators. The remaining 10% of the generated electricity is obtained from nuclear power plants. The first three sources of energy are discussed briefly in this unit; nuclear power is discussed in unit 38. This unit concentrates on the principles of generator operation because once chemical or physical energy is changed to mechanical energy, the generator converts this energy into usable electrical energy.

## Steam Generating Plants

In modern plants using coal as the primary source of energy, the coal is pulverized into a fine flour-like powder. The coal powder is then forced under pressure through burner nozzles into the combustion chamber of a furnace. Air is automatically added to the powdered coal in the chamber to cause it to burn and produce heat.

An interesting application of the principles of charged particles is seen in the objectionable gases which sometimes are given off by the smokestacks of the generating station. The smoke particles are given a positive charge and are then passed through highly charged negative plates. In this manner, the negatively charged particles are held until they can be removed.

*Converting Heat to Steam.* Heat as a form of energy is of little value until it is used to convert water into steam. Water passes through an assembly of tubes in a boiler. Hot gases pass around the tubes and the heat of the gases changes the water within the tubes into steam. This steam, however, is not hot enough to do useful work. Thus, it is passed from the boiler to another series of tubes in a superheater. It is known that at higher temperatures, greater pressures are possible. As applied to this example, the greater the pressure, the greater is the amount of work that can be performed by the steam. The heat produced by burning pulverized coal can heat the steam in a super-heater assembly to 1000°F at a pressure of 2000 lb./in.$^2$.

*The Steam Turbine Converts Steam to Mechanical Energy.* Steam at a pressure of 2000 lb./in.$^2$ and a temperature of 1000°F has little value until it produces mechanical energy. The *steam turbine* is the machine that converts the heat energy in the steam into rotating mechanical energy. In very simple terms, the steam turbine acts on much the same principle as that used in the design of a windmill. For a windmill, the wind blowing on the slanted blades causes the blades to revolve.

Instead of just four blades mounted on a shaft, as in the windmill, the steam turbine has many blades curved around a rotor. In addition, the steam turbine has a series of blades (called buckets) spaced along the rotor. Each series of blades is called a stage. In a conventional steam turbine, there are as many as twenty different stages through which the steam passes until its energy is spent.

The superheated steam (at 1000°F) hits the blades at speeds up to 1200 miles per hour. While the steam passes through the twenty stages (hitting 5000 blades in less than 1/30 of a second as it passes), the pressure of the steam drops from 2000 lb./in.$^2$ to less than one lb./in.$^2$. In giving up this energy, the steam causes the rotor of the turbine to turn and thus produces mechanical energy.

Within a turbine, the blades at each stage vary in shape and size. The steam enters the high-pressure side where the blades are the smallest. At the end of the stages, the steam is exhausted from the low-pressure side where the blades are the largest. In this manner, the driving force of the superheated steam is carried equally along the entire rotor. The exhaust steam is condensed into warm water which is then carried back to the boiler tubes so that the cycle can be continued.

## CONVERTING THE ENERGY OF FREE-FALLING WATER TO MECHANICAL ENERGY

*Hydroelectric plants* must be located in areas where two conditions are satisfied: (1) a constant and adequate source of water and (2) free-falling water. The free-falling water replaces the coal or oil as the energy source. Mechanical kinetic energy is created when the stored up potential energy of the water rushing down *penstocks* hits the buckets of a *hydroturbine*. The greater the height of water above the turbine blades, the greater is the pressure with which the water hits the blades.

This high-velocity water, falling through an elevation difference of about 1000 feet, is directed through a nozzle against the blades of the hydroturbine waterwheel. The water pressure forces the blades to rotate the shaft to which they are attached. In this manner, the potential energy of the water is converted into mechanical energy.

*Comparison of Steam and Hydroelectric Plants.* Almost two-thirds of the electrical power generated is produced by steam plants because they are more dependable and can be located near the load centers. Hydroelectric plants depend on the weather and the flow of a large amount of water in streams. Both of these factors are unpredictable. Most hydroelectric plants are supplemented by steam units which are used to produce the required energy whenever the supply of water is insufficient.

### Diesel Operated Plants

Diesel engines driving generators and alternators produce about 5% of the total electrical energy generated in the United States. A less expensive (than coal or oil) type of fuel oil can be used in these diesel engines. The fuel-driven diesel engine turns a crankshaft to produce mechanical energy. This crankshaft is connected to and rotates a generator shaft.

## GENERATING ELECTRICITY (CONVERTING MECHANICAL TO ELECTRICAL ENERGY)

### Principles of the Elementary Generator

Any electrical conductor which moves in a magnetic field produces an electric current. An ammeter connected to such a conductor shows the presence of current in one direction as the conductor cuts down through magnetic lines of force, figure 28-2. When the conductor is moved back up through the lines of force, the ammeter needle registers a current in the opposite direction.

If there is no movement of the conductor, or if it moves in a direction parallel to the magnetic

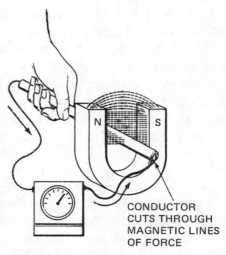

CONDUCTOR CUTS THROUGH MAGNETIC LINES OF FORCE

Fig. 28-2 Producing an induced current

lines of force, the ammeter reading is zero. The movement of the ammeter needle indicates that an electromotive force is produced. This electromotive force is known as an *induced voltage* because it is the movement of a conductor through the lines of force that cause or induce the voltage.

The amount or magnitude of this electromotive force is increased when either the conductor speed or the density of the magnetic field is increased. In other words, the induced electromotive force in a conductor is proportional to the number of the lines of force which it cuts in a given period of time.

The following section presents a simple explanation of the principles which govern the operation of a generator. The *generator* is a machine that converts to electrical energy the mechanical energy of the turbine to which it is connected. To make this conversion, a generator has a series of stationary electromagnets mounted in a rotary frame which surrounds a rotating core called an armature. The electromagnets create the strongest magnetic field possible. The conductor cutting through this field is not a single conductor, but rather is a number of conductors mounted on an iron core armature which is rotated through the magnetic field, figure 28-3.

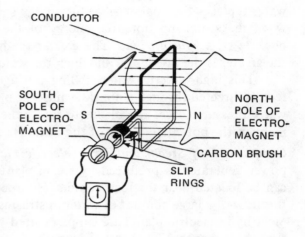

Fig. 28-3

## The Alternating-current Generator

A great number of conductors cutting through the magnetic field at a high speed produce a high voltage. Figure 28-4 illustrates the principles of generator operation. Only one set of magnetic field poles is shown in this example. In addition, a single conductor wire is shown rather than conductors mounted in an armature. One end of the conductor wire is connected to one *slip ring,* as shown in figure 28-3; the other end of the conductor is connected to a second slip ring.

An external circuit is connected to these slip rings through carbon *brushes.* The two terminals attached to the carbon brushes can then be tapped so that the induced voltage can be transmitted to any location where it is needed. (However, the voltage transmitted may not be used directly because the amperage and voltage may be greater than is desirable for practical use.)

As shown in figure 28-4A, the conductor in the vertical position cuts through a limited number of magnetic lines of force. As the conductor rotates clockwise 90°, it cuts the greatest number of lines of force in the same time, figure 28-4B. The induced voltage, therefore, begins at zero and reaches a maximum at 90°. From 90° to 180°, figure 28-4C, the number of lines of force cut decreases from the maximum number at 90° and the induced voltage decreases again to zero at 180°.

The induced electromotive force and the current can be measured at each step of this process. As the conductor continues to turn clockwise beyond 180°, the current

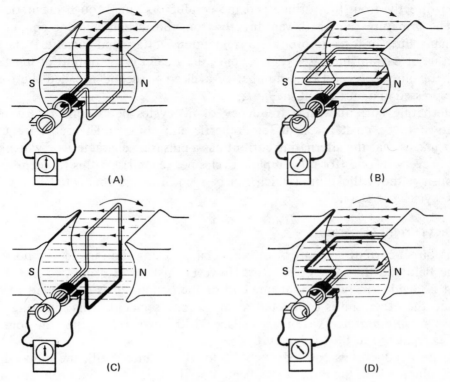

Fig. 28-4

in the external circuit is reversed. The zero reading at 180°, figure 28-4C, again reaches a maximum value at 270°, figure 28-4D, and returns to zero at 360°.

Thus, for every complete rotation of the conductor through 360° in a magnetic field, the induced electromotive force and current varies from zero at 0° to a maximum at 90° and back to zero at 180°. It then reverses from zero at 180° to a maximum at 270° and again back to zero at 360°.

A *waveform* of this voltage for a complete revolution of the conductor is shown in figure 28-5. This waveform differs from the one for direct current in that a dc voltage maintains the same voltage polarity output. In the case of ac electricity, the voltage alternates from plus to minus. Thus, this current is called *alternating current* because it is the alternately positive and negative alternating voltage that causes the electrons to flow in this manner.

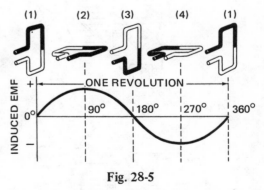

Fig. 28-5

In the case of a generator, the single conductor and the single set of electromagnets is replaced by many conductors mounted on an armature and many electromagnets mounted on the frame surrounding the armature. In most large generators, the magnetic field revolves while the conductors are mounted on the frame. There are other generators, however, in which the magnetic field is stationary and the conductors revolve. Regardless of the construction of the generator, electrical energy is produced when the conductors cut through magnetic lines of force. The armature shaft is driven directly from the turbine and the conductors cut through magnetic lines of force. The armature shaft is driven directly from the turbine and the conductors cut across many lines of force to produce a tremendous voltage and current. If the number of reversals of current increases each second, the greater is the output of the generator for each revolution (*cycle*). The number of cycles per second is called the *frequency* and is expressed in a unit called the *hertz*.

Alternating current with a frequency of 60 cycles per second or 60 hertz has been shown to be most practical for domestic and commercial use. A frequency of 60 hertz means that the alternating current has a pulsating characteristic which reverses 120 times every second (for 60 complete cycles per second). At this frequency, electric lamps burn without flickering (which can be seen in lamps connected to a voltage at 25 or 40 cycles per second).

### The Direct-current Generator

The direct-current generator can be explained using the same set of poles for the magnetic field and the same conductor. However, instead of using two slip rings, one slip ring is divided in two so that one end of the conductor is connected to one segment, and the other end is connected to the second segment. This split ring design is known as a *commutator*. Two brushes spaced 180° apart ride against the commutator to carry current out to the load circuit.

As the conductor is rotated from 0° to 90°, figure 28-6B, the induced current in the conductor is in the direction indicated on the meter. The induced current builds up from zero to a maximum at 90° and then drops back to zero at 180°, figure 28-6C. When the conducting loop is rotated to 270°, a maximum voltage and current are

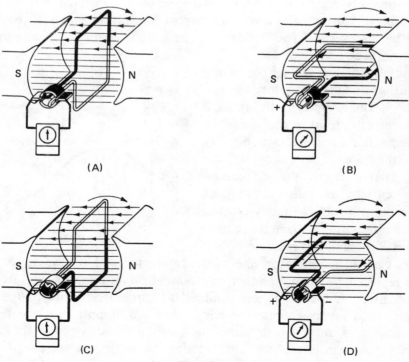

(A)

(B)

(C)

(D)

**Fig. 28-6**

again induced, figure 28-6D. Although the current in the rotating loop of wire alternates, (note the current directions relative to the white and black sides of the loop), the sliding of the commutator halves from one brush to the other means that the current in the external circuit is in the same direction at all times. Thus, the induced voltage in the rotating loop reverses frequently, but the external current values at 90° and 270° are in the same direction. The induced emf as shown in figure 28-7 illustrates the pulsating character of the direct current.

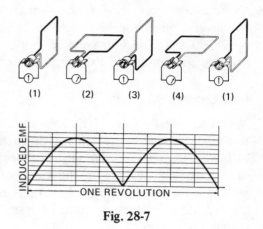

(1)  (2)  (3)  (4)  (1)

**Fig. 28-7**

## TRANSMITTING AND DISTRIBUTING ELECTRICITY

### Step-up Transformers for Transmission

The maximum potential difference at which an electric generator delivers energy is usually 15 000 volts. This voltage can be *stepped up* to higher values so that it is more economical to transmit the energy through the power lines to the points of operation. A standard generated emf of 13 200 volts can be stepped up as high as 250 000 volts (depending on the distance over which it must be transmitted and the amount of energy available).

The increasing or stepping up of the voltage is accomplished by *step-up transformers*. Such transformers are installed near generating stations and look like a battery of steel tanks. The transformer has three main parts: a *primary winding* to which the generated electrical energy is applied, a *secondary winding* which is tapped to provide the output voltage, and the laminated iron core around which both the primary and secondary windings are wound, figure 28-8.

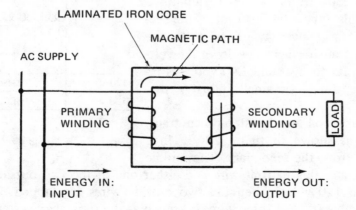

**Fig. 28-8**

The transformer transfers electrical energy from one coil to the other by magnetic induction. *Magnetic induction* is the result when a voltage applied to the primary winding sets up a voltage in the secondary winding.

The amount of voltage induced in the secondary winding depends on the ratio of the primary and secondary windings. For example, if the electrical energy received at the primary winding of a transformer is 13 200 volts, and there are five times as many windings around the laminated iron core at the secondary winding, then the output is 13 200 x 5 = 66 000 volts. The output is determined by the ratio of the turns in the primary coil compared to the number of turns in the secondary coil.

### Line Losses

The question may be asked as to why it is necessary to transform a low voltage to a high voltage to transmit it through power lines. The answer is simple. Regardless of how efficient the conductors of electricity are, they still offer resistance to the flow of electrons. The greater the value of current, the more heat is generated and the greater is the loss in electrical energy. Since the energy is equal to voltage x current, the greater the value of the voltage, the smaller is the current required to produce the same amount of energy.

### Step-down Transformers for Distribution

As the length of the transmission lines increases, more emf is lost in forcing the current to the point of use. The amount of energy is controlled by stepping up the transmission voltage. However, such a high transmission voltage is dangerous. To distribute the energy safely once it reaches the point of use, the high voltage is stepped down to 2300 or 4000 volts. Transformers are installed in *substations* to step down the voltage to practical value.

Another common type of transformer is usually mounted on a pole or in a transformer vault below the surface of the ground. This transformer further reduces the voltage to that supplied to 120/240-volt, three-wire services used for lighter industrial and domestic applications.

### Three-wire System of Distribution

Current at 2300 volts enters a step-down transformer. In this type of transformer, the secondary coil contains fewer windings than the primary coil. The transformer steps down the voltage from 2300 volts to 120 and 240 volts, as required for home, farm, or small industry use, figure 28-9.

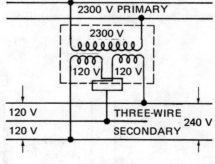

Fig. 28-9

The secondary coil of the step-down transformer is wired as shown in figure 28-9. Three wires are tapped from the secondary. The middle wire is neutral. Using this neutral wire and either one of the two outside wires, it is possible to obtain 120 volts. When the two outside wires are tapped, 240 volts are obtained. This combination of three wires is known as the *three-wire system*.

A three-wire service is brought into a building at a point called the *service entrance*. This service entrance is a switching center where the electrical energy is first sent through a meter which measures the amount of energy used. From the meter, three wires are brought into a main distribution center. The number of circuits required branch off from the distribution center. A protective device is included in the center to cut off the service if a short circuit develops in the electrical equipment.

This protection may be a simple fuse or a circuit breaker which is reset each time the shorted circuit is repaired for a normal operation.

Figure 28-10 shows how 4000 volts is transformed into 120/240 volts, how this voltage is brought into a building through a service entrance, and how the voltage is brought safely through a distribution center to the various types of circuits.

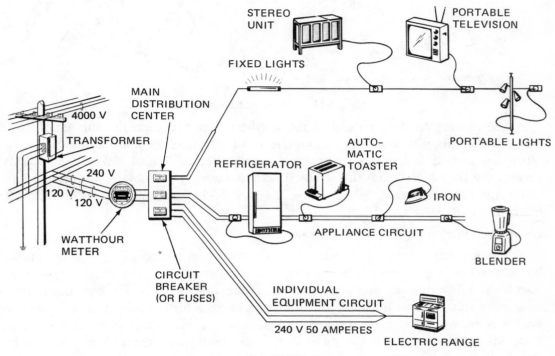

**Fig. 28-10**

## MEASURING AND PAYING FOR ELECTRICAL ENERGY

The rate of doing work is expressed as power. Power = work ÷ time or energy ÷ time. In electricity, the common unit of power is the watt. One *watt* represents the power used when one ampere of current is pushed through a circuit by a pressure of one volt. In other words, watts = volts x amperes. A motor which draws two amperes of current under a pressure of 120 volts requires 240 watts of power. However, the energy consumed in electrical circuits is measured in watthours. The *watthour* is an energy measurement. Energy = power x time; watthours = watts x hours.

For example, if 200 watts of power are needed and energy is used at this rate for five hours, the energy consumed is 1000 watthours.

**Watts = Volts x Amperes**

**Watthours = Volts x Amperes x Time (hours)**

### The Watthour Meter

Electrical energy is measured by a device known as a *watthour meter*. The watthour meter indicates the number of watthours of energy consumed to operate the electrical equipment attached to the circuits in one building.

A watthour meter contains four dials. Each dial is marked from 0 to 9, figure 28-11. Since the most practical unit of measurement in this situation is thousands of watthours, the watthour meter dials indicate the number of kilowatthours (kwh) consumed. One kilowatt is equal to 1000 watts. In other words, one kilowatthour is the energy expended at the rate of one kilowatt (or 1000 watts) over a period of one hour.

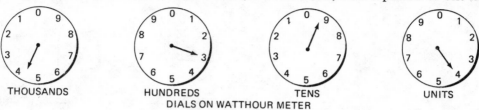

THOUSANDS     HUNDREDS     TENS     UNITS

DIALS ON WATTHOUR METER

**Fig. 28-11 Dials on watthour meter**

The reading on the first dial of the watthour meter registers the number of single units. When this dial makes one complete revolution, the second dial indicates tens. When the second dial makes one complete revolution, the third dial reads hundreds; finally, the fourth dial reads thousands. Taken as a whole, the meter dials indicate kilowatthours of energy used. The reading in figure 28-11 indicates that 4.294 kilowatthours of energy were used.

## Determining Costs for Electrical Energy

The amount of energy consumed during any given period of time is determined by subtracting the previous reading from the present reading. The difference in readings is the kilowatthours of energy used. A sliding rate or cost per kilowatthour determines the amount that must be paid by the consumer for the energy used. The first charge is known as a minimum charge and must be paid regardless of whether the minimum kilowatthour value of energy is used. The minimum charge is used to pay the expenses of the power company for the installation and maintenance service. Added to the minimum charge is the cost of the additional kwh of energy used.

Although the rates charged differ in various localities, the method of computing the electricity bill is the same. In some areas the statement of charges sent to the consumer gives two rates: the lower or *net rate* applies when the bill is paid by a certain date; the higher or *gross rate* applied after that date.

For example, assume that the power company rates are as follows:

Service charge (first 15 kwh or less) . . . . . . . . . . . . . . . . . $2.00
Next 45 kwh per month . . . . . . . . . . . . . . . . . . . . . . . . . $0.08 per kwh
Next 60 kwh per month . . . . . . . . . . . . . . . . . . . . . . . . . $0.06 per kwh
Additional kwh per month . . . . . . . . . . . . . . . . . . . . . . $0.036 per kwh

To apply these rates, assume that the difference between two watthour meter readings (a month apart) is 820 kilowatthours. Thus, the cost of this service is $2.00 for the first 15 kwh, $3.60 for the next 45 kwh, $3.60 for the next 60 kwh, and $25.20 for the remaining 700 kwh at the rate of $0.036 kwh. The total cost of the 820 kwh of electrical energy is $34.40.

This unit briefly shows how one form of energy is converted into one or more different forms until it can be tapped as electrical energy to do useful work. Ultimately, a 120- or 240-volt line supplies electrical circuits which operate lamps and electrical equipment. In all instances, the amount of energy used must be metered and its cost to the consumer computed.

———————————— SUMMARY ————————————

- The generating station changes the energy of falling water, oil, coal, or other fuel into electrical energy.

- Steam generating plants use fuel to convert water into superheated steam.

  The heat energy of steam at a tremendous pressure and temperature is transformed into rotating mechanical energy in a steam turbine.

- Hydroelectric plants transform the stored potential energy of water into mechanical energy.

- The elementary generator operates on the principle that an induced electromotive force and current is produced when a conductor cuts through magnetic lines of force.

  The generator converts rotating mechanical energy into electrical energy.

- A pulsating current and electromotive force is produced by a rotating conductor armature cutting through a magnetic field set up by stationary electromagnets mounted on a frame.

- Alternating current is that type of electrical energy in which the current changes direction twice in each cycle from positive to negative values. The alternating current pulsates at a frequency of a given number of cycles per second.

- The direct-current generator produces pulsating electric current continuously in one direction.

- The voltage of generated electricity is stepped up by a transformer to provide economical operation before the electricity is sent over high-voltage lines.

- A step-up transformer has a laminated core, a primary coil tapped to the generator leads, and a secondary coil with more turns than the primary coil. At the secondary coil, the electrical energy is increased by magnetic induction.

- High-voltage current is stepped down near the point of distribution to an operating voltage of 2000 volts.

- In the final stages of stepping down a high voltage, a three-wire system brings electrical energy to the service entrance of a building.

- The three-wire system permits the operation of electrical devices and appliances at 120 volts (obtained by tapping one outside wire and the neutral center wire) or 240 volts (across the two outside wires).

- The watt is the standard unit for measuring electric power. The watthour and kilowatthour measure energy.

## ASSIGNMENT UNIT 28 GENERATION, TRANSMISSION, AND DISTRIBUTION OF ELECTRICITY

### PRACTICAL EXPERIMENTS WITH ELECTRICAL ENERGY

**Equipment and Materials Required**

| Experiment A<br>Characteristics of Transformers | Experiment B<br>Measuring Electrical Energy |
|---|---|
| Laminated iron core | Watthour meter |
| Spool of conductor wire | Lamp circuit (two lamps) |
| Voltmeter | Two 100-watt bulbs |
| Low-voltage transformer | Connecting insulated wires |

## EXPERIMENT A.  STEP-UP AND STEP-DOWN TRANSFORMERS

**Procedure**

1. Form a primary coil for a transformer by winding 24 turns of conductor wire around one part of a laminated iron core.
2. Form a secondary coil around the same core but on the opposite side. Wind 12 turns of conductor wire for this coil.
3. Place a low-range voltmeter in a circuit with the secondary coil.
4. Connect the two ends of the conductor wire of the primary coil to a low-voltage transformer.
5. Adjust the voltage regulator on the transformer so it delivers 10 volts. Then insert the 110-volt transformer into a 110-volt line. Note the reading on the voltmeter. Prepare a table for Recording Results and record this reading.
6. Remove the plug from the power source. Unwind the secondary coil until there are just eight turns of wire remaining on the core. Repeat step 5 and record the voltage at the secondary coil.
7. Repeat step 6 using six turns of wire and again determine the voltage.
8. Increase the number of turns on the secondary coil to 24, 36, and 48 turns respectively. Conduct three voltage tests in each case:  (a) with the transformer voltage at 10 volts, (b) at 15 volts, and (c) at 20 volts. Record the voltmeter readings for each test.

**Recording Results**

| | Primary Coil | | Secondary Coil | |
|---|---|---|---|---|
| | Turns | Volts | Turns | Volts |
| A | | | 12 | |
| B | 24 | 10 | 8 | |
| C | | | 6 | |
| D | 24 | 10<br>15<br>20 | 24 | |
| E | 24 | 10<br>15<br>20 | 36 | |
| F | 24 | 10<br>15<br>20 | 48 | |

## Interpreting Results

1. From the voltages recorded for experiments A, B, and C (see the table for Recording Results), determine whether the transformer is a step-up or step-down type of transformer. Give the characteristics of the transformer identified.

2. Determine what type of transformer is indicated from the data at parts D, E, and F of the table. (a) Make a simple sketch of this transformer. (b) Label the parts of the transformer. (c) Explain briefly how the transformer works.

## EXPERIMENT B.  MEASURING ELECTRICAL ENERGY

### Procedure

1. Connect a watthour meter to a 110-volt power source and then connect it to a lamp circuit with two 100-watt bulbs and a switch.

2. Note the reading on the four dials of the watthour meter.

3. Close the circuit for six minutes. Note the reading on the watthour meter. Subtract this reading from the previous reading to obtain the number of watthours consumed. Record this value.

4. Close the circuit again and take readings at the end of 12 and 18 minutes, respectively. Determine the energy consumed at the end of each time interval and record this value.

### Recording Results

|   | Watthour Meter Reading | |
|---|---|---|
|   | Time | Watthours Consumed |
| A | 6 min. | |
| B | 12 min. | |
| C | 18 min. | |

### Interpreting Results

1. Use the formula given to compute the number of watthours of electrical energy consumed by the two bulbs for the 6, 12, and 18 minute intervals.

2. Compare the computed energy with that recorded on the watthour meter. Explain why the values are the same or different.

## PRACTICAL PROBLEMS ON THE GENERATION, TRANSMISSION, AND DISTRIBUTION OF ELECTRICITY

For statements 1 to 10, determine which are true (T) and which are false (F).

1. Almost two-thirds of the electricity generated in the United States is produced by hydroelectric plants.

2. The heat energy of fuels can be used to convert water into steam.

3. Superheated steam at a low temperature and low pressure produces the greatest amount of mechanical energy.

4. The superheated steam can be passed through the rotor blades of a turbine to produce mechanical energy in usable form.

5. Hydroelectric plants depend on a constant and adequate source of free-falling water.

6. An electromotive force can be induced when a conductor cuts across a magnetic field.

7. The magnitude of an induced electromotive force remains constant regardless of the position of the conductor.

8. The induced electromotive force in a conductor does not depend on the number of the lines of force cut by the conductor in a given period of time.

9. A generator is used to convert mechanical energy into electrical energy.

10. The stationary electromagnets mounted in a generator frame are intended to create as weak a magnetic field as possible.

Supply the word or phrase to complete statements 11 to 22.

11. A high voltage is produced when many _____ cut through a magnetic field.

12. The conductor on a generator armature is connected to a _____ to harness the induced current.

13. A conductor on a generator cuts the greatest number of magnetic lines of force at _____ and _____ degrees.

14. A conductor cuts the fewest number of magnetic lines of force at _____ and _____ degrees.

15. A _____ voltage fluctuates from a minimum value to a maximum value twice during each cycle.

16. The graph of an induced electromotive force is called its _____ .

17. The pattern of current flow in an alternating-current circuit alternates from _____ to _____ values.

18. Frequency refers to the number of _____ per _____ .

19. The slotted slip ring on a direct-current generator is known as a _____ .

20. The induced current in a dc generator is greatest at _____ and _____ degrees.

21. Direct current reverses itself _____ in each revolution.

22. The _____ of induced current in a dc generator is always the same.

Select the correct word or phrase to complete statements 23 to 32.

23. The maximum pressure at which electrical energy is produced by a generator is approximately (120) (2300) (15 000) volts.

24. Stepping up a voltage means (increasing) (decreasing) the induced electromotive force for transmission.

25. A transformer transfers electrical energy by (conduction) (magnetic induction) (movement across lines of force).

26. The ratio of windings between the primary and secondary coil determines the (amount) (direction) of voltage produced by a transformer.

27. To transmit 13 200 volts from the generator over high-power lines at 132 000 volts, there must be (five times as many) (ten times as many) (the same number of) turns on the secondary coil as there are on the primary coil.

28. The (smaller) (greater) the pressure, the smaller the current required to produce the same amount of energy.

29. The voltage in a three-wire service system is usually (120/240) (2300/4600) (240/480) volts.

30. Electrical power is measured in units of (amperes) (watts) (volts) (watthours) (coulombs).

31. A kilowatthour is equal to (1000 watts) (1000 watthours) (100 watts).

32. Electrical energy consumed is measured on a (watthour meter) (voltmeter) (ammeter) (wattmeter).

33. Determine the gross and net charges for electrical energy for quantities A, B, C, D, and E at the rates given in the table.

| Kilowatthours Consumed | A | B | C | D | E |
|---|---|---|---|---|---|
| | 12 | 125 | 320 | 460 | 695 |

| Demand | Rate | |
|---|---|---|
| | Gross | Net |
| First 15 kwh | $1.90 | $1.90 |
| Next 35 kwh | $0.08/kwh | $0.064/kwh |
| Next 55 kwh | $0.06/kwh | $0.052/kwh |
| Additional kwh | $0.04/kwh | $0.034/kwh |

34. Determine the cost of burning 20 75-watt lamps for five hours at the rate of four cents per kilowatthour.

35. What current is required to operate an electrical device rated at 1200 watts at a voltage of 240? At 120 volts?

36. Compute the cost of operating all of the appliances indicated in the table for a period of one year at the rate of $0.052 per kilowatthour. Determine the average cost per month.

| | Appliance | Capacity in Watts | Estimated Yearly kwh |
|---|---|---|---|
| A | Electric blanket | 200 | 150 |
| B | Clothes dryer | 4500 | 560 |
| C | Clock | 4 | 36 |
| D | Dishwasher | 1800 | 435 |
| E | Attic fan | 365 | 280 |
| F | Oil burner | 795 | 1020 |
| G | Range | 10 200 | 1250 |
| H | Radio | 100 | 100 |
| I | Television | 320 | 300 |

37. A small electric motor draws 6.5 amperes at 110 volts. Determine the power rating in watts and in kilowatts.

38. List eight important steps in the generation, transmission, and distribution of electrical energy.

# Unit 29 Application of Electricity to Machines and Lighting

Electricity is a convenient and comparatively inexpensive method of transmitting and distributing energy to any place it is needed. However, electrical energy must be changed to another form of energy at the point of application. Electric motors are the machines that convert electrical energy to mechanical energy to do useful work. Motors can range from small thimble-sized ones of fractional horsepower to very large motors with capacities of more than 45 000 horsepower.

Basically, motors are of two types: direct-current and alternating-current motors. The principles of operation and the advantages and disadvantages of both types of motors are treated in this unit.

## DIRECT-CURRENT MOTORS

The dc motor and the dc generator are very similar. The essential difference between the two machines is in their use. The dc generator converts mechanical energy into electrical energy and the dc motor takes electrical energy and changes it to mechanical energy at the place where it is needed. Both the dc generator and the dc motor depend on the cutting of magnetic lines of force and the principles of magnetism for their operation.

The simplest direct-current motor has a set of permanent magnets which produce opposing poles, a single-loop armature connected to the two halves of a commutator, and two brushes spaced 180° apart from each other. Essentially, a dc motor diagram looks like the one for a dc generator. The magnetic field around the permanent magnets is called the *field flux*. This term is in contrast to the *conductor flux* which refers to the flux produced by the action of a current through the single wire loop.

## CURRENT-CARRYING CONDUCTOR

Two other principles must be considered before the action of a wire loop in a magnetic field can be completely understood. The first principle concerns what happens to a current-carrying conductor when it is placed in a magnetic field. Figure 29-1A shows the north to south direction of the field flux and the counterclockwise direction of the conductor flux.

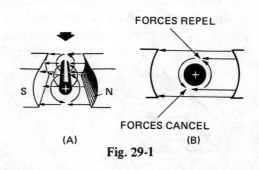

FORCES REPEL

FORCES CANCEL

(A)    (B)

**Fig. 29-1**

Figure 29-1B shows that both the conductor and the field flux are in the same direction on the top side of the conductor. On the bottom side, however, the field flux and the conductor flux tend to cancel each other. The result of these unequal forces tends to force the conductor downward and out of the field of the magnet.

## Direction of Rotation

The second principle concerns the direction of rotation. The direction of current in the conductor illustrated in figure 29-2A is reversed. The result is shown in figure 29-2B where the forces on the top of the conductor cancel each other. At the same time, the forces at the bottom repel one another while the conductor is moving in a clockwise direction.

The direction in which the conductor is moving can be determined easily (when the current direction is known) by applying *Fleming's Right-hand Motor Rule*. This rule states that if the forefinger of the right hand points in the direction of the field flux and the middle finger points in the direction of the current in the conductor, the thumb indicates the direction in which the conductor moves. Figure 29-3 shows the positions of the fingers at right angles to each other.

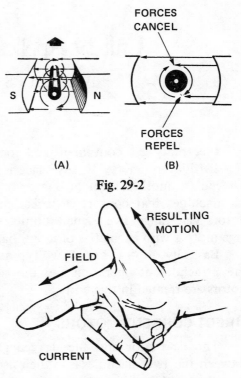

Fig. 29-2

Fig. 29-3  **Right-hand motor rule**

## Principles of Operation

As current is applied to the single loop through the brushes and the commutator, the loop tends to rotate counterclockwise due to the forces explained previously, figure 29-4A.

When the loop reaches a vertical position, figure 29-4B, the circuit is broken because the brushes rest on the insulated portion of the commutator between the two commutator segments. Although there is no current in the loop for an instant, the momentum of the motion of the loop carries it past this neutral point.

Once the loop is past the vertical position, current again flows in the loop, but in reverse direction. As a result, there is a reversal in the direction of the conductor flux. Thus, the magnetic forces continue to cancel and repel each other and the loop continues its rotation in one direction, figure 29-4C.

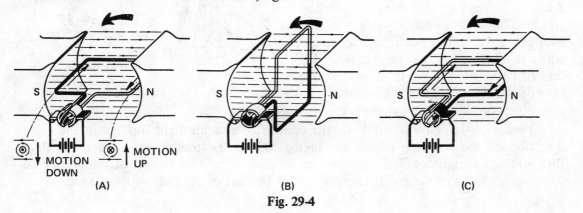

Fig. 29-4

The turning force (torque) of the loop in this illustration is extremely small. For practical applications, electromagnets are used for the field poles and the strength of the conductor flux is also increased.

The graph in figure 29-5, resulting from one revolution of a single loop conductor shows two peaks and two valleys. The torque is greatest at 90° and 270° and has a minimum effective value at 180° and 360°.

The pulsating torque shown in figure 29-5 is undesirable for practical applications. This effect can be improved by increasing the number of loops and the corresponding number of commutator segments.

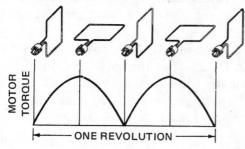

Fig. 29-5 Torque graph
of single-loop armature

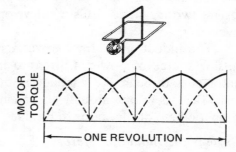

Fig. 29-6 Torque graph
of double-loop armature

The torque graph in figure 29-6 shows the effect of doubling the number of wire loops and commutator segments. The single wire loops used to describe the basic principles of motor operation are duplicated on commercial motors by an armature, figure 29-7. The direction of rotation of a dc

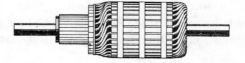

Fig. 29-7 Armature

motor can be reversed by reversing either the field connections or the armature connections. Basically, dc motors are used to provide varying speeds with flexibility of speed control within a desired range.

## ALTERNATING-CURRENT MOTORS

The major portion of the alternating type of electrical energy generated is used for ac motors. Alternating-current motors are preferred over dc motors for general use because they depend on fewer parts for their operation, are well adapted for constant speed applications, and can be operated on either a single-phase or a multiphase ac line. There are two main types of ac motors: (1) single phase and (2) polyphase. For applications requiring small amounts of power and having comparatively light loads, the single-phase motors are used. These motors are used in the home, on the farm, and in industry because they are less expensive to manufacture and the need for three-phase service is eliminated. Three-phase motors are used for heavier applications.

Large alternating-current generators generally contain three sets of coils (rather than one set). As a result, they generate three separate alternating currents which are 120° out of step. That is, the currents take turns in reaching their maximum values at equal intervals, as shown in the sine-wave graphs in figure 29-10, page 309. A three-phase power system is economical in terms of wiring and equipment. The energy of the three generated currents can be delivered on three wires and used by three-phase motors.

### One- and Three-Phase Induction Motors

Portable electrical equipment such as drills, saws, buffers, fans, and refrigerators commonly use a single-phase motor that runs on single-phase ac. There are two basic types of single-phase ac motors: (1) the induction motor and (2) the series motor (whose brushes and commutator resemble those of a dc motor).

The two main operating parts of any motor are the stator and the rotor. The *stator* is the stationary part of the motor; the *rotor* is the revolving part. The stator is housed in a cast steel frame. An end shield is attached to each end of this frame. The end shields house the bearings that support the rotor shaft. Regardless of the motor design, two magnetic fields are developed in every motor with the result that the rotor revolves.

The induction motor receives its name from the fact that the rotor requires no outside source of power. Current is induced in the rotor as the rotating field of the stator cuts through the conductors of the rotor. As rotor current is induced, a magnetic field develops around the conductors. This magnetic field interacts with the magnetic field of the stator to produce a torque which causes the rotor to turn.

*Single-phase Induction Motors.* For a single-phase induction motor, the stator has only one set of running windings. The pulsating currents in the winding induce currents in the rotor. In turn, the induced rotor current produces a magnetic field that is opposite in polarity to that of the stator. These opposing fields tend to turn the rotor 180° from its starting position. As a result, the rotor will not turn. Although a single-phase motor will run once it is started, some type of starting device must be provided. Once the rotor is rotating, it continues to turn in the same direction for almost 180° for each alternation of the stator field. The rotor increases in speed until a maximum speed is reached.

The starting device used with a single-phase motor may be added to the stator circuit so that a rotating field is generated. Once the motor is started, the starting windings are cut out of the circuit. Since the motor is now running on single phase, it is no longer necessary for the extra winding to continue to draw starting current.

*Three-phase Induction Motors.* Three sets of windings, spaced 120° apart, are mounted on the stator of a three-phase induction motor. When three-phase ac current is applied to these windings, a magnetic field is generated in each set of windings. When these fields are combined, a rotating magnetic field is produced.

*Principles of Generating a Rotating Magnetic Field.* The three sets of windings in a three-phase motor are out of phase. Thus, as shown in figure 29-8, the current flows through line 1 first, then through line 2, through line 3, and back to line 1. The magnetic field generated by any one of these phases at any instant depends on the current through that phase at the same instant. In one revolution, the three magnetic fields shift through one complete cycle; this cycle continues over and over as long as the motor is running.

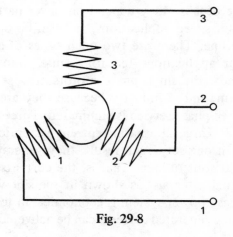

**Fig. 29-8**

The three-phase stator wiring which produces this rotating field is shown more accurately in figure 29-9. Figure 29-10 further defines the rotating field which is common to all types of polyphase motors.

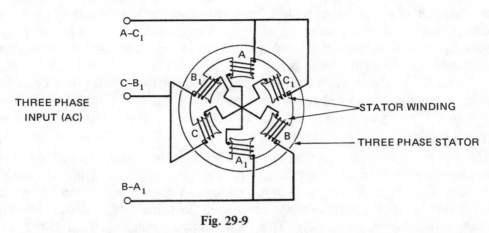

**Fig. 29-9**

*Waveforms of Current and the Magnetic Field of a Stator.* The waveforms shown in figure 29-10 are lettered A, B, and C to represent the three alternating magnetic fields generated on the stator during a complete cycle of three-phase current. As shown, one revolution or cycle of current is divided into six parts. Each part is equal to 60°. The waveforms show graphically how alternating current pulsates from ( + ) to ( – ) values and how each phase of a three-phase current is 120° out of phase.

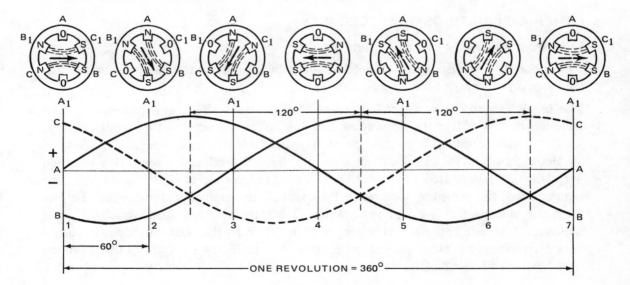

Note the circular drawings above the waveforms. Each drawing is lettered with the same ac input letters as the schematic wiring diagram shown in figure 29-9. Each drawing shows: (1) the polarity of each phase at 60° intervals, (2) the resultant magnetic field of the stator, and (3) the direction of the magnetic field.

*Development of Rotating Magnetic Field.* Starting at position (1) in figure 29-10, the value of phase C is positive; phase A has a zero value; and phase B has a negative value. Thus, the phases are out of phase with each other. The polarities of the fields of the three sets of windings of the stator are shown by the letters (N) for north, (S) for south, and (0) for no polarity. The dotted lines show the magnetic fields created and the arrow indicates the direction of the resultant magnetic field.

At a point 60° later, phase A now has a positive value, phase C has a zero value, and phase B has a negative value. The stator drawing at position (2) indicates that current flows from the north pole at phase A to the nearest south pole at phase B and from $B_1$ to $A_1$. Both of these fields are equal in amplitude and the direction of the resultant magnetic field is as shown.

At 120° of the cycle, the value of phase A is positive, phase C has a negative value which is equal to A, and phase B has a zero value. Current flows from the north pole of A to C (its nearest south pole), and from $C_1$ to $A_1$. There is no polarity at windings $B_1$ and B.

***Effect of the Interaction of the Magnetic Fields on the Rotor.*** As each phase is traced at intervals of 60° through one complete cycle, the magnetic fields and the resultant magnetic field direction are as shown. One cycle occurs in a fraction of a second. The end result of applying three-phase ac current to three sets of stator windings is to create a rotating magnetic field. This field induces a magnetic field in the rotor. The interaction of the two magnetic fields causes a torque to be set up in the rotor, causing it to turn.

In summary, the motor transforms electrical energy into mechanical energy which is used to drive machines, tools, equipment, and instruments to do productive work. Regardless of the construction features of dc and ac motors, the principles underlying their operation are the same.

## OTHER APPLICATIONS OF ELECTRICITY

### Light from Electricity

Electricity is also used to produce light in incandescent lamps, vapor and gas lamps, and fluorescent tubes.

***The Incandescent Lamp.*** Electric current has a heating effect as it passes through a conductor. The better the conductor and the larger the size of the conductor, the smaller is the amount of heat produced. For light applications, the heating effect of an electric current is desirable. For example, light is produced when a fine tungsten wire (*filament*) is heated to a white heat or *incandescence* at 2800°C. At this high temperature, the tungsten wire normally oxidizes and gradually burns away. To prevent this action, the filament is enclosed in a glass envelope from which all air is removed. An inert gas, such as argon, is then added to the glass envelope. In such a closed environment, electrical energy is converted to light energy using the incandescent tungsten filament as the medium.

***Vapor- and Gas-filled Lamps.*** At times, the light from a lamp is emitted by a vapor or gas. The term *vapor* means that although the substance in the tube is a liquid or solid at normal temperature, when it is heated the substance produces a vapor that is *luminous* because it gives off light. *Gas lamps* are filled with substances which are in a gaseous state at normal temperatures. The red light in display signs is produced by neon gas. Other colors are obtained by using other inert gases.

The sodium vapor lamp with its amber-colored light and the mercury vapor lamp (violet-blue light) are two common types of lamps that depend on the vaporizing of sodium and mercury respectively. In many sodium lamps, some neon gas is included to act as a starter in the vaporization process. When a high voltage is applied to the terminals of a lamp, some of the electrons of the gas travel between the two electrical terminals at the same time that each charged particle of gas moves toward the oppositely charged terminal. Light is given off as countless numbers of atoms are hit and set into motion.

Only two of the many kinds of vapor lamps are treated in this brief description. Another type of lamp is the *ultraviolet* lamp. This is a mercury vapor lamp that has a tungsten filament as a starter. Such a lamp is valuable as a sterilizer because it gives off much of its energy in wavelengths that are valuable for their germicidal effects.

*Fluorescent Lamps.* The fluorescent lamp operates as the result of the action of a chemical powder mixture inside the tube which glows or *fluoresces* when it is excited by the absorption of high-energy waves. The fluorescent lamp produces from two to four times as much light per watt as does the ordinary incandescent lamp. Another advantage of the fluorescent lamp is that the light output per circuit is high. As a result, the number of circuits required for an installation can be decreased. The amount of heat given off by a fluorescent lamp is less. In addition, the longer tube permits shadows to be overcome because of the uniform brightness achieved.

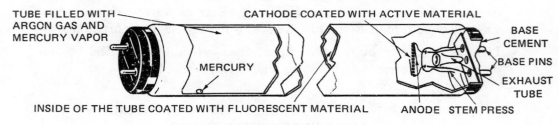

Fig. 29-11

The construction and operation of a fluorescent tube is comparatively simple. Figure 29-11 shows the essential parts of the lamp as a cap on each end of a glass tube. Terminals inside the caps connect to the ends of a tungsten filament. The tube is filled with the inert gas argon, and a small quantity of mercury which produces a vapor when heated.

As a current is passed through the tube, the bombardment of atoms and electrons produces high-energy, ultraviolet waves which cause the chemical powder of the tube to fluoresce. The color of the light produced depends on the chemical makeup of the powder mixture.

### Heat from Electricity

Electricity is also useful when it is converted into heat energy such as in electric furnaces, household heat ranges, toasters, and heating units.

——————————————————— SUMMARY ———————————————————

- Dc motors convert electrical energy into mechanical energy.
- Rotation is produced in a dc motor by the interaction of the opposing conductor flux and the field flux. The greatest torque occurs at 90° and 270°; the least torque is at 180° and 360°.

- Undesirable pulsating torque is overcome by increasing the number of conductor wires and commutator segments.

- Ac motors are used in single-phase or multiphase (polyphase) circuits.

- Induction motors operate on induced currents in a rotor that are opposite to those in a stator.

  The interaction of the two fields produces a torque on the rotor which causes it to turn.

- Single-phase induction motors require a starting device which cuts out automatically as the motor reaches speed.

- Three-phase motors operate on the principle that a rotating magnetic field on a stator induces a field in the rotor.

- Waveforms of a three-phase ac input circuit show that the waveform for any one phase is 120° out of phase with the next phase.

- The incandescent lamp depends on electricity to heat a filament to incandescence, at which point it gives off light waves.

- Gas and vapor lamps operate on the principle that as a gas or vapor becomes charged at a high voltage and electrons move between oppositely charged terminals, light is given off when the electrons hit atoms.

- Light is produced in the fluorescent lamp when the chemical coating on the tube fluoresces as it absorbs high-energy waves from the gas.

## ASSIGNMENT UNIT 29  APPLICATION OF ELECTRICITY TO MACHINES AND LIGHTING

### PRACTICAL EXPERIMENTS WITH MOTORS

**Equipment and Materials Required**

| Experiment A Electromagnetic Induction | Experiment B Direct-current Motors | Experiment C Alternating-current Motors |
|---|---|---|
| Horseshoe magnet with stand Bar magnet Wire-wound core Conductor wire Galvanometer | Simple experimental direct-current motor Two dry cells Wire to make connections | Small single-phase ac motor prepared for disassembly Tools to disassemble the motor Small simple three-phase ac motor |

## EXPERIMENT A.  ELECTROMAGNETIC INDUCTION

### Procedure:  Induced Currents in a Conductor

1. Connect the ends of a single conductor wire to a galvanometer. (A galvanometer measures the presence of an electrical charge.)

2. Hold the wire midway between the poles of the mounted horseshoe magnet. Observe the movement (if any) of the galvanometer needle. Record the action in a table for Recording Results.

3. Move the conductor wire slowly downward through the magnetic field of the magnet. Note the galvanometer reading and record this value at D in the table.

4. Bring the conductor wire slowly back up through the same field. Again note the position and reading on the galvanometer. Record this value (at U).

5. Repeat steps 3 and 4, cutting the lines of force at a slightly faster rate. Record any difference in readings from the slow movement.

6. Increase the speed again and repeat steps 3 and 4. Record the reading for the downward and upward movement in the table.

7. Double the strands of wire and repeat steps 2 – 6. Record the readings in the appropriate blanks of the table.

8. Triple the number of strands of wire in the conductor and repeat steps 2 – 6. Record all readings as in previous steps.

## Recording Results

| Conductor Wire | | At rest | Reading and Direction | | | |
|---|---|---|---|---|---|---|
| | | | | Movement through Field | | |
| | | | | Slow | Medium | Fast |
| A | Single | | D | | | |
| | | | U | | | |
| B | Double | | D | | | |
| | | | U | | | |
| C | Triple | | D | | | |
| | | | U | | | |

## Interpreting Results

1. How does the speed of a conductor through a magnetic field affect the induced voltage in the conductor? Prove this answer in terms of the recorded data.

2. Justify the statement that the direction of a moving conductor influences the direction of the induced current in the conductor.

3. Prove or disprove that an increase in the number of conductors cutting a magnetic field increases the total induced current.

## Procedure: Electromagnetic Induction in a Coil

1. Attach the two bare ends of a coil to a galvanometer. Plate one end of a permanent magnet in the coil. While the magnet is at rest, check the position of the galvanometer dial and note the reading, if any. Prepare a table for Recording Results.

2. Remove the magnet. Then move the north pole of the magnet slowly into the coil. Note the direction and magnitude of the induced current. Record these observations in the table.

3. Bring the magnet slowly out of the coil. Again note the direction and reading of the induced current in the table (at U).

4. Repeat steps 1 – 3 with the south pole of the magnet.

5. Push and pull the bar magnet in and out of the coil quickly. In each case, note and record the approximate reading and its direction on the galvanometer.

## Recording Results

| | Polarity | Reading and Direction of Induced Current in Coil | | | |
| --- | --- | --- | --- | --- | --- |
| | | At Rest | | Slow | Fast |
| **A** | North | | D | | |
| | | | U | | |
| **B** | South | | D | | |
| | | | U | | |

## Interpreting Results

1. The needle of a galvanometer points in one direction when the magnetic field moves in one direction and vice versa. Prove or disprove this statement in terms of the recorded results.

2. Explain briefly the effect of speed on the amount of induced current in the coil.

## EXPERIMENT B.  DIRECT-CURRENT MOTORS

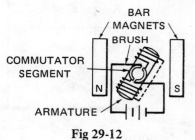

Fig 29-12

### Procedure

1. Connect one terminal of a battery to one brush terminal of a simple experimental direct-current motor having permanent magnets.

2. Connect the second battery terminal to the second brush.

3. Note the speed and direction in which the motor runs.

4. Reverse the terminals. Again note the motor speed and direction.

5. Disconnect the battery and place two batteries in series in the circuit. Recheck the speed and direction of the motor.

6. Reverse the terminals on the current source and note any change in the speed or direction of the motor.

### Interpreting Results

1. State the effect on armature speed that an increase in electrical pressure (emf) produces. Explain.

2. What effect does changing the direction of current flow have on the direction of rotation?

## EXPERIMENT C.  ALTERNATING-CURRENT MOTORS

### Procedure

1. Disassemble a small single-phase ac induction motor.
   CAUTION: *Make certain beforehand that the power source is disconnected.*

2. Make a simple schematic sketch of the stator and label the parts.

3. Make a simple drawing of the rotor and label the parts.

4. Reassemble the motor according to the manufacturer's specifications. Check the wiring and insure that the rotor turns freely. The assembly is to be checked before applying power to observe the operating characteristics of this motor.

5. Repeat steps 1 – 4 with a simple, small polyphase motor.

### Interpreting Results

1. Describe briefly the function of each part of the rotor and the stator of the single-phase induction motor.

2. Briefly explain how a rotating magnetic field is produced in a three-phase induction motor and what causes the rotor to turn.

## PRACTICAL PROBLEMS WITH ELECTRICAL MACHINES AND LAMPS

### Dc and Ac Machines

Select the correct word or phrase to complete statements 1 to 10.

1. The dc generator converts (electrical) (mechanical) (light) energy to another form.

2. The dc motor converts (electrical) (mechanical) energy into (electrical) (mechanical) energy.

3. The magnetic field around the stationary magnets of a dc motor is known as the (field) (conductor) flux.

4. The flux produced in a single-wire loop when current is passed through it is called (field) (conductor) flux.

5. The armature of a dc motor contains many conductor wires to overcome the (constant) (pulsating) torque conditions.

6. Ac motors are preferred for (constant) (varying) (extra high) speed applications.

7. The (single-phase) (polyphase) ac motor is widely used on small machines in the home.

8. Motors depend on the principle of a (rotating) (stationary) magnetic field.

9. A rotating magnetic field on a motor depends on the current source being (in phase) (out of phase).

10. The interaction of magnetic fields in a motor causes a (rotating) (nonrotating) torque to be set up in the rotor.

11. State what effect the following changes have on the magnitude of an induced magnetic field.

    a. Decreasing the field strength of the stator 50%.

    b. Doubling the number of wires in the armature.

    c. Increasing the conductor speed 25%.

    d. Increasing the number of field poles on the stator.

12. Match the items in Column I with the conditions in Column II.

| Column I | Column II |
|---|---|
| a. Principle of a dc motor | 1. The one stator winding does not produce a rotating field. Therefore, a secondary field 90° out of phase with the stator field is needed to start the motor. |
| b. Principle of a dc motor commutator | 2. The magnetic field of each phase is 120° out of phase. However, at any instant, the fields combine to make one resultant field. |
| c. Induction ac motor | 3. The armature current is reversed when the brushes contact another segment of the commutator. This causes the armature field polarity to change and produces continuous armature rotation. |
| d. Three-phase motors | 4. The current through the armature coil causes the armature to become a magnet. The poles of this magnet are attracted to the field poles of opposite polarity, producing a rotating torque. |
| | 5. The rotor does not require an external source of current. Current is induced by action of a rotating magnetic field cutting the rotor conductors. |

13. Describe briefly what is meant by the statement that the interacting magnetic fields in motors produce a rotating torque.

14. Give four reasons why electric motors are preferred to gasoline and steam engines in industry.

## Light from Electricity

For statements 1 – 10, determine which are true (T) and which are false (F).

1. The prime purpose of an inert gas in the ordinary incandescent lamp is to color the light.

2. Incandescence is produced in an electric lamp by the heating effect of electricity on a tungsten filament.

3. Vapor lamps depend on the gas emitted from a liquid or solid substance (when heated) to give off light.

4. The color of the light produced by a gas-filled lamp does not depend on the gas used.

5. Gas and vapor lamps are used for purposes other than to provide light.

6. The light output for the fluorescent lamp is the same per watt as for the incandescent lamp.

7. The initial cost of fluorescent lamps and fixtures is less than for the incandescent lamp and regular sockets.

8. Less heat and a more uniform light is distributed by the fluorescent lamp as contrasted with the incandescent lamp.

9. A chemical powder inside the fluorescent tube fluoresces as argon gas and mercury vapor receive an electrical charge.

10. The life of the lamp is increased by sealing a lamp filament in an enclosed glass shell, evacuating the air, and inserting an inert gas.

11. List two advantages of using fluorescent lamps.

12. State two applications in which an incandescent lamp is more practical than a fluorescent lamp.

13. Examine a cutaway section or drawing of an early incandescent lamp. Compare this lamp with a modern lamp, listing three improvements made in the new lamp versus the old.

14. Examine carefully either a cutaway model or a drawing of a gas- or vapor-filled lamp.

    a. Make a simple sketch of the lamp.

    b. Label the important parts.

    c. Describe briefly the function of each part.

# Achievement Review
## MAGNETISM AND ELECTRICAL ENERGY

### STATIC ELECTRICITY AND THE ELECTRON THEORY

Select the letter in Column II which describes the corresponding numbered item in Column I.

| Column I | Column II |
|---|---|
| 1. An electron | a. Composed of a heavy nucleus and one or more electrons traveling in orbit(s). |
| 2. Free electrons | b. Has an excess of protons (lack of electrons). |
| 3. A negatively charged part | c. Electrons near the nucleus which cannot be forced out of their orbits. |
| 4. A positively charged part | d. A negative charge in an orbit around a nucleus. |
| 5. An atom | e. Has an excess of electrons. |
| | f. Electrons that are easily forced out of their orbits. |

Supply the correct word or words to complete statements 6 to 9.

6. Electrical charges may be transferred by _____ , _____ , and _____ .

7. According to the electron theory, current flows from _____ to _____ .

8. Current flow is controlled by _____ , _____ , and _____ .

9. The _____ is a laboratory device for detecting electrical charges.

10. Give the scientific reason why some materials are good conductors.

11. a. Name three materials that are used by power companies as insulators.
    b. Explain briefly why each of these materials is a poor conductor.

### MAGNETISM

For statements 1 to 4, determine which are true (T) and which are false (F).

1. The intensity of the magnetic field remains the same regardless of the position or distance from the field.

2. Nonmagnetic materials affect the lines of force.

3. A shield is provided by a magnetic material which diverts the magnetic lines of force.

4. Magnetic materials offer greater resistance to magnetic lines of force than air.

5. Name the nonmagnetic materials in the following list:

| a. copper | c. wood | e. plastics | g. brass |
|---|---|---|---|
| b. glass | d. steel | f. aluminum | h. paper |

Define each of the following terms or conditions.

6. Magnetic flux density.

7. Force between two magnetic poles.

8.  Permeability.
9.  Modern concept of magnetism.
10. Domains.
11. Unmagnetized domains.
12. Magnetic domains.
13. Curie point.

## ELECTROMAGNETISM

Select the correct word or phrase to complete statements 1 to 3.

1.  Reversing the direction of current flow (has no effect on) (reverses the direction of) the magnetic flux.
2.  The amount of current, material and number of turns used in a current-carrying coil (affects) (does not change) its magnetic strength.
3.  Flux density (decreases) (increases) (is not affected) according to nearness to the current-carrying conductor.
4.  Describe the effect each of the following has on the strength of an electromagnet.

    a. Number of turns in the current-carrying coil.
    b. Increasing the current.
    c. Using a hardened steel core as compared with a soft iron core.
    d. Using a soft iron core as compared with air.

5.  Make a simple sketch of a machine or device which depends for its operation on an electromagnet.

    a. Letter the important parts.
    b. Indicate the direction of current flow, the flux direction, and the poles.
    c. Explain briefly how the electromagnet actuates the device.

## ELECTRICITY IN MOTION (DC)

In a single sentence, define each of the following electrical terms:

1.  Voltage
2.  Ampere
3.  Electromotive force
4.  Current flow
5.  Ohm
6.  Resistor

7.  Explain in a single sentence the effect each of the following conditions has on current flow.

    a. Changing from a copper wire to a nichrome wire (or other type of resistance wire).
    b. Increasing the length of the conductor.
    c. Decreasing the cross-sectional area of the conductor.
    d. Varying the temperature of the conductor.

8.  Make a simple sketch of each of the four different circuits indicated, labeling all important parts.

    a. A series circuit
    b. A parallel circuit
    c. An open circuit
    d. A broken circuit

9. Compute the missing values as indicated in the table for each of the four circuits given, to the nearest whole number.

| | Voltage (E) | Current (I) | Resistance (R) |
|---|---|---|---|
| A | 117 | 20 | |
| B | 117 | | 8.5 |
| C | | 20 | 11.7 |
| D | | 30 | 14.67 |

## ELECTROCHEMICAL SOURCES OF ELECTRICAL ENERGY

1. Compute the missing values for amperage, ampere-hour capacity, or hours of service for the four fully charged batteries given in the table.

| | Amperes | Battery Life (Hr.) | Ampere-hours |
|---|---|---|---|
| A | 12 | 10 | |
| B | | 20 | 160 |
| C | 9 | 220 | |
| D | 6.4 | | 192 |

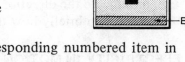

2. List six rules which should be observed to obtain maximum battery life.

3. a. Name the six parts shown on the cross-sectional sketch of the dry cell battery.

   b. Mark the positive plate (+) and the negative plate (−).

Select the letter in Column II which describes the corresponding numbered item in Column I.

### Column I

4. Parts of a primary cell
5. Service of a primary cell
6. A secondary cell
7. Service of a storage cell

### Column II

a. A series of separated positive and negative electrodes suspended in an electrolyte and capable of being recharged without replacing parts.

b. Heavy loads for short periods of time.

c. A number of primary cells connected in parallel.

d. Intermittent light loads for short periods of time.

e. Two different cell plates and an electrolyte.

## GENERATION, TRANSMISSION, AND DISTRIBUTION OF ELECTRICITY

Complete statements 1 to 4 by indicating the correct words or phrases.

1. The steam turbine operates by sending superheated steam at a temperature of _____ and a pressure of _____ through a series of blades until the steam is exhausted at the low-pressure end (pressure is less than one pound).

2. Hydroelectric plant operation depends on two major factors: (a) _____ and (b) _____ .

3. The magnitude of an emf is increased by increasing (a) _____ or (b) _____ .

4. A voltage applied at a primary winding sets up a voltage in a secondary winding by _____ .

5. What does the statement "alternating current with a frequency of 60 hertz" mean?

6. Determine the gross and net charges for the electrical energy required to operate all of the appliances in the table at the power rates given.

| Appliance | Monthly kwh Consumed | Demand (kwh) | Rate | |
|-----------|----------------------|--------------|------|-----|
| | | | Gross | Net |
| Oil burner | 80 | First 12 | 1.80 | 1.80 |
| Range | 100 | | /kwh | |
| Refrigerator | 10 | | | |
| Radio | 6 | Next 36 | 0.08 | 0.07 |
| Television | 30 | | | |
| Clothes Dryer | 48 | Next 60 | 0.06 | 0.05 |
| Clock | 3 | | | |
| Lighting | 33 | Additional | 0.05 | 0.042 |

7. List the steps which show how (a) an alternating current is reversed, and (b) how this current pulsates in a complete cycle of 360°.

## APPLICATION OF ELECTRICITY TO MACHINES AND LIGHTING

Supply the correct words or values to complete statements 1 to 12.

1. The four most important parts of the dc generator are: _____ , _____ , _____ , and _____ .

2. _____ in a dc motor is produced by the interaction of the opposing conductor and field fluxes.

3. The magnetic field around the permanent magnets of a motor is known as the _____ .

4. The greatest torque obtainable for a single-wire loop generator occurs at _____° and _____°.

5. Increasing _____ and _____ compensates for the undesirable pulsating torque condition when a single-loop conductor and a pair of brushes are used.

6. Ac motors are preferred to dc motors because: (a) _____ and (b) _____ .

7. The single-phase motor is widely used because it (a) _____ , and (b) _____ .

8. In a (an) _____ motor, currents are induced in the rotor that are opposite to the currents in the stator.

9. The _____ of a three-phase ac input circuit shows that any one phase is 120° out of phase with the next phase.

10. The _____ lamp contains a chemical powder which glows as the particles absorb high-energy waves from the enclosed gas.

11. _____ lamps contain a substance in the envelope which is either a liquid or a solid at normal temperatures.

12. In the _____ lamp, the filament is brought to a white heat to give off light waves.

13. a. Name three of the most common types of electric lamps.
    b. List one advantage or disadvantage of each type of lamp.

14. Study the specification plate on either a small dc or ac motor.

    a. List each of the items which are included on the plate.
    b. Explain in one sentence the meaning of each item or term.

15. Give three advantages to show why, when electricity is available, an electric motor may be preferred over an internal combustion engine as a source of power.

# Section 5
# Heat Energy and Heat Machines

## Unit 30 Heat Energy-Its Effects and Its Measurement

Heat is one of the most essential elements for human survival. Body heat is produced as the chemical energy in food is consumed. Heat also results when various forms of energy are converted into useful work.

### SOURCES AND NATURE OF HEAT ENERGY

The radiant energy of the sun provides a primary source of heat as it is absorbed and put to work. Electrical energy, too, is converted into heat energy by machines such as resistance furnaces, the common toaster, and resistance coils. Atomic energy is converted into heat when heavy atoms break up with the loss of mass or when light atoms combine and also lose mass (giving off heat and energy).

These are but a few examples of the way in which heat can be obtained or released by four major methods: (1) chemical action, (2) changing electrical energy to heat, (3) developing heat from work as friction develops from motion, and (4) by nuclear changes. In each instance, the form of the energy that disappears is replaced by heat energy. Thus, heat is a form of energy.

There are other phenomena that are as interesting and important when heat is to be explained. Every solid, liquid, and gas is said to be made up of particles which are capable of moving. Scientists have proved that as a material in the form of a solid, liquid, or gas takes on heat, the motion of the particles increases. As a body cools, the motion decreases. Heat then is the kinetic (internal) energy possessed by a body.

Heat is also described as a form of energy capable of flowing between two bodies which are at different temperatures. The addition of heat energy to a body increases the kinetic energy of motion of its molecules.

### Factors Affecting Heat Energy

Heat is also a measure of molecular motion. In solids, molecular motion is in the form of vibration in which the particles never move far from a fixed position. This freedom of motion is greater for liquids than solids because the motion is limited by the container only. For uncontained gases, the molecular motion is unlimited, figure 30-1, page 324.

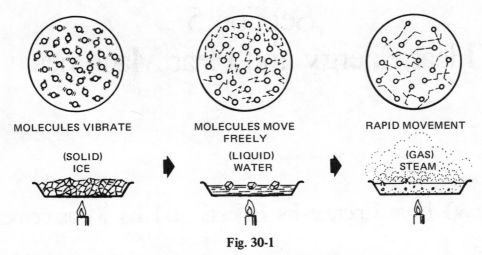

MOLECULES VIBRATE MOLECULES MOVE FREELY RAPID MOVEMENT

(SOLID) ICE (LIQUID) WATER (GAS) STEAM

Fig. 30-1

In each case, heat is transferred from the molecules of the hot body (more heat energy) to the molecules of the cold body (less heat energy) by intermolecular collisions. The fast-moving molecules of a hot body collide with the slower-moving molecules of a cold body. Some of the energy of the fast-moving molecules is used to speed up the slow-moving molecules.

The total kinetic energy of a body is a measure of the heat of that body. Three factors must be considered in any study of heat energy: (1) the mass of the body as determined by the number of particles or molecules; (2) the average activity of the particles; and (3) the specific heat of the material in the body. The term *specific heat* is defined later. In other words, the heat energy of a body depends on the speed, the weight, and the number of molecules in motion and transferring heat in a given time.

## TEMPERATURE AND ITS MEASUREMENT

The first step in measuring the quantity of heat in a body is to measure the temperature change of the body. There are several types of temperature measuring instruments which are used in three basic systems of temperature measurement.

### Measuring Instruments

*Thermometers.* The most obvious effect of heat or its absence is that a body becomes hotter or colder. The addition or removal of heat produces changes in temperature which can be measured. The instrument commonly used for measuring ordinary temperatures is called a *thermometer*. Most thermometers consist of a glass tube with a fine orifice, a small bulb which acts as a reservoir, and a fluid. The thermometer works on the principle that a liquid expands as it is heated and contracts as it cools. The glass tube is graduated so that any increase or decrease of the temperature, as reflected by the raising or lowering of the fluid, can be read directly.

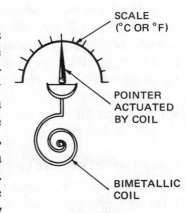

SCALE (°C OR °F)

POINTER ACTUATED BY COIL

BIMETALLIC COIL

Fig. 30-2

The *metal thermometer* is another type which works on the principle of the expansion of metals due to heat. Strips of two unlike metals are secured together to form a coil, figure 30-2. The difference in the unit expansion of the two metals causes the coil to move a pointer with any change in temperature. This change is read on a scale calibrated in degrees.

The *thermograph* combines the features of a metal thermometer and a recording mechanism to record permanently variations in temperature.

***Pyrometers.*** The *pyrometer* is widely used in the metal industry to measure and control temperatures within precise limits. A pyrometer is needed because the fluids and glass used in ordinary thermometers cannot withstand the tremendous temperatures used in heat treating, ore reduction and refining, and fabricating processes. These temperatures reach 3000°F.

***The Thermocouple.*** The simplest type of pyrometer combines a thermocouple and a millivoltmeter. A *thermocouple* consists of two wires of dissimilar metals. One set of the wire ends is welded together and protected by a tube covering. When this end is heated, a small electric current is generated by the difference in electromotive force between the two wires. This current is measured by connecting the opposite ends (the cold junction) of the thermocouple to a millivoltmeter. Any difference in temperature is reflected by the amount of current flowing from the hot junction of the thermocouple to the cold junction at the millivoltmeter, figure 30-3.

***Thermoelectric Current.*** Electric current produced as a result of heating is called *thermoelectric current.* The following description explains how this type of electric current is generated. Within the boundaries of each grain in a metal, there is a gaslike atmosphere of free electrons. This atmosphere exerts a vapor pressure which resembles that of gases or that above the surface of liquids. As heat is applied, the movement of the atoms and their electrons increases with a resulting increase in the electron vapor pressure.

At the hot junction of the thermocouple, the two dissimilar metals have markedly different electron

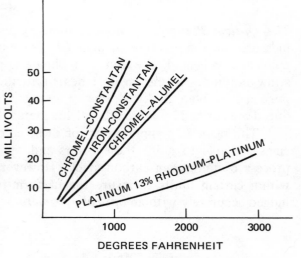

Fig. 30-3 **Relationship of temperature difference (°F) to millivolts of emf**

vapor pressures so that electrons flow from one metal to the other. There is less flow of electrons at the cold junction. Thus, the electromotive force which develops causes the electrons to flow from the hot junction to the cold junction of the thermocouple. Since the millivoltmeter used in this type of application has a scale calibrated to read in degrees Fahrenheit, direct readings can be taken at high temperatures.

Although any two dissimilar metals can be used in a thermocouple, certain combinations of metals are preferred. The metals used depend on the conditions under which the thermocouple must operate. For low temperature ranges (300° – 600°F), copper and constantan are desirable; iron and constantan are used for a range of 0° – 1300°F; and chromel and alumel are used for a range of 600° – 2000°F. Chromel, alumel, and constantan are alloys. The melting temperatures of these metals are:

chromel, 2800°F; alumel, 2600°F; and constantan, 2450°F. Other combinations are used in industry as shown in figure 30-4. The figure also indicates the polarities of the various metallic combinations.

| Thermocouple Materials | Polarity | Range in °F. |
|---|---|---|
| Chromel<br>Alumel | Positive<br>Negative | 600° – 2000° |
| Iron<br>Constantan | Positive<br>Negative | 0° – 1300° |
| Chromel<br>Constantan | Positive<br>Negative | 0° – 1000° |
| Platinum, 13% Rhodium<br>Platinum | Positive<br>Negative | 1300° – 2700° |
| Copper<br>Constantan | Positive<br>Negative | 300° – 600° |
| Platinum, 10% Rhodium<br>Platinum | Positive<br>Negative | 1300° – 2700° |

Fig. 30-4

*The Optical Pyrometer.* Modern industry uses a method called *color inspection* to measure the temperature of a metal. For example, a polished piece of steel changes color as it is heated. The first color change appears near 400°F when the steel becomes straw colored. As the steel is heated further, this color darkens into brown, then purple, then blue, dark red, cherry, orange, yellow and, finally, white. The colors produced are formed by the oxides of the metal as it is heated above room temperature.

The range of temperatures for these colors is shown in figure 30-5. Many technicians are able to judge these colors and temperatures with accuracy, such as in the process of tempering carbon steels. However, when it is necessary to control a heat within certain limits to insure the uniformity of the product, these colors must be judged accurately with an *optical pyrometer.*

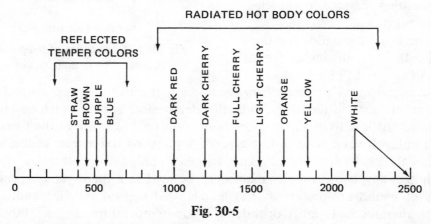

Fig. 30-5

The brightness with which heat is radiated from a hot object depends on the temperature of the object. The optical pyrometer works on the principle of color matching the filament in the instrument against the color of the heated object or furnace. The color of the filament is regulated by the amount of electrical current supplied. The correct temperature is reached and read on the instrument when the brightness of the filament in the pyrometer matches the color of the object. If the filament is darker in color than the object, then the filament is colder than the object; a brighter filament color indicates that it is hotter than the object.

Portable optical pyrometers are an accurate means of determining temperature. They are especially useful in steel mills, foundries, heat-treating departments and in any other applications where it is necessary to measure high temperatures accurately from a distance using a portable instrument.

## Systems of Temperature Measurement

*Celsius and Fahrenheit Scales.* There are two basic systems for measuring changes in temperature: Celsius and Fahrenheit. In both systems there are two fixed points which indicate (1) the temperature at which ice (pure water ice) melts anywhere in the world, and (2) the temperature at which pure water boils at a standard pressure of 760 millimeters.

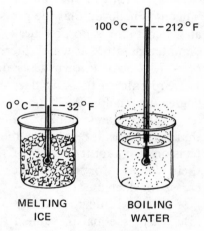

The Celsius and Fahrenheit scales were adopted many years ago as a convenient way of indicating temperature. Temperature is measured by a unit called the *degree*. The point at which ice melts is known as zero degrees on the Celsius (C) scale and 32 degrees on the Fahrenheit scale. The boiling point of pure water at the standard pressure is fixed at 100°C and 212°F, figure 30-6.

**Fig. 30-6**

Thus, the difference between the melting point of ice and the boiling point of water is 100° in the Celsius system and 180° in the Fahrenheit system. This means that each degree change of temperature on the Fahrenheit scale is equal to five-ninths of a degree on the Celsius scale. Note, too, that the Fahrenheit scale begins with the number 32.

A reading on either scale can be converted to the other scale by using the formulas,

$$°C = (°F - 32°) \frac{5}{9}$$

$$°F = (\frac{9}{5} °C) + 32°$$

For example, a reading of 100°C can be changed to the Fahrenheit equivalent as follows:

$$°F = (\frac{9}{5} °C) + 32°$$

$$°F = (\frac{9}{5} \times 100°) + 32° = 212°$$

This same reading (212°) can be changed back to a Celsius reading as follows:

$$°C = (°F - 32°) \frac{5}{9}$$

$$°C = (212° - 32°)\frac{5}{9} = 100°$$

If decimals are preferred, these formulas can be used:

$$°F = 1.8°C + 32°, \text{ and}$$
$$°C = (°F - 32°) \, 0.56$$

**The Kelvin Absolute Temperature Scale.** The Celsius and Fahrenheit scales indicate relative temperature. *Absolute temperatures* can be measured only when the scale starts at the real zero of temperature. This point occurs when there is no heat and thus there is no degree of heat. Absolute zero is the basis of a third temperature scale known as the *Kelvin* scale (named after the British scientist who formulated the scale).

It has been proved that regardless of the gas used, the pressure of the gas decreases 1/273rd of the value at 0°C for each degree that the gas is cooled below 0°C. Since it is believed that the pressure exerted by a gas is caused by heat, the temperature at which there is no pressure exerted is the same temperature at which there is no heat, or –273°C. All known gases undergo this 1/273rd rate of pressure decrease for each degree drop below 0°C.

Before gases liquefy, it is also true that at a constant pressure there is an expansion (or contraction) of 1/273rd of their volume for each degree Celsius increase (or decrease).

The unit degree on the Kelvin absolute scale is equal in value to the Celsius degree. To convert from the Celsius scale to the Kelvin scale, simply add 273° (since the zero reading on the Kelvin scale is located 273° below 0°C). Fahrenheit readings can be converted to absolute readings by either changing into (C) units first, or by using the formula:

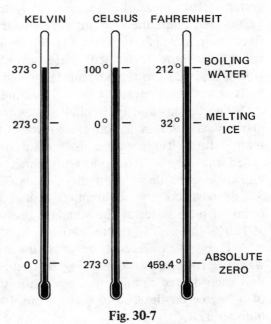

Fig. 30-7

$$°K = (°F + 459.4°) \times \frac{5}{9}$$

For example, 32°F is determined on the Kelvin scale as follows:

$$°K = (32° + 459.4°) \times \frac{5}{9} = 273°$$

## MEASURING HEAT ENERGY

Heat energy depends on the temperature, composition of the material, and the amount of fuel or other energy sources involved. Some system of measurement is needed to determine how efficiently heat energy is produced or consumed. Measurements are also needed to plan the amount of material needed to supply known quantities of heat. Since heat produces work, a temperature change can be measured directly.

### The Calorie (Metric System)

There are two units of heat measurement: (1) the calorie (an SI metric unit) and (2) the British thermal unit (Btu), the commonly used English unit. A *gram-calorie* is the quantity of heat absorbed or released when one gram of pure water is raised or lowered one degree Celsius. When 10 grams of water are raised 10°C, 100 gram-calories of heat are absorbed. The *kilogram-calorie* is a unit 1000 times larger than the gram-calorie.

## The British Thermal Unit (English System)

The British thermal unit (Btu) is the amount of heat exchanged when one pound of water undergoes a temperature change of one degree Fahrenheit. If the temperature of 10 pounds of water is cooled 10 Fahrenheit degrees, then 100 Btu of heat are released. The heat energy of one Btu is also equal to 778 foot-pounds of work. This relationship permits calculations to be made of the amount of work or mechanical energy produced by a given amount of heat energy.

## Specific Heat Values

*Heat Coefficients.* Materials vary in their ability to absorb and exchange heat. If equal weights of steel, lead, copper, aluminum, and water are heated through equal temperature changes, each material absorbs a different amount of heat. This procedure shows that water requires the largest amount of heat. Such a result suggests that water can be used as a standard of comparison of the specific heat of different materials.

Specific heat is useful in computing the amount of heat required to raise the temperature of any material. The specific heat is the number that expresses the fractional number of Btus exchanged when one pound of a material changes one degree Fahrenheit in temperature.

*Determining Heat Values. Specific heat* is the numerical ratio of heat energy required to raise or lower the temperature of a given weight of a material a required number of degrees as compared to the amount of heat energy needed to raise or lower the temperature of the same weight of water the same number of degrees.

Tables are available which show the specific heat of common materials. A few examples are shown in Table 30-1. To find the heat required to raise or lower a temperature of a material, multiply the weight of the material by the temperature change; then multiply this product by the specific heat of the material. For example, if the internal stresses in a 100-pound steel casting are to be relieved, the amount of fuel required to raise the temperature from 80°F to 880°F can be computed as follows:

| Material | Specific Heat (Approximate) |
|---|---|
| Machine Oil | 0.40 |
| Aluminum | 0.21 |
| Glass | 0.19 |
| Steel | 0.12 |
| Copper | 0.10 |
| Brass | 0.09 |

**Table 30-1**

$$(100 \text{ lb.}) \times (880° - 80°) = 80\,000$$

Specific heat of steel (from table 30-1) = 0.12

Heat required = 80 000 x 0.12 = 9600 Btu

The value in Btu (9600 Btu) is then divided by the Btu of a given quantity (by weight or by volume) of fuel. If coke is used as the fuel and coke produces 3200 Btu per pound, then 9600 ÷ 3200 = 3 pounds of coke must be burned to obtain the required amount of heat.

*The Calorimeter.* In the laboratory, the heat values of fuels and foods are usually determined by a device known as a *calorimeter,* figure 30-8. The field of experimentation and study dealing with these heat values is known as *calorimetry.*

The calorimeter works on the principle that heat is exchanged between two bodies having unequal temperatures. That is, heat moves from the one body having the higher temperature to the other body at the lower temperature until both temperatures are equal. Since energy can neither be created nor destroyed, the heat gained by the colder body is equal to the heat lost by the warmer body. The calorimeter is the device used to measure experimentally the amount of heat gained or lost.

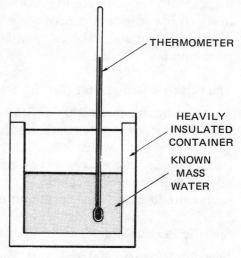

THERMOMETER

HEAVILY INSULATED CONTAINER

KNOWN MASS WATER

Fig. 30-8 Calorimeter

---

## SUMMARY

- Heat energy can be obtained by converting chemical, electrical, mechanical, or nuclear energy.

- Heat energy flows between two bodies at different temperatures — from the hot body to the cold body.

- The heat energy of a body is determined by its mass, the average activity of its molecules and its specific heat.

- Thermometers and pyrometers are instruments which measure one effect produced by heat; namely, a change in temperature.

- A thermocouple combined with a millivoltmeter (calibrated to read in degrees) is used for high-temperature measurement.

- Optical pyrometers depend on the color matching of a filament with a hot object. The temperature reading is obtained from a millivoltmeter which records the amount of current required to produce the matching filament temperature.

- There are two comparative systems for measuring temperature: the Celsius and the Fahrenheit systems.

  Temperature readings may be converted from one scale to the other by the formulas:

  $$°C = (°F - 32°) \frac{5}{9}$$

  $$°F = (\frac{9}{5} °C) + 32°$$

- The Kelvin temperature scale starts at absolute zero or the point at which there is no heat. At –273°C, or 0°K, or –459.4°F, there is no pressure and no heat. These are absolute zeros of temperature.

- The calorie is a metric unit of heat measurement. The gram-calorie measures the amount of heat absorbed or released in changing the temperature of one gram of pure water one degree Celsius.

- The Btu is the amount of heat exchanged when the temperature of one pound of pure water is raised or lowered (at a standard pressure) one degree Fahrenheit.

- Specific heat is a number ratio which expresses the heat energy required to raise or lower the temperature of a given weight of a body one degree Fahrenheit as compared to that of an equal weight of water.

- The calorimeter is a laboratory device which measures the amount of heat exchanged between two bodies of unequal temperature.

## ASSIGNMENT UNIT 30  HEAT ENERGY:  EFFECTS AND MEASUREMENT

## PRACTICAL EXPERIMENTS WITH HEAT, ITS EFFECTS, AND MEASUREMENT

### Equipment and Materials Required

| Experiment A<br>Thermometers | Experiment B<br>Pyrometers | Experiment C<br>Specific Heat |
|---|---|---|
| Two glass tubes with orifices of two different sizes | Two dissimilar metals (with high melting points) | Four small billets of four different metals (same weight) |
| Small quantity of mercury for the tubes | Brazing materials and equipment | Four metal containers with same weight of water |
| Beaker for heating water | Small furnace | Four thermometers |
| Ring stand | Connecting lead wires | Small furnace with thermo-couple and tongs |
| Bunsen burner | Eight temperature cones (two each for 600°F, 700°F, 800°F, and 900°F) | Temperature cones (four each for 275°F, 250°F, and 200°F) |
| Ice tray and ice | Millivoltmeter | |
| | Low-range optical pyrometer | |

## EXPERIMENT A.  MEASURING TEMPERATURE: THERMOMETERS

### Procedure

1. Heat each glass tube and form a small bulb at one end of each tube to serve as a reservoir.

2. Allow the tubes to cool. Add enough mercury to one tube to fill the bulb and about one inch of the tube.

3. Pack cracked ice in a tray and insert a tube. Mark the lowest point on the tube that the mercury drops to as the melting point of ice.

4. Remove the tube and suspend it in a beaker of water at room temperature.

5. Heat the water slowly until it boils. Then carefully mark the highest point to which the mercury rises. This is the boiling point of water.

6. Repeat steps 2 – 5 with the second tube.

7. Allow both tubes to cool. Then divide the space on both tubes between the freezing and boiling points. Roughly calibrate one tube for Celsius degrees and the other tube for Fahrenheit degrees (at five- or ten-degree intervals).

8. Place the bulb and lower part of the stem of each thermometer in a beaker of water at room temperature. Record the reading on both scales.

9. Heat the water slowly. When the temperature reaches 145°F and 190°F, record the temperatures on the thermometer with the Celsius scale for each Fahrenheit value, respectively.

### Interpreting Results

1. Does the level of mercury or other liquid in a thermometer continue to rise the longer the thermometer stays in boiling water? Does the level continue to fall the longer the thermometer stays in ice? Briefly explain each action.

2. Prove that $°C = (°F - 32°) \times 5/9$ from the results of the experiment.

3. What effect, if any, does the size of the tube orifice or the amount of mercury have on the measurement of temperature? Explain.

## EXPERIMENT B. PYROMETERS

### Procedure: Making a Thermocouple

1. Wind any two fairly high-melting point metals together at one end and braze that end.

2. Connect the two open ends in a circuit with a millivoltmeter.

3. Place the welded end (hot junction) in a small furnace. Heat this junction and note whether the thermoelectric effect on the two wires produces a small current, as measured on the millivoltmeter.

4. Shut down the furnace and allow the thermocouple to cool.

5. In the furnace, place four temperature cones (pellets) with low melting points. Prepare a table for Recording Results and record the rated temperature at which the cones should melt.

6. Relight the furnace. The hot junction of the thermocouple is to be near the cone with the lowest melting point.

7. Record the reading on the millivoltmeter when the first cone melts; then indicate the reading for the cone with the second lowest melting point, then the reading for the third cone, and finally that of the fourth cone.

8. Draw a rough temperature scale on paper and paste this scale on the outside of the instrument. This scale shows the temperatures at which the different temperature cones melt.

9. Place four more cones in the furnace. These cones are to have the same melting points as the original set of cones.

10. Heat the furnace. Again, note on the prepared instrument scale the temperature at which each cone melts.

11. Check each melting point by color matching the cones with a commercial optical pyrometer. Note each reading on the instrument in the table.

## Recording Results

| Cone | Rated Temperature | Temperature | |
|------|-------------------|-------------|---|
| | | Millivoltmeter Scale | Optical Pyrometer |
| A | | | |
| B | | | |
| C | | | |
| D | | | |

## Interpreting Results

1. Prove or disprove the statement that heating two dissimilar metals produces a thermoelectric effect.

2. The same amount of heat applied to the hot junction of a thermocouple always produces the same amount of electrical energy. This energy can be read directly on an instrument as temperature. Support this statement by referring to the results of the experiments.

3. Did the results obtained with the optical pyrometer check with the homemade millivoltmeter temperatures and with the manufacturer's temperature ratings for the cones (within 10% accuracy)? Explain.

## EXPERIMENT C. DETERMINING SPECIFIC HEAT

### Procedure

1. Check the weight of each of four small metal billets. The billets should all weigh the same.

2. In each of four containers, place an amount of water equal to the weight of each billet (the weights of the water will be the same for each container since the billets have the same weights).

3. Prepare a table for Recording Results. Indicate the type of metal in each billet, the weight of the billet, the equal weight of water in each container, and the temperature of the water.

4. Heat the first metal piece to 300°F.

5. Remove the billet from the furnace and quench it in the first container of water until the temperature of the metal drops to 275°F. Remove the metal billet.

6. Measure the temperature to which the water is raised and record this value.

7. Place the metal piece back in the water until its temperature drops to 250°F and 200°F, respectively. Each time, remove the piece, measure the temperature of the water, and record this temperature.

8. Repeat steps 4 – 7 with metal pieces B, C, and D. In each instance, record the changes in temperature for the metal billets and the water. In each case, start with the water at room temperature.

## Recording Results

| | Metal | Metal | | | Water | | | |
|---|---|---|---|---|---|---|---|---|
| | Metal | Weight | Temp. Heated | Amount of Heat Lost | Weight | Starting Temp. | Temp. Heated | Amount of Heat Gained |
| A | | | | 1<br>2<br>3 | | | 1<br>2<br>3 | | |
| B | | | | 1<br>2<br>3 | | | 1<br>2<br>3 | | |
| C | | | | 1<br>2<br>3 | | | 1<br>2<br>3 | | |
| D | | | | 1<br>2<br>3 | | | 1<br>2<br>3 | | |

## Interpreting Results

1. Determine the amount of heat lost by each of the metal billets and the amount of heat gained by the water.

2. Compute the specific heat of each metal.

3. Compare the computed specific heat value with the specific heat given for each metal in a prepared table. Account for any variations between the computed specific heat and the separate values given in a table.

## PRACTICAL PROBLEMS ON HEAT ENERGY AND ITS MEASUREMENT

## SOURCES OF HEAT ENERGY AND MEASURING INSTRUMENTS

Add a word or phrase to complete statements 1 to 8.

1. Heat energy is produced or released by the conversion of: _____ , _____ , _____ .

2. Changes in heat are produced in solids, liquids, and gases by _____ of particles in that body.

3. Heat is a form of energy which can flow between two bodies at _____ temperatures.

4. The molecular freedom of motion is _____ in fluids than it is in solids.

5. Heat transfer always takes place in the direction of _____ to _____ temperatures; never the reverse.

6. Heat energy depends on three major factors: _____ , _____ , _____ .

7. When the fast-moving molecules of a heated material collide with the slower-moving molecules of a cooler body, some of _____ of the fast-moving molecules is released.

8. The simplest method of measuring the quantity of heat is to start with a measurement of _____ .

Select the correct word or phrase to complete statements 9 to 15.

9. The (pyrometer) (thermocouple) (thermometer) is used to measure ordinary temperatures.

10. The metal thermometer works on the principle of the effect of temperature on the expansion and contraction of (a liquid in a tube) (two dissimilar metals).

11. The (pyrometer) (thermometer) is used in industry to measure temperatures ranging up to 3000°F.

12. Thermoelectric current is produced when heat is applied to the (hot) (cold) junction of two (like) (unlike) metals.

13. The (optical pyrometer) (thermocouple) is used to measure high temperatures by color matching.

14. A steel part heated to a straw color is heated to (a higher) (a lower) (the same) degree of temperature as compared to another steel part at an orange color.

15. In color matching with an optical pyrometer, when the color of the filament is brighter than the heated object, this is an indication that the object is at (a colder) (a hotter) (the same) temperature as compared to the temperature of the filament in the instrument.

Select the letter corresponding to the word or value which best completes statements 16 to 23.

16. The instrument which measures high temperatures is a (A) galvanometer (B) pyrometer (C) barometer.

17. (A) One (B) Two (C) Three (D) Five metals are needed to make a thermocouple.

18. The electrical units used to measure the emf produced by heating a thermocouple are (A) volts (B) megavolts (C) millivolts (D) microvolts.

19. An iron-constantan thermocouple can be used to measure a temperature of (A) 1200°F (B) 1800°F (C) 2400°F .

20. A light straw color appears on a steel part being tempered at (A) 250°F (B) 400°F (C) 650°F (D) 1000°F .

21. A pyrometer measures the (A) amount of heat (B) hardness (C) magnetism (D) level of temperature.

22. A thermoelectric effect is produced by (A) the movement of free electrons when a body is heated (B) a chemical change in a body (C) the magnetic properties of metals (D) recrystallization of the metals.

23. For temperatures within the range of 2200° – 2700°F, the thermocouple to be used is a combination of (A) chromel-alumel (B) platinum rhodium-platinum (C) iron-constantan.

## Systems of Temperature Measurement

1. Match the item in Column I with the correct condition in Column II.

<table>
<tr><th>Column I</th><th>Column II</th></tr>
<tr><td>a. The Celsius and Fahrenheit systems of temperature measurement.</td><td>(1) There are two fixed points on the temperature scale: (1) to indicate the melting point of ice, and (2) to show the boiling point of pure water at 760 mm of pressure.</td></tr>
<tr><td>b. The Kelvin temperature scale.</td><td>(2) Subtract 32° and multiply by 5/9.</td></tr>
<tr><td>c. Changing a temperature measurement from a Celsius reading to a Fahrenheit reading.</td><td>(3) Add 459.4 to the temperature reading and multiply by 5/9.</td></tr>
<tr><td>d. Changing a reading on the Fahrenheit scale to an absolute reading on the Kelvin scale.</td><td>(4) This system measures absolute temperature starting with -273 °C (the temperature at which there is no heat).</td></tr>
<tr><td></td><td>(5) Multiply the reading by 9/5 and add 32 ° to the answer.</td></tr>
</table>

2. Compute the missing equivalent values on the Fahrenheit, Celsius, or Kelvin scales for the temperatures given in the table. Use the applicable formula and show all computations.

| | Temperature Readings | | |
|---|---|---|---|
| | °F | °C | °K |
| A | 8 | | |
| B | 95 | | |
| C | | 0 | |
| D | | 100 | |
| E | | | 0 |
| F | | | 100 |

## Measuring Heat Energy

1. Determine the gram- or kilogram-calories of heat lost or gained in producing the temperature changes on metal parts A, B, C, and D. The specific heat of steel is 0.1.

| | Weight of Body | Temperature Changes °C | | Gram- or Kilogram-calories |
|---|---|---|---|---|
| | | From | To | |
| A | 1 gram | 0 | 100 | |
| B | 1 gram | 75 | -25 | |
| C | 100 grams | 8 | 72 | |
| D | 175 kilograms | 132 | 118 | |

2. a. Compute the Btu exchanged for the temperature changes of the liquids and solids indicated in the table at A, B, C, and D.

   b. Convert the Btu value to foot-pounds of work.

| | Liquid or Solid | Weight | Temp. Changes °F | | Btu | Ft.-lb. of Work |
|---|---|---|---|---|---|---|
| | | | From | To | | |
| **A** | Water | 20 lb. | 32 | 82 | | |
| **B** | Oil | 40 lb. | 78 | 228 | | |
| **C** | Steel | 60 lb. | 75 | 450 | | |
| **D** | Copper | 1 ton | 75 | -32 | | |

3. Prepare a table like the one illustrated.

   a. List in the materials column three common metals at A, B, and C, and the two nonmetals at D and E.

   b. Indicate the specific heat of each material. Locate the specific heat in the Appendix or in a Metals Handbook.

   c. Indicate the order in which the materials have the capacity to raise the temperature of an equal weight of water the greatest number of degrees. Arrange in descending order from 1 (greatest) to 5.

| | Material | Specific Heat | Capacity to Heat Water |
|---|---|---|---|
| **A** | | | |
| **B** | | | |
| **C** | | | |
| **D** | | | |
| **E** | | | |

4. Three bids are received for delivering quantities of coal. Determine which of the three bids (A, B, or C) is the best to accept in terms of energy cost when the properties of each coal are identical except for variations in the Btu rating.

| Bid | Btu Per Lb. | Cost Per Ton (2000 Lb.) |
|---|---|---|
| A | 12 900 | $60.00 |
| B | 12 600 | $57.00 |
| C | 13 200 | $63.00 |

# Unit 31 Expansion, Transfer of Heat, and Change of State

Nearly all solids expand when heated and contract when cooled. The amount of expansion and contraction varies with different materials. This fact is important because the design and construction of all parts, mechanisms, and devices are affected by changes in temperature. Every rail, bridge, underground tube, engine cylinder head, bearing — every conceivable device — is produced with tolerances on all working dimensions. This design allowance provides for expansion and contraction of all parts so they can operate without interruption within prescribed limits of temperature changes.

## EXPANSION BY HEAT

### Coefficient of Expansion in Solids

Expansion due to heat results from the motion of molecules. Heat applied to a material causes the molecules to vibrate faster, thus increasing the distance between the molecules. The more heat applied to the material, the greater is the molecular activity and the greater is the distance between molecules. This increased distance between molecules accounts for the expansion of the material.

In solids, expansion takes place in each direction and is called *linear expansion*. This expansion is measurable and its rate is called the *coefficient of linear expansion*. This coefficient refers to the increase in length (in fractional parts of an inch) for each inch of length of a material for each Fahrenheit degree that its temperature is increased. The coefficient is expressed by a formula as follows:

$$\text{Coefficient of Linear Expansion } (E_c) = \frac{\text{Change in Length } (L_c)}{\text{Original Length } (L_o) \text{ x Change in Temperature } (T_c)}$$

For example, a 10-inch bronze bar measures 10.010 inches when heated 100°F. The coefficient of expansion can be determined by substituting values in the previous formula

$$E_c = \frac{0.010''}{10'' \text{ x } 100°} = 0.000\ 010$$

The coefficient of linear expansion $(E_c)$ of bronze is (0.000 010). The formula for $(E_c)$ may be used any time three of the four values are known. Coefficient tables, similar to the one shown in Table 31-1, are available for most of the commonly used solids.

In actual practice, all dimensions (not only linear dimensions) are affected by expansion and/or contraction. One of the most striking examples of contraction occurs in the foundry. Patterns and molds are made oversize to compensate for the rate of shrinkage for a particular metal. As the molten metal in a casting solidifies, it shrinks to a required size.

### Expansion of Liquids and Gases

Two important points should be emphasized about the expansion of liquids and gases:

- The coefficient of volume expansion of a liquid increases with a rise in temperature.

- Water has the peculiar property of contracting when heated from 0°C to 4°C, the range over which it has the greatest density.

| Material | Coefficient of Expansion |
|---|---|
| Aluminum | 0.000 012 |
| Brass | 0.000 010 |
| Bronze | 0.000 010 |
| Copper | 0.000 009 |
| Cast iron | 0.000 007 |
| Lead | 0.000 016 |
| Steel | 0.000 006 |
| Glass | 0.000 004 |

**Table 31-1**

Gases expand more rapidly than either solids or liquids. The expansion or contraction rate of a gas (as stated in the description of absolute zero), is 1/273rd of its volume at 0°C for each Celsius degree of temperature change. The effect of temperature change on the volume or pressure of a gas is found by applying either Boyle's or Charles' Law.

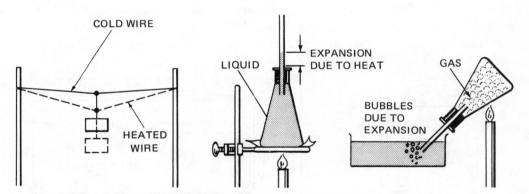

**Fig. 31-1 Heating effect on solids, liquids, and gases**

## TRANSFER OF HEAT

### Heat Transfer by Conduction

Heat energy always flows from an area of high intensity to one of low intensity, or from hot to cold. In this process, the speed of the molecules at the cold end is increased by collision between molecules or by the bombardment of molecules as heat is absorbed. The process continues until the same temperature is reached. At this temperature, all of the parts have the same average molecular activity. This method of transferring heat from molecule to molecule is known as *conduction*.

For example, a heated steel part transfers heat to the water in which it is quenched. That is, the fast-moving molecules on the outer surface of the steel transmit the heat to the slower-moving molecules of the water. As a result, the speed of the water molecules increases. As the steel cools to the temperature of the water, the rate of collision of the molecules decreases.

Although the outer surface of the steel is cool, the center of the steel part is still hot. The rapid motion of the molecules in the interior is transmitted progressively to the outer surface by conduction. The collision of a fast-moving molecule with a slow-moving one causes one to lose speed with a subsequent gain in the speed of the other molecule until a uniform temperature of all molecules is reached.

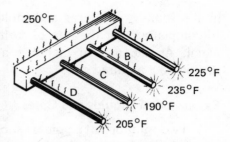

Fig. 31-2 Metals differ in conductivity

While most metals are said to be good conductors of heat, different metals differ in their property of conductivity, as shown in figure 31-2. As compared to metals, most liquids, gases, and nonmetallic solids are partial *insulators*; gases are among the best insulators. This property of gases explains why glass, wool, Rockwool, and asbestos packing are such effective insulators. The air trapped in the small cells of these materials acts as an auxiliary insulator to increase the effectiveness of the materials.

### Heat Transfer by Convection

Convection is another method of transferring heat by the circulation of heated portions of a fluid. As a gas or a liquid is heated from contact with a hot surface, the heat expands the fluid and causes it to become less dense. The cooler and denser fluids surrounding this fluid push the less dense portion of the fluid upward. This method of transmitting heat by the upward currents (*convection currents*) caused by heating is called *convection,* figure 31-3.

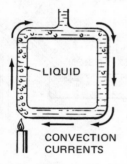

Fig. 31-3 Convection currents

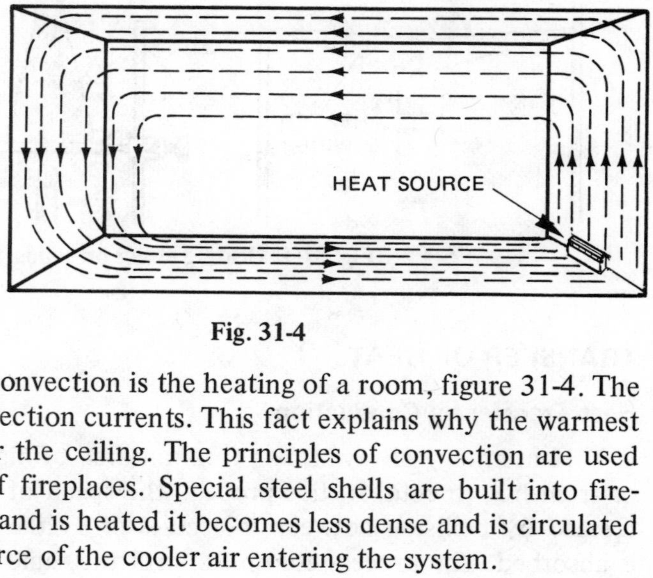

Fig. 31-4

One of the best examples of convection is the heating of a room, figure 31-4. The heat is transferred upward by convection currents. This fact explains why the warmest air in a room is usually found near the ceiling. The principles of convection are used to increase the heating capacity of fireplaces. Special steel shells are built into fireplaces. As cool air enters the shell and is heated it becomes less dense and is circulated into the area to be heated by the force of the cooler air entering the system.

### Heat Transfer by Radiation

A third method of transferring heat is known as *radiation.* This process transfers heat by energy waves. Unlike both conduction and convection (which require solids or fluids to transmit the heat energy), the heat waves caused by radiation move freely through space. The ordinary incandescent lamp is an example of heating by radiation. Within the lamp, all of the air is evacuated and there is no connection between the filament and the outer glass shell. The incandescent heat cannot be transmitted either by conduction or convection; but, it is transmitted by heat waves (radiation). These heat waves or rays are also called *infrared radiation.*

***Reflecting, Transmitting, Absorbing Heat Rays.*** Heat rays can (1) be reflected as they strike a body, (2) penetrate the material, or (3) be absorbed. Heat absorption and reflection by a material depend on certain characteristics of the material. For example, material usually has good absorption properties when it is dark and rough. The surface area and temperature of a material also determine the rate at which a body loses heat. That is, the greater the surface area and temperature, the greater is the cooling rate.

Color and finish also affect the reflection of heat rays. The smoother the surface finish and the brighter and lighter the color of the surface, the better are the conditions for reflecting heat rays.

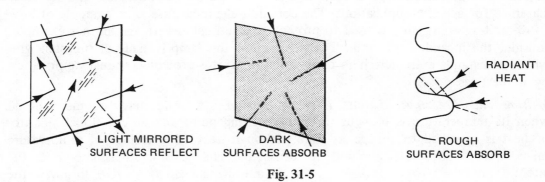

LIGHT MIRRORED
SURFACES REFLECT

DARK
SURFACES ABSORB

RADIANT
HEAT

ROUGH
SURFACES ABSORB

**Fig. 31-5**

In industry, the temperature of surfaces can be measured with a *radiation pyrometer.* This instrument operates on the principle that heat radiation from an object is proportional to the fourth power of its absolute temperature ($0°C + 273°$). Radiant heat energy is absorbed in the radiation pyrometer. This radiation is focused on one junction of a thermocouple which then indicates the temperature.

To summarize, radiant heat has many unusual properties. For example, if radiant heat strikes a mirrored surface, it may be reflected. However, if the radiant heat hits an opaque object that casts a shadow, it may be stopped. Also, as radiant heat passes through materials which transmit light, it does not heat such materials to any extent. In other words, it is only when a material interferes with the free passage of heat waves that it is heated. More and more use is being made in industry of radiant heat as a means of drying and baking fluids on surfaces, such as the exterior finish paint on an automobile.

## Insulation Against Heat Losses

Insulating materials are used to reduce heat losses due to conduction. Such non-metallic substances as ceramic clays and asbestos are good heat insulators for high temperatures. Cork, wood, and glass are better suited as insulators for moderate and low temperatures. Air is the best insulator, if its tendency to circulate is prevented. It is for this reason that good insulators are manufactured with small cells or air pockets. These cells provide dead air space to insure low heat conductivity. Heat losses due to radiation can be reduced by using polished surfaces to reflect the radiant heat. Other uses of insulation to retard heat losses and heat transfer are found in devices such as refrigerators, cold storage rooms, and deep freeze units.

## CHANGE OF PHYSICAL STATE

The change of state from a solid to a liquid or from a liquid to a gas is a heat-absorbing process. The temperature at which a solid changes to a liquid is known as its *melting point*. The temperature at which a liquid changes to a solid is its *freezing point*. As a point of interest, note that the melting and freezing points are one and the same temperature. Pure substances have a definite melting point. Mixtures of solids usually have a melting range. When this range is reached, the solid softens before it becomes a liquid.

### Change from a Solid to a Liquid State

An understanding of melting points is important to the chemist, physicist, and engineer for this knowledge is used to check the purity of substances and to select materials for specific applications. The common electrical fuse is an example of how a low-melting point alloy is used to protect an electrical circuit. As another example, consider the incandescent lamp. The filament of this lamp is made of tungsten wire which withstands extremely high temperatures — if it is protected against vaporization by an inert gas.

*Melting and Freezing: Latent Heat of Fusion.* A solid may become a liquid when its temperature is brought up to its melting point and an additional quantity of heat is absorbed to change its state, although no further change of temperature takes place. The heat required to cause the change of state is known as the latent (concealed) heat of fusion. By definition, the *latent heat of fusion* is the amount of heat required to change one pound of a substance already at its melting point from the solid to the liquid state without changing the temperature.

The fact that the melting and freezing points of a material are the same is explained by the theory that the heat of fusion acts to break down the forces of cohesion between the molecules of the material. These cohesive forces consist of the molecular potential energy which must be overcome. Once these forces are overcome, additional heating increases the kinetic energy of the molecules and the temperature continues to rise.

When the process is reversed and energy is removed, the liquid begins to lose the potential energy stored in the fusion process. The forces of cohesion come into action and solidification takes place at the freezing point. Note that the temperature does not change again until the whole mass of the material reaches the freezing point.

In other words, the latent heat of fusion is a number which indicates the amount of heat required to melt a unit mass of a substance with no change in temperature. The value for the latent heat of fusion can be given either in calories per gram or Btu per pound. The heat of fusion of ice is 80 calories/gram or 144 Btu/lb. Thus, at 0°C, 144 Btu of heat energy are required for each pound of ice if a change of state to a liquid at 0°C is to occur. Other common heats of fusion in calories per gram are: iron, 35; copper, 42; zinc, 28; and silver, 21 (each value is rounded to the nearest whole number).

### Change from a Liquid to a Gaseous State

*Evaporation (Vaporization).* Liquids that are exposed to air evaporate or vaporize. While heat speeds up this change in state, there are four other major factors that determine how fast the process takes place:

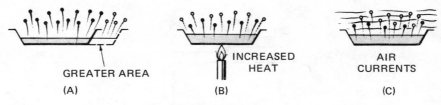

GREATER AREA
(A)

INCREASED HEAT
(B)

AIR CURRENTS
(C)

**Fig. 31-6**

- The greater the area of the liquid surface, the faster the liquid evaporates.
- Any increase in temperature increases the rate of evaporation; heat increases the speed of the molecules and the fast-moving molecules escape.
- The condition of the space surrounding the liquid determines whether the saturation point is reached or if it can absorb more of the evaporated liquid.
- The liquid itself determines the rate of evaporation; that is, water evaporates more slowly than alcohol, gasoline or ether, at the same temperature.

*Saturation Point and Condensation.* When no additional evaporation seems to be taking place and just as many molecules return to the liquid as leave it, then this point is known as the *saturation point.* The saturation point is a condition of equilibrium that is usually associated with a liquid contained in a closed vessel.

The return of molecules from the gaseous state to the liquid state is called *condensation.* The process of condensation is the reverse of the evaporation process.

*Latent Heat of Vaporization.* A considerable amount of heat is required to change the state of a liquid at its boiling point to a vapor without any additional change in temperature. The required heat in this case is called the *latent heat of vaporization.* This value is the amount of heat (in calories or Btu) required to change one gram or one pound, respectively, of a liquid (already at its boiling point) into a vapor at the same temperature.

When the process is reversed and the vapor condenses, the *latent heat of condensation* is numerically equal to the latent heat of evaporation. The latent heat of vaporization of water at its normal boiling point is 540 cal/g or 970 Btu/lb. The heats of vaporization for other common materials are: mercury, 65 cal/g; ethyl alcohol, 204 cal/g; ammonia, 327 cal/g; and sulfur, 362 cal/g. Note that 540 cal/g are required to change water from a liquid state to a gas as contrasted with 80 cal/g required to change water from a solid to a liquid state.

## COOLING EFFECTS OF VAPORIZATION

One of the most practical applications of the cooling effect of vaporization is the refrigerator. Most refrigerators include a gas (a refrigerant that liquefies easily), a cooling chamber, a compression pump, cooling coils, a cooling device, and a storage compartment for the gas. The refrigeration unit also contains two main pressure areas: the low-pressure area in which the cooling occurs and the high-pressure area in which the gas is liquefied.

The schematic drawing in figure 31-7 shows the cooling effects of vaporization and the principles on which many refrigeration units operate. The refrigerant is stored under pressure in a tank (T). As the cycle begins, the compression motor creates a low-pressure area between the expansion jet (A) and the pump (C). This sudden vaporization of a liquid at high pressure to a gas at low pressure produces an extremely cold gas which cools the refrigeration unit (B).

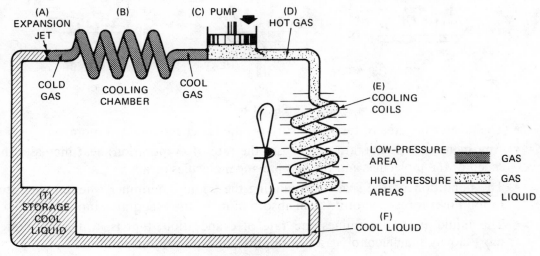

**Fig. 31-7**

The heat surrounding the cooling chamber warms the gas so that when it reaches the pump it is cool gas. When it is pressurized at the pump, it becomes a hot gas (D). As the pressure continues and builds, the gas molecules move faster and the gas gets hotter. As the hot gas at a high pressure moves past the cooling coils (E), it is cooled suddenly and changes to a cool liquid at (F). The cool liquid (still under pressure) then enters the refrigerant tank (T) for the cycle to continue. Thus, the refrigeration process depends on evaporation to produce a cooling effect.

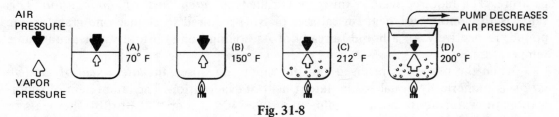

**Fig. 31-8**

*Boiling Point Affects Vapor Pressure.* Water at room temperature has a very low vapor pressure as shown in figure 31-8A. As the temperature of the water increases, the vapor pressure increases, figure 31-8B, until the liquid boils at air pressure, figure 31-8C, and large numbers of molecules are released as a vapor. By reducing the air pressure, figure 31-8D, the vapor pressure is decreased and the water begins to boil at a lower temperature. This action is explained by the fact that the liquid boils when it has absorbed enough heat energy to permit its molecules to overcome the pressure of the air molecules which are preventing the liquid molecules from escaping. The boiling point, therefore, is the temperature at which the vapor pressure of a liquid is equal to the air pressure.

### Change from a Solid to a Gaseous State and from a Gaseous to a Solid State

*Sublimation.* Solids such as dry ice, camphor, and iodine crystals become gases without going through the liquid state. This type of change is called *sublimation.*

To summarize, there are five common changes of state:

- melting, where a solid becomes a liquid;
- freezing, where a liquid becomes a solid;

- boiling, where a liquid becomes a gas;
- condensing, where a gas becomes a liquid; and
- sublimation, where a solid changes directly into a gas or a gas into a solid.

──────────────────────── SUMMARY ────────────────────────

- The coefficient of linear expansion is the fractional amount of expansion of a material for each Fahrenheit degree which its temperature is raised.

- Allowances must be made in the design of all devices to compensate for changes in size resulting from changes in temperature.

- Liquids expand more than solids; gases expand more than either solids or liquids.

- Heat energy is transferred in a solid by conduction.
  - The rapid motion of heated molecules is transmitted to the slower-moving cold molecules. The speed of the molecules is equalized when all reach the same temperature.

- Metals are usually good conductors; liquids, gases, and nonmetallic solids are partial insulators.

- Heat can be transferred in liquids and gases by conduction and convection currents.

- Upward currents are produced as heat expands a fluid and it becomes less dense. The cooler and denser surrounding fluids push the less dense portion upward.

- In radiation, heat is transferred by energy waves which move freely in space.
  - Heat rays may be reflected, they may penetrate a body, or they may be absorbed.
  - Color, surface finish, surface area, and temperature affect the absorption or reflection of heat rays.

- The heat required to change the state of a material from a solid to a liquid without changing the temperature of the material is known as latent heat.
  - The heat of fusion is given in Btu for a definite weight of material.

- Evaporation depends on surface area, temperature, condition of the space surrounding the liquid, and the kind of liquid.

- The saturation point is a condition of balance in a contained vessel where just as many molecules return to the liquid as leave it.

- The latent heat of vaporization refers to the amount of heat in Btu/lb. required to change the state of a liquid at its boiling point to a vapor without undergoing any additional change in temperature.

- The boiling point of a liquid is the temperature at which the vapor pressure of the liquid is equal to the air pressure.

# ASSIGNMENT UNIT 31  EXPANSION, TRANSFER OF HEAT, AND CHANGE OF STATE

## PRACTICAL EXPERIMENTS WITH EXPANSION, CONDUCTION, CONVECTION, AND RADIATION

### Equipment and Materials Required

| Experiment A Expansion | Experiment B Conduction | Experiment C Convection | Experiment D Radiation |
|---|---|---|---|
| Ball and ring<br>Bimetallic bar<br>Bunsen burner | Copper, aluminum, steel, and brass rods (each rod is to be the same in diameter and length, about 18 in.)<br>Three bricks<br>Sealing wax<br>Lead shot<br>Bunsen burner | Square pyrex tube<br>Bits of paper<br>Bunsen burner<br>Funnel<br>Thermometer with a stopper to fit the funnel<br>Ether (1 oz.) | Incandescent lamp<br>Electric heater<br>Radiometer<br>Light-colored, smooth metal plate<br>Dark-colored, rough-surfaced metal plate<br>Two similar surfaces of nonmetallic materials |

## EXPERIMENT A.  HEAT EXPANSION

### Procedure

1. Test the ball and ring when cold to see if one fits into the other.

2. Heat the ring and keep trying the ball until it fits. When the ring and ball are assembled together and are cooled, try to remove the ball.

3. Reheat the ring until the ball can be removed.

4. Heat the end of a bimetallic bar that is supported on one end. Note the direction in which the bar bends.

5. Allow the suspended bimetallic bar to cool. Note the effect that cooling has on the shape of the bar.

6. Pack ice around the supported end of the bar and again note what happens to the shape of the bar.

### Interpreting Results

1. Describe briefly what effect heating has on the expansion and contraction of metals.

2. Do all metals expand at the same rate? Justify the answer in terms of what was observed with the bimetallic bar.

3. Name two common household devices that may be operated by a bimetallic strip.

## EXPERIMENT B.  CONDUCTION

### Procedure

1. Place four metal rods of the same diameter and length on three bricks so they fan out from a center point like a set of spokes.

2. Attach lead shot or pieces of paper on the end of each rod at regular intervals. Sealing wax may be used to attach the pieces.

3. Heat one end of each of the four rods at the same time.

4. Note when the shot drops from each of the rods and watch the heat as it travels along each rod.

### Interpreting Results

1. Is heat conducted at the same rate through different metals? Prove the answer in terms of the observed results.

2. Is heat energy like electrical energy in that it is instantaneous in moving from the hot end to the cold end? Base the answer on the results of this experiment.

## EXPERIMENT C.  CONVECTION AND CONVECTION CURRENTS

### Procedure

1. Fill a square tube with colored water and mount the tube on the ring stand.

2. Add small bits of paper to the water to make it easier to observe the action.

3. Heat the tube at one end. As bubbles appear, note the direction and quantity as they form.

4. Remove the source of heat. Note the decreased quantity of bubbles and the speed of currents.

5. Insert a thermometer in a pyrex funnel with a stopper and mount the assembly on the ring stand.

6. Fill the funnel partly full with water. Note the temperature reading on the thermometer. Adjust the height of the thermometer if necessary.

7. Add an ounce or less of ether to the top of the water. Cap the ether bottle and remove it to a safe place.

8. Light the ether with a taper and check the temperature carefully when all of the ether has burned.

### Interpreting Results

1. Explain the principle of convection currents and convection. Did the results of the experiment prove or disprove this theory?

2. Why does the thermometer reading scarcely rise in spite of the hot flame from the burning ether? Show by a simple sketch how convection currents affected the temperature.

## EXPERIMENT D.  RADIATION AND ITS CONTROL

### Procedure

1.  Mark two distances (X) and (Y) with chalk on a laboratory table. Place a heat source at a third spot.

2.  Place a radiometer at distance (X) from a lighted incandescent lamp. Prepare a table for Recording Results. Indicate the effect of the heat from the lamp on the vanes of the radiometer.

3.  Move the radiometer to distance (Y) and again note the effect on the speed with which the vanes move at this distance.

4.  Insert a smooth polished metal plate between the lamp and the radiometer at distances (X) and (Y). Record the results.

5.  Note the temperature of the shield in each instance.

6.  Replace the smooth plate with a rough, dark metal plate and again note the effect on the radiometer and the temperature of the metal plate.

7.  Repeat steps 4 and 5 with two nonmetallic surfaces like the two metal plates. Record the effect of these plates on the heat waves and the temperature of the surfaces.

### Recording Results

| Condition | | Distance | Radiometer Speed | | | |
|---|---|---|---|---|---|---|
| | | | Slow | Med. | Fast | None |
| Incandescent lamp | | X | | | | |
| | | Y | | | | |
| Lamp and metallic plate | Bright | X | | | | |
| | | Y | | | | |
| | Dull | X | | | | |
| | | Y | | | | |
| Lamp and nonmetallic material | Bright | X | | | | |
| | | Y | | | | |
| | Dull | X | | | | |
| | | Y | | | | |
| Electric heater | | X | | | | |
| | | Y | | | | |
| Heater and metallic plate | Bright | X | | | | |
| | | Y | | | | |
| | Dull | X | | | | |
| | | Y | | | | |
| Heater and nonmetallic material | Bright | X | | | | |
| | | Y | | | | |
| | Dull | X | | | | |
| | | Y | | | | |

**Interpreting Results**

1. Prove or disprove the statement that the greater the distance traveled by waves from a radiant heat source, the less is the heat energy at the greater distances.

2. Based on the observed results with the radiometer, explain what effect each of the following has on the radiation of heat waves:

   a. a light, smooth-surfaced metal plate;

   b. a dark, rough-surfaced metal plate;

   c. a light-colored, smooth, nonmetallic surface; and

   d. a darkened, rough, nonmetallic surface.

3. Name three factors which influence and control the amount of radiant heat energy that may be transferred.

## PRACTICAL PROBLEMS ON EXPANSION, TRANSFER OF HEAT, AND CHANGE OF STATE

### Expansion by Heat

Add a word or phrase to complete statements 1 to 10.

1. Most solids _____ when heated and _____ when cooled.

2. The design of all parts and mechanisms must include _____ on all dimensions to compensate for _____ during operation and _____ on cooling.

3. Heat applied to a material causes the molecules to _____ .

4. The expansion of a material when heated is caused by _____ .

5. The coefficient of linear expansion refers to _____ .

6. _____ dimensions are affected by linear expansion.

7. The volume expansion of a liquid is _____ than for a solid.

8. The expansion rate of gases is _____ of its volume at 0°C for each Celsius degree of temperature change.

9. Gases expand _____ rapidly than either a solid or liquid.

10. Water _____ when heated from 0°C to 4°C.

11. Select four common metals.

    a. List the metals and find the coefficient of expansion of each one in a table.

    b. Which metal expands most for a given temperature change?

    c. Which metal expands the least?

12. Give three industrial applications where knowledge of the coefficient of linear expansion is used in shrink fitting parts together.

13. Name two places where expansion joints must be provided. Show how the amount of expansion may be computed.

14. Four parts (A, B, C, and D) are to be machined to the dimensions indicated. The difference in temperature between the precision instrument at room temperature and the work is also given. Compute the dimension to which each part should be machined so that when the part operates at room temperature, it fits accurately.

| | Part | Required Dimension (Inch) | Temp. Difference Between Instrument and Work | Dimension to Machine (Inch) |
|---|---|---|---|---|
| A | Steel | 10.000 | 100°F | |
| B | Bronze | 10.000 | 100°F | |
| C | Cast iron | 5.000 | 75°F | |
| D | Glass | 0.800 | 20°F | |

## Transfer of Heat:  Conduction, Convection, Radiation, Insulation

For statements 1 to 10, determine which are true (T) and which are false (F).

1. Heat always flows from an area of low intensity to one of high intensity.

2. The greater the heat, the greater is the speed of the molecules.

3. The transfer of heat from molecule to molecule occurs by the process of convection.

4. Fast-moving molecules colliding with slow-moving ones produce no effect on the heat transmitted.

5. The speed of the molecules in three different metals with different coefficients of expansion is the same if they are heated through the same temperature range.

6. Nonmetallic solids and most gases are poorer conductors of heat than metals.

7. Transfer of heat by circulating upward currents in fluids is known as convection.

8. Convection currents are formed when hotter and less dense fluids push the colder and heavier portion upward.

9. Radiation is the phenomenon of transferring heat by energy waves.

10. Energy waves caused by radiation move freely through solids.

Select the correct word or phrase to complete statements 11 to 16.

11. The air trapped in the cells of insulators (does not affect) (increases) (decreases) the efficiency of the insulator.

12. A hot air system of heating depends on (conduction) (convection).

13. A dark rough surface (absorbs) (reflects) heat rays.

14. A light-colored smooth surface (absorbs) (reflects) heat rays.

15. Heat losses may be reduced by using (insulating) (conducting) materials.

16. Radiant heat (heats) (does not heat) materials which transmit light and through which it may pass.

17. Study a cutaway section or drawing of a thermos bottle. Name five specific design features and describe how each feature tends to reduce the heat loss.

18. Secure samples of three different types of insulating materials.
    a. Describe briefly the principle on which each one works.
    b. Give the main advantage and disadvantage of each material.
19. Using a sketch, describe the way in which heat is conducted through a solid.
20. Explain why heat cannot be transmitted through a metallic solid by convection or radiation.

## Change of State

Select the correct word or phrase to complete statements 1 to 8.

1. The melting point is that temperature at which (a solid changes state to a liquid) (a liquid changes to a gas).
2. The amount of heat needed to change the state of a material from a solid to a liquid is known as the latent heat of (fusion) (vaporization).
3. The heat of fusion is (greater than) (less than) (the same as) the heat of vaporization.
4. The process by which molecules change state from a gas to a liquid is known as (condensation) (evaporation) (boiling).
5. (Condensation) (Sublimation) (Vaporization) is the process where a solid changes state by going directly to a gas and the reverse, from a gas to a solid.
6. The temperature at which the vapor pressure of a liquid is equal to the air pressure is its (condensation) (evaporation) (boiling) point.
7. A liquid under high pressure that vaporizes at a low pressure produces a (cooling) (heating) (no effect) on the temperature.
8. A refrigerant gas subjected to high pressure and a high cooling action (becomes a liquid) (remains a gas).
9. Determine the heat required or the heat to be removed to produce the state of change indicated in the table at A through J.

|   | Weight (Lb.) | State | Temperature | Change To | Btus Required |
|---|---|---|---|---|---|
| A | 1 |  |  | Water |  |
| B | 50 | Ice | 32°F | Water |  |
| C | 1 |  |  | Water at 82°F |  |
| D | 50 |  |  |  |  |
| E | 1 | Water | 212°F | Steam |  |
| F | 50 |  |  |  |  |
| G | 1 | Steam | 212°F | Water |  |
| H | 25 |  |  | Water at 162°F |  |
| I | 1 | Water | 32°F | Ice |  |
| J | 35 |  |  | Ice at 2°F |  |

10. Locate a schematic drawing of a domestic or industrial refrigeration unit. Make a simple sketch of the important parts and explain how the process is carried out.

# Unit 32  Heat Engines and Turbines

Heat engines convert the chemical energy of fuel into mechanical energy. One of the earliest types of heat engines was the steam engine. More than any other single factor, the steam engine made it possible to industrialize in many countries of the world and to bring about a high standard of living and production.

Steam engines have been performing work since the eighteenth century. One of the first types of steam engines was developed around 1700 by Thomas Newcomen. This highly inefficient engine depended upon the force of steam in a cylinder to push a piston. To return the piston to its starting position, the steam was condensed by a cold water spray. The combination of the condensation and atmospheric pressure pushed against the piston.

James Watt, a Scottish inventor, perfected the first practical steam engine and is credited with its invention. From 1750 to 1900 the steam engine was the principal source of mechanical power. Since then, the steam engine has been replaced by more efficient steam turbines and internal combustion engines. Five of the major types of engines that are covered in this unit include the popular gasoline and diesel engines, gas turbines, jet engines, and rockets. The mechanical equivalent of heat and the measurement of engine efficiency are also explained in detail.

## PRINCIPLES OF STEAM ENGINES AND STEAM TURBINES

All heat engines operated on one common principle. In each case, a fuel is burned to give additional kinetic energy to particles of a gas. The increase in heat causes the molecules of a gas to move faster. As these molecules strike a piston or turbine blade, part of the energy is absorbed by the blades which then move.

### The Steam Engine

Three things happen to heat energy in the steam engine: (1) energy is transferred to the movable piston, (2) the heat energy lost by the gas results in a gain in heat by the piston or blade, and (3) as energy is lost, the gas temperature drops. Steam engines and turbines depend on the force of expanding steam (when water changes to steam, it expands to about 1700 times its original volume and a tremendous pressure develops).

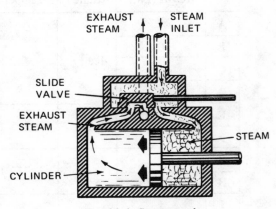

Fig. 32-1  Steam engine

As shown in figure 32-1, a sliding valve permits steam to enter the chamber when the piston is at the start of the cycle. The force of the expanding steam pushes the piston to the left end of the cylinder. Any steam on the left side of the piston is forced out the exhaust port by the action of the piston. At the same time, the slide valve moves to the right and uncovers the left-hand inlet. Steam then enters from the left side of the engine and pushes the piston back to the right to exhaust any steam remaining on the right side of the piston. The cycle continues and produces the reciprocating action of a simple, single-cylinder steam engine.

This back-and-forth movement of the piston is changed to rotating motion by using the force of the piston to drive an eccentric crank. This crank is attached to a shaft. A heavy flywheel mounted on the shaft gives it momentum so that the rotation can continue without any hesitation at the beginning or end of a stroke. Mechanical energy in the form of the rotation of the shaft is then transferred through drive belts or other devices to machines.

Figure 32-1 shows the simplest type of steam engine, the single-cylinder engine. More cylinders can be added and better controlling devices can be used to replace the simple slide valve. These improvements increase the capacity of the engine and provide smoother operation. A *condenser* can be added in a closed system to cool the escaping steam below its condensation point. The condenser saves the boiler water and creates a high vacuum which, in turn, causes a greater difference in pressure between the two sides of the piston. As a result, the work output is increased. Steam engines are rugged in construction and are easy to repair and maintain.

### The Steam Turbine

The steam turbine has replaced the steam engine because of its better operating characteristics and greater efficiency and compactness. The turbine rotor, in which thousands of curved blades are inserted, is moved by directing a flow of steam at an angle against a set of rotating blades, figure 32-2. A similar set of stationary blades is mounted in the crankcase housing. As the energy of the steam exerts a force against the rotor blades, some of the energy is used at each stage until the

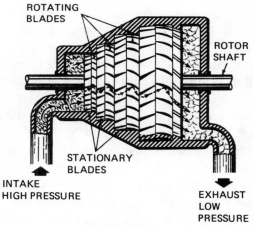

Fig. 32-2

steam loses almost all of its energy. The steam enters the turbine as superheated steam. As the steam loses some of its heat energy and gets cooler, the blades become correspondingly larger to compensate for this condition.

The point at which the superheated steam enters the turbine is known as the high-pressure side. The exhaust side of the turbine is known as the low-pressure end. The efficiency of the turbine is increased by extracting the remaining heat at the exhaust side through the use of heat exchangers in which the remaining steam is condensed. This action produces a vacuum which creates an even greater difference in pressure between the input and output sides of the turbine. Thus, the steam passes through the turbine at a maximum speed and gives off as much of its kinetic energy as possible.

The steam turbine is especially valuable where flexibility of speed control is essential. It has many advantages over the cumbersome gear drives which are used with other types of power drives. The steam turbine is widely used in steam-electric generating plants, aboard large ships to drive generators to produce electrical energy, and where there are heavy power demands.

## PRINCIPLES OF INTERNAL COMBUSTION ENGINES

The term *internal combustion* describes an engine in which the fuel is burned inside a chamber of the engine. Contrast this operation with that of the steam engine where the hot gas is transferred to the engine from a boiler. There are four common types of internal combustion engines: the piston-type gasoline engine used in automobiles, the diesel-type engine used for heavier loads, the gas turbine, and the jet engine.

### The Gasoline Engine

The operation of gasoline engines depends on the principle of the compressibility of gases. When a gas is ignited, it expands greatly to produce extremely high pressures. In a gasoline engine, a mixture of gasoline and air enters the cylinder as a gas. Within the closed cylinder, a piston compresses the gas from a pressure of about 80 lb./sq. in. to 150 lb./sq. in. As the pressure increases, the gas molecules become

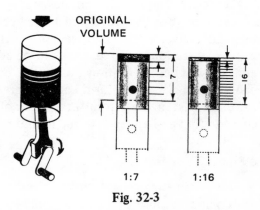

Fig. 32-3

heated until at the top of the stroke (greatest pressure), the temperature of the mixture is almost 600°F. At this point, the mixture is ignited by an electrical discharge from a spark plug. The mixture explodes with tremendous force and the piston is driven downward by the rapid expansion of the gases as they burn. The temperature of the gases ranges momentarily up to 4500°F. The force exerted against the piston by the expanding gases is changed to a torque (rotating force) by a crankshaft which is turned by the piston.

Near the end of this downward power stroke, the exhaust valve opens. Thus, while there is still pressure in the cylinder, the exhaust gases are eliminated through the exhaust manifold, exhaust pipe, and muffler.

In every cylinder cycle, there are four distinct successive steps that produce the power in an internal combustion engine: (1) intake, (2) compression, (3) power, and (4) exhaust. These steps occur repeatedly in the same sequence in each cylinder, figure 32-4. The cycle is regulated by such parts as the intake and exhaust valves and the carburetor. For example, the carburetor mixes and supplies fuel in the correct proportion to make an explosive mixture at the proper moment.

The steps shown in figure 32-4 indicate how the energy of the fuel mixture is converted into heat energy. Since the heat energy decreases toward the end of the power stroke, it is exhausted when it has done the maximum amount of work. The heat energy is converted into mechanical energy by the action of the piston turning a crankshaft. The rotary motion of the crankshaft may be used to drive machinery directly; or, the rotary motion may be changed through gear mechanisms, to produce other types of desired motion.

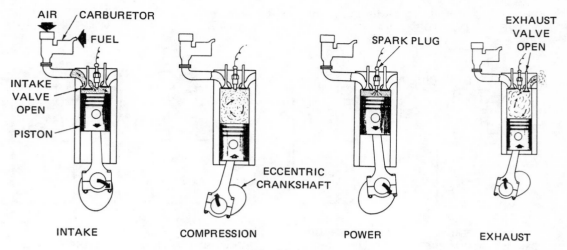

Fig. 32-4

Single-cylinder engines are not practical where large amounts of power are required and smooth operation is desirable. For this reason, it is common practice to design gasoline engines with four, six, eight, twelve, or more cylinders. Such combinations produce an engine having smoother operation because each piston has a smaller fraction of work to do, thus reducing stresses and strains on all of the operating parts.

The gasoline engine used on aircraft operates on the same principle. However, instead of having the cylinders inline (one behind the other), or dividing the cylinders into two sets placed at an angle to each other, airplane engines are usually *radial* in design. This means that the cylinders are spaced apart at a fixed number of degrees (like the spokes of a wheel) around the crankshaft on which the propeller is mounted.

### The Diesel Engine

For a heated substance, the heat energy depends on the number of molecules present, their velocity, and the mass of the substance. The speed of the molecules is controlled by temperature. The basic difference in operation between the diesel and gasoline engines is one of temperature and pressure. The compression ratio of the diesel engine is considerably higher than that of the gasoline engine. The high compression ratio results from the fact that on the intake stroke only air is admitted into the cylinder. The air is then compressed to a high degree during the compression stroke and reaches an extremely high temperature and pressure. At the top of the stroke, a fuel injector sprays a fine mist or fog of diesel oil into the combustion chamber and the fuel ignites immediately. Under these conditions of high temperature and pressure, a spark ignition system is not required. The burning of the combustible mixture causes an even greater increase in the temperature and pressure, producing a tremendous force. The stroke becomes the power stroke and is followed by the exhaust stroke to remove the burned gases. It is interesting to note that without spark plugs an electrical system is not required. Thus, greater efficiency is possible because of the greater range of temperatures through which the engine operates.

Disadvantages of the diesel engine include its weight, starting procedure, and obnoxious fumes. The weight has been reduced slightly by the perfection of a two-stroke cycle engine which supplies power half of the time as compared to the four-stroke cycle engine which supplies power only one-quarter of the time. The two-stroke cycle

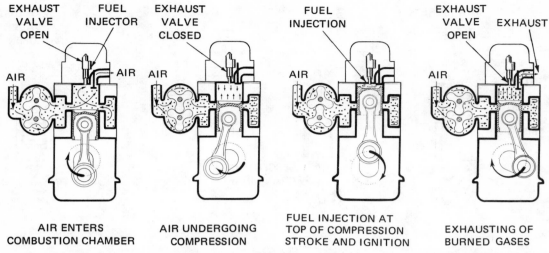

**Fig. 32-5** Intake, compression, ignition, and exhaust phases of a two-stroke cycle diesel engine

combines the intake and exhaust strokes by using the incoming air to scour and flush out the burned gases through an arrangement of inlet and exhaust valves. A special air compressor provides the air since there is no suction intake stroke as such. The air intake is followed by compression, fuel injection, ignition, and the power stroke.

Diesels use low-priced fuels. Thus, diesel engines are widely used where heavy power demands are made, such as for locomotives, ships, power plants, and other types of large equipment.

## Gas Turbines

The gas turbine is an engine. It is designed as a complete unit to convert heat energy into another form of usable energy, such as mechanical or electrical energy, or the high-speed thrust of a jet engine. The basic patents for gas turbines were issued over 160 years ago.

The gas turbine engine has a minimal number of moving parts and limited vibration. This type of engine produces more power for its weight than do conventional piston engines. The gas turbine is highly reliable and has a high operating efficiency. Gas turbines operate on fuel oil or natural gas. They are especially adaptable for industrial power generation service and for surface and air transportation (including supersonic aircraft). Further economy is achieved with gas turbines by reusing the hot exhaust gases for heating and other equipment and processes that require heat energy.

Gas turbine engines basically consist of three major sections: (1) a gas generation section, (2) a power conversion section, and (3) an inlet section. The gas generation section contains a compressor, a combustor, and the turbine. The power conversion section includes various types of exhaust nozzles and one or more additional turbine wheels for certain applications. The inlet section contains ducts and other equipment. These ducts are located in front of the compressor to control the amount and pressure of the incoming air and the engine thrust. Different kinds of gas turbines are produced by changing the design, construction and functions of the inlet and exit (exhaust) components.

*Newton's Laws of Motion.* Gas turbines operate according to the principles of Newton's second and third laws of motion:

> **Second Law of Motion: Force is equal to the mass accelerated.**

$$F = m \cdot a$$

> **Third Law of Motion: For every action there is an equal and opposite reaction.**

For gas turbine jet engines, the *action* is called the *jet exhaust* and the *reaction* is known as the *thrust.*

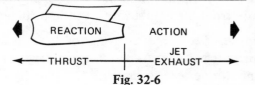

Fig. 32-6

### Operation of the Gas Generator

*Compressor Section.* The three main components of the gas generator are the compressor, combustor (burner), and the turbine. The mass required by Newton's second law of motion consists of the air that is sucked into the gas turbine from the surrounding air of the inlet. Stationary blades (stators) and rotating blades (compressor wheels) are contained within the first part of the generator. The intake air is compressed and forced into the generator by the rotor blades. As the air is pulled in, it is diffused by the stationary blades and is directed into the rotating blades. This airflow produces an increased pressure at each successive stage. As the air is compressed it requires less space and there is an increase in its temperature and velocity. To compensate for these changes, the compressors and the housing are gradually reduced in size. The number of compressors in the turbine gas generator depends on the output requirements.

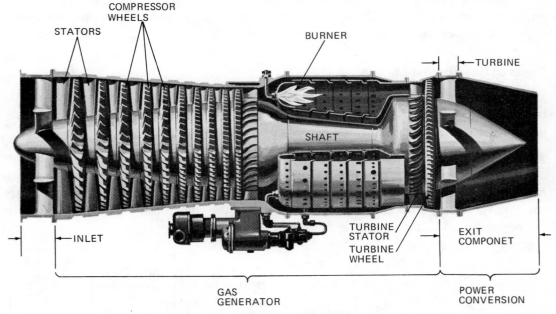

Fig. 32-7 Major components of the gas turbine engine

*Combustor Section.* Although the compressed air is at a high temperature and velocity, these conditions are still not adequate to operate a gas turbine engine. A still higher heat energy is needed. The additional energy is obtained by heating the air in the combustor section. Two conditions exist at this stage: first, the compressed air contains a large amount of oxygen and so supports combustion with a very hot flame; and second, the compressed air moves at a great velocity and tends to blow out the flame or fire (flameout). To prevent flameout, a diffuser is used to slow down the air velocity.

After passing through the diffuser, some of the air is mixed with fuel and burned. The remaining air is directed through the combustor where it expands due to the high heat energy produced by the burning fuel. In this process, the heat energy is transferred to the larger mass of the compressed air. A nozzle guide vane directs the hot gases from the combustor into the turbine section.

*Turbine Section.* The force of the gas at a high velocity against the turbine wheel causes it to turn. As a result, a rotary motion is produced in the drive shaft. The turbine nozzle acts to increase the velocity of the hot gases and thus causes an increase in the energy by which the turbine wheel is turned. The exhaust gases leave the turbine at high speeds. This heat energy can be utilized by designing different exit components to meet varying needs.

## Power Conversion

After the exhaust gases pass through the turbine wheel, the remaining energy is used in two principal ways. In aircraft applications, an exhaust nozzle (which operates on the same principle as the nozzle on a garden-type water hose) increases the exhaust gas velocity. It is this high velocity of the exhaust gases that produces the forward thrust of the aircraft.

The second important method of harnessing the remaining energy of the exhaust gas is by the addition of more turbine wheels. These wheels use more of the heat energy to turn the drive shaft which powers a gear box, propellers (as in the turboprop engine), or other mechanical and electrical equipment.

## Types of Gas Turbine Engines

Changes in the design of both the inlet and exit (exhaust) components of gas turbines have produced a family of engines that are called *turboengines*. Some of the most widely used turboengines are described in the following paragraphs.

*Turbojet Engines.* The turbojet engine combines the principles of the gas turbine and a jet engine. For this type of engine, the greatest amount of heat energy is used to operate the compressor. The remaining heat energy (derived from the fuel) becomes the kinetic energy of a stream of gas molecules shooting from an exhaust opening at the opposite end of the engine.

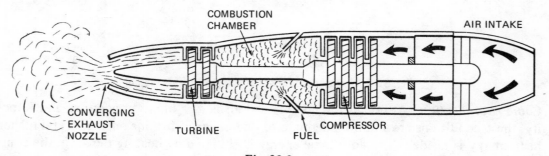

Fig. 32-8

The pressure pushing these molecules backward is the same as the pressure that drives the object forward. This situation is another application of Newton's *Third Law* of Motion which states that for every action there is an equal and opposite reaction.

In the turbojet engine, air is taken in through the *divergent subsonic inlet duct,* as shown in figure 32-9. After passing through the compressor and combustor stages, and through the turbine wheel, added thrust is provided at the exhaust nozzle. The exhaust air is expanded to a high velocity through power conversion by the addition of a *converging exhaust nozzle.*

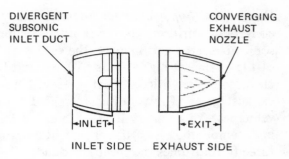

Fig. 32–9 Turbojet Engine

*Turbofan Jet Engine.* The compressor, combustor, and turbine components of the gas generator are also used in the turbofan jet engine, figure 32-10. A large fan wheel and inlet duct are added to the inlet side. The fan wheel increases the engine thrust and reduces fuel consumption. At the exhaust stage, a turbine wheel is added ahead of a converging exhaust nozzle. Additional thrust is created by the expansion of air at high speed through the converging nozzle.

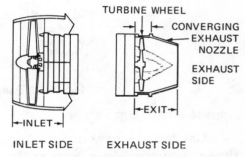

Fig. 32–10 Turbofan Jet Engine

*Turboprop Engine.* The turboprop engine has a reduction gear box added at the inlet side, figure 32-11. The gear train rotates a propeller at an rpm value which insures the maximum thrust. Additional turbine wheels are mounted at the exit (exhaust) stage to rotate the fan, propeller, and the power take-off shaft. Again, the gas generator components process the incoming air to produce the required pressure, velocity, and temperatures.

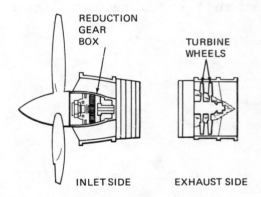

Fig. 32–11 Turboprop Engine

*Turboshaft Engine.* This type of engine is similar to the turboprop engine, with the exception of the propeller. A reduction gear box, figure 32-12, is added to the turboshaft engine to produce rotary shaft motion. This motion may be used to drive an alternator to produce electrical energy, or to provide rotary motion for applications such as machines, trucks, trains, ships, helicopters, and other devices, where a reliable source of mechanical or electrical energy is required.

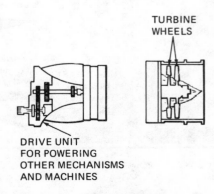

Fig. 32-12 Turboshaft engine

*Ramjet and Supersonic Jet Engines.* The ramjet and the supersonic jet engine have a converging diffuser-type inlet duct, figure 32-13, which is designed to slow supersonic air speeds and increase the pressure. The air is rammed into the intake or throat of the engine at high speeds. A large exit burning chamber containing a diverging nozzle provides additional heat energy. This energy increases the velocity of the hot exhaust gases. The resulting increased thrust allows the aircraft to attain supersonic velocities. An auxiliary device is needed to start the cycle by which air is forced under high temperature into the engine. Ramjet and supersonic jet engines can attain speeds three to four times the speed of sound.

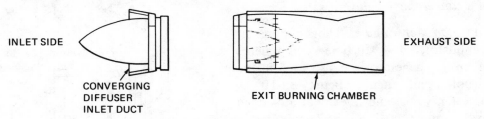

Fig. 32-13  Ramjet and supersonic jet engine

## Applications of Gas Turbines

Gas turbines have numerous applications in propulsion and power generation as shown in figure 32-14. Note in the figure the range of performance of gas turbine engines in terms of speed, power capacity, and operation at altitudes ranging from sea level to rarified atmospheric levels.

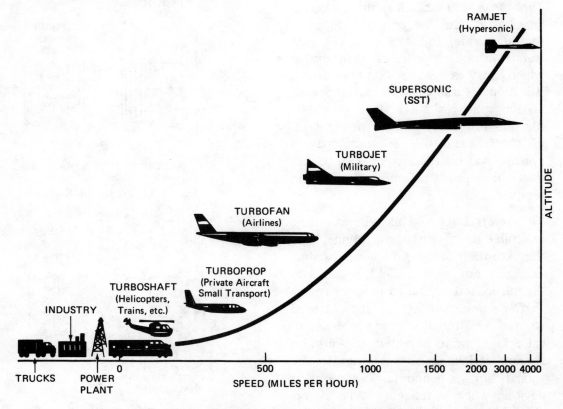

Fig. 32-14  Examples of gas turbine applications

## Rocket Engines

Rocket engines also operate according to Newton's Third Law (action and reaction). The rocket engine differs from the jet engine in that rockets carry their own supply of fuel and oxygen and are independent of air intake. In fact, the higher the rocket ascends from the earth, the faster is its speed. At altitudes where the air is rarified, there is less air resistance and the ambient temperature is much lower. Thus, the exhaust gases of the rocket produce more thrust.

The rocket engine operates by changing the potential chemical energy of the fuel into the kinetic energy of motion. The thrust of a rocket is produced by the rapidly expanding gases of the combustion process. These gases flow through the exhaust nozzle to produce the opposite reaction needed to push the rocket forward.

Rocket engines are classified into two main groups: those using liquid propellant and those using solid propellant. Liquid propellant rocket engines have a greater impulse and range and are used for long-range applications, such as intercontinental ballistic missiles. Solid propellant rockets are used for short-range missiles and aircraft boosters.

*Liquid Propellant Rocket Engines.* This type of rocket engine includes a propellant fuel system, a control system, and a rocket thrust chamber. The propellant consists of a fuel and an oxidizer which are mixed in the correct proportions in the injector. The mixture is fed to the rocket thrust (combustion) chamber where it is converted into extremely hot burning gases. The gases then flow through a divergent exhaust nozzle to produce an accelerated thrust.

The speed of the burning gases at the nozzle throat is designated by a number which is the ratio of the velocity of an object to the velocity of sound at that point. This ratio is known as the Mach number. A speed of Mach 1 is approximately 1130 ft./sec. Thus, the burning gases at the nozzle throat have a speed in excess of 4000 ft./sec. which is equivalent to Mach 3.5. At the diverging section, the expanding gas velocity increases to a speed of 8000 ft./sec.

*Solid Propellant Engines.* The design and construction of solid propellant rocket engines are simpler than those of liquid propellant rocket engines. Rocket engines using solid propellants require less servicing and have a high reliability factor. The fuel for this type of engine consists of solid propellant units (grains) molded into special shapes to produce the required burning characteristics. The rate and duration of burning, and the operating pressure and thrust are determined by the construction, size, shape, and the exposed burning surface of the grains. The propellant burning rate is stated as the number of inches of propellant burned per second.

The thrust and burning time ranges have been established for solid propellant rocket engines. The rockets are grouped in the restricted burning class when the thrust is delivered over a longer time span; rockets are in the unrestricted burning class when a greater thrust is produced for a shorter period of time. For example, propellants classified for restricted burning typically produce a thrust range of from 100 to 10 000 pounds over a time span of from 4 to 120 seconds. The higher thrust of unrestricted propellant grains is in the range of 5000 to more than 100 000 pounds over a 0.05- to 10-second time range.

## MECHANICAL EQUIVALENT OF HEAT; ENGINE EFFICIENCY

The input to a steam engine is heat energy. The output of a steam engine is expressed in foot-pounds of work. The efficiency of such an engine is known as its *brake thermal efficiency.* This efficiency is the ratio of the potential work which a given amount of energy should produce to the actual output. However, before such comparisons can be made, it is necessary to know the relationship between mechanical work and heat energy.

In the mid-nineteenth century an English physicist, James P. Joule, measured the amount of work required to raise the temperature of one pound of water one degree Fahrenheit. In his experiments, Joule measured the change in temperature of a known quantity of water to obtain the work output. He then obtained the work input by measuring the distance through which known weights moved to produce the work output. Joule's work and that of other experimenters using more controlled conditions and precise measuring instruments, formed the basis for establishing that the mechanical equivalent of heat is 778 foot-pounds of work per Btu, or 42 700 gram-centimeters of work per calorie.

It is now possible to compute the operating efficiency of heat engines by comparing the input in terms of Btu and the work output in foot-pounds.

### Brake Thermal Efficiency of Heat Engines

The *brake thermal efficiency* of heat engines is another way of saying that the efficiency of the engine is equal to the brake horsepower x 33 000 foot-pounds of work per minute divided by the product of the quantity of the fuel used in each minute x the heat value of the fuel x 778 foot-pounds of work. This statement can be expressed as a formula as follows:

$$\text{Brake Thermal Efficiency (BTE)} = \frac{\text{Brake Horsepower} \times 33\,000}{\substack{\text{Quantity of Fuel} \times \text{Heat Value} \times 778 \\ \text{(lb./min.)} \qquad \text{(Btu/lb.)} \quad \text{(ft.-lb.)}}}$$

As an example of the use of this formula, assume that a steam engine develops 26 h.p. using coal that yields 13 000 Btu per pound. The engine uses 60 pounds of coal per hour. These values can be substituted in the formula to yield:

$$\text{BTE} = \frac{26 \times 33\,000}{(1 \text{ lb./min.} \times 13\,000 \text{ Btu/lb.}) \times 778 \text{ ft.-lb.}} = 8.5\%$$

In other words, the steam engine in the example has an operating efficiency of 8.5% — this value indicates how inefficient the steam engine really is. If all of the heat energy of one Btu can be converted without loss to mechanical energy, then this energy could raise 778 pounds a distance of one foot. In actual practice, the efficiency of steam and internal combustion engines ranges from 6% to 38% with averages of 23% for steam engines, steam turbines, and condensing engines; 28% for gas engines; 32% for shaft-driving gas turbines; and 35% for diesel engines. As improvements are made in engine design, the efficiency of these engines increases. However, in each instance the number of Btu lost is greater than the number of Btu converted to do useful work. Figure 32-15 compares the five major types of engines.

ENGINE EFFICIENCY, PERCENT

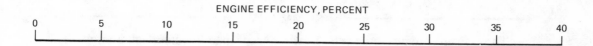

SIMPLE STEAM ENGINE – 6 TO 8%

STEAM TURBINES, CONDENSING ENGINES – 16 TO 33%

GASOLINE ENGINES – 25 TO 32%

DIESEL ENGINES – 32 TO 38%

GAS TURBINE ENGINES – 32 TO 36%

| Major Types of Engines | Btu of Useful Work | Btu Lost |
|---|---|---|
| Steam engine | 1110 to 1480 | 17 020 to 16 390 |
| Steam engine and condensing steam engine | 2960 to 5550 | 15 540 to 12 950 |
| Gasoline engine | 4060 to 5200 | 14 440 to 13 300 |
| Diesel engine | 5920 to 7040 | 12 580 to 11 460 |

**Fig. 32-15**

────────────────── **SUMMARY** ──────────────────

- The steam engine converts heat energy into mechanical energy. The energy of expanding heated gas molecules is transferred in a chamber to a number of pistons, causing them to move. The loss in the heat energy of the gas results in heat gained by the moving parts.

- Condensers used on heat engines save boiler water and create a high vacuum. The vacuum, in turn, increases the pressure differential to increase the work output.

- Steam turbines have a greater operating efficiency than steam engines.
  - Mechanical energy is produced by the action of heat energy in the form of superheated steam against many stages of rotor blades.

- Internal combustion engines operate on the principle of the compressibility and expansion of gases.

- Every cycle of an internal combustion engine includes (1) fuel intake, (2) compression, (3) ignition and power, and (4) exhaust.

- Diesel engines operate at a higher temperature and pressure than gas engines. When fuel is injected into a cylinder containing air under high compression, the fuel ignites without an electrical charge and produces mechanical energy.

- Gas turbine operation depends on two of Newton's Laws of Motion: (1) action and reaction and (2) force equals the mass accelerated. Gas turbine engines consist of three major sections: the gas generator, the inlet, and the power conversion area (exhaust).
  - The gas generator contains three main components: the compressor, the combustor, and the turbine.
  - The exhaust section of the gas turbine may contain one or more turbine wheels and a specially shaped exhaust nozzle.

- Turbojet, turbofan jet, turboprop, turboshaft, ramjet, supersonic jet, and other types of gas turbine engines use the same gas generator system.
  - Each type of gas turbine engine requires different designs for the inlet and exhaust sections and components.

- The turbojet engine does the same work as the jet engine. However, added thrust is gained by accelerating the hot exhaust gases through a nozzle. The stream of hot gases at high speed produces the propulsive force.

- The ramjet and supersonic jet engines depend on ramming air into a constricted area. The temperature and pressure of the air are increased tremendously as it flows through the gas generator section.
  - Additional thrust is developed to produce supersonic velocity with the use of an extended exit burner chamber and a diverging nozzle.

- Rocket engines are air-independent. Thrust is produced by changing the chemical energy of a propellant into rapidly expanding exhaust gases.
  - The engine using liquid propellant produces hot exhaust gases by burning a properly proportioned liquid fuel and oxidizer.

— The solid propellant rocket engine depends on the construction and burning characteristics of the propellant grain.

• The design of the engine and the propellant grain characteristics determine whether restricted or unrestricted grain burning is required, depending on the thrust and distance ranges required.

• The mechanical equivalent of work is 778 foot-pounds per Btu or 42 700 gram-centimeters per calorie.

• Brake thermal efficiency = $\dfrac{\text{Brake horsepower x 33 000}}{\text{Amount of fuel x Heat value x 778}}$.

## ASSIGNMENT UNIT 32  HEAT ENGINES AND TURBINES

### PRACTICAL STUDIES OF STEAM ENGINES, TURBINES, AND INTERNAL COMBUSTION ENGINES

**Equipment and Materials Required**

| Experiment A<br>Steam Engines | Experiment B<br>Steam Turbines | Experiment C<br>Gasoline or<br>Diesel Engines | Experiment D<br>Jet Engines |
|---|---|---|---|
| Model of steam engine with cutaway section of steam chest<br><br>Or<br><br>A visual aid showing the sectional view of the internal construction of the steam engine | Homemade steam turbine<br><br>Or<br><br>Sample turbine blade<br><br>Or<br><br>Sectional view of an enlarged drawing showing the construction features of a steam turbine | Cutaway model of the engine head, valve, piston, and crankshaft assembly<br><br>Or<br><br>Enlarged sectional drawing showing the principal parts of the engine | Disassembled jet engine drawing showing all parts<br><br>Or<br><br>Disassembled engine |

### EXPERIMENT A.  STEAM ENGINES

**Procedure**

1. Study the construction features of the steam chest and slide valve mechanism on either a model steam engine or a cutaway sectional drawing of a steam engine.

2. Make two simple sectional sketches of the steam engine. One sketch is to show the power stroke and the other sketch is to show the return stroke. Label the important parts of the engine on each sketch.

**Interpreting Results**

1. State one fundamental law of science which controls the operation of the steam engine.

2. List the steps showing how the heat energy of steam is converted to mechanical energy in the steam engine.

## EXPERIMENT B.  STEAM TURBINES

**Procedure**

1. Set up the steam turbine apparatus so that the steam escaping from an opening in the top of a metal container hits the angularly shaped blades cut into a circular disc (multivaned rotor).

2. Heat the partially filled container with water and vary the amount of steam produced by adding or removing heat.  Note the action on the rotor.

3. Study the shape and size of a sample rotor blade and a cutaway sectional drawing showing the internal construction features of a steam turbine.

4. Make a simple sketch showing the stationary and revolving blades at the different stages of the steam turbine engine.

5. Trace the path and action of the superheated steam from the high-pressure side of the low-pressure exhaust side of the engine.

**Interpreting Results**

1. Identify two basic principles of science governing the operation of the steam turbine.

2. Explain why the blades of the steam turbine vary in size and shape at each stage.

## EXPERIMENT C.  GASOLINE ENGINES

**Procedure**

1. Examine the condition of each cylinder of a cutaway model of a gasoline engine. Note the position of the piston, valves, connecting rod, and crankshaft.

2. Prepare a table similar to the one shown for Recording Results.  Use as many vertical columns as there are cylinders in the model.

3. Make simple sketches in the appropriate places in the table to show the positions of the piston, valves, connecting rod, and crankshaft.

4. In each case, indicate what part of the cycle is represented, i.e., intake, compression, power, or exhaust stroke.

5. Turn the crankshaft one-half turn and again note the same conditions and positions of the parts listed.  Record these positions in the table and identify what part of the stroke is represented.

6. Repeat steps 3 and 4 at one, one and a half, and two revolutions.

## Recording Results

| | Cylinders | | | | | | | |
|---|---|---|---|---|---|---|---|---|
| | 1 | 2 | 3 | 4 | 5 | 6 | 7 | 8 |
| Start of Cycle | | | | | | | | |
| 1/2 Revolution | | | | | | | | |
| 1 Revolution | | | | | | | | |
| 1 1/2 Revolutions | | | | | | | | |
| 2 Revolutions | | | | | | | | |

## Interpreting Results

1. Identify the scientific principle on which the operation of the gasoline engine depends.

2. Describe briefly what takes place within a cylinder during the four parts of a cycle in a gasoline engine.

## EXPERIMENT D. TURBOJET ENGINES

### Procedure

1. Examine a cutaway section of either an actual turbojet engine that may be available for display or an assembly drawing showing the major parts and mechanisms.

2. Make a simple sketch of the engine for the purpose of identifying the important parts.

### Interpreting Results

1. List two scientific principles on which the turbojet engine operates.

2. Trace the action that takes place from the air intake to the exhaust of hot gases. Explain the function of each step.

3. Explain how added thrust is obtained by utilizing the exhaust gases.

## PRACTICAL PROBLEMS WITH HEAT ENGINES AND TURBINES

### Steam Engines and Turbines

Select the correct word or phrase to complete statements 1 to 6.

1. The simplest and oldest industrial heat engine is the (gas turbine) (steam engine) (steam turbine).

2. The steam engine converts (mechanical) (heat) energy to (mechanical) (heat) energy.

3. In the Newcomen engine, the piston was returned to its starting position by (the momentum of a flywheel) (condensing the steam at the end of the stroke with a cold water spray).

4. Steam engines depend on the force of (expanding steam) (decreased temperature).

5. The (reciprocating) (rotary) motion of a piston on a steam engine is converted to a (reciprocating) (rotary) motion of a flywheel.

6. A smoother operating steam engine is produced when the number of cylinders is (limited to one) (increased).

For statements 7 to 12, determine which are true (T) and which are false (F).

7. Steam turbines are preferred to steam engines because of their greater operating efficiency, better characteristics, and compactness.

8. The heat energy of superheated steam is greater than that of the exhaust steam in a turbine.

9. The efficiency of a steam turbine is not affected by heat exchangers.

10. Condensing the heat at the output side of a turbine creates a pressure differential which steps up the operating efficiency of the turbine.

11. The steam turbine and steam engine are two examples of internal combustion engines.

12. The size and shape of the turbine blades is the same throughout all of the stages from the high-pressure side to the low-pressure side.

13. Name three changes that take place in the heat energy in a steam engine.

14. List two main disadvantages of steam engines.

15. Describe briefly the difference between an internal combustion engine and an external combustion engine. Give two examples of each type of engine.

16. State two differences between a water-driven turbine and a steam turbine. Name the laws of science which govern the operation of the water-driven turbine.

**Internal Combustion Engines**

1. The _____ engine depends on the laws governing the compressibility of gases for its operation.

2. The greater the compression in an engine, the _____ is the temperature of the compressed fuel.

3. The energy of a fuel mixture under pressure is converted from _____ energy to _____ energy.

4. Heat energy _____ toward the end of a power stroke.

5. That part of a stroke where fuel and air are introduced into the cylinder is called the _____ .

6. The _____ stroke begins when the compressed fuel mixture is ignited.

7. The _____ the number of cylinders, the smoother is the engine performance.

8. The two basic differences between the principles of operation of a diesel engine and a gasoline engine are _____ and _____ .

9. Diesels have a higher compression ratio because only _____ is admitted into the cylinder on the intake stroke.

10. _____ injected into a cylinder of a diesel during its compression stroke _____ from the intense heat of compression.

11. The operating efficiency of a diesel is _____ than that of a gasoline engine.

12. The fuel costs for a diesel as compared to those for a gasoline engine are _____ .

Select the correct word or phrase to complete statements 13 to 24.

13. The gas turbine converts (heat) (mechanical) (electrical) energy to (heat) (mechanical) (electrical) energy.

14. The (steam) (gas) turbine requires an auxiliary power supply to start the cycle.

15. Rotary motion is produced in the gas turbine by (compressing air and igniting it with a fuel spray under pressure) (compressing an oil spray and igniting it).

16. The heat energy of a burning fuel mixture in a gas turbine acting on the (turbine) (compressor) blades is converted into mechanical energy.

17. Newton's Third Law of Motion, when applied to a turbojet engine, means that (the pressure exerted by the kinetic energy of a stream of gas molecules escaping from the exhaust is the same as the pressure that drives the object forward) (the air pressure in the combustion chamber is needed to burn the fuel).

18. The (air) (gasoline) is highly compressed in the compressor of the turbojet engine.

19. The thrust for a turbojet engine in flight is produced by (the turbine) (the remaining heat energy of the gas).

20. The ramjet engine operates by combining (an auxiliary device to start the cycle and the ramming of the air into the engine intake at high speed) (a turbine with another heat engine).

21. The ramjet depends on the kinetic energy of hot gases escaping under great pressure to develop thrust for propulsion (during flight) (at low speeds on takeoff).

22. The ramjet is adaptable to flight in (the atmosphere) (the atmosphere and a vacuum) (a vacuum).

23. The rocket engine is adaptable to flight in (the atmosphere) (the atmosphere and a vacuum) (a vacuum).

24. One major difference between a rocket engine and either a turbojet or ramjet engine is that the rocket engine (depends on an atmospheric air supply) (is air independent) (converts an external air supply).

25. List four applications or types of heat engines that depend for their operation on Newton's Third Law of Motion.

26. Describe briefly how any two types of heat engines operate.

27. Cite an example of a situation in which a turbojet engine may be used and one in which a propeller type of airplane engine may be used.

28. Explain why the efficiency of the diesel engine is greater than that of the gasoline engine.

29. Explain briefly how the forward thrust of a jet engine is obtained.

30. Make a simple schematic drawing of a ramjet engine or a gas turbine engine.

   a. Name the important parts.

   b. List two scientific principles on which the engine operates.

   c. Describe one complete cycle of operation of the engine.

31. Explain briefly how a gas turbine is started. What is the function of the exhaust gases?

## Mechanical Efficiency of Heat; Engine Efficiency

Use the formula for brake thermal efficiency to solve problems 1 and 2:

$$\text{BTE} = \frac{\text{Brake horsepower x 33 000}}{\text{Amount of fuel x Heat value x 778}}$$

1. A steam turbine and condensing engine develops 195 brake horsepower. The engine uses 150 pounds of coal per hour. The Btu rating of the coal is 12 000 Btu per pound. Determine the operating efficiency of the engine.

2. A diesel engine burns a volume of fuel equivalent to 15 pounds per minute. The fuel has a value of 2200 Btu per pound. The brake horsepower of the engine is 260. Determine the engine efficiency.

3. Match each item in Column I with the correct condition in Column II.

| Column I | Column II |
|---|---|
| a. Brake thermal efficiency | 1. The useful work (in Btu) is less than the Btu lost. |
| b. Mechanical equivalent of heat | |
| c. Heat engines | 2. The ratio of actual work output to the potential work output of an engine. |
| | 3. The useful work (in Btu) is greater than the Btu lost. |
| | 4. 778 foot-pounds of work per Btu of heat energy. |

# Achievement Review
## HEAT ENERGY AND HEAT MACHINES

### HEAT ENERGY: ITS EFFECTS AND ITS MEASUREMENT

For statements 1 to 9, determine which are true (T) and which are false (F).

1. Heat energy is produced whenever the form of chemical, electrical, mechanical, or nuclear energy is changed.

2. Any increase in the molecular movement within a liquid, solid, or gas results in heat.

3. Heat energy does not depend on the speed, weight, or number of molecules in motion at any given time.

4. The metal thermometer, clinical thermometer, and pyrometer are used in the same way under the same conditions.

5. Thermoelectric current is generated by the difference in electron vapor pressure of two dissimilar metals when heated.

6. The thermocouple-pyrometer combination is a portable device for measuring high temperatures by color matching.

7. Heat energy does not depend on temperature, composition of material, and amount of fuel involved.

8. The calorimeter measures the quantity of heat exchanged between two bodies of unequal temperature.

9. Iron-constantan, platinum-rhodium, and chromel-alumel alloys are used in thermocouples for temperature around 600°F.

10. (a) Give the boiling point of water at standard atmospheric pressure on the Celsius, Fahrenheit, and Kelvin temperature scales.

    (b) What is the freezing point of water on these same scales?

    (c) List the three formulas that may be used to convert a temperature reading on one scale to its corresponding value on either of the two other scales.

11. (a) What is the basic unit of heat measurement in the metric system?

    (b) Show by example how this unit is used in heat measurement.

12. (a) Explain briefly what is meant by a Btu.

    (b) Give the equivalent value of work for one Btu.

13. Refer to a Metals Handbook to select one material each having a specific heat within 0.005 of the values for A, B, C, and D given in the table.

| | Specific Heat | Material |
|---|---|---|
| A | 0.122 | |
| B | 0.097 | |
| C | 0.208 | |
| D | 0.192 | |

14. Find the heat required to raise the temperature as shown in the table of materials A, B, C, and D, according to the given data.

| Material | Weight | Temperature Change | Specific Heat | Heat Required |
|----------|--------|--------------------|---------------|---------------|
| A | 100 lb. | 32°F to 532°F | 0.12 | |
| B | 1 ton | 0°C to 100°C | 0.45 | |
| C | 250 g | 0°K to 100°K | 0.11 | |
| D | 275 lb. | 57°K to 307°K | 0.072 | |

## EXPANSION, TRANSFER OF HEAT, AND CHANGE OF STATE

Determine the description in Column II which corresponds with the correct term given in Column I.

| Column I | Column II |
|----------|-----------|

**Column I**

1. Linear expansion
2. Coefficient of linear expansion
3. Flow of heat energy
4. Convection
5. Latent heat of vaporization
6. Latent heat of fusion

**Column II**

a. A transfer from molecule to molecule from a high intensity to a low intensity until the same temperature is reached.

b. The Btu per pound needed to change the state of a boiling liquid to a vapor without any change in temperature.

c. The heat required to change one pound of a material at its melting point from the solid state to the liquid state without increasing its temperature.

d. The temperature at which the vapor pressure and the air pressure are equal.

e. The fractional amount a material expands for each F° its temperature is increased.

f. Heat rays may be reflected, absorbed, or may penetrate a body.

g. The expansion which takes place in all directions in a solid.

h. A heat transfer method for gases and liquids in which less densely heated portions cause a movement to a more densely heated portion.

Supply a word, phrase, or value to complete statements 7 to 14.

7. Vaporization refers to the _____ .

8. Rate of evaporation depends on (a) _____ ,
(b) _____ , and (c) _____ .

9. The _____ is that point at which the number of molecules leaving a liquid is equal to the number of molecules returning to it.

10. The latent heat of vaporization is expressed in terms of _____ in the English system and _____ in the metric system.

11. _____ is the process of changing state from a gas to a solid or from a solid to a gas.

12. The return of molecules from the gaseous to the liquid state is known as _____ _____ .

13. The gas turbine has three major sections: (a) _____ ,
    (b) _____ , and (c) _____ .

14. The turboprop engine develops its thrust from the _____
    and the _____ .

15. State the heat energy functions of the stationary blades (stators) and the rotating blades (compressor wheels) in the compressor section of a gas generator for a turbine engine.

16. Indicate the functions of (a) the turbine stator and (b) the turbine wheels in the turbine section of the gas generator.

17. Cite two devices which use the cooling effects of vaporization.

18. Compute the heat energy needed to produce the changes indicated in the table for materials A, B, C, and D according to the data given.

|   | Material | Latent Heat of Fusion | Specific Heat | Weight | State | Change To | Heat Energy Required |
|---|----------|-----------------------|---------------|--------|-------|-----------|----------------------|
| A | Steel | 35 cal/g | 0.12 | 100 g | Solid at 1468°C | Liquid at 1668°C | |
| B | Copper | 42 cal/g | 0.10 | 20 kg | Liquid at 2408°F | Solid at 302°F | |
| C | Water | 80 cal/g | 1.00 | 100 g | Liquid at 77°F | Solid at –23°F | |
| D | Water | 144 Btu/lb. | 1.00 | 2 tons | Solid at –10°C | Liquid at 25°C | |

## HEAT ENGINES AND GAS TURBINE ENGINES

Select the correct word, phrase, or value to complete statements 1 to 10.

1. Steam turbines depend on (the reciprocating motion of a slide valve) (burning compressed gases in a cylinder) (the force created by expanding steam).

2. The steam turbine has (greater) (less) (the same) flexibility of control and efficiency as the steam engine.

3. The vacuum created to extract the remaining heat at the output side of a steam turbine (has no effect) (decreases) (increases) the energy output of the engine.

4. (Internal combustion engines) (Steam engines) (Steam turbines) operate due to the compression, explosion, and expansion of a fuel mixture.

5. Diesel engines operate at (the same) (lower) (higher) temperatures and pressures than gasoline engines.

6. The operating parts of a gas turbine include: (a carburetor, compression-combustion chamber, and a drive mechanism) (an intake, steam chest, and exhaust) (a turbine rotor, combustion chamber, compressor, and air intake).

7. The kinetic energy of the stream of exhaust gases in a turbojet engine (has no effect) (is a disadvantage to) (adds to) the thrust produced by the engine.

8. The thrust for propulsion in a ramjet engine is produced by (injecting fuel and burning it under tremendous air pressure) (creating and burning an explosive mixture in a four-stroke cycle) (passing live steam through a series of turbine blades).

9. The (brake thermal efficiency) (horsepower) of a heat engine equals

$$\frac{\text{Brake Horsepower x 33 000}}{(\text{Qty. Fuel x Heat Value) x 778}}$$

10. The average efficiency of steam engines is (18%) (23%) (35%) (75%) as compared with a (5%) (35%) (60%) (80%) efficiency for diesel engines.

11. Determine the missing values for engines A, B, and C from the data given in the table.

| Engine | Brake Horsepower | Heat Value (Btu/lb.) | Brake Thermal Efficiency | Quantity Fuel (/min.) |
|--------|------------------|----------------------|--------------------------|------------------------|
| A | 200 | 2200 | | 100 lb. |
| B | | 1000 | 30% | 1 kg |
| C | 520 | | 10% | 2000 g |

12. For each of the five functions listed, name the major components which carry on these functions in gas turbine jet engines.

   (a) air intake

   (b) compression

   (c) combustion

   (d) power

   (e) thrust

13. Describe (a) the different inlet and power sections and (b) their functions for (1) the turbofan jet engine and (2) the turboshaft engine.

# Section 6
# Light Energy

## Unit 33  Light: Sources, Application, Control, and Measurement

Light is one of the most important forms of energy. The light energy of the sun is responsible for plant growth. During this growth process, the potential energy of food is stored. The radiation from the sun also supplies heat energy to warm the earth and sustain human life. In another form, light may cause motion because light exerts a very small push when it strikes an object. When light strikes chemicals, such as those on a photographic plate, a change takes place. When light strikes some metals, an electric current is produced because the light imparts enough energy to cause the free electrons to move out of their orbits. Electrical energy is produced by the action of light on so-called solar cells and batteries. Most importantly, it is light which affects the eye structure to cause sight.

### SOURCES OF LIGHT

There are five principal sources of light: the sun, fire or combustion, incandescence, electrical discharge, and phosphorescence. One of the earliest forms of artificial light was a fire. Later, oil lamps, candles, and gas light were used to produce light by combustion. Improved lighting resulted when the incandescent electric light bulb was invented. In this type of bulb, a metal (tungsten) filament is electrically heated to incandescence to produce light. In the gas-filled lamp, an electric current causes the gas to give off light. Modern fluorescent lamps provide improved light sources. Substances such as radium and uranium also give off light and are known as radioactive substances.

Light is of value to the extent that it may be controlled and measured through an understanding of its nature.

### THE NATURE OF LIGHT

Three conditions are necessary before an object that is not giving off its own light can be seen: (1) the object must have a source of light in its environment; (2) the light from the source must strike the object; and (3) the light must be reflected from the object to the eye. Thus, in a room that is totally dark and free of dust particles, no light will be seen. However, if there is a light source and if the room contains particles of chalk dust, some of the light will bounce off the chalk dust, hit the object, and will be reflected from the object to the eye.

Light travels in straight lines. When light hits an object, the light may be reflected from the object, absorbed by the object, or it may pass through the object. If the object is smooth (a mirror, for example), the light is reflected in a regular pattern and is known as *regular reflection,* figure 33-1. When light hits a rough surface, the surface irregularities cause the light to spread out in an irregular pattern in all directions. The light is said to be diffused and the pattern is called *diffuse reflection.*

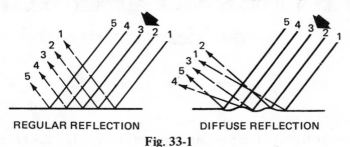

REGULAR REFLECTION          DIFFUSE REFLECTION

Fig. 33-1

## Translucent and Opaque Materials

If a beam of light passes through an object, then the material of which the object is made is said to be *translucent* or *transparent.* If a piece of paper is held in the path of a light beam, the light passes through the paper and spreads out. In this case, the paper is translucent. Glass and clear cellophane are examples of transparent materials. Most of the light beam passes through such materials and the beam retains its original shape. When an object does not permit a light beam to pass through it, the object is said to be *opaque.* In all of the instances given, the light travels in a straight line. If the object is white, most of the light is reflected. If the object is black, the light is absorbed and converted into heat energy.

The fact that light moves in straight lines accounts for shadows. If the light source is small and intense, a sharp and dark shadow is produced. It is for this reason that lamp bulbs are frosted. The light from the frosted bulbs is diffused and there is less contrast to create eyestrain.

The speed of light was first determined in 1676. A Danish scientist, Olaf Roemer, computed the speed of light to be 192 000 miles per second. In 1902, Albert Michelson, an American scientist, measured the speed of light to be 186 285 miles per second. This great speed accounts for the fact that people see things at what appears to be the instant they happen.

## The Wave Motion Theory of Light

Scientists have proposed a number of theories about light. One of the earliest theories was that a light wave is similar to a water wave in which a continuous series of crests and troughs is produced. As these waves strike a smooth surface they rebound and retain their original shape. This theory states that light travels in waves and the waves have different frequencies.

The waves of light can neither be seen nor touched as waves. But, whenever waves are used to describe light, the words wavelength and frequency are used, figure 33-2. The distance from one point on a wave to a corresponding point on another wave is its wavelength. The lengths of the waves that produce light vary from 0.000 032 4 in. (0.000 81 millimeter) to 0.000 014 4 in. (0.000 36 millimeter). *Frequency* refers to number of waves that move past a point in a given time.

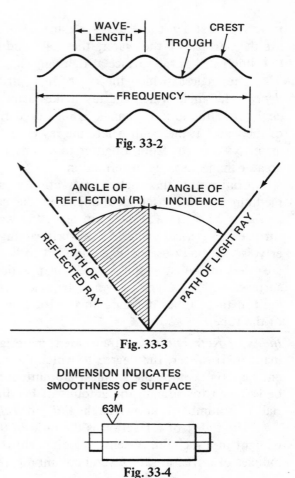

Fig. 33-2

Fig. 33-3

Fig. 33-4

### Reflection of Light

*The Law of Reflection.* When a ray of light shines on a smooth surface and is reflected, the light bounces off the surface at the original angle, figure 33-3. The angle at which the ray hits the object is the *angle of incidence.* Thus, in this case, the angle of incidence is equal to the *angle of reflection.*

The *Law of Reflection* states that the angle of incidence of a reflected ray of light is equal to its angle of reflection.

*Diffuse Reflection.* As light is reflected from a rough surface the rays are scattered or diffused. This effect is due to the different angles at which the rays are reflected. The wavelength of a reflected wave of light is about 0.000 02 in. (or 1/50 000th in.).

If a machined surface has irregularities which are spaced farther apart than the dimension of the wavelength, the light is diffused. This fact is used to determine the surface finish of a machined part by visually comparing the part being machined with another standard of known surface irregularity. Drawings of machined parts include dimensions which indicate the degree of smoothness required in microinches, figure 33-4.

### Refraction of Light

Whenever light passes from air or from one transparent material to another, the light is bent according to fixed laws. The Law of Reflection no longer applies under these conditions. The *Law of Refraction* applies when a transparent material bends or changes the direction of light rays.

For example, consider the rays of light in figure 33-5. The rays hit the water at five different places and at varying angles. For each of the rays, the light

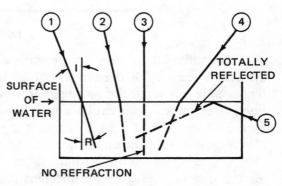

Fig. 33-5

travels in a straight line until the surface of the water is reached. Once the light rays hit the surface of the water, they all bend and take a different direction except ray (3) which strikes the surface at right angles.

The angle at which the ray of light strikes the surface is called the *angle of incidence*. The angle which the ray makes after it is bent by passing into the water is the *angle of refraction*. This angle is measured from a perpendicular drawn to the surface of the water at the point where the ray enters the water. For a ray of light passing from air into water (or from a lighter to a denser medium), the angle of incidence is always greater than the angle of refraction.

Conversely, the angle of refraction is greater when light rays pass from a denser medium into a lighter medium, as in the case where light passes from water to air. In figure 33-5, ray (5) does not leave the water because the angle of incidence on the surface is a *critical angle*. At this critical angle, the ray is totally reflected. This property is especially useful when applied to light tubes or fiber optics. These light tubes are used in industry and in medical practice to bring light to machine parts and to surfaces which otherwise would be inaccessible to a light source. The bending of the light in these tubes depends on the total reflection of light rays bouncing off the walls of the tube.

*Index of Refraction.*  All transparent materials bend or refract rays of light in the same manner. However, the degree to which the refracted rays are inclined differs for different materials. Since the smallest amount of bending occurs in a vacuum, a number can be assigned to indicate the amount of bending that takes place in each different material. This number is known as the *index of refraction.*

The index of refraction is the ratio of the speed of light in a medium to the speed of light in a vacuum. A denser medium slows light more than a less dense medium. The indexes of refraction for several common materials are shown in table 33-1.

| Material or Condition | Index of Refraction |
|---|---|
| Vacuum | 1.000 0 |
| Air | 1.000 3 |
| Water | 1.33 |
| Quartz (fused) | 1.46 |
| Glass | 1.53 |
| Diamond | 2.47 |

**Table 33-1**

The value for air may vary depending on the temperature of air (since temperature affects the density of the air).  Cold air is denser than hot air; therefore, cold air has a higher index of refraction. As light travels from a cold layer of air to a hot layer, the rays are refracted. This explains why objects above a hot road or stove appear to shimmer. This bending of light as it passes through layers of air at varying temperatures is also responsible for mirages which appear over heated surfaces.

Figure 33-6 shows what takes place when a ray of light passes from air through materials having different indexes of refraction. These sketches illustrate the *Law of Refraction* which states that a ray of light bends closer to a perpendicular

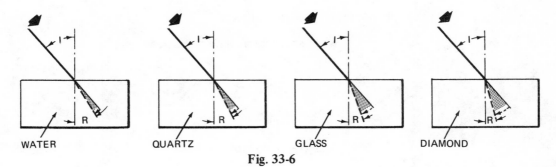

WATER  QUARTZ  GLASS  DIAMOND

Fig. 33-6

to the surface for a transparent material with a higher index of refraction than it does for a material with a lower index of refraction. In other words, angle (R) in figure 33-6 is less for the diamond than it is for the water, assuming that the angle of incidence (I) is the same in each case.

### Diffraction of Light

Energy can be transmitted from one place to another by particles or by waves. One characteristic difference between the two methods is described by the phenomenon of diffraction. Since light consists of waves, *diffraction* refers to the ability of light waves to bend around the edges of obstacles in their path, figure 33-7.

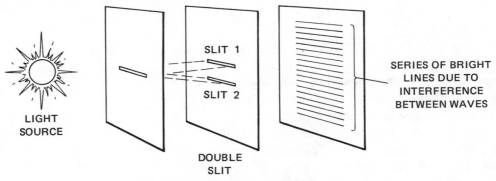

LIGHT SOURCE

SLIT 1

SLIT 2

DOUBLE SLIT

SERIES OF BRIGHT LINES DUE TO INTERFERENCE BETWEEN WAVES

Fig. 33-7 Light transmitted by waves

When light waves pass through a small opening or pass the edge of an obstacle, they are bent into a region not directly exposed to the light source. The image thus formed by the diffracted light has bright and dark fringes surrounding it. If the object is considerably larger than the wavelength of the light that illuminates it, diffraction is not important because a clear image is formed. However, for an object whose dimensions are smaller than the wavelength of the light illuminating it, there is a great deal of diffraction. Thus, the object cannot be seen clearly because the image is distorted or it may not be seen at all.

The wavefronts (which are perpendicular to the light rays) of a beam of unobstructed light produce secondary wavelets that interfere in such a way as to produce new wavelets that are exactly like the original ones. Whenever a part of a wavefront is obstructed, the shadow region is reached by only a part of the secondary wavelets from the initial wavefront.

***Destructive and Constructive Interference.*** Light and dark fringe patterns are the result of interference between secondary wavelets. *Destructive interference* of the wavelets,

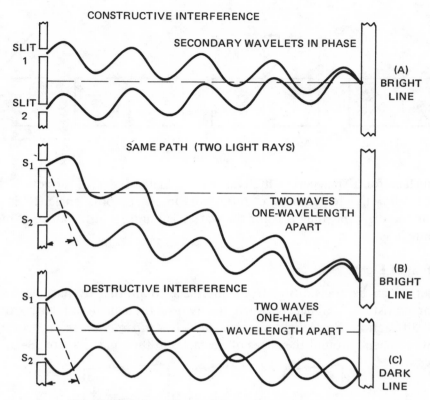

Fig. 33-8 Constructive and destructive interference

figure 33-8C, produces darkness and indicates that there is no wave activity. Bright areas are the result of constructive interference. *Constructive interference* takes place when the paths of two rays are the same, (as in figure 33-8A), or when they differ by any number of whole wavelengths such as twice the wavelength or three times the wavelength (as in figure 33-8B).

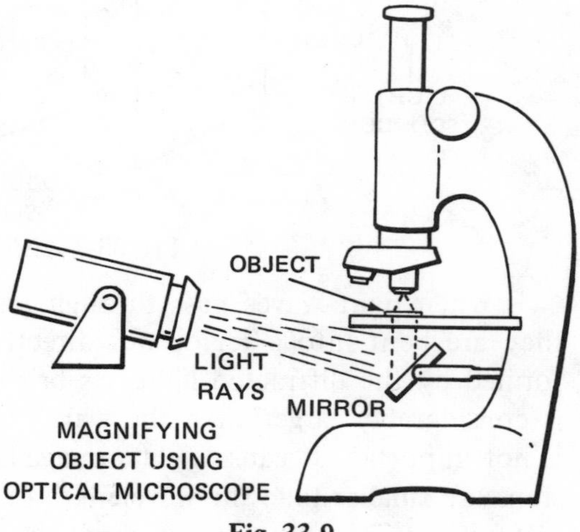

Diffraction must be considered in the design and use of optical and other measuring instruments due to possible magnification limits. Such instruments are designed to have a particular *resolving power*. This value expresses the capability of the device to produce well-defined separate images. Resolving power is the useful magnification range of a microscope using visible light and as limited by diffraction. Beyond this range, further magnification would result in magnifying the diffracted fringes around the edges of minute objects.

The 200-inch diameter concave mirror of the world's largest telescope on Mt. Palomar in California illustrates this point. Despite the extreme limits of accuracy to which the mirror was finished, it is still limited to the measurement of features on the moon's surface that are slightly more than 150 feet apart.

Fig. 33-9

## APPLICATION OF REFRACTION PRINCIPLES TO LENSES

A ray of light traveling through a perfectly flat plate of glass emerges from the glass at the same angle as it entered the glass, but it is displaced a short distance. A similar ray passing through a triangular glass prism is bent so it returns toward a baseline, figure 33-10. In both instances, each path can be predicted by the law of refraction.

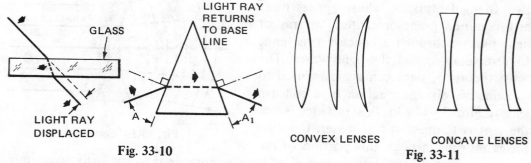

Fig. 33-10

Fig. 33-11

Glass plates which are used to help control light so that it can produce useful work are called *lenses*. The flat plate and the triangular prism illustrated in figure 33-10 are two types of lenses. Other common lens types include convex or converging lenses and concave or diverging lenses, figure 33-11. A simple way of distinguishing between these two types of lenses is to think of *convex* as being thicker at the center and *concave* as being thinner in the center. There is no limit to the size and degree to which a lens can be curved. Most optical, projection, and other instruments, which depend on lenses for their operation, use a combination of lenses.

### Types and Functions of Lenses

*Convex Lenses.* The cross section of a convex lens shows it to consist (approximately) of two triangular prisms with a common base, figure 33-12. As light strikes one edge of the lens, it is bent as it passes through the lens and is refracted toward the base. The rays at the thinnest part of the lens move faster and are refracted more than the portion of the light wave passing through the center part of the lens because the light rays are retarded more in the thicker area. The rays converge toward a single *focal point*, (F). The shape and quality of the lens determines whether all of the light rays are refracted through this focal point.

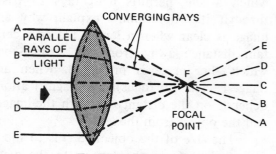

Fig. 33-12 Convex lens

The converging of the light rays through a convex lens can also be explained by the theory of light waves. As shown in figure 33-13, the light waves hit the center (C), or the thickest portion of the lens, first.

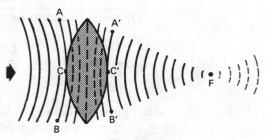

Fig. 33-13

Since the light rays are slowed down as they pass through the glass, the rays at the thinner outer edges move through the lens and travel from points (A) and (B) to points (A') and (B') during the same time interval that the light passes through the center (C to C'). The light ray thus curves back on itself and converges at the focal point (F).

*Concave Lenses.* A concave lens can also be viewed as consisting of (approximately) two triangular prisms whose apexes meet at a common point or center. Any ray of light passing through a concave lens tends to diverge or spread the light waves. The wave theory of light can again be used to explain this diverging action of a concave lens, figure 33-14. In this instance, since the lens is thinnest at the center, the light waves moving through this portion of the

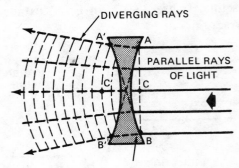

Fig. 33-14  Concave lens

lens are retarded for a smaller period of time as compared to the light waves moving through the thicker portions of the lens. Thus, the light waves emerging from the lens move outward in ever-increasing waves and the light spreads apart or diverges. These simple principles of convergence and divergence are the basis of all optical instrument and lens design.

### Determining Magnification of Images

The appearance of an object as seen by the light passing through a lens is called an *image*. There are limits to the way in which a lens permits light rays to pass through it. These limits explain why an image is clear when a lens is placed a certain distance away from the surface on which the image is projected. When an image is clear (in focus), the magnification depends on the distance of both the image and the object from the lens.

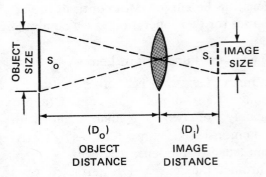

Fig. 33-15

The size of the object ($S_o$) is to the size of the image ($S_i$) as the distance of the object ($D_o$) from the lens is to the distance of the image ($D_i$), figure 33-15. This statement can be expressed as a formula as follows:

$$\frac{\text{Distance Object to Lens }(D_o)}{\text{Distance Image to Lens }(D_i)} = \frac{\text{Size Object }(S_o)}{\text{Size Image }(S_i)}$$

$$\frac{D_o}{D_i} = \frac{S_o}{S_i}$$

For this reason, an object six feet tall and eight feet from a camera lens appears only 1 1/2 inches high when photographed on a sensitized film which is two inches from the lens.

$$\frac{D_o}{D_i} = \frac{S_o}{S_i}$$

$$\frac{8'}{2''} = \frac{6'}{X}$$

$$\frac{96''}{2''} = \frac{72''}{X}$$

$$144'' = 96X$$

$$1\ 1/2'' = X$$

### Calculating Image Distance

The curvature of a lens determines its focal length. Recall that the *focal length* is the distance range through which an object appears with its sharp and distinguishing features in focus. This focusing distance can be determined by using the following expression:

$$\frac{1}{D_o} + \frac{1}{D_i} = \frac{1}{\text{Focal length}}$$

The image and object distances are correct for a lens of given focal length if the dimensions fit correctly in this formula.

The importance of lenses cannot be overstressed. The physical, biological, chemical, and other sciences depend on the magnifying power of lenses to enlarge the structure of materials and living things to a point where each may be studied. Diseases have been brought under control through the use of the microscope. Cameras, telescopes, and other optical measuring instruments and devices depend on the kind and placement of lenses for the projection of images and their control. In each application, one or more lenses are combined with the properties of light to produce useful work.

## ILLUMINATION

### Measuring Illumination

Light-producing power is measured in *candlepower*. The unit of candlepower was adopted many years ago when the candle was the principle source of light. Originally, the standard candlepower measured just what the term indicates: the light-producing power of a single candle. One candlepower is the unit illumination on a surface one foot from the light power of one candle. Although this standard is satisfactory for crude measurements, it is not dependable for measurements that must be consistently accurate.

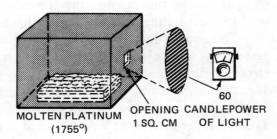

MOLTEN PLATINUM (1755°)  OPENING 1 SQ. CM  CANDLEPOWER OF LIGHT

**Fig. 33-16**

The modern standard of measuring illumination is based on the light-producing power of a fixed quantity of platinum heated to its melting point at 1755°C. If this metal is heated in a closed container and the light thus produced is allowed to escape through a hole one centimeter square, the resulting illumination is equal to 60 candlepower, figure 33-16. While the platinum remains at its melting temperature, the light-producing power remains constant and is a dependable standard.

## Conditions Which Affect Illumination

By observing light sources, it can be seen that the closer an object is to the source, the greater is the amount of illumination on the object. Within limits, as the light source is moved away from the object, the area over which the light falls is greater, but the light becomes less intense on that area. The amount of light is also affected by the surface behind the light source and the surface of the object. As stated previously, a white smooth surface reflects light and a rough dark surface tends to absorb light. These two conditions help determine the amount of illumination given off by a light-producing source.

## Distance and Illumination

Experiments have shown that the intensity of illumination from a light source varies inversely as the square of the distance from the source. In other words, the intensity of illumination decreases as the distance from a light source increases. The intensity may be computed by squaring the distance and using the product as a denominator of a fraction whose numerator is one:

$$\text{Intensity} = \frac{1}{d^2}$$

This mathematical process is a restatement of the *Law of Inverse Squares*. Thus, the intensity of the light two feet from a light source is one-fourth of its value at one foot. At three feet, the intensity is one-ninth or $(1/3^2)$ its value at one foot. If the light at one foot has a value of 60 candlepower, the intensity at two feet is 15 candlepower (60 x 1/4), and at three feet the intensity is 6 2/3 candlepower.

## Measuring Intensity by Instruments

The candlepower can be computed when the distance and intensity are known. At times, this method is impractical as, for example, in photography. The intensity of light can be measured in the laboratory using a bunsen photometer. This gauge is used to compare the brightness of the light source to a light having a known rating.

The most common instrument for measuring illumination is the familiar light-meter. The *lightmeter* works on the principle that light produces minute amounts of electrical current. The greater the intensity of light, the greater is this electrical current. The current, in turn, operates a dial which indicates on a calibrated scale how many foot-candles of light are present. This reading is quick to obtain and is fairly accurate. Therefore, the lightmeter is widely used in industry and for many domestic purposes.

## Area of Illumination

Another interesting fact about illumination is that while the intensity varies inversely as the square of the distance from the light source, the area of illumination increases as the distance from the light source increases. The area increases in proportion to the original area and the square of the distance from the original area. If the light shining through a one-inch square

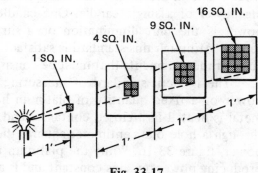

Fig. 33-17

opening in a plate one foot from a light source is measured on a second plate two feet from the original light source, the new area of illumination is a square two inches on a side or four square inches (4 sq. in.).

If this area is three feet from the light source, then an area of illumination of one square inch on the first plate is increased to nine square inches at the third plate. In each case, the area of illumination is proportional to the square of the distance from the light source.

The distance from the light source and the area of illumination are important conditions as they form the basis for measuring light and applying it to do useful work.

### Candlepower Output for Industrial Needs

The amount of illumination required in industry depends largely on the nature of the work to be performed. The greater the precision needed in machining or assembly operations, the higher is the range of illumination needed. A comparison of such ranges is given in table 33-2.

| Kind of Work | Foot-candles (Illumination Range) |
|---|---|
| Rough (Foundry, Construction) | 10–20 |
| Bench Work (Filing, Rough Assembly, Standard Operations) | 15–50 |
| Fine Assembly and Fine Operations | 40–65 |
| Precision Machining and Instrument Work | over 60 |

Table 33-2

As progress is made in the design and manufacture of light sources, both the candlepower per watt and the efficiency values increase. The open gas flame, one of the early sources of light, was only 0.04% efficient in producing 0.02 candlepower per watt. The efficiency of the carbon filament lamp reached 0.40%.

With each new lamp development, such as the tungsten incandescent lamp, the sodium arc lamp, and the fluorescent lamp, the efficiency of operation increased to 1.80%, 10%, and 12% to 16%, respectively. Thus, it is now possible to obtain the type, quantity, and conditions of illumination required for exacting industrial and scientific research conditions.

Better illumination is also a key to safety. Industrial, commercial, and residential hazards are decreased as a result of the reduction of light glare and the availability of the light needed to perform machine, bench, and other operations safely.

### Polarization:  The Control of Light

A light wave consists of oscillating electrical and magnetic forces that are perpendicular to each other and to the direction of propagation. Thus, a light wave is made up of oscillating fields. Through the process of *polarization,* the oscillations of a wave motion may be confined to a definite pattern. Polarization is a characteristic of transverse waves; longitudinal waves are not affected.

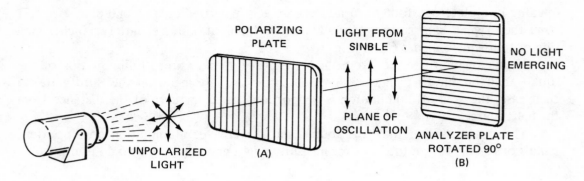

**Transverse Nature of Light Waves**

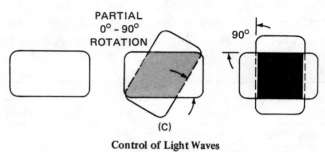

**Control of Light Waves**
**Fig. 33-18**

In the process of polarization, only those waves whose vibrations are parallel to the vertical plane of a polarizing plate can pass through the plate, figure 33-18. All other vibrations are blocked. If a second screen (an analyzing plate) is placed so that its vertical pattern is rotated 90 degrees from the position of the polarizing plate (the plane of the analyzer is perpendicular to the incoming waves), the light waves are stopped and no light emerges. Thus, polarization is used in this instance to control the light waves and their intensity.

Polarization plates are made from a number of substances whose crystalline structures have different indexes of refraction for light with different planes of polarization. Plates made from materials such as quartz, calcite, and tourmaline transmit light in a single plane of polarization only.

Figure 33-18A shows that as unpolarized light passes through a polarizer (a plate of tourmaline or some other crystal), the light beam emerging from the plate is *plane polarized* and has a reduced intensity. The remainder of the waves are absorbed and reflected within the plate. The second plate at (B) in the figure serves as an analyzer. The intensity and amount of light passing through the second plate is controlled by rotating the plate. The intensity of the light is gradually reduced from 0° to 90° where relatively no light is transmitted, figure 33-18C.

***Applications of Polarization Principles.*** The term *Polaroid*® refers to a polarizing material that is produced artificially. This material has the property of being able to transmit light within a single plane of polarization. The development of polarized sheets has advanced the practical applications of polarization. Polarized sheets consist of a thin layer of iodosulfate crystals (aligned by a strong electric field) sandwiched between two plastic sheets. The light intensity can be controlled by using two polarized sheets.

Polaroid® glass is used in vehicle windshields to reduce glare – from the sun, headlights, and other light sources. Polaroid® lenses used in welding shields and safety glasses provide eye protection in manufacturing situations where intense heat and emerging light rays are produced. The stress patterns in plastic models of tools, machines, and processes can be analyzed using polarized light. For example, the interference fringes that are created by the polarized light show stress concentrations and other operating characteristics.

Polaroid® filters are used in many common applications. In photography, for example, these filters are used to adjust the light intensity and control the glare from reflected light.

The applications just listed are only a small sampling of the vast number of practical uses of polarization phenomena which are being made or will be made by the consumer, and by business and industry.

## LASERS: CHARACTERISTICS, PRINCIPLES, AND APPLICATIONS

Two of the new scientific terms that have emerged in the second half of this century are maser and laser. The *maser* was developed in 1954 as a device for producing coherent waves in the microwave region. Each letter of maser designates part of its function: **M**icrowave **A**mplification **S**timulated **E**mission of **R**adiation. Instead of light waves, the operation of the maser is based on electromagnetic radiation having very short wavelengths (radio waves or microwaves).

One of the most important applications of the maser is in the measurement of time intervals. An extremely accurate frequency standard was established. The error in time measurement is presently estimated as one second in 1000 years!

When the principles of the maser were extended into the areas of light and optics in 1960, the resulting optical maser was called a *laser*. Laser is the shortened version of **L**ight **A**mplification by **S**timulated **E**mission of **R**adiation. The light beam produced by a laser device has a number of remarkable and very valuable properties as follows:

- for all practical purposes, the light is monochromatic.

- the light is coherent; that is, the waveforms are all in phase with each other. When interference patterns are needed, they can be produced by using two separate laser beams or by placing two slits in a laser beam.

- the light beam is highly *collimated.* This means that the angular divergence is extremely limited. An important application of this property was made during the Apollo 11 moonflight. A mirror was placed on the moon to reflect a laser beam sent from earth. The collimation of the laser beam was so small that the beam was detected when it returned to earth after traveling more than half a million miles.

- the laser beam is the most intense light source. The magnitude of the energy density of some laser beams is equal to the energy density of a hot object at a temperature of $10^{30}$ K.

- the laser beam can be continuous (CW) or it can be a pulsing beam. These different types of beams serve different functions. CW lasers have lower power outputs and better optical properties than pulsed lasers. However, pulsed lasers generate extremely high power outputs over short intervals (in the order of milliseconds).

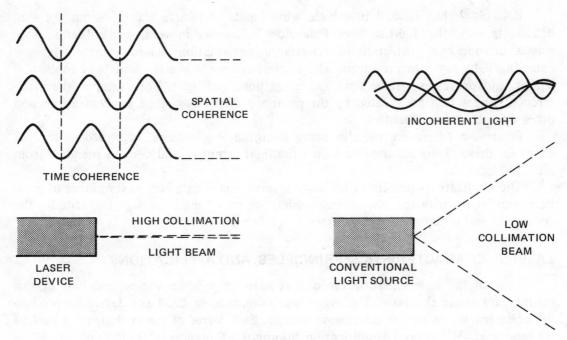

**Fig. 33-19 Properties of a laser beam**

## Classification and Principles of Lasers

To simplify the description of how a laser works, a comparison should be made with the principles of ordinary light emission. A photon of light energy is emitted when an electron falls from a higher to a lower energy level within an atom. This action takes place when the atom is excited by an external source of energy, such as heat or electricity. This excitation causes electrons to be raised to a higher energy level from which they fall back randomly to a lower level. This random electron motion produces an *incoherent light emission.*

If a material is to have suitable energy levels for light emission, it must have a metastable state. Electrons in a *metastable state* are trapped at an energy level so that they do not immediately and spontaneously drop to lower energy levels. The importance of the metastable state lies in the fact that an atom can exist at a metastable energy level longer, as shown in figure 33-20B, before radiating than it can if it is at

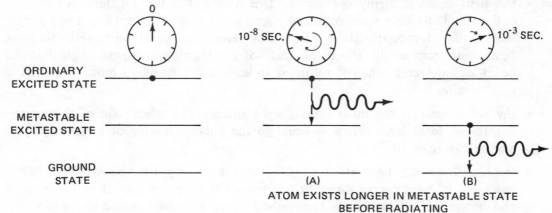

**Fig. 33-20**

an ordinary energy level, figure 33-20A. Electrons can remain in a metastable state for an interval ranging from a few microseconds to several milliseconds. The time interval must be long enough for a stimulated emission to occur.

Electrons in a laser material are raised to the metastable energy levels by a so-called pumping system that supplies the required additional energy. When the number of electrons trapped in the metastable energy level exceeds those in the lower energy level, a *population inversion* exists. Laser action is possible under these conditions. The pumping systems used may be an optical device, accelerating electrons from an electron gun, or chemical reactions.

Within the laser material, the excitation of the electrons inside a region known as a *resonant cavity* produces monochromatic light. The resonant cavity has two parallel mirrors: one fully silvered mirror is mounted at one end and the other mirror, only half silvered, is mounted at the other end. Some of the light produced will pass through the walls of the resonant cavity, but other light waves will strike the mirrors, figure 33-21. The distance between the mirrors is adjusted to set up standing waves. The resonant cavity also contains atoms that can be raised to the energy level of the cavity radiation. The cavity radiation stimulates the emission of light waves from the cavity atoms. A coherent light beam results because the energies of both the stimulating and the emitted photons are the same and the two are in phase.

Laser light is produced from solid, liquid, and semiconductor devices. The structure and functions of three basic types of laser devices are treated in this unit.

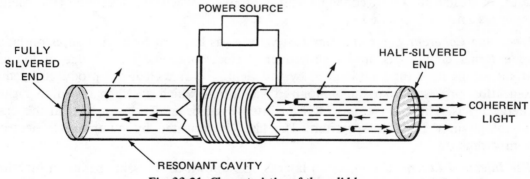

Fig. 33-21 **Characteristics of the solid laser**

***Ruby Lasers.*** The rubies used in lasers are synthetic gem stones produced by fusing aluminum and chromium oxides to form large crystals. This common form of laser (using a solid material) was the device which first successfully generated coherent light.

The most common ruby laser produces a pulsed beam; however, this device can be made to generate a continuous light beam. A very high energy pulse of light of several milliseconds duration is delivered by a pulsed ruby laser. This electrical energy is built up in capacitor banks and is released as a pulse of energy.

A ruby laser system includes an energy pumping source (power supply), an optical coupling, the laser material (ruby), and a cooling system. These components are shown in a simple schematic form in figure 33-22, page 390. The two ends of the ruby laser are flat, parallel, and silvered. The end that is slightly silvered permits the resulting light beam to move out of the laser.

A flash tube containing xenon or other gas is wound around the ruby end of the laser. As a pulse of electricity is passed through the flash tube, the gas is ionized and

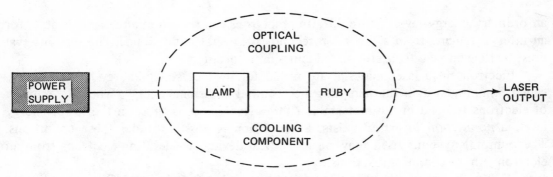

**Fig. 33-22 Components of a typical ruby laser system**

glows briefly. The chromium atoms in the ruby crystal absorb the light from the xenon gas. As a result, the electrons of these atoms are raised to a higher energy level. As the electrons fall back to their *ground state* (the state at which the atom has the least possible energy), they emit photons. The wavelengths of these photons are in the red region of the spectrum. This red light forms the cavity radiation which is reflected between the ends of the laser. Other chromium atoms are stimulated by this cavity radiation to emit photons of the same wavelength. As the photons become more numerous and as they trigger still more chromium ions into emitting radiation, the photon stream continues to build up. Finally, the intensity of the photon stream is such that a pulse of monochromatic, spatially coherent light bursts from the partially silvered end of the laser. A steady stream of coherent light is produced when the flash tube (the energy pump) is activated repeatedly.

*Producing Coherent Light in a Gas Laser.* In a gas laser, such as a helium-neon laser, an external circuit sets up a high-frequency electromagnetic field. Some of the gas atoms in the laser tube are ionized by this field. The free electrons produced by this ionization process strike the un-ionized gas atoms which are then raised to higher energy levels. The neon atoms emit an intense red beam. The laser light in this case is not all in the visible spectrum. The laser is activated by pumping or electron bombardment.

*The Injection Laser.* The injection laser is also known as the semiconductor junction diode (injector) laser. One advantage of this type of laser is that to modulate the device, the output current can be varied rather than varying the laser output beam. One disadvantage of the injection laser is that the laser beam output is less coherent and directional than the light from a gas laser or a solid laser. Another problem arises because the device must be maintained at the temperature of liquid nitrogen to insure continuous operation.

In the injector laser, extremely high current is passed between the terminals of a semiconductor diode. Coherent light is emitted along the line defined by the semiconductor junction. The light emission is incoherent until the intensity of the current is increased to the point that coherent light is produced.

The semiconductor junction is the narrow plane which cuts the typical injector laser block along its long axis. The opposite sides of the block are parallel and finely polished to reflect light back into the laser. This feature causes coherent light to be emitted in parallel rays from the sides of the block. Electric current is applied to opposite sides of the rectangle. In other words, the current flow is perpendicular to the semiconductor junction.

## Applications of the Laser

The light beam leaving a laser can be concentrated even more by focusing devices. In this manner, the power of laser beams can be brought up to several million watts.

Laser light is used in industry to cut metal, drill fine holes, and spotweld complex components. In addition, it is used in precision machining processes and measurement and in research into the crystalline structure of metals. Lasers are used in meteorological instruments to detect atmospheric aerosols and to study the fine-scale motion and structure of the atmosphere.

In other sciences, the intense and precise nature of lasers makes it possible to control and modify the path of chemical reactions and thus lead to the development of new materials and chemicals. The monochromatic quality of the light produced by lasers provides a highly accurate means of measuring lengths and frequencies in the physical sciences. In biological applications, a laser beam can be focused through a high-powered microscope to a spot 0.0002 in. in diameter; this precision gives scientists the capacity to explore inside a living cell. In medicine, lasers are used in operations to repair the retinas of eyes, in cancer surgery in the brain, and in other very delicate surgical procedures.

The use of lasers to extend the capability of communications is almost infinite. Scientists estimate that a single laser beam has the capacity to carry simultaneously all the communications functions that now travel by microwaves or along wires.

## Laser Photography

Laser technology is being used to produce three-dimensional pictures called *holograms*. In conventional photography, a two-dimensional image of an illuminated three-dimensional object is recorded on a light-sensitive film. In holography, light from a laser is split by a mirror which is half reflecting and half transparent. One part of the light thus split goes directly to a film on which the hologram is to be produced.

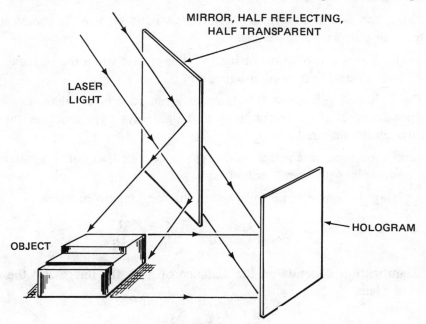

**Fig. 33-23 Producing a hologram**

Another part of the light is directed to the object to be photographed. This light is reflected back from the object to the film. When the two beams reach the film they interfere. This interference makes it possible for the film to record at each point both the intensity and the relative phase of the light.

A three-dimensional image of the object is projected when laser light is shown on or through the hologram at the proper angle. Different perspectives of the same image can be produced by changing the relative position of the hologram with respect to the laser light hehind it. This holographic technique has been applied to motion pictures and television. Holography is also being used in data processing to increase the storage capacity of memory elements for computers.

## SUMMARY

- There are five principal sources of light energy: the sun, fire and combustion, incandescence, electrical discharge, and phosphorescence.

- Light travels in straight lines. Light may be absorbed, reflected, or it may pass through an object which it strikes.

- When light is reflected in a regular pattern, a regular reflection is produced. In diffused reflection, the light spreads out in an irregular pattern.

- A wavelength denotes the distance from one point on a light wave to the corresponding point on the next wave. Wavelength is expressed as a linear measurement.

- Frequency refers to the number of waves that move past a point in a given interval of time.

- The angle of incidence of a reflected ray of light is equal to its angle of reflection.

- Refraction is the change in direction of light rays as they pass into or through transparent materials.

  – The angle of refraction of light rays is greater when the light ray passes from a dense to a less dense medium.

- The index of refraction is a number given to a material to indicate the amount of bending (refraction) that light rays experience as they pass through the material.

- Lenses are used to control light rays: a convex lens converges light rays; a concave lens diverges the light rays.

- An image is clear when it falls within the focal length of a lens.

$$\frac{1}{\text{Focal Length}} = \frac{1}{D_o} + \frac{1}{D_i}$$

- Magnification depends on the distance of both the image and the object from a lens.

$$\frac{D_o}{D_i} = \frac{S_o}{S_i}$$

- Light energy is measured in candlepower units. One candlepower refers to the unit illumination on a surface one foot from the light power of one candle.

- The intensity of illumination decreases and the area of illumination increases as the square of the distance from the light source.

- Diffraction describes the ability of light waves to bend around the edges of obstacles in their paths. Diffraction limits the power of resolution of an optical system to produce separate, clear images of nearby objects.

- Polarization relates to the process of confining the oscillations of a wave motion to a definite pattern. A polarized beam is one whose vibrations occur in only a single direction which is perpendicular to the direction of travel of the beam.

- Laser is the term derived from Light Amplification by the Stimulated Emission of Radiation.

- A laser light has extremely valuable properties.  The light is monochromatic, coherent, collimated, has an exceptionally high magnitude of energy, and can be either a continuous beam or a pulse beam.

- The ruby, gas, and semiconductor injector lasers are three common laser devices.

- Holography as a new concept in photography utilizes a laser film called a hologram. Viewing laser light through a hologram (on which interference patterns are formed) at the proper angle provides a three-dimensional view of the object.


## ASSIGNMENT UNIT 33  LIGHT: SOURCES, APPLICATION, CONTROL, AND MEASUREMENT

### PRACTICAL EXPERIMENTS ON THE MEASUREMENT AND CONTROL OF LIGHT AND LENSES

### Equipment and Materials Required

| Experiment A | |
| --- | --- |
| Reflection | Refraction |
| Optical disc apparatus with adjustable shutter | Light source (fine flashlight) |
| | Test stand |
| Mirror | Beaker with water |
| Light source with rheostat control and stand | Optical disc apparatus |
| Protractor | Large protractor or graduated disc |
| | Glass prisms (rectangular and triangular) |

| Experiment B<br>Lenses | Experiment C<br>Light Measurement | Experiment D<br>Polaroid® Lenses |
|---|---|---|
| Optical disc apparatus with light source and stand<br>Convex and concave lenses (two different sizes and curvatures of each) | Inverse squares light measurement apparatus<br>Light source<br>Rule or measuring tape<br>Lightmeter | Three Polaroid® lenses to fit light box<br>Light box with slide to receive lenses<br>Lightmeter |

## EXPERIMENT A. REFLECTION

### Procedure

1. Place a mirror on the horizontal centerline of the optical disc apparatus.

2. Adjust the light source and the shutter on the apparatus so that a ray of light strikes the mirror at the intersection of the vertical and horizontal centerlines.

3. Note the angle of the light ray with relation to the normal line (vertical centerline). Sketch and record this angle in the table for Recording Results.

4. Read the angle of the reflected ray directly on the optical disc (the angle can also be measured with a protractor). Record the angle and add this line to the sketch.

5. Repeat steps 1–4 but change the angle of the entering light ray.

6. Turn the disc with the mirror attached so that the normal line is 45° from the vertical centerline.

7. Pass a light ray through the shutter so that it again strikes the mirror at the center point.

8. Measure, record, and sketch the angles of incidence and reflection. Show the relationship of these angles to the normal line.

9. Repeat steps 7 and 8 with another light ray hitting the mirror at a different angle. Sketch the results.

### Recording Results

| | Entering<br>Light Ray<br>∡ Incidence (I) | Reflected<br>Ray<br>∡ Reflection (R) | Sketch (I) and (R)<br>in Relation to<br>Normal Line (a-b) |
|---|---|---|---|
| A | | | a · b |
| B | | | a · b |
| C | | | b / a |
| D | | | b / a |

### Interpreting Results

1. Compare the angle of incidence with the angle of reflection and the sketches at A, B, C, and D in the table.

2. State briefly two laws governing light that may be developed from the data recorded and the observations made in this experiment.

## B. REFRACTION
### Procedure

1. Shine a fine light ray down to the surface of a beaker of water at a number of angles.

2. Observe and sketch what happens as each light ray passes from air through the water.

3. Position the light ray so that it shines up through the water at several different angles.

4. Sketch what happens when the angle of incidence is such that the ray does not leave the water.

5. Mount a rectangular prism or plate on the horizontal centerline of the optical disc apparatus.

6. Direct a ray of light on the horizontal surface of the glass plate or prism at an angle of 90°. Repeat this step by shining the light at 45° to the plate or prism.

7. Observe and sketch the angle and position of the light ray as it passes from air through the glass at 90° and 45°.

8. Pass the light rays at a 45° angle through the glass into the air.

9. Sketch the angle and position of the light ray in the table.

10. Repeat steps 5–9 using other shapes of transparent solids and other liquids.

### Recording Results

|   | Condition | Sketch of Rays |
|---|-----------|----------------|
| A | From air through water | |
| B | Ray does not leave water | |
| C | Ray at 90° to horizontal — from air through glass | |
| D | Ray at 45° to horizontal — from air through glass | |
| E | Ray at 45° to horizontal — from glass into air | |

### Interpreting Results

1. Compare the indexes of refraction of glass and water with air.

2. State briefly the law of refraction as it applies when light rays (a) move from a material with a high index of refraction to one with a lower index and (b) from a material with a lower index of refraction to one with a higher index of refraction.

3. Explain why a light ray at right angles to a material passes through the material without bending.

## EXPERIMENT B. CONCAVE AND CONVEX LENSES

### Procedure

1. Mount a convex lens in a vertical position on the horizontal centerline of the optical disc.

2. Adjust the shutter on the apparatus and shine a light ray on the lens from the point shown in the table for Recording Results at (A).

3. Sketch the path of the light rays in a table for Recording Results.

4. Replace the lens with another convex lens having a different curvature.

5. Change the position of the light rays shining on the lens.

6. Sketch the shape and characteristics of the resulting image at (B) of the table.

7. Repeat steps 1–3 with two different convex lenses as at ($A_1$) and ($B_1$) in the table.

8. Replace the convex lens with a concave lens.

9. Start with a light ray in a horizontal position. Repeat steps 1–7 using concave lenses and record the results at C, $C_1$, D, and $D_1$ of the table.

10. Add a convex lens to the concave lens to form a combination of lenses. Pass the light ray through both lenses from two positions (E and $E_1$).

11. Sketch the light rays in the table.

### Recording Results

Characteristics of Images Produced by Lenses

## Interpreting Results

1. State two physical laws demonstrated by these experiments with concave and convex lenses.

2. Explain why a ray of light bends when passing from a less dense medium (air) to a more dense object (glass).

## EXPERIMENT C. MEASUREMENT OF LIGHT

### Procedure

1. Adjust the four panels on the inverse square apparatus so that the distance between each panel is the same as that from the light source to the first panel (one foot apart).

2. Note the intensity of the light at the first panel. Read the amount of illumination at the panel with a lightmeter and record the reading in a table for Recording Results.

3. Remove the center section from the second panel (B) and note the area and intensity of the illumination on Panel B. If necessary, adjust the position of the panel so that the light falls within the ruled area.

4. Measure and record the dimensions of the illuminated area. Also, note the lightmeter reading at the second panel. Record this value.

5. Determine the brilliance and record the light intensity and area of illumination for the third and fourth panels (C and D), respectively.

### Recording Results

| | Observed Brilliance | Lightmeter Reading | Area |
|---|---|---|---|
| Panel A | | | |
| Panel B | | | |
| Panel C | | | |
| Panel D | | | |

### Interpreting Results

1. Use the recorded data to prove or disprove the statement that the intensity of the illumination decreases as the square of the distance from the light source.

2. Compare the areas of illumination on each of the four panels.

3. Give the formula which can be used to determine these areas of illumination.

4. Compute the area of illumination from the measured dimensions and check against the areas computed by formula.

## EXPERIMENT D. EFFECT OF POLAROID® LENSES ON LIGHT RAYS

### Procedure

1. Read the intensity of light one foot from a light box using a lightmeter. Prepare a table for Recording Results and record the reading.

2. Insert one Polaroid® glass plate in the light box so that the rays of light pass from the source through the Polaroid® lens.

3. Note and record the light intensity.

4. Place a second Polaroid® glass plate in the light box and rotate it so that the pattern of the glass is 90° to that of the first plate.

5. Read and note any variations in the intensity of the light shining through the two plates. Record this value.

6. Repeat the procedure by adding a third Polaroid® glass plate.

### Recording Results

|   | Condition | Foot-candles of Illumination |
|---|---|---|
| A | Light — direct | |
| B | One Polaroid® lens | |
| C | Two Polaroid® lenses (at right angles) | |
| D | Three Polaroid® lenses | |

### Interpreting Results

1. From the observations made, describe what effect one, two, and three Polaroid® lenses have on light intensity.

2. Use the recorded data to prove that light waves are absorbed as they pass through a Polaroid® lens, with a resulting reduction in the intensity of the illumination.

## PRACTICAL PROBLEMS ON THE MEASUREMENT AND CONTROL OF LIGHT AND LENSES

### The Nature of Light and Light Paths

1. List three principal sources of light.

For statements 2–7, determine which are true (T) and which are false (F).

2. An object may be seen when light strikes the object and is reflected.

3. An object may be seen when there is no light reflected from the object to the eye.

4. An object may be seen in a totally dark, dust-free room.

5. As a light ray hits an object, it may be absorbed, reflected, or it may pass through the object.

6. Regular reflection results when light hits a rough surface and the rays spread out in an irregular pattern.

7. Frosted electric light bulbs provide reflection by diffusion.

Select the correct word or phrase to complete statements 8 to 15.

8. The reflection produced by rays of light striking an irregular object and spreading irregularly in all directions is (regular) (diffuse) reflection.

9. An object which permits light to shine through it is said to be (opaque) (translucent).

10. Shadows are the result of light rays traveling in (straight) (curved) (irregularly shaped) lines.

11. The speed of light is (220 000) (186 285) (1500) miles per second.

12. A wavelength is (the distance from one wave to a corresponding point on the next wave) (the number of waves in a given time).

13. The frequency of a light wave (is approximately 1/50 000th of an inch) (refers to the number of waves in a given time).

14. The index of refraction of air is (greater than) (smaller than) (the same as) that of glass.

15. The index of refraction of a diamond is (2.47) (1.00) (1.33) (0.050).

Match the description in Column II with the term in Column I to which each refers (16 to 20).

| Column I | Column II |
|---|---|
| 16. Angle of incidence | a. The spreading out or diffusing of light rays in an irregular pattern. |
| 17. Angle of reflection | b. The ability of light waves to bend around the edges of obstacles in their paths. |
| 18. Refraction | c. The angle a bent ray makes with a perpendicular to the surface at a point where the ray enters the object. |
| 19. Angle of refraction | d. The point at which the angle of incidence of a light ray reaches the critical angle and the light ray does not leave the material. |
| 20. Total reflection of a light ray | e. The bending of light rays as the path or direction is bent or changed in passing from one material to another. |
| 21. Diffraction | f. The paths of two light rays are the same or differ by any whole number of wavelengths. |
| 22. Constructive interference | g. The numerical value (standard) used to indicate the amount of bending that takes place in different materials. |
| | h. The angle at which a light ray is reflected from an object. |
| | i. The angle at which a light ray hits an object. |
| | j. There is no wave activity. |

## Application of Lenses

1. List four common types of lenses.

2. Give three industrial applications of diverging lenses or mirrors.

3. Name three industrial uses of converging lenses or mirrors.

Add a word or phrase to complete statements 4 to 10.

4. A _____ lens is a curved lens which is thinner at the center than at the outer edges.

5. A _____ lens is thicker at the center than at the outer edges.

6. Light rays striking a lens with converging properties are refracted toward the _____ and converge toward a common _____ .

7. Light waves are affected by the _____ of the material through which they pass.

8. Light rays passing through a concave lens are _____ .

9. A (an) _____ is the reflection of an object through a lens.

10. The _____ is the lens range through which an object appears clear and in focus.

11. Determine the distances and sizes of objects or images indicated by the missing dimensions in the table at A to F.

$D_o$ = Object Distance     $S_o$ = Size of Object

$D_i$ = Image Distance     $S_i$ = Size of Image

|   | $D_o$ | $D_i$ | $S_o$ | $S_i$ |
|---|---|---|---|---|
| A | 16 ft. |  | 8 ft. | 2 ft. |
| B |  | 12 in. | 6 in. | 4 in. |
| C | 12 in. | 8 in. |  | 1/2 in. |
| D | 8 in. | 2 in. | 0.800 in. |  |
| E | 2 in. | 10 in. |  | 0.200 in. |
| F | 6.25 in. | 7.5 in. |  | 0.200 in. |

12. Compute the focal length for each object and image distance given in the table at (A) and (B). Find the missing dimensions at (C) and (D).

|   | Object Distance $D_o$ | Image Distance $D_i$ | Focal Length |
|---|---|---|---|
| A | 6 in. | 3 in. |  |
| B | 5 ft. | 3 ft. |  |
| C | 12 in. |  | 6 in. |
| D |  | 4.5 in. | 3 in. |

For statements 13 to 19, determine which are true (T) and which are false (F).

13. The farther away an object is moved from a light source, the greater is the amount of illumination on the object.

14. Relatively speaking, the farther away an object is from a light source, the greater is the area of illumination.

15. Surface roughness has no effect on the amount of light reflection.

16. The amount of illumination is equal to the candlepower divided by the distance from the light source, squared.

17. The bunsen photometer is not a convenient portable device for measuring illumination.

18. The area of illumination is proportional to the square of the distance from a light source.

19. The foot-candles of illumination available in industrial settings affect the quality and quantity of production and the personal safety of the workers.

Select the correct word or phrase to complete statements 20 to 25.

20. A machine spindle turning at 500 revolutions per minute appears to the operator to turn (faster) (slower) (at the same speed) when the illumination is increased from 20 to 60 foot-candles.

21. Of the early sources of light, the carbon filament lamp is (more) (less) efficient than the fluorescent lamp.

22. For instrument work and precision machining, (10) (50) (100) (300) foot-candles of illumination are needed.

23. Rough construction work requires the (same) (less) (more) candlepower than is required for fine assembly operations.

24. Light may be controlled with Polaroid® lenses whose crystalline structure (is arranged the same as fine glass) (is such that the crystals are arranged in parallel columns).

25. Two Polaroid® lenses turned 90° to each other permit a (greater) (lesser) amount of light to shine through than a single lens whose crystals are in one direction.

26. List four useful industrial applications of Polaroid® lenses.

27. Compute the candlepower needed to deliver the required illumination for the type of work and conditions shown in the table at A, B, C, and D.

|  | Type of Work | Required Illumination (Foot-candles) | Distance from Work | Required Candlepower (Foot-candles) |
|---|---|---|---|---|
| A | Foundry | 20 | 20 ft. | |
| B | Rough Machining | 40 | 10 ft. | |
| C | Fine Assembly | 60 | 5 ft. | |
| D | Precision Machining | 70 | 3 ft. | |

## Lasers: Characteristics, Principles, and Applications

1. State briefly the principles that relate to the production of a laser beam. Secure a technical manual or scientific article that deals with any one of the basic types of laser devices.

2. Make a simple schematic of the laser device and label the important parts.

3. State how the laser device produces a beam of light. Name two different fields in which lasers can be applied.

4. Name a practical application (other than those cited in this unit) for each of the two fields listed in question 3.

5. State the function served by the laser light beam in one of the applications.

6. Identify the advantage of using (or the need for) a laser light beam as compared to any other device or technique previously used for this same function (question 5).

# Unit 34  Principles, Control, and Applications of Color

Light waves are part of an electromagnetic spectrum of waves, figure 34-1, in which the wavelengths vary from over 20 miles in length to the billionth part of an inch and smaller. Note in the figure that radio waves have long wavelengths and heat waves, ultraviolet, X rays, gamma and cosmic rays have shorter wavelengths. The combination of all wavelengths makes up the *electromagnetic spectrum.*

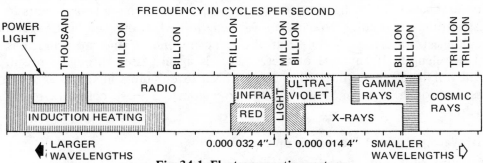

Fig. 34-1 Electromagnetic spectrum

The light waves which are visible to the eye are represented by only a small section of the electromagnetic spectrum. This section is known as the *visible spectrum.*

To the left of the visible spectrum, there is a series of wavelengths which are longer than those for red light. These wavelengths of light are shorter than the wavelength of violet light. These shorter wavelengths are known as *ultraviolet waves* and also produce heat. The infrared and ultraviolet wavelengths do not fall in the range of visible light.

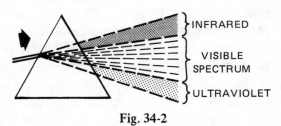

Fig. 34-2

## PRINCIPLES OF COLOR

### Newton's Experiments With Color

Newton is credited with devising a method of separating light into its component colors. He passed a beam of sunlight through a triangular prism of glass. The light beam was refracted toward the base of the prism. The light emerging from the prism, figure 34-3 (page 404), was not a parallel beam. Rather, it was dispersed and broken up into seven colors:  red, orange, yellow, green, blue, indigo, and violet.

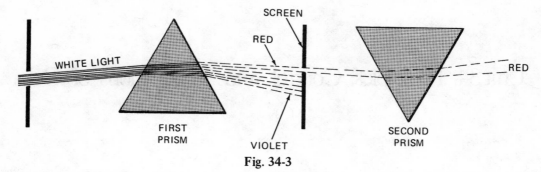

**Fig. 34-3**

*The Solar and Visible Spectrums.* Newton then passed the separate beams of the individual colors through a screen so each one in turn fell on a second prism. In no case was any one of the seven colors dispersed further. The seven colors comprising the *solar spectrum* are known as the *visible spectrum* because they are visible to the eye. When the visible spectrum is passed through another prism, the seven separate colors again come together and emerge as a white ray.

To summarize, white light can be broken up into the basic seven colors in the visible spectrum. Each color depends on wavelength. Red, at one end of the spectrum, has the longest wavelength. Violet, at the opposite end of the spectrum, has the shortest wavelength. If any one of these colors is again passed through a prism, the color will not break down into other colors. If all seven colors in the solar spectrum are passed through a prism, it is possible to recombine these colors to reproduce white light.

### Effect of Wavelength and Frequency on Color

In figure 34-3, note that the red light is bent or refracted less than the violet light. The difference between one color and another is the difference between the wavelength and frequency. Since the speed of light remains constant (186 285 miles per second), it follows that each different color has a different frequency.

The rate of vibration of red light is nearly 375 trillion vibrations each second. Its wavelength is almost 26 millionths of an inch. Red has the longest wavelength of the visible spectrum, table 34-1, and consequently, the lowest frequency. At the other end of the color spectrum, indigo has a wavelength of 16 millionths of an inch. The frequency of indigo is almost double (26/16 or 13/8) that of red light. Table 34-1 lists the wavelengths of the colors in the solar spectrum.

| Primary Color | Wavelength | | |
|---|---|---|---|
| | Inches | Millimeters | Angstroms |
| Deepest Red | 0.000 032 4 | 0.000 81 | 8100 |
| Red | 0.000 026 | 0.000 65 | 6500 |
| Orange | 0.000 023 3 | 0.000 583 | 5830 |
| Yellow | 0.000 022 | 0.000 55 | 5500 |
| Green | 0.000 020 5 | 0.000 51 | 5100 |
| Blue | 0.000 018 | 0.000 45 | 4500 |
| Indigo | 0.000 016 | 0.000 4 | 4000 |
| Deepest Violet | 0.000 014 4 | 0.000 36 | 3600 |

**Table 34-1**

In all cases, the product of the frequency of a color times its wavelength gives the constant velocity of light. Thus, the wavelengths are inversely proportional to the frequency. Colors with lower frequencies have longer wavelengths.

## PRODUCING COLORS BY ABSORPTION AND REFLECTION

### Colors by Absorption or Subtraction

Colors are produced and seen by absorption and reflection. If a red glass is placed between a prism and a screen as shown in figure 34-4A, and a beam of white light is passed through the prism, then all of the colors of the spectrum are absorbed except the red which emerges and shines on the screen. Similarly, if a piece of yellow glass is placed between the prism and the screen, only yellow light shines on the screen, figure 34-4B. In figure 34-4C, the purple glass cuts out the yellow, orange, and green wavelengths but permits the red, blue, indigo, and violet wavelengths to shine through.

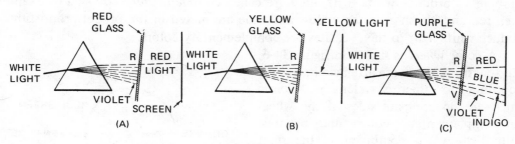

Fig. 34-4

The process of color absorption (subtraction) has a great value in industry. By taking away or absorbing any of the fundamental colors in the visible spectrum, it is possible to mix different colors of lights, paints, or other substances in varying proportions to obtain an infinite variety of shades and hues.

### Colors by Reflection

A colored opaque object absorbs all of the spectrum colors except the one that matches the object. This color is reflected. For example, the blue surface shown in figure 34-5A reflects only blue and absorbs all of the remaining colors of the spectrum. The red, orange, yellow, green, indigo, and violet rays are absorbed by the object and are converted into heat energy. A red object will absorb all of the spectrum colors except red which it reflects. Again, with a purple surface, the orange, yellow, and green are absorbed while the red, blue, indigo, and violet are reflected.

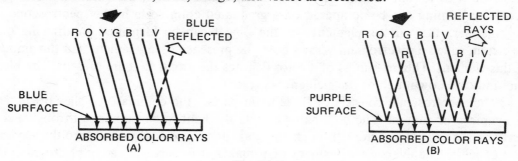

Fig. 34-5

In actual practice, there are few colored objects which reflect only one color. Generally, very small amounts of all of the spectrum colors are reflected from every object. The color of an object depends on the color of the light illuminating the object as well as the color reflected by the object.

The rainbow is an example of color reflection in nature. The white light of the sun is reflected through water droplets which act as prisms. The light rays (sunlight) are refracted and dispersed into the colors of the spectrum.

### Color Combinations

All colors can be made by mixing the red, green, and blue colors in the right proportions. These colors are known as the *primary colors*. When rays of each of the three colors overlap, the area where the three colors come together appears in white light.

White light can also be produced by arranging the seven basic colors in the spectrum on a disc and rotating the disc rapidly. When any two colors of these seven colors combine to produce white light, they are called *complementary colors*. For example, when red and bluish-green or yellow and blue are mixed in the proper proportions to produce white light to the eye, they are complementary colors.

A *color wheel* is shown in figure 34-6. The spectrum colors are arranged in a circle according to their wavelengths. Colors which are on opposite sides of the wheel are complementary colors and produce white light when combined in the right proportions.

An object is black when it absorbs all of the light rays (no light is reflected). An object is white when it reflects all of the light rays. An object is red when it absorbs all of the rays except red, which it reflects. The same principle applies to objects of other colors. Thus, to reemphasize a previous statement, the color of an object depends on (1) the kind of light that falls on it, and (2) the wavelengths of the light it reflects.

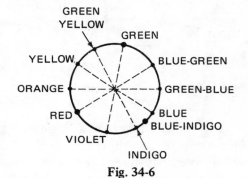

Fig. 34-6

## ANALYZING MATERIALS THROUGH COLOR

### The Spectrum Applied to the Spectroscope

The *spectroscope* is an instrument used by scientists, technicians, and others, to identify different materials and elements. Since all elements burn with a different color, or with a combination of colors in differing amounts, the color or colors produced by burning can be compared on a graduated color scale in the spectroscope. In this manner, the various elements in the object can be identified. A drawing of a spectrum, called a spectrogram, shows both the presence of an element and the amount of this element. The brightness of a color denotes the atoms that are present. The black lines indicate how much of the element is present.

The spectroscope consists of three main parts: the collimator, a glass prism, and an eyepiece or camera. The collimator directs the light to be examined through a convex lens. The light continues through the glass prism and is dispersed into the separate rays of light. The eyepiece or camera then makes it possible to see or photograph the separate rays of light so they can be compared with a spectrum.

When white light from the burning of a bright lamp or white liquid is examined in the spectroscope, a continuous spectrum of all of the colors from red to violet appears. In the case of the solar spectrum shown in figure 34-7, the black lines indicate the absorption of light in the sun's atmosphere. The black lines on the spectrogram of white light from the sun are known as Fraunhofer lines (named for the German physicist who discovered them).

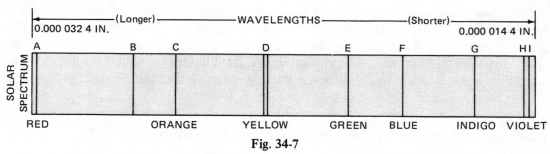

Fig. 34-7

Every element in a heated gaseous or vapor state has its own identifying spectrum lines. This fact is important because it enables scientists to identify the elements comprising a material. For example, when sodium burns a very bright yellow flame is produced. This color always falls within a certain identified region of the spectrum.

If a metallurgist wants to determine if a stainless steel specimen contains chromium or nickel, a small amount of the material is vaporized to produce light. The spectrum of the light thus produced can be photographed, and then compared with the known spectra of chromium and nickel. Figure 34-8A shows the spectrum of stainless steel; the drawing at (B) is the spectrum of chromium; and that of nickel is shown at (C). Note that all of the lines in the chromium spectrum and those in the nickel spectrum match with the lines in the stainless steel spectrum. This matching indicates that the stainless steel part contains both chromium and nickel.

Scientists also use the spectroscope to analyze the light from stars and to learn about new elements found in nature.

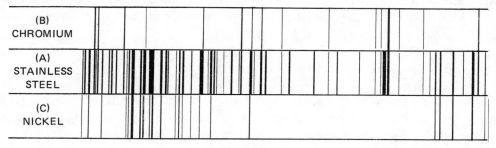

Fig. 34-8

## Analyzing Colorful Elements

The recording spectrometer is an instrument which analyzes chemical elements and determines the quantity of each contained by metals. The metal is heated to a high enough temperature to produce light. The light is broken up by the instrument into its component wavelengths which are read on a recording unit. In this manner, the types of elements and the quantity of each present in a metal sample can be determined.

The recording spectrometer combines Sir Isaac Newton's principles of color with the discovery that each element gives off a characteristic color pattern when subjected to intense heat energy. The combination of these principles and good photographic film in the spectroscope (spectrophotometer) has provided scientists and technicians with valuable information about the various elements and the amounts of these elements contained in different metals.

In the recording spectrometer, the photographic and visual interpretative steps of the spectrophotometer are eliminated. Photomultiplier electronic tubes scan the color waves produced by an intensely heated metal sample and change the wavelengths into electric currents. These currents operate a recording unit on which the results are read directly. Furnace operators and other technicians in metal industries find that this instrument quickly yields laboratory reports on the elements and the amounts contained in a metal prior to tapping the molten metal. Using this instrument, several elements may be analyzed in less than one minute.

## COMMERCIAL APPLICATIONS OF LIGHT AND COLOR PRINCIPLES

The perception of color is one of the most important human senses by means of which all objects may be identified. In this process, the eye is the medium for transmitting the sensations which light rays make on the brain. Metals, plastics, woods, textiles, foods, and a limitless number of materials and objects depend on color for identification and use.

### Value of Color in Industry and Science

A few examples are given as follows to show how color is used. The chemist and biologist use simple color tests to analyze the contents of solutions. One application of such an analysis is that the condition of the human bloodstream can be diagnosed. The metallurgist uses color to identify different metals and study the grain structure of these metals through photomicrographs.

Metals can be heat treated and tempered accurately because specialized instruments control the temperature of the process by color matching. The physicist and other scientists identify the elements in various substances by vaporizing the material and color matching the resulting light wavelengths.

### Use of Color for Safety

The use of color in the environment of workers (on walls and machines) has a tremendous effect on improving the quality of workmanship, reducing eye strain, decreasing industrial accidents, and on worker morale. Certain colors produce the sensation of restfulness; other colors are used to identify safe areas or emergency locations. Yellow is commonly used to mark safety lanes in plants. Red generally indicates danger zones. In some industries, systems of colored lights provide communication between sections of a production line extending for great distances.

In the home, color again is an important consideration for safety and for comfort. Color is the basis for interior decoration. The fact that any color may be produced by combining the primary colors of red, green, and blue is used in color television. The colors received in the television camera at the studio are duplicated on the television set in the home. The process is so rapid that the separate images of red, green, and blue are blended together to produce an image in full color.

## Color Printing

Most color printing is done with a series of printing plates and colored inks. If four basic colors are to be printed, four separate photographs are taken using a fine mesh screen and a color filter which separates out one color at a time. The screen breaks the object into a series of dots which photograph as the light and dark parts of the object. The color filter makes it possible to photograph those parts of the object in one color. Four negatives are produced — one each for a different color. On each negative, the light and dark spots on the object appear as a pattern of dots which vary in darkness.

A set of plates is made from these negatives for printing. On each plate only those dots appear which are to be printed in that color. The plate for yellow is used first. This plate produces a pattern of light and dark yellow dots. The sheet printed in yellow is then run through the printing press a second time, using the plate on which the pattern is formed by all of the parts which are solid red or consist of red and other colors. After this press run, both the yellow and the red colors are printed.

If part of the object is solid yellow, then this part appears only on the yellow plate. Those parts of the object that require the mixing of one or more colors are printed as a series of dots, one overlapping the other until the desired color is obtained. The process is continued with the plates for the third and fourth colors.

The result of the combination of dots of different colors is a solid-appearing blend of the original colors of the object. The individual dots of color are usually extremely fine and are not visible to the naked eye.

This same technique of color printing applies regardless of the number of colors needed and the kind of printing used. In the case of black and white printing, only one plate is made. A screen is used in photographing the object. The light and dark effects are produced by the fine mesh screen which breaks the object into a limitless number of dots. The amount of light absorbed by these dots determines how light or dark each one will appear when etched on the printing plate. A negative of this object, which may be taken directly from a photograph, is called a *halftone*.

Fig. 34-9

## LIGHT AND COLOR PRINCIPLES APPLIED TO PRECISION MEASUREMENT

Measurement with a degree of accuracy to a millionth part of an inch is possible with devices called *optical flats*. This method of linear measurement is the most accurate way of checking for errors in parallelism, flatness, and the size of precision gauge blocks. Measurements made by the use of the principles of light wavelengths represent a simple measuring method.

## Optical Flats

The *optical flat* is a flat lens having an extremely accurate surface and an ability to transmit light. Optical flats are made of a semiquartz material and are accurate to within one millionth of an inch (for a perfect plane surface) for a master flat. Working flats have a surface accurate to within 0.000 002 in. to 0.000 005 in.

### Light Interference

When light strikes an optical flat, part of the light is reflected from the top surface, part of the light is transmitted through the optical flat and is reflected from its lower surface, and part of the light is reflected from the work surface.

Any interference between the light rays reflected from the bottom of the optical flat and the top of the work

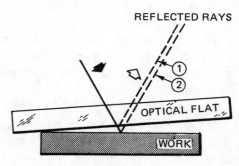

Fig. 34-10

appears as dark bands. These bands are the result of the *principle of light interference.* Unless the two surfaces are perfectly flat, the part of the light beam that is reflected from the surface of the work must travel a greater distance as shown by ray (2) in figure 34-10. Thus, when the reflected ray rejoins the ray reflected from the optical flat, ray (1), the rays are no longer in step. When one ray is half a wavelength out of step with the other ray, the interference between the rays causes alternate dark bands to appear.

## MEASURING WITH OPTICAL FLATS AND INTERFERENCE BANDS

The interference bands resulting from the use of an optical flat can be made to have a definite pattern. The wavelength of daylight is 20 millionths of an inch. Thus, when daylight is used with an optical flat, the interference bands appear at every ten-millionth of an inch of surface variation from a perfectly flat position. This result is due to the fact that interference occurs at multiples of one-half the wavelength of the light.

Although daylight may be used for measurements, it is more practical to use a monochromatic light. *Monochromatic light* is light of a single color which has a different wavelength than daylight. If the wavelength of a monochromatic light is 23.4 millionths of an inch, the difference between two interference bands is one-half the wavelength or 11.7 millionths of an inch.

The optical flat is widely used to check the flatness and parallelism of an object. If a surface is perfectly flat, the interference bands are evenly spaced and parallel to the line of contact between the surface and the optical flat, figure 34-11. While such surfaces are in nearly perfect contact, there is an air space between them. Any error in the surface being checked can be measured by the amount of deviation in the interference pattern formed.

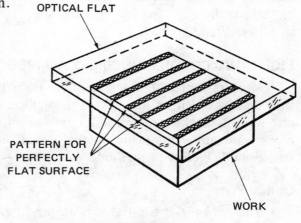

Fig. 34-11

### Common Patterns of Measurement

A few common patterns for interference bands using daylight are shown in figure 34-12. At (A), the regular and parallel bands show that the surface being measured is perfectly flat. The curvature of the bands at (B) indicates that the center of the work is higher than the edges by about 0.000 007 in. (the curvature is about 7/10 of the distance between two interference bands or 0.7 of ten-millionths).

The pattern at (C) indicates that the work is perfectly flat except for the edges which are worn about 0.000 004 in. (four-tenths of the distance between two inter-

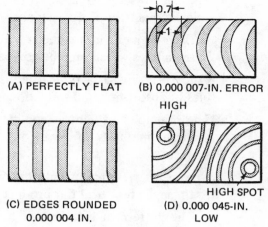

(A) PERFECTLY FLAT    (B) 0.000 007-IN. ERROR

(C) EDGES ROUNDED      (D) 0.000 045-IN.
    0.000 004 IN.           LOW

**Fig. 34-12**

ference bands). The pattern at (D) shows that there are two high spots on the work. Since there are nine bands between the centers of the high spots, the valley between the two high spots is 9/2 x ten-millionths or 45-millionths of an inch deep.

Many other patterns can be formed. Each variation in pattern has a definite meaning to the technician because it indicates how accurate a surface is and the points on the surface which are not flat. Although optical flats are still used with precision gauge blocks to measure the accuracy of parts, these measurements can also be done on an instrument known as an *optical comparator*. This instrument permits very exacting measurements to be made of the positions of the bands as compared to the estimations made of the fractional parts of bands using optical flats.

The system of interchangeability of parts, which is a factor in the improved standards of living in industrial nations, depends largely on the degree of accuracy to which master instruments, gauges, and tools can be produced. Modern systems of interchangeable manufacture to extremely fine limits of accuracy are possible because it is possible to apply a fundamental scientific principle of light to a perfectly flat optical plate.

---

## SUMMARY

- The visible spectrum includes seven basic colors: red, orange, yellow, green, blue, indigo, and violet. All of these colors are found in white light.

- Ultraviolet rays appear at one end of the visible spectrum and infrared rays appear at the other end. These light rays are not visible.

- The difference between colors depends on the wavelength and the frequency of each color.

- Colors are produced by absorption and reflection. The color of an object depends on the color illuminating the object and the wavelengths of the light reflected by the object.

- Light rays of the primary colors (red, green, and blue) produce white light.

- Complementary colors refer to those combinations of two of the seven basic colors which produce white light.

- The spectroscope is used to determine both the presence of an element and the quantity of the element.

  - Each element in a gaseous or vapor state has its own identifying spectrum lines which may be compared.

- Color printing, as a typical industrial application, depends on the principles of absorption and reflection in the preparation of screened plates.

- The principles of light as applied to optical flats provide one of the most accurate methods of measuring flatness, parallelism, and the size of a surface.

  - Any interference of half a wavelength between the light rays reflected through an optical flat produces dark colored bands.

  - The pattern of interference bands indicates the degree and location of an error.

## ASSIGNMENT UNIT 34 PRINCIPLES, CONTROL, AND APPLICATIONS OF COLOR

### PRACTICAL EXPERIMENTS WITH COLOR AND ITS APPLICATION

#### Equipment and Materials Required

| Experiment A<br>Light and Color | Experiment B<br>The Spectroscope | Experiment C<br>Optical Flats |
|---|---|---|
| Hand or power color rotating device<br><br>Multicolor disc (all colors of the spectrum)<br><br>Tricolor disc (red, green, and blue)<br><br>Optical disc apparatus<br><br>Light beam (white, red, green, and blue)<br><br>Glass prism<br><br>Screen | Spectroscope<br>Bunsen burner<br>Steel wire in holder<br>Sodium chloride<br>Potassium chlorate<br>Hydrochloric acid and beaker<br>Lithium chloride | Monochromatic light source<br>Optical flats (set)<br>Precision gauge blocks (small set)<br>Precision ground flat plate |

## EXPERIMENT A. PRODUCING LIGHT AND COLOR

### Procedure

1. Mount the multicolored disc on the spindle of the rotating device.
2. Rotate the disc. Prepare a table for Recording Results (Table A) and note the color produced.
3. Replace the multicolored disc with the tricolored disc.
4. Rotate the disc rapidly and note the color(s) produced.

5. Repeat steps 3 and 4 using any other combination of colors.

6. Pass a white light ray through a narrow opening on the optical disc apparatus so that it hits a triangular prism and travels through to a screen.

7. Prepare a table for Recording Results (Table B) and record the colors which are reflected on the screen.

8. Repeat steps 6 and 7 using a red light ray, then a green one, and then a blue ray.

## Recording Results

| | Color on Disc | Color(s) Produced |
|---|---|---|
| A | Multicolor | |
| B | Tricolor | |

Table A

| | Color Light | Reflected Color(s) |
|---|---|---|
| A | White | |
| B | Red | |
| C | Green | |
| D | Blue | |

Table B

## Interpreting Results

1. Review the results of the rotating multicolor disc and the reflected light experiments.

2. Prove or disprove the statement that white light is produced when an object reflects all of the colors in the spectrum.

3. Based on the results of the experiment, defend the principle that when the primary colors are mixed in the proper proportions, all other colors can be produced.

## EXPERIMENT B. IDENTIFYING MATERIALS WITH THE SPECTROSCOPE

### Procedure

1. Light and adjust the bunsen burner so that an impression of the bright yellow flame is seen in the spectroscope.

2. Look through the eyepiece and note the colors formed on the spectroscope scale.

3. Readjust the flame until it is a hot blue flame. Clean the steel wire by first dipping it in a solution of hydrochloric acid and then heat it.

   CAUTION: Keep the acid away from the skin and clothing to prevent skin irritation and the eating away of cloth fabric.

4. Prepare a table for Recording Results for any of four different salts.

5. Dip the clean steel wire in sodium chloride (or any other salt). Hold the wire in the flame and note the color produced and the scale reading on the spectroscope.

6. Record the data in the prepared table; then clean the wire.

7. Dip the wire in calcium chloride (or other salt). Heat the wire and repeat step 5. Again note the color of the flame and compare the location of the color on the spectroscope scale.

8. Repeat the experiment with potassium chlorate and lithium chloride (or any two different substances).

## Recording Results

| | Substance | Spectroscope | |
|---|---|---|---|
| | | Color | Scale Reading |
| A | Sodium chloride | | |
| B | Calcium chloride | | |
| C | Potassium chlorate | | |
| D | Lithium chloride | | |

## Interpreting Results

1. Compare the colors and the scale readings as recorded for the spectroscope experiments with those shown in a metals handbook.

2. If the colors and readings do not agree, repeat the experiment and justify the results.

3. Describe briefly the principles on which the spectroscope operates. Use simple sketches to show the relationship of the parts of the spectroscope.

## EXPERIMENT C. APPLYING COLOR AND LIGHT TO PRECISION MEASUREMENT

### Procedure: Testing for Flatness

1. Place a clean gauge block on a flat surface.

2. Wring an optical flat to this surface.

3. Place the assembly under a monochromatic light. Note in a table for Recording Results how the light rings are formed.

4. Repeat steps 1–3 with each of the gauge blocks.

### Procedure: Testing for Accuracy and Parallelism

1. Wring a master gauge block of the required size to an optical flat.

2. Repeat step 1 with the gauge block to be measured.

3. Place a second clean optical flat on the two gauge blocks.

4. Prepare a table for Recording Results and sketch the pattern of the interference bands.

5. Repeat the procedure using a master gauge of the required size to measure any other gauges or parts. In each case, record the results.

## Recording Results

| Gauge Block | Pattern of Interference Bands |
|:---:|:---|
| A | |
| B | |
| C | |

## Interpreting Results

1. When testing for flatness with optical flats and light, what interference band patterns indicate that a surface is perfectly flat? That a surface is not flat?

2. How many gauge blocks in the set are perfectly flat?

3. Indicate the stamped sizes of the gauges which are not perfectly flat. (If none, indicate accordingly.)

4. When checking the accuracy of a size, what patterns do the interference bands make? Show the patterns with a simple sketch.

5. Show how optical flats and gauge blocks may be used to check the accuracy of a round part. Use a sketch.

## PRACTICAL PROBLEMS WITH COLOR
## PRINCIPLES, CONTROL, AND APPLICATIONS

### Principles of Color

Select the correct word or phrase to complete statements 1 to 6.

1. The wavelengths of electromagnetic waves vary from (1 mile) (20 miles) (1000 miles) to the (thousandth) (millionth) (billionth) part of an inch.

2. The shortest wavelengths are for (induction heating) (cosmic rays) (infrared rays).

3. The longest wavelengths are for (ultraviolet) (X-ray) (radio) waves.

4. The electromagnetic spectrum consists of (the colors in the solar spectrum alone) (the combination of all wavelengths).

5. The visible spectrum includes the seven basic colors (alone) (and the infrared waves) (and the infrared and ultraviolet waves).

6. The (longer) (shorter) wavelengths as compared to those in the visible spectrum represent the ultraviolet waves.

Add a word or phrase to complete statements 7 to 11.

7. Infrared and ultraviolet waves produce _____ .

8. The seven basic colors into which white light can be dispersed are: _____ , _____ , _____ , _____ , _____ , _____ , and _____ .

9. _____ has the longest wavelength; _____ the shortest.

10. The difference between two colors is the difference between the _____ and _____ of each.

11. The wavelength of red rays is _____ millionths of an inch.

12. Indicate on the line drawing of a solar spectrum, the lengths (in millionths of an inch) of colors A, C, D, E, F, G, and H.

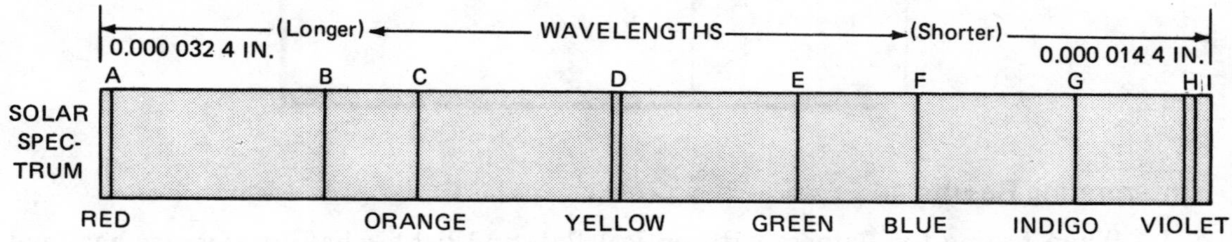

Fig. 34-13

## Producing Colors by Absorption and Reflection

Add a word or phrase to complete statements 1 to 8.

1. The primary colors are _____ , _____ , _____ .

2. Any two colors which, when combined, produce white light are _____ colors.

3. The color of an object depends on _____ and _____ .

4. Through the process of color _____ , lights, paints, inks, and other substances, can be mixed in varying proportions to obtain a desired shade or color.

5. _____ will shine through a red glass prism into which white light is passed.

6. A colored opaque object absorbs all of the spectrum colors except _____ which it reflects.

7. A rainbow is produced by refracting and dispersing the light rays of the sun into the _____ colors.

8. An object appears _____ when it absorbs most of the light rays.

9. Explain why the same fabrics or other materials appear to be differently colored when seen under artificial and natural lights.

## Analyzing Materials Through Color

For statements 1 to 8, determine which are true (T) and which are false (F).

1. All elements burn with a different color or combination of colors in differing amounts.

2. Cobalt burns with a dark blue light; chromium, a green light; nickel, a light blue light; and sodium, a yellow light.

3. Color is used by scientists, inventors, and technologists to analyze living things and nonliving matter.

4. The amount of light absorbed through a photographic plate has no effect on the light and dark spots printed for a shaded object.

5. For every shade of a color to be printed, a series of different shades of ink must be used.

6. The spectrograph, spectroscope, and recording spectrometer depend on color principles to operate.

7. The spectroscope produces a photographic film which indicates the elements and quantity of each found in a metal.

8. In a modern foundry, it takes hours using the recording spectrometer to analyze several elements in a metal being tested.

9. Select any four chemical elements.
   a. List these in the prepared chart.
   b. Use a handbook to determine the identifying color of each element.
   c. Determine (approximately) the scale reading or wavelength for each element.

| | Element | Color | Wavelength |
|---|---|---|---|
| A | | | |
| B | | | |
| C | | | |
| D | | | |

## Applications of Color and Light Principles

1. Identify five materials by the color characteristics of each.

2. Name four sciences (other than those included in this unit) which depend on the principles of color and light to identify different conditions or materials.

3. Select any piece of industrial equipment that is power driven and used for some type of production. Indicate the colors that the different parts may be painted (a) to reduce eye strain and (b) to serve as safety signals.

## Light and Color Principles Applied to Precision Measurement

Match the term in Column I with the correct description in Column II.

| Column I | Column II |
|---|---|
| 1. Optical flat | a. A single color light with a different wavelength than daylight. |
| 2. Light interference | b. A semiquartz substance machined to an accurate and perfectly plane surface. |
| 3. Interference bands | |
| 4. Monochromatic light | c. A system of interchangeable parts manufactured to extremely fine precision limits. |
| | d. The difference between two light rays reflected from a translucent plane surface. |
| | e. A pattern which appears when reflected rays from the surface being measured are out of step with the reflected rays from the perfectly plane optical surface. |

5. Explain briefly why a monochromatic light is better than daylight when precision measurements involving interference bands are to be made.

# Achievement Review
## LIGHT ENERGY

### LIGHT: SOURCES, APPLICATION, CONTROL, AND MEASUREMENT

Add the correct word or phrase to complete statements 1 to 10.

1. Two of the three conditions needed to see an object are (a) _____ and (b) _____ .

2. When light hits an object, it may be _____ or _____ .

3. Diffuse reflection means _____ .

4. Light travels _____ at a speed of _____ .

5. The number of light waves that move past a point at a given time is the _____ of the light.

6. The _____ of a reflected light ray is equal to its _____ .

7. The angle of refraction of a light ray is _____ when the ray passes from a dense medium to a less dense medium.

8. The ratio of the speed of light in a medium to its speed in a vacuum is its _____ _____ .

9. The light rays passing through a convex lens move faster at the _____ than at the _____ .

10. Light rays from a concave lens _____ .

11. In a single sentence define each of the following terms:

    a. Focal length
    b. Candlepower
    c. Intensity of illumination
    d. Area of illumination

12. Arrange the following lamps in the correct order ranging from the most efficient lamp to the least efficient: (a) sodium arc, (b) incandescent, (c) fluorescent, and (d) carbon filament.

13. Arrange the following jobs according to the foot-candles of illumination required (in the order of least to most).

    a. Foundry work
    b. Concrete pouring
    c. Instrument making
    d. Tractor tread assembly
    e. Motor winding
    f. Machining operations

14. a. What effect does polarization have on light rays?
    b. Cite one industrial application of a Polaroid® lens.

15. Make a simple sketch to show graphically how various lenses affect light rays: (a) a diverging lens, (b) a converging lens, and (c) a combination of a diverging and converging lens.

### PRINCIPLES, CONTROL, AND APPLICATIONS OF COLOR

Supply the correct word or phrase to complete statements 1 to 7.

1. Every material in a gaseous or vapor state has its own _____ .

2. The recording spectrometer is used in industry to _____ .

3. Color is a useful property in materials for (a) _____ and (b) _____ .

4. Colored paints are used in manufacturing plants for (a) _____ and (b) _____ .

5. Interference bands result when _____ .

6. When light interference bands from two surfaces are evenly spaced and are parallel to the line of contact, _____ .

7. The curvature in light interference bands indicates _____ .

8. Describe briefly how variations in a single color are produced in printing.

9. Name two of the three actions that may happen when a light ray hits an optical flat.

Match the term in Column I with the corresponding description in Column II.

| Column I | Column II |
|---|---|
| 10. Visible spectrum | a. A series of wavelengths greater than those for red light, which produce heat. |
| 11. Infrared waves | |
| 12. White light | b. The product of the frequency of a color and its wavelength. |
| 13. Variations in colors | c. A small section of the electromagnetic spectrum containing the wavelengths of the seven basic colors. |
| 14. Velocity of light | d. A combination of any three primary colors. |
| | e. The difference between the wavelengths or the frequencies of the colors. |
| | f. The color of a material produced by burning. |
| | g. Composite of the seven basic colors in the visible spectrum. |

For statements 15 to 23, determine which are true (T) and which are false (F).

15. Ultraviolet light has a longer wavelength than ordinary visible light.

16. A concave lens can be used to concentrate light at a single point.

17. Polarized light is used to examine structural parts for defects or strains.

18. White light is light of a single wavelength.

19. The colors formed on a metal surface as it is heated are due to the formation of oxides.

20. Light interference bands can be produced by combining light waves of either the same or different wavelengths.

21. Light waves of longer wavelengths have higher frequencies.

22. Light waves which vibrate in only one plane are said to be polarized.

23. Optical flats are used to make very precise measurements.

24. Explain briefly how color plays an important part in determining the composition of any material.

25. Refer to the table on tempering and heat colors in the Appendix.

    a. Give the temperature in °F for these tempering colors: straw, full blue, deep straw, and bronze.

    b. Give the temperature in °C for the following heat colors: white, cherry, lemon, salmon, and orange.

    c. Give the °F temperatures for dark cherry, light yellow, bright cherry, and dark orange.

# Section 7
# Sound Energy

## Unit 35  Sound Properties, Production, and Acoustics

Each technological advance in transportation, communication, new areas of experimentation, development, and production increases the need for a greater understanding of the nature, transmission, and control of sound. An important area of research is concerned with sound as it relates to safety and health. The effects of certain types of sounds and noises affect individual health, hearing, and industrial output. In addition, national security requires sophisticated equipment to detect, identify, and record sound.

These applications are only a few of the many instances in which an understanding of the principles of sound is valuable. This unit and the next cover the nature of sound and how it is produced, transmitted, recorded, and reproduced.

### NATURE OF SOUND

Sound is vibration traveling through matter in the form of a wave motion. In this process, the original vibrations of an object cause vibrations in the gas, liquid, or solid with which they are in contact. Thus, sound can be transmitted through any form of matter; but, sound cannot be transmitted through a vacuum.

In general, air is the most important of all of the materials which transmit sound. Air is the medium through which sound is brought to the human ear. Certain other materials such as steel, water, and gases are also good transmitters of sound. However, soft objects such as fabrics, fibrous wool, and wood products tend to absorb sound waves and deaden them.

### The Speed of Sound

Numerous experiments have shown that sound travels at a speed of 331 meters per second or almost 1090 feet per second through air at 0°C. The speed of sound increases two feet per second with each degree Celsius rise in temperature. The speed of sound in air at 100°C is equal to 1090 ft./sec. at 0°C + 200 ft./sec. = 1290 ft./sec. at 100°C.

Sound travels almost 4700 ft./sec. in water and 16 400 ft./sec. through steel. In other words, sound travels approximately 15 times as fast through steel and four times as fast in water as it does in air. Another way of comparing the speed of sound in different materials is to consider that sound travels almost 1/5 of a mile per second in air, almost a mile per second in water, and slightly over three miles per second through steel.

The speed of sound can be determined using the formula for velocity:

$$\text{Velocity} = \frac{\text{Distance}}{\text{Time}}$$

$$V = \frac{D}{T}$$

This formula must be used with care because the velocity (speed) of sound varies for different materials and according to temperature.

## THE PRODUCTION OF SOUND

### Sound Waves

Sound is produced by a series of vibrations which move longitudinally along the length of the wave. A simple way to study the production of a sound is to use a tuning fork. This device is a heavy pronged piece of metal with a handle. A slight pressure applied to the prongs causes them to vibrate and produce a sound.

*Compression.* As the prongs move back and forth (in outward and inward directions), the molecules of air near the prongs also move. On the outward stroke the air near the prongs is compressed as shown in figure 35-1. This air under high pressure pushes against the air next to it and starts a *compression* or *high-pressure wave.* This wave continues to travel away from the prongs at a speed determined by the medium through which it is moving. A pressure wave, or the condition of compression, is shown in figure 35-1A.

*Rarefaction.* As the tuning fork prongs return to their original position and then continue to move inward beyond that point, a low-pressure area is created by the inrushing air, figure 35-1B. In this area, a *low-pressure* or *rarefaction wave* is created. The rarefaction wave also moves outwardly from the prongs at the same speed as the compression wave because the medium is the same, figure 35-1C.

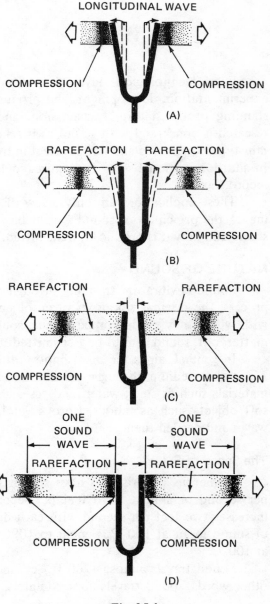

Fig. 35-1

Each time the prongs move outward they create a high-pressure compression wave. As the prongs move inward, a low-pressure rarefaction wave is created.

Thus, for each complete cycle of the prongs (back-and-forth movement), a compression wave and a rarefaction wave are formed, figure 35-1D. This combination of waves represents a single sound wave.

The number of sound waves produced for any interval of time is the combined number of compressions and rarefactions produced in sequence in a specified time.

A sound wave can be illustrated as a series of light and dark concentric circles, figure 35-2. The figure shows three characteristics of sound waves:

1. Sound waves travel outwardly in all directions at the same speed. As a result, the sound wave is shaped like an expanding ball.

2. The sound wave includes a series of compressions and rarefactions.

3. The compressions and rarefactions and the resulting sound decrease as the distance from the source increases.

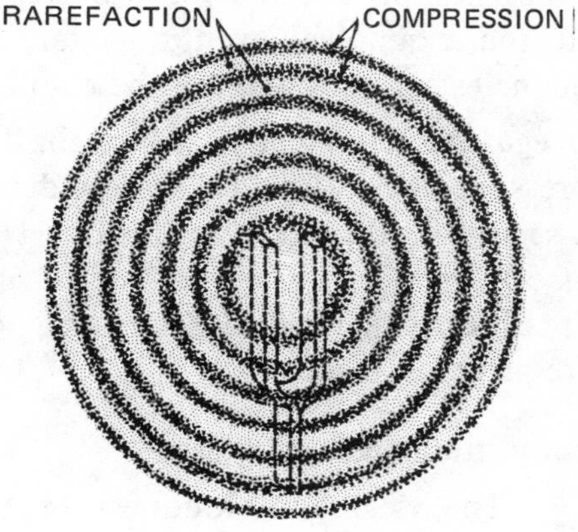

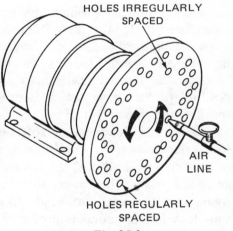

Fig. 35-2

## Pitch and Frequency

Different vibrations cause different sounds. Some sounds have high tones and others have low tones. The word *pitch* is used to indicate the difference between the high tones and the low tones.

*Regular Frequencies.* The pitch of a sound depends on the number of compressions in a given period of time. For example, as air is blown through equally spaced holes on a rotating disc, figure 35-3, the pitch changes as the speed of the disc is varied. When the disc revolves at a slow speed and the holes are regularly spaced, a low-pitched musical sound is produced by the vibrations set up as the air alternately strikes a hole and then the rotating disc. As the speed of the disc is increased, more vibrations are produced and a high-pitched shrill sound results.

Fig. 35-3

*Frequency* means the number of vibrations per second required to produce a tone, figure 35-4. For example, if there are 50 holes on the rotating disc in figure 35-3, and the disc turns at 20 revolutions per second, a jet of air directed through the holes produces a frequency of 1000 vps (vibrations per second), or hertz (Hz).

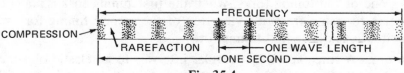

Fig. 35-4

If a tuning fork vibrates at 1000 vps (Hz), the tone is identical to the one produced by the disc rotating at 20 revolutions per second. In other words, the two tones are identical because the frequencies of vibration are the same. Higher tones are produced by higher frequencies and lower tones are produced by lower frequencies. In the examples given thus far, the frequencies are considered to be regular.

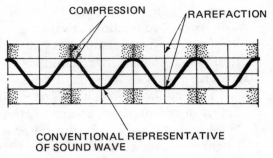

Fig. 35-5

*Irregular Frequencies.* When the frequency of sound is *irregular* — that is, one compression at 1/500th of a second is followed by the next compression at 1/400th of a second, followed by a third at 1/485th of a second, there is no fixed frequency such as 500 vps. As a result, sound having an irregular frequency is noise. Thus, the difference between musical tones and noise is the difference between regular and irregular frequencies, respectively.

## Amplitude and Loudness

The degree of loudness or faintness of a sound depends on four factors; (1) the distance from the sound; (2) the state of the medium through which the sound is transmitted; (3) the amplitude of the original vibration; and (4) the frequency of the vibration. *Amplitude* is the distance between the normal position to the final position of an object which is producing specific vibrations, figure 35-6.

Since the loudness or softness of a sound depends on the amplitude, the larger

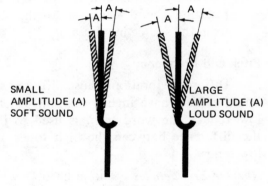

Fig. 35-6

or greater the amplitude, the louder is the sound. Figure 35-6 shows two variations in amplitude, using the tuning fork to produce the sound waves. Note in part A of the figure that the small amplitude (A) of vibration of the tuning fork produces a faint sound. A louder sound is produced at part B of the figure because the amplitude of vibration is greater. Sounds having too low a frequency or too high a frequency may be outside the range of hearing of the normal human ear.

## Resonance

*Induced Vibrations.* Under certain conditions, a vibrating object will produce vibrations of the same frequency in another object. Such vibrations are called *induced vibrations* or *sympathetic vibrations.* To demonstrate induced vibrations, two tuning forks can be mounted a few feet apart with a ping-pong ball suspended from a test stand so that it touches one of the tuning forks. When the first tuning fork is made to vibrate, the ball will start to swing from the vibration of the second tuning fork when it responds to the vibrations of the first tuning fork.

*Forced Vibrations.* A force applied at regular intervals to an elastic object causes the object to vibrate with the same frequency as the applied force. Such a vibration is known as a *forced vibration*. The purpose of forced vibrations is to cause a sound producer to set into vibration more air than the small vibrating object itself can affect. Typical applications of this principle include the cone of a loudspeaker, a violin body, or a piano sounding board. In each instance, the responding surface is designed so that it has no natural vibration of its own in the frequencies with which it is to be used. This condition is necessary to prevent exceptionally loud sounds at the required frequency.

***Natural and Applied Frequencies: Resonance.*** When the natural frequency of vibration of an object is the same as the frequency of an applied force, the amplitude of vibration is increased. This increase is due to the fact that sympathetic vibrations are produced and the response is called *resonance*. Resonance takes place when the natural and applied frequencies are identical.

Resonance can be demonstrated with a tuning fork, a beaker of water, and a glass tube which is open at one end and closed at the other end. As the prongs of the tuning fork vibrate and one end of the tube is moved up and down in the water to vary the column of air, there is one height at which the air in the tube resonates with the vibration of the tuning fork and a loud sound results. This resonance is explained by the fact that the vibrating fork induces vibrations in the air in the column.

***Resonance of a Closed Tube.*** The previous demonstration showed that at the proper height, the vibrations of the air column and the vibrations of the tuning fork are in step and resonance occurs.

The glass tube in this demonstration is known as a *resonance tube* because its length influences the tone produced. The best resonance occurs when the length of a resonance tube is equal to one-quarter of a wavelength. Thus, a high-pitched tone having a high frequency requires a longer resonance tube.

Figure 35-7 illustrates what happens as a tuning fork vibrates over a resonance tube. As the lower prong of the tuning fork moves to the dotted

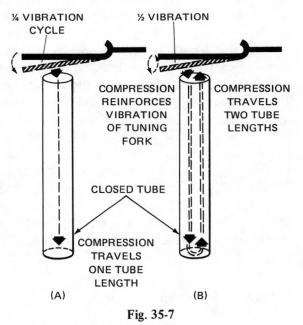

Fig. 35-7

position as shown at (A), it sends a compression wave down the tube. This compression wave moves to the bottom of the tube, strikes the bottom, is reflected by the water, and returns up the tube. If the compression reaches the prong of the fork at the instant the prong returns to its original position (as shown by the solid black outline), the compression will match the vibration of the tuning fork and resonance will occur.

This movement of the prong represents only one-half a cycle as the prongs must still spring back beyond the original position. Since the motion of the vibrating prong is one-half of a vibration (producing one-half wavelength), resonance takes place in a closed tube when the length of the air column is one-fourth the wavelength of the sound waves produced.

In the time required by the tuning fork to move through one-half of its cycle, the compression moves down the tube, strikes the surface of the water, and is reflected back to the prong again. This compression travels two lengths of the tube. If the compression is to reinforce the amplitude of the sound, it must move in the same direction as the prong. If the upward moving compression reaches the prong at the instant the prong is moving downward, the compression opposes the vibration of the prong and the amplitude of the sound is not reinforced.

Thus, resonance causes an increase in the sound when the ratio of the tube length traveled by the sound during one vibration of the tuning fork is 4:1, that is, four tube lengths to one vibration. In other words, at resonance, four lengths of a resonating tube are equal to one wavelength produced by the tuning fork.

*Resonance of an Open Tube.* In an open tube, the best resonance occurs when the length of the tube is one-half the wavelength of the sound it reinforces. As a compression is sent down through an open tube and reaches the open bottom, it spreads out rapidly into the surrounding air. As a result, a rarefaction is started back through the tube. Thus, the reflection is a rarefaction rather than a compression. To create resonance, the rarefaction must reach the prong at the rarefaction phase of its cycle.

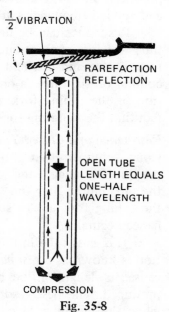

If the length of the resonance tube is any whole number multiple of one-half the wavelength, resonance takes place. Consider an open tube length which is twice the size found to be equal to the wavelength of a sound. The tuning fork makes two whole vibrations while the compression travels down the tube and the rarefaction returns up the tube. While the tube length is in multiples of one-half the wavelength, resonance takes place. However, the longer the tube, the weaker is the resonance.

Fig. 35-8

Resonance is important in increasing sound intensities and in reproducing sound. The principles of resonance are applied to musical instruments, to industrial instruments which test the soundness of a material by resonance, and to electrical amplifying devices.

## KINDS OF SOUNDS

The most ingenious mechanism for hearing and distinguishing many kinds of sounds is the human ear. The outer ear is so sensitive that the compressions and rarefactions of sounds cause it to vibrate in time to the vibrations. These vibrations are transmitted to a structure called the cochlea in the ear. Within the cochlea is a flexible membrane containing many nerve pathways leading to the brain.

As vibrations reach the cochlea, the membrane begins to vibrate. Different parts of the membrane vibrate to the different frequencies of the original sound. As each part of the membrane responds to a different frequency, such as 312 vps in one location, 245 vps in another location, and 478 vps in a third location, each of these parts sends impulses via the nerves to the brain. Thus, it is possible to distinguish one tone from another.

## Limitations of Hearing

The human ear is capable of hearing sounds within a frequency range of 16 vps to 20 000 vps. The hearing range of animals and birds extends to 80 000 vps and more. As a result, they are capable of hearing a greater range of sounds than a human being.

Today, scientists are experimenting with ultrasonic waves which are sound waves of high frequency ranging from 20 000 vps to more than 200 000 vps. Ultrasonic sound waves are extremely important in modern aviation, medicine, and other branches of science.

## Wave Interference

*Interference* refers to the condition existing when two sound waves meet and one wave partially or completely cancels the other wave. This condition is created whenever the compression of one set of sound waves is trying to push the ear drum in at the same time that the rarefaction of a second set of waves is tending to move the ear drum outward. Thus, the compression cancels the rarefaction and the net result is that nothing is heard.

*Beats.*  When two regular sounds having slightly different frequencies interfere with each other, there are instantaneous moments of silence and loudness. The resulting effect is a *beat*. A beat occurs whenever a rarefaction interferes at any half wavelength with a compression. If the amplitudes of the frequencies are equal during the interference, there is momentary silence, figure 35-9.

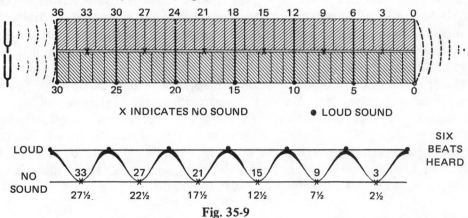

Fig. 35-9

Whenever the compressions are in step and reinforce each other, the moments of silence are followed by moments of loudness. Beats result when two vibrations of slightly different frequencies reach the ear at the same time. This number of beats per second represents the difference between the frequencies of two sounds or notes. Thus, if two notes have frequencies of 160 vps and 164 vps, respectively, there will be four beats per second. The number of beats per second increases as the difference between two regular frequencies increases; it follows that when the difference between two regular frequencies decreases, the number of beats per second will also decrease.

## Musical Tones

The principles of sound are extremely important in music. All musical instruments are constructed of materials having enough elasticity to allow them to vibrate easily at a definite and regular rate to produce specific tones. The three major groupings of instruments — string, wind, and percussion — all depend on regular vibrations. It has been found that when the tone frequencies are in a simple ratio to each other, such as 3:2, or 5:4, harmony results. On a piano keyboard, the sounds produced range from 28 vps at one end (low tones) to over 4000 vps at the other end (high tones). Harmony results when keys are struck that are in a simple ratio to each other. For a piano, there is a simple ratio (4:5) between middle C (256 vps), the next E key to the right (320 vps), and the next G key (384 vps). When these three keys are struck together, they produce a harmonious sound.

## SOUND ACOUSTICS

### Reflection of Sound Waves

A sound wave travels as a longitudinal wave away from the vibrations causing it. When this wave strikes a surface, the sound may be deadened by contact with a sound-absorbing material, or it may be reflected from a nonabsorbing material. One practical application of the principle of sound reflection is a sonar device. Such a device helps to determine how close a ship or other object is to a second object. In sonar, sound waves are sent out from one object to strike another object. The distance between the objects is determined by measuring the time interval required for the sound to return to the source.

Since the sound travels to the object and returns, the distance from the object is one-half the speed of sound through the medium being measured multiplied by the time in the same unit. In a room 100 feet long and at a temperature of 60°C, sound reaches from the front to the back of the room in 1/20 of a second and is reflected back in another 1/20 second.

Under certain conditions, reflection is helpful as in the case of mining operations where the depth reached by a charge is measured by sound reflection. When sound is reflected, but the time interval causes a confusing effect, these sound reflections are called *reverberation*. When there is a considerable reverberation in a room, it is said to have poor acoustical properties. Reverberation can be overcome by altering the construction of a room and/or using sound-absorbing materials. These materials absorb most of the sound waves and reflect only a few waves.

### Refraction of Sound Waves

Sound travels at a constant speed in all directions if the medium through which it is transmitted is uniform and at rest. A medium at rest refers to variations in temperature and other conditions of the medium. When the medium is not uniform, the direction of travel of sound waves changes and the speed of the sound is greater in one part of the medium than in another. This changing or bending of sound is known as refraction.

As an example of refraction, consider a sound in open air. The sound travels uniformly in all directions if the air is at rest. However, such is rarely the case because air is seldom of uniform temperature for any distance and almost never at rest. As a result, the sound waves are refracted. The sound travels through the moving medium at a speed which is relative to the medium and the position from which it is heard. Sound carries a greater distance with a wind as compared to the distance it travels against a wind.

## Absorption of Sound Waves

Some of the regular motion in a sound wave is changed to an irregular motion as the wave passes from one medium to another. This change in wave motion reduces the amount of sound transmitted and is described by the word absorption. *Absorption* is the ability of a material to absorb some of the energy of the sound wave and change this energy to heat energy. Materials which are good absorbers of sound are usually fibrous or porous materials such as fabrics, felts, wall boards, and cork. Some of the energy of sound waves is changed in the pores of these materials to heat energy, thus reducing the sound transmitted.

## Insulation and Absorption

A sound that begins in a room and is reflected from the walls persists until its intensity is such that it is no longer audible. This repeated reflection is known as reverberation. Some reverberation is desirable; the absence of all reverberation makes a room sound dead or hollow.

Reverberation is undesirable when speech sounds or musical sounds persist as a result of reflection after the next sound is made. If the walls of a room or a container in which the sound is made are good reflectors of sound, some of the sound reflections can be diminished by covering the walls with a sound-absorbent material. The sound waves are converted into irregular motion in the absorbent materials and there is limited sound energy passing through them.

## Insulating Materials

When possible, heavy stone or masonry walls are used to separate rooms if it is undesirable for sound to pass from one room into an adjoining room. In other cases, tight walls prevent sound waves from passing through them. However, although the waves may not pass through the walls, they frequently set up a vibration in the wall which causes sound vibrations in the air in the next room.

Another way of insulating against sound is to separate two surfaces by an insulating material so the vibrations of one wall are not transmitted to another. One method of insulating is to use separate studding for each surface and insure that there is no mechanical contact between the studs on the different surfaces. Soft insulating material can then be added between the outside surface and the studs. Some sound transmission is eliminated by this method of insulating. If the surfaces are nailed through the insulating material, some vibrations are transmitted through the nails.

Another method of partially insulating against sound for walls that vibrate easily is to add fillers between the walls such as rock wool and other highly sound-absorbent materials. Any material that tends to reduce the sound vibrations may be used. Part or all of the wall sections also can be covered with rough plaster, draperies, felt, or compressed fiberboard to provide sound-absorbent surfaces.

## Reverberation Time

It was pointed out previously that a certain amount of reflection is needed if a voice is to carry. Although some reverberation is desirable for speaking rooms, there are many places in business and industry where the reverberation of machinery and other manufacturing sounds causes nervous strain and inefficiency.

The recommended *reverberation time* for an auditorium is from one to two seconds. This period is the time required for a sound to die away until it is inaudible. Reverberation time is expressed by a formula as follows:

$$\text{Reverberation Time (T)} = \frac{\text{0.05 Volume of Room (in cu. ft.) (V)}}{\substack{\text{Total Absorption of all Materials} \\ \text{in the Room (A) (in sq. ft.)}}}$$
$$\text{(in seconds)}$$

The total absorption of all materials in the room is found by multiplying the area of each material by its coefficient of absorption and adding each of these separate values together.

The coefficient of absorption represents the fraction of sound energy that a material absorbs for each reflection. For example, for sounds of medium pitch, ordinary plaster absorbs 3.4% of the sound for each reflection. Carpets absorb from 15% to 20% of the sound for each reflection; draperies absorb from 40% to 75%; and marble absorbs 1%. By varying the materials used and the construction features of a room, it is possible to control the reflection of sound and the reverberation time.

## MEASUREMENTS APPLIED TO SOUND ENERGY

### Frequency, Wavelength, and Velocity

Frequency, wavelength, and velocity can be computed by using the formula:

$$\text{Wavelength (W)} = \frac{\text{Velocity (V)}}{\text{Frequency (F)}}$$

For convenience, the formula can also be written in two other forms:

$$\text{Frequency (F)} = \frac{\text{Velocity (V)}}{\text{Wavelength (W)}}$$

$$V = W \times F$$

By substituting two values in one of the three forms of the formula, it is possible to determine the third value. For example, if the speed of radio waves is 186 000 miles per second and a radio station broadcasts at 744 kilocycles, the wavelength can be determined. The value of 744 kilocycles is equal to 744 000 vps because the prefix *kilo* means 1000 and each cycle is one complete vibration.

$$W = \frac{V}{F}$$

$$W = \frac{186\ 000}{744\ 000} = \frac{1}{4} \text{ mile}$$

The same formulas apply to sound waves. If the wavelength of an object vibrating at the rate of 240 vps in a room at 55°C is to be found, first determine the velocity of sound.

$$V = 1090 + 110 = 1200 \text{ ft./sec.}$$

$$W = \frac{1200}{240} = 5 \text{ feet}$$

Thus, with a frequency of 240 vps and a velocity of 1200 ft./sec. the wavelength is 5 feet.

### The Power (Intensity) of Sound

The loudness, power, or intensity of sound waves can be measured. Loudness is produced by variations in pressure. The pressure of the loudest sound is about 1 000 000 times as great as that of the softest audible sound. Above the audible range of hearing, added pressure causes the sensation of pain rather than hearing.

The power of sound is measured on a scale of intensity ratios. For example, if the intensity of a sound is increased 100 times, the sound seems twice as loud; with an increase of 1000 times, the sound seems three times as loud; and at 10 000 times, the sound seems four times as loud.

Note in the first instance that 10 x 10 is 100. This expression can be written mathematically as $10^2$. Similarly, 10 x 10 x 10 = 1000 = $10^3$ and 10 x 10 x 10 x 10 = 10 000 = $10^4$, and so on. Thus, the difference in intensities of sounds can be stated as the difference in the exponents of 10. If the sound is increased 100 times or $10^2$, the gain in power equals the exponent, two. When the sound is increased 1000 times ($10^3$), the gain is three.

The exponent of ten is also called a *bel,* in honor of Alexander Graham Bell. In practice, a *decibel* is used to describe sound intensity because the bel is too large a unit of measurement. Since the prefix *deci* means one-tenth, there are ten decibels in one bel. The decibel is the smallest difference in sound intensity which the human ear can distinguish.

A comparison of common sounds in terms of their intensities can be made, figure 35-10. A barely audible sound, which is said to be on the *threshold of hearing,* is rated at 0 decibels; ordinary conversation is rated at 60 decibels; a stamping plant at 100 decibels; and the loudest sounds at 170 decibels. Sound intensities above 120 decibels produce a feeling of pain. The safest continuous noise level of sound is below 70 decibels.

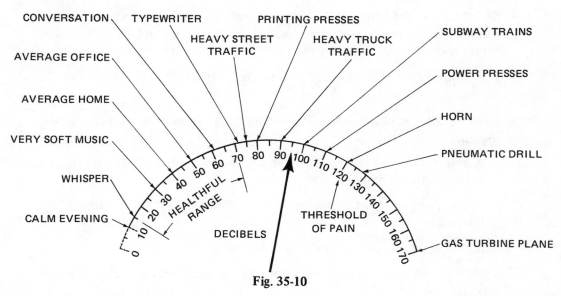

Fig. 35-10

The intensity of sound is measured with a sound-level meter which transforms sound energy into electrical energy. The microphone on the meter converts sound energy into feeble electrical impulses which, in turn, are amplified to move an indicator. The indicator shows the decibel reading on a graduated scale of the instrument.

---
## SUMMARY
---

- Sound is produced by the vibration of any material in an elastic medium. Sound is heard through the auditory nerves which transmit the vibrations and communicate the sensation to the brain.

- Sound waves travel at the rate of approximately 1090 ft./sec. at 0°C, and increase 2 ft./sec. for each degree Celsius rise in temperature.

- A sound wave consists of a high-pressure compression wave and a low-pressure rarefaction wave.

  - A wavelength is the distance between two successive compressions or rarefactions.

- Sound waves produce different sounds depending on the number of vibrations per second (frequency). The greater the frequency, the higher is the pitch.

- Musical tones result from sound waves of regular frequency, noise is due to sound waves of irregular frequencies.

- The quantity of sound depends on the distance, the material through which the sound is transmitted, the amplitude of the sound, and its frequency.

- Vibrations may be induced or forced.

- Resonance takes place when the natural and applied frequencies of vibration are identical.

- At resonance, the length of a closed resonating tube is one-fourth its wavelength. For an open tube, the greatest resonance occurs when the tube length is one-half its wavelength.

- The frequency range of audible sounds is from 20 Hz to 20 000 Hz (cycles/sec.).

- Beats result when two vibrations of different frequencies reach the ear together.

  - A rarefaction interfering at one-half a wavelength with a compression produces momentary silence.

  - A compression reinforcing another compression produces momentary loudness.

- Sound may be reflected, refracted, or absorbed.

- The frequency, wavelength, and velocity of sound waves are related by the formula: $W = V \div F$.

- The intensity of sound is measured in units of one-tenth a bel or decibels.

  - Sounds vary in intensity as the difference of the exponents of 10.

  - The scale on sound-level meters ranges from 0 to 170 decibels.

# ASSIGNMENT UNIT 35  SOUND PROPERTIES, PRODUCTION, AND ACOUSTICS

## PRACTICAL EXPERIMENTS WITH SOUND AND ACOUSTICAL TREATMENT

### Equipment and Materials Required

| Experiment A<br>Pitch | Experiment B<br>Frequency | Experiment C<br>Resonance |
|---|---|---|
| Small motor and variable speed control<br><br>Sound disc with two circles of holes, one contains regularly spaced holes and the other has irregularly spaced holes<br><br>Air source and control valve<br><br>Tachometer<br><br>Stop watch | Tuning fork and scriber<br><br>Chalked rectangular plates (approx. 4" x 24") | Tuning fork with sounding hammer<br><br>Beaker<br><br>Graduated hollow tubes (one to slide in the other) |

| Experiment D<br>Insulation | Experiment E<br>Absorption |
|---|---|
| Sound box with double opening | Sound box with double opening |
| Matched bells or chimes | Matched bells or chimes |
| Switch and power source | Switch and power source |
| Single panels for the sound box (sheet metal; heavy felt; 1/4-in., 1/2-in., 3/4-in. plywood, composition board, cork, or any other materials available for testing) | Material liners for the five sides of a sound box (sheet metal, acoustical plaster, plaster board, acoustical celotex, plywood, cork, linoleum, rubber tile) |
| Double-surfaced panels for sound box constructed with rigid spacers (sheet metal, plywood, composition board, cork, or any other suitable materials) | |
| Double-surfaced panels with surfaces insulated from each other using the same face panels | |
| Double-surfaced panels with surfaces insulated from each other and with fillers added such as sawdust, plaster, and cork | |

## EXPERIMENT A.  PITCH

### Procedure:  Regular Frequency

1.  Direct a small stream of air into the outer circle of regularly spaced holes on the sound disc.

2.  Revolve the motor slowly and check its speed with a tachometer for ten seconds.

3. Prepare a chart for Recording Results as shown.

4. Determine the number of revolutions per second of the disc.

5. Determine the frequency by multiplying the number of revolutions per second by the number of holes in the disc circle. The product is the number of vibrations per second or frequency.

6. Indicate whether the tone of the sound produced has a low, medium, high, or shrill pitch.

7. Increase the speed of the motor. Calculate the frequency and determine the pitch. Record both the frequency and the pitch.

8. Repeat step 7 and increase the speed a third and a fourth time. Record the results at C and D in the table.

9. Use the same frequencies as recorded for A, B, C, and D and increase the amount of the airflow.

10. Indicate in the spaces at E, F, G, and H respectively, the pitch produced.

**Recording Results**

| | Frequency (vps) | Pitch | | Pitch |
|---|---|---|---|---|
| A | | | E | |
| B | | | F | |
| C | | | G | |
| D | | | H | |

**Interpreting Results**

1. Using the recorded data, prove or disprove the statement that the tone depends on the number of vibrations per second.

2. State the conditions under which tones having a low pitch and those having a high pitch are produced.

**Procedure: Irregular Frequency**

1. Direct the airstream against the inner circle of irregularly spaced holes on the the sound disc.

2. Rotate the disc at various speeds (very slow, slow, medium, and fast) using a control mechanism.

3. Prepare a table for Recording Results and note in each case the pitch of the sound produced.

4. Increase the amount of air and repeat steps 2 and 3 (using speeds A, B, C and D).

5. Determine if there are any variations in the characteristics of the sounds produced. (Enter the results in the table at E, F, G, and H.)

## Recording Results

|  | Speed | Characteristics of Sound Produced |  | Variation of Sound Produced |
|---|---|---|---|---|
| A |  |  | E |  |
| B |  |  | F |  |
| C |  |  | G |  |
| D |  |  | H |  |

## Interpreting Results

1. Explain any variations between the sounds produced by regular compression and irregular compression.
2. Based on the results of the experiment, explain in scientific terms the factors which control the production of noise.

## EXPERIMENT B.  FREQUENCY

### Procedure

1. Select a tuning fork with a marking device (scriber) attached.
2. Set the prongs of the tuning fork in motion so that they vibrate slowly.
3. Trace the frequency pattern by placing the marking device on the chalked metal surface and moving the fork slowly on the rectangular plate.
4. Determine the amount of time required to move the tuning fork lengthwise along the plate.
5. Repeat steps 2-4 with the prongs of the tuning fork vibrating at a faster speed.
6. Trace the frequency pattern longitudinally along a second chalked plate in the same time interval determined in step 4.
7. Repeat steps 2 and 4 a third time with the prongs of the tuning fork vibrating at an even greater speed.
8. Prepare a table for Recording Results. Count the number of waves on each plate and record these values.
9. Indicate the pitch produced by the tuning fork in each case.

### Recording Results

|  | Number of Waves | Pitch |
|---|---|---|
| A |  |  |
| B |  |  |
| C |  |  |

## Interpreting Results

1. State the relationship between frequency and pitch as shown by this experiment.

## EXPERIMENT C. RESONANCE

### Procedure: Closed End Resonating Tube

1. Place one end of a hollow graduated tube in a beaker of water so that one end of the tube is sealed.
2. Tap a tuning fork so that its prongs vibrate.
3. Hold the vibrating tuning fork over the open end of the tube.
4. Vary the length of the tube by moving it up and down in the water until the vibrations reinforce each other and resonance occurs.
5. Note the graduation on the tube at which the compressions reinforce each other.
6. Continue to test the effect of tube length on sound. Increase the length and record the results in a prepared table.
7. Decrease the tube length and record the result.

### Recording Results

|   | Condition | Result |
|---|-----------|--------|
| A | At Resonance Length | |
| B | Greater Length | |
| C | Smaller Length | |

### Interpreting Results

1. Explain briefly how resonance occurs in a closed tube.
2. Based on observation and the recorded results, prove or disprove the statement that resonance occurs when the natural frequency of the tuning fork is identical to the applied frequency of the resonating tube.

### Procedure: Open Resonating Tube

1. Slide a second hollow tube over the first and again tap the tuning fork to start it vibrating.
2. Hold the vibrating tuning fork over one end of the tube and adjust the length of the open tube until resonance occurs.
3. Measure the length of the tube.
4. Increase the tube length and note any changes in the sound produced.
5. Decrease the tube length and again note any changes in the sound produced.

### Interpreting Results

1. Explain briefly the principles and conditions necessary for resonance.

2. Compare the length of the resonating tube needed to produce resonance in the enclosed tube with the length of an open tube.

3. State the relationship between the wavelength of an open tube as compared with a closed tube.

## EXPERIMENT D. INSULATION

### Procedure

1. Test the bells in both compartments to insure that the sounds are identical.

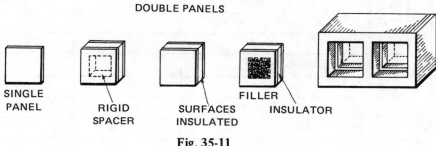

Fig. 35-11

2. Leave one opening bare and cover the other with a panel of sheet metal.

3. Sound the bell in each compartment. Prepare a table for Recording Results and note all differences in sound.

4. Replace the sheet metal panel with a heavy felt panel. Sound the bells and again note and record any difference in sound.

5. Continue to test the remaining single panels, rating each material in the table for its ability to transmit sound or insulate against sound.

   NOTE: Use both sides of the sound box whenever it is difficult to distinguish between the insulating qualities of any two panels.

6. Test the double-surfaced panels constructed with rigid spacers. Classify each type of panel as to its ability to transmit or insulate against sound.

7. Repeat the test using double-surfaced panels with the face surfaces insulated from each other. Record the results.

8. Test the double-surfaced panels which are insulated from each other and filled with different materials. Record the results.

9. Rate each material and construction in the table for insulating properties. Use the numeral (1) to indicate the material with the best insulating properties, (2) the next best, (3) the third best, and so on.

## Recording Results

| | | | Material and Construction | Insulating Properties | |
|---|---|---|---|---|---|
| | | | | Sound Intensity | Numerical Rating |
| Double-surface Panels with | Single Panels for Sound Box | A | Sheet Metal | | |
| | | B | Heavy Felt | | |
| | | C | 1/4" Plywood | | |
| | | D | 1/2" Plywood | | |
| | | E | 3/4" Plywood | | |
| | | F | Composition Board | | |
| | | G | Cork | | |
| | Rigid Spacers | H | Sheet Metal | | |
| | | I | 1/4" Plywood | | |
| | | J | Composition Board | | |
| | | K | Cork | | |
| | Surfaces Insulated | L | Sheet Metal | | |
| | | M | 1/4" Plywood | | |
| | | N | Composition Board | | |
| | | O | Cork | | |
| | Sawdust Filler | P | Sheet Metal | | |
| | | Q | 1/4" Plywood | | |
| | | R | Composition Board | | |
| | | S | Cork | | |
| | Plastic Filler | T | Sheet Metal | | |
| | | U | 1/4" Plywood | | |
| | | V | Composition Board | | |
| | | W | Cork | | |
| | Cork Filler | X | Sheet Metal | | |
| | | Y | 1/4" Plywood | | |
| | | Z | Composition Board | | |
| | | $A^1$ | Cork | | |

## Interpreting Results

Based on the results of the experiment, identify:

1.  Three materials that transmit sound best. Arrange these materials in decreasing order with the material having the best transmission properties first.

2.  Three materials that insulate best. Again arrange these materials with the best insulator listed first.

3. The type of construction that provides the best insulation against sound.

4. The best material and construction to insulate against sound.

## EXPERIMENT E.  ABSORPTION

### Procedure

1. Test the bells in both compartments to see that they are matched.

2. Place the sheet metal liner in one side of the sound box so it covers all five sides.

3. Prepare a table for Recording Results as illustrated.

4. Ring both bells and determine whether the sounds from the lined compartment are louder or softer than those from the bare one. Record the findings.

5. Replace the sheet metal liner with an acoustical plaster liner and repeat step 4.

6. Repeat the test using the six other liners in turn, and any other materials that should be tested.

7. Review the eight materials in the table. Rate each one on a numerical scale where (1) indicates the material which absorbs the greatest amount of sound and (8) is the material absorbing the least amount of sound.

### Recording Results

|   | Material in Liner | Sound-absorbing Property | Rating |
|---|---|---|---|
| A | Sheet Metal | | |
| B | Acoustical Plaster | | |
| C | Plasterboard | | |
| D | Acoustical Celotex | | |
| E | Plywood | | |
| F | Cork | | |
| G | Linoleum | | |
| H | Rubber Tile | | |

### Interpreting Results

1. List the characteristics of those materials which possess the best qualities for absorbing sound.

2. Based on the experiment, what are the characteristics of those materials which reflect sound the best?

## PRACTICAL PROBLEMS ON SOUND
## PROPERTIES, PRODUCTION, AND ACOUSTICS

### Nature of Sound

Select the correct word or words to make statements 1 to 5 correct.

1. Sound may be transmitted through (solids) (a vacuum) (liquids) (gases).

2.  (Fibrous wool and wood products) (Cork) (Steel) (Water) (Cloth) are poor materials for transmitting sound.

3.  Sound travels faster in (water) (a vacuum) (steel) than through air.

4.  Sound travels at the rate of approximately (1090 ft./sec.) (2000 ft./sec.) (186 000 miles/sec.) through air at 0°C.

5.  Sound travels about fifteen times as fast through (a vacuum) (water) (steel) as it does through air.

**Production of Sound**

Match each term in Column I with the correct description in Column II.

| Column I | Column II |
|---|---|
| 1.  Compression | a.  The difference between high tones and low tones. |
| 2.  Rarefaction | b.  A low-pressure air wave created by in-rushing air. |
| 3.  Single sound wave | c.  A vibrating object that induces vibrations in another object. |
| 4.  Pitch | d.  The number of vibrations per second (vps) required to produce a tone. |
| 5.  Frequency | e.  A high-pressure air wave which moves outwardly from a sound producer. |
| | f.  A complete cycle consisting of a rarefaction and a compression. |

Add the term or quantity needed to complete statements 6 to 15.

6.  The degree of loudness (quantity) depends on the _____ of the original vibration.

7.  The greater the amplitude of vibration, the _____ is the sound.

8.  Vibrations of regular frequency produce _____ sounds.

9.  The pitch of a sound becomes _____ as the number of compressions in a given time increases.

10.  The difference between musical tones and noise is the difference between _____ and _____ frequencies.

11.  _____ vibrations result when a small vibrating object sets a larger volume of air into motion to produce a greater sound.

12.  The _____ of vibration is increased when the natural frequency of an object and the frequency of the applied force are the same.

13.  _____ occurs when the natural and applied frequencies are identical.

14.  When the length of a closed tube is equal to _____ its wavelength, resonance takes place.

15.  The best resonance occurs in an open tube when its length is _____ the wavelength of sound.

16. Compute the best resonance length for tubes A, B, C, and D, according to the information given in the table. Use 1090 ft./sec. as the velocity of air.

|   | Sound Frequency (vps) | Temperature | Type Resonating Tube | Best Resonating Length |
|---|---|---|---|---|
| A | 302.5 | 60°C | Closed | |
| B | 60 | 25°C | Closed | |
| C | 302.5 | 60°C | Open | |
| D | 4950 | 80°C | Open | |

17. Explain briefly how beats are produced. Use a diagram.

## Kinds of Sounds and Measurements

For statements 1 to 6, determine which are true (T) and which are false (F).

1. Compressions and rarefactions cause the inner ear to vibrate at the different frequencies of the original sound.

2. Resonance is the result of rarefaction canceling a compression.

3. The frequency range of human hearing is between 5 vps and 200 000 vps.

4. Beats result from instantaneous moments of silence and loudness of two regular sounds with different frequencies.

5. Two notes having regular frequencies of 32 and 36 vps, respectively, have four beats.

6. Harmony results when tone frequencies are in simple ratios, such as 5 1/2 : 4 1/3 or 13 : 11, to each other.

7. Compute the wavelength in air, water, and steel for each frequency and temperature given in the table.

|   | Frequency (vps) | Temperature (°C) | Medium | Wavelength (ft.) |
|---|---|---|---|---|
| A | 120 | 55 | Air | |
| B | 1090 | 0 | Air | |
| C | 120 | 55 | Water | |
| D | 1090 | 0 | Water | |
| E | 120 | 55 | Steel | |
| F | 1090 | 0 | Steel | |

8. The scale on a sound-level meter indicates that a sound of 40 decibels is 10 X louder than one at 30 decibels; a sound at 50 decibels is 10 X louder than one at 40 decibels; and so on.

   a. Select any three sounds. Either measure the intensity of each one with a sound-level meter or determine the intensity from a table.

   b. Tell what the relationship of intensities is between the lowest decibel reading and the other two readings for the three sounds.

## Sound Acoustics

Match each term in Column I with the correct definition in Column II.

|  | Column I |  | Column II |
|---|---|---|---|

Column I

1. Sound Reflection
2. Reverberation
3. Sound Refraction
4. Sound Absorption

Column II

a. The effect of a time lag in the return of sound waves which causes confusing sounds.

b. The ability of a material to change the energy of a sound wave to heat energy, thus reducing the amount of sound transmission.

c. Identical natural and applied frequencies of vibration which increase the amplitude of vibration.

d. A sound wave that travels as a longitudinal wave away from the vibrations causing it, strikes a nonabsorbing material, and returns.

e. The changing or bending of sound by a medium that is not uniform and is in motion.

5. Arrange the following materials in the order in which each may be used as sound-insulating material: felt, plywood, composition board, cork, sheet metal, and marble.

|  | Material | Sound Insulating Rating |  |
|---|---|---|---|
| A |  |  | Best |
| B |  |  | ↑ |
| C |  |  |  |
| D |  |  |  |
| E |  |  | ↓ |
| F |  |  | Poorest |

6. Prepare a Rating Table similar to the one illustrated. Give a numerical rating to each of the double-surfaced constructions to indicate its ability to insulate against sound. Numeral (1) is to indicate the best material and construction, (2) the next best, and so on, to (25) for the poorest combination of material and construction.

|  | Material | Double Wall Construction With |  |  |  |  |
|---|---|---|---|---|---|---|
|  |  | Rigid Spacers | Surfaces Insulated | Surfaces Insulated With | | |
|  |  |  |  | Sawdust Filler | Rock Wool Filler | Cork Filler |
| A | Sheet Metal |  |  |  |  |  |
| B | Acoustical Plaster |  |  |  |  |  |
| C | Concrete Block |  |  |  |  |  |
| D | Wood |  |  |  |  |  |
| E | Composition Board |  |  |  |  |  |

**Rating Table**

7. Use a reference table of absorption coefficients and indicate the percentage of sound energy absorbed at each reflection for each material given in the table.

| | Material | Percent Absorption Coefficient |
|---|---|---|
| A | Draperies | |
| B | Acoustical Plaster | |
| C | Marble | |
| D | Steel | |
| E | Open window | |
| F | Wood | |
| G | Carpet | |
| H | Celotex (1/2'') | |

8. Determine either the reverberation time (T), room volume (V), or total absorption area (A), as indicated by the blank spaces in the table, for rooms A, B, C, and D.

| Room | T (sec.) | V (cu. ft.) | A (sq. ft.) |
|---|---|---|---|
| A | | 200 000 | 2000 |
| B | 1.5 | 30 000 | |
| C | 0.75 | | 3000 |
| D | | 5000 | 2500 |

9. The walls of the four rooms indicated in the table as A, B, C, and D are constructed of the materials shown. Determine either the total absorption area (A), the volume (V), or the reverberation time (T) for the missing values.

| Room | Sound-Absorbing Material | Coefficient of Absorption | A (sq. ft.) | V (cu. ft.) | T (sec.) |
|---|---|---|---|---|---|
| A | Concrete | 2% | | 2000 | 5 |
| B | Hair Felt (1'') | 60% | 5000 | | 0.5 |
| C | Brick | 5% | 20 000 | 60 000 | |
| D | Regular Plaster | 3.5% | | 7000 | 2 |

10. A laboratory whose volume is 30 000 cubic feet has a reverberation time of 1.5 seconds when empty. If each worker has a sound absorption capability equal to 5 square feet (with coefficient of 1), and there are 20 technicians working in the laboratory, what is the reverberation time?

# Unit 36 Transmitting, Recording, and Reproducing Sound

## TRANSMITTING AND REPRODUCING SOUND

One of the greatest advances in the field of communications started in 1876 when Alexander Graham Bell demonstrated how sound could be transmitted and received. Bell's first experimental telephone was a variation of a telegraph sender and receiver. Bell made some slight changes in the standard telegraph system in which the movement of a telegraph key actuated an electromagnet to transmit signals over telegraph wires (to be heard as duplicate clicks on a sounder).

### The Early Telephone

The transmitter and receiver of the first telephone were alike. In general, each device consisted of a permanent magnet and a strip of iron which was attached to a drumhead, figure 36-1. As sound waves reached the receiver, the vibrations produced by the compressions and rarefactions of the waves caused the drumhead to vibrate. This vibration varied the size of the air gap between the poles of the electromagnet, resulting in a variation in the number of lines of force passing through the electromagnetic leg of the permanent magnet.

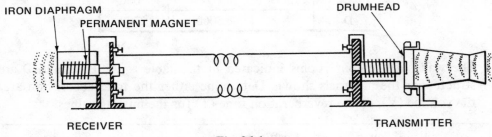

Fig. 36-1

This variation in current induced similar variations in current in the receiver. As a result, the drumhead in the receiver duplicated the vibrations of the transmitter. Although Bell's first receiver worked fairly well, the transmitter did not have the sensitivity to variations in sound required for good transmission.

### The Modern Telephone

In 1877 Edison invented a more sensitive carbon transmitter. Since that time, the carbon transmitter has been improved until its present construction, as shown in figure 36-2 (page 445), includes carbon granules, a carbon disc, a thin diaphragm, and other parts. Vibrations hitting a thin diaphragm push a carbon disc into the carbon granules in a container or box. One wire is connected to the granule box and another wire is attached to the diaphragm.

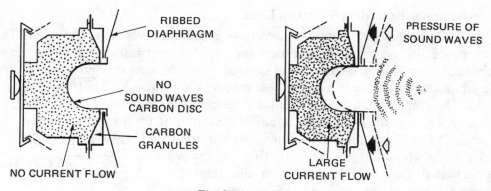

Fig. 36-2

The pressure of the carbon disc on the granules determines the amount of the current flowing through the transmitter. Variations in the pressure of the disc are determined by the vibrations of the sound waves striking the diaphragm. Any variation in air pressure on the transmitter is carried electrically to the receiver where the original vibrations in the transmitter are duplicated and changed back to sound waves.

Figure 36-3 shows how the transmitter sends out electrical currents that vibrate at the same rate as the sound waves which strike the transmitter. The receiver produces sound waves that vibrate with the same frequency as the electrical currents that it receives. Thus, the transmitter-receiver combination changes the original sound to an electrical current and then changes this current back to sounds which duplicate the sounds spoken into the transmitter, figure 36-3.

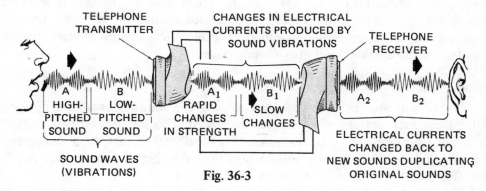

Fig. 36-3

There are limitless variations to telephone circuitry. Many improvements have been made in the application of the principles of sound to reproduction and transmission, transmission lines, and the amplification of sound through radio transmitters.

Significant improvements in the design and construction of telephone cables greatly increased the transmission capacity of telephone companies. In 1877, each telephone was supplied by a single set of wires. Now, large cables containing over 2000 pairs of telephone wires supply many telephones.

## SOUND RECORDING AND REPRODUCTION

Thomas A. Edison is credited with making the first sound recording. Edison was able to use the vibrations of his voice to create sympathetic vibrations which were scratched onto a piece of metal foil.

### Mechanically Recording Sound on Discs

The mechanical method of recording sound waves is widely used. The sound waves are first converted to electrical energy. This electrical energy is then transformed through a crystal element into an equivalent mechanical energy which is used to cut a varying spiral groove in a disc. These grooves are formed by a cutting head.

The transformation from sound to electrical energy takes place through a microphone and (usually) a speech amplifier. The output from this microphone /amplifier combination feeds the cutting head to control the manner in which the groove is cut. The cutting needle is actuated by a flexible diaphragm vibrating at the same frequency and amplitude as the original sound. Thus, a master record (disc) is produced. Other records can then be cut from the master disc or sound tapes can be produced.

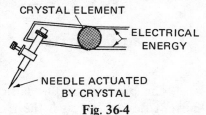

CRYSTAL ELEMENT

ELECTRICAL ENERGY

NEEDLE ACTUATED BY CRYSTAL

Fig. 36-4

### Mechanical Reproduction

To reproduce the sound from a disc, a pickup head transforms the mechanical energy obtained from the varying grooves of the disc to an equivalent electrical energy. When a phonograph record is played, the process of cutting is reversed. The grooves of the record cause the needle to vibrate at the same frequency and intensity as those of the original sound. Since the needle is connected to a diaphragm which is free to vibrate, the vibrations produced are a duplicate of those recorded.

The phonograph needle touches a crystal (cartridge) which instantaneously changes the amount of electrical current. These changes in electrical currents are amplified by electronic tubes and changed into sounds of the same frequency and amplitude as those originally recorded. The electrical energy is strengthened by an amplifier whose output is fed into a loudspeaker.

### Sound Films

Sound can also be recorded by changing the sound energy into light energy. A visible frequency pattern of sound can be traced. The sound waves are converted into electrical vibrations by a microphone. These vibrations are then converted into a changing light pattern which is photographed on film. The width of the light and dark areas on the film varies according to the sound.

To reproduce the sound recorded on the film, the film is run through a projector. As light shines through the film, the sound pattern (light energy) is converted into electrical energy. This electrical energy, in turn, controls the sounds produced in a loudspeaker.

A great deal of experimentation is continuing to improve the quality of reproduced sound. At the higher frequencies in the hearing range, it is difficult for a recording needle, loudspeaker, or other transmitting device, to vibrate freely. In addition, the problem of reproducing very high-pitched and extremely low-pitched sounds at the same time is also being studied.

### Magnetic (Tape) Sound Recorders

Another type of recorder-reproducer unit is the tape recorder. The recording process is illustrated in figure 36-5A (page 447); the playback operation is shown

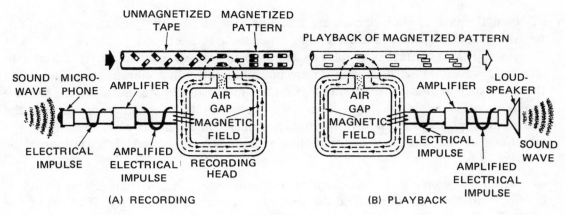

**Fig. 36-5**

in figure 36-5B. In the recording portion of the system, the sound impulse is changed to an electrical impulse through the use of a microphone.

The electrical impulse is strengthened by an amplifier and then is fed into a recording head. This head consists of a coil wound on a soft iron core having a small air gap. Variations in the electrical impulse produce a pulsating magnetic field in the recording head. As a thin, unmagnetized tape is passed over the air gap, the magnetic lines of force pass through that section of the tape and a magnetic pattern of sound is induced on the tape.

### Magnetic Sound Reproducer

When the magnetic (tape) sound recorder is used as a playback device to reproduce the sound, the process is reversed. The magnetized tape produces a pulsating magnetic field in the recording head as it passes over the air gap in the core. This pulsating magnetic field cuts the coil and induces electrical impulses in the coil. These impulses are strengthened by an amplifier and are transformed by the loudspeaker into sound waves.

Magnetic tape recorders and reproducers provide: (1) a completely portable unit; (2) instant playback; (3) excellent fidelity with a minimum of background noise; (4) a recording where a portion of the entire spool can be removed and new parts inserted; and (5) a tape that can be reused by passing it through a demagnetizing head to remove the magnetic pattern.

## SOUND TRANSMISSION AND REPRODUCTION

### Radio Communication

Sound waves can travel at a speed of 1090 feet per second (approximately) for short distances. These same sound waves can also be made to travel at the speed of light (186 000 miles per second) through any medium by changing the form of energy from sound energy to electrical energy. In this process, a faithful reproduction of the frequencies of the sound waves is needed. The audio unit which converts sound waves to electrical waves at the sending point is called a *microphone*. These electrical waves are converted back to sound waves by another device known as a *loudspeaker*.

***The Microphone Converts Sound Waves to Electrical Waves.*** All microphones convert the mechanical variations in air pressure due to sound to an equivalent electrical wave of equal frequency. Microphones are available in a variety of styles to suit many different uses. Both portable and stationary types are available.

In addition, contact microphones are used to convert mechanical vibrations into electrical waves. These microphones are placed directly on a vibrating medium other than air, such as musical instruments. Contact microphones are useful where the surrounding noise levels are high. Two other common types of microphones are the throat microphone and the lip microphone. These microphones react directly to the vibrations of the throat or lips and thus exclude any other outside variations of air pressure.

*Frequency Response and Sensitivity of a Microphone.* When transmitting sound, it is important that the quality of the reception approach the quality (fidelity) of the original sound at the point of transmission. This quality is made possible to some degree by insuring that the microphone is sensitive to a wider range of frequencies than the receiver. Microphone frequency usually ranges from 16 to 16 000 cycles; the average home receiver has a range of 50 to 12 000 cycles.

Sound waves have been described as mechanical vibrations of air at frequencies that are audible. It is interesting to note that normal air pressure must be increased by only a minute amount (about one-millionth of one percent) to create an audible sound. Sounds that are so loud as to produce a painful sensation increase the air pressure about one-tenth of one percent. Thus, it is apparent that if there is to be good reproduction fidelity, a microphone must be sensitive to these fine variations in sound for the entire hearing range.

It must be remembered that since the intensity of a sound wave decreases with the distance from the source, the sensitivity of the microphone also decreases as the distance from the source of the sound to the microphone increases. The output of a microphone is usually rated by the American Standards Association in terms of a unit of atmospheric pressure known as *dynes per square centimeter.*

*Characteristics of Microphones.* One basic type of microphone is the *carbon microphone.* This device operates on the principle that the resistance of a quantity of carbon granules will vary as the pressure exerted upon it varies. This is the same principle on which Edison based his carbon transmitter. Figure 36-6 shows that any variation in the pressure exerted by the diaphragm varies the pressure exerted on the carbon granules. Thus, as sound waves strike the diaphragm and cause it to vibrate, the variations in the vibrations cause variations in the pressure exerted on the carbon granules. The variations are controlled by the intensity and frequency of the sound waves and produce changes in the current flow in the circuit. Therefore, the output of this device is an alternating voltage whose magnitude and frequency correspond to the sound waves.

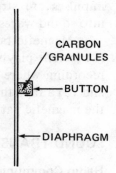

Fig. 36-6

The carbon microphone is valuable because it is: light in weight, low in cost, rugged in construction, portable, and has a high sensitivity.

*Crystal Microphones.* Sound waves can also be converted into an electric current by another type of microphone which uses a crystal. In a *crystal microphone,* thin crystal slabs are clamped together with electrical contact made by cementing a thin sheet of tin foil on both faces of the crystal. To create a sensitive microphone, the crystal is secured to the diaphragm. As the diaphragm vibrates, the sound waves cause the crystals to vibrate and produce an alternating voltage at the terminals of the crystals. This voltage has a frequency and intensity which corresponds to the sound waves.

Crystal microphones are used in public address systems, on recording equipment, and in broadcasting systems. This type of microphone is also light in weight, rugged, easily maintained, and requires no external power source.

## Other Types of Microphones

*Capacitor Microphone.* Three other types of microphones are used to convert sound energy into electrical energy for transmission purposes. A *capacitor microphone* has a diaphragm that is separated from a backplate by a spacer made of an insulating material, figure 36-7. Sound waves cause the diaphragm to vibrate. This vibration changes the air gap and thus the voltage between the capacitor plates.

In this manner, an electrical equivalent of the sound waves is obtained. Although a capacitor microphone is not very sensitive, it has good uniformity. As a result, this device is useful for sound measurement, but is not practical for radio or public address systems.

*Dynamic Microphone.* The *dynamic microphone* is a moving coil type of microphone, figure 36-8. The coil wire is attached to the back of the diaphragm and is free to move back and forth in a strong radial magnetic field. This action induces a voltage in the coil. The induced voltage is an electrical reproduction of the sound waves.

Dynamic microphones are of rugged construction so that they will withstand the effects of temperature, moisture, and vibration. Dynamic microphones may be used indoors or outdoors for broadcasting, public address systems, and for recording.

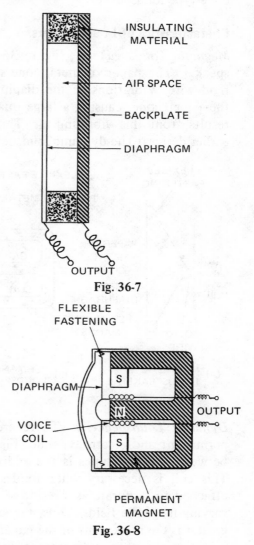

Fig. 36-7

Fig. 36-8

*Velocity Microphone.* The *velocity microphone* is a variation of the moving coil type of microphone. A strip of metal replaces the moving coil and vibrates in a magnetic field. This single conductor (a thin corrugated metal strip) is suspended between the poles of a permanent magnet. As sound waves cause this conductor to vibrate and cut the magnetic lines of force, a voltage is induced in the strip. The voltage is proportional to the pressure and frequency of the sound waves. In other words, the force is proportional to the velocity of the air set in motion by the sound waves.

The velocity microphone is directional; that is, it transmits only those sounds occurring in front of it. Another characteristic of this microphone is its good frequency response.

## SOUND REPRODUCTION

The reproduction of sound is the reverse of the transmission process. In some cases, electrical energy is converted to sound energy directly. At other times, electrical energy must be changed to mechanical energy first and then to sound energy. One of the most common devices used in this process of sound reproduction is the cone-type loudspeaker.

### Characteristics of Loudspeakers

*Magnetic Loudspeakers.*   Figure 36-9A indicates the main parts of a magnetic loudspeaker. The paper or metal cone speaker is attached to the metal diaphragm with a rigid wire. Vibrations of the diaphragm are transmitted to the large surface area of the paper cone, causing a large quantity of air to be set into motion. A large sound results from this vibrating air. This cone-type speaker arrangement was one of the earlier designs. The dynamic loudspeaker is a more recent development.

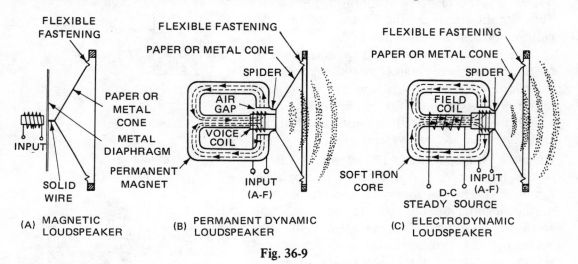

Fig. 36-9

*Dynamic Loudspeakers.*   There are two common types of dynamic loudspeakers: permanent and electrodynamic. Figures 36-9A and B show that the only difference between the two types is the addition of a field coil in the electrodynamic speaker. This coil is necessary when loud sounds are required. The voice coil and the cone attached to it vibrate as a result of variations in the strength of both the fixed and the varying magnetic fields. Thus, the greater the strength of the fixed magnetic field, the greater is the movement of the paper cone and the louder is the resulting sound.

For the permanent magnet dynamic loudspeaker, the strength of the permanent magnet and the volume output of the loudspeaker become weaker in time. The electromagnet dynamic loudspeaker provides a strong magnetic field which does not decrease with time. This electrodynamic loudspeaker derives its name from the fact that an electromagnet is substituted for the permanent magnet. This type of loudspeaker is widely used because of its constant strength and wide range of power output.

### Sound Amplifiers

*Sound amplifiers* are used to increase the current and voltage output of a radio receiver, microphone, or similar device. Depending on its use, a sound amplifier is also known as a speech amplifier, phonograph amplifier, public address amplifier, or microphone amplifier. In each case, the amplifier is used to reproduce speech and music with a good degree of fidelity.

***Sound Horns.***   Sound transformation can also be accomplished by coupling a dynamic loudspeaker to a horn. The *horn* transforms sound energy at a high pressure and low velocity to sound energy at a low pressure and high velocity. The horn contour is such that there is a smooth and continuous increase in the cross-sectional area.

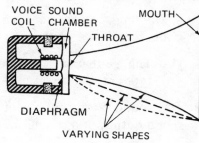

Fig. 36-10

\* \* \* \* \* \* \* \* \* \*

Although this unit covers only a few of the many devices and instruments for recording, transmitting, and reproducing sound, it does describe the principles of operation of these devices and gives applications of equipment in wide use.

─────────────────── SUMMARY ───────────────────

- Sound vibrations are transformed in a telephone receiver into electrical energy by varying the pressure exerted on carbon granules by a thin vibrating diaphragm.
  - Electrical impulses produced in the transmitter are duplicated on the receiver where electrical energy is converted to sound energy.

- Edison is credited with the first recording of sound on metal foil by reproducing sympathetic vibrations.

- Mechanical recording (disc method) requires the transforming of sound waves to electrical energy. The electrical output controls a cutting head which produces a master disc.
  - The process of recording is reversed in playback where the mechanical energy is changed back into sound waves.

- Sound is also recorded on sound films, sound tapes, on light recorders, and so on.

- Sound may be recorded on a tape recorder by feeding electrical impulses into a magnetic recording head. These variations in output produce a pulsating magnetic field which induces variations in the sound pattern of the tape.

- Sound waves can be changed into radio waves traveling great distances at a speed of 186 000 miles per second.
  - Radio waves are changed back to sound waves at the receiving point.

- The microphone converts mechanical variations in air pressure (sound waves) into electrical waves of equal frequency.
  - The common types of microphones include carbon, crystal, capacitor, and velocity types.

— The amplifier increases electrical energy output and quantity of reproduction without changing fidelity.

— The loudspeaker changes electrical waves to sound waves by duplicating the sound pattern through the vibration of a large surface area.

— The permanent and electrodynamic loudspeaker are two common types of speakers.

- The sound horn and various sound amplifiers are used to increase the sound output of mechanical, electrical, and light energy devices.

## ASSIGNMENT UNIT 36 TRANSMITTING, RECORDING, AND REPRODUCING SOUND

### PRACTICAL EXPERIMENTS WITH SOUND EQUIPMENT
### Equipment and Materials Required

| Experiment A<br>Transmitting and Reproducing Sound | Experiment B<br>Recording and Reproducing Sound |
|---|---|
| Early model of telephone transmitter and receiver<br>Modern transmitter and receiver<br>OR<br>Schematic or phantom view drawing of early and modern telephone sets | **Mechanical Reproduction of Sound**<br>Mechanical recording unit (disc method)<br><br>OR<br><br>Sectional and cutaway drawings and specifications for a cutting head |
| **Experiment C**<br>**Transmitting Sounds (Radio)** | **Electrical Recording and Reproduction** |
| Carbon, crystal, capacity, dynamic, velocity, or other types of microphones<br>OR<br>Construction (sectional) drawings and specifications of each type of microphone | Tape sound recorder, complete with accessories<br><br>OR<br><br>Sectional drawings of the recording heads of different types of recorders |
| **Experiment D**<br>**Sound Reproduction** | **Light Recording and Reproduction** |
| Magnetic loudspeaker<br>Permanent magnet dynamic loudspeaker<br>Electromagnet dynamic loudspeaker<br>Sound amplifier | Strip of sound film with projection equipment, or operating diagram of a sound film projector<br><br>Magnifying glass |

## EXPERIMENT A. TRANSMITTING AND REPRODUCING SOUND: COMPARING THE EARLY AND MODERN TELEPHONE

### Procedure

1. Connect an early telephone set or model.

2. Talk into the receiver. Note the action of the drumhead on the transmitter.

3. Listen to the quality of voice reproduction and the action of the drumhead.

4. Connect a modern telephone set for two-way conversation and talk through the transmitter.

5. Note the fidelity of sound reproduction and intensity as compared with the early model.

6. Study the internal construction of a modern telephone. If the actual telephone is not available, use a schematic, cutaway, or exploded view drawing.

## Interpreting Results

1. Compare the sound quality and quantity of an early telephone and a modern telephone receiver.

2. Make a simple sketch of a modern telephone transmitter and receiver.

   a. Label the important parts.

   b. Explain briefly how the telephone works.

## EXPERIMENT B.  RECORDING AND REPRODUCING SOUND

### Procedure:  Mechanical Reproduction of Sound

1. Examine the main parts of the cutting head of a disc recorder.

2. Determine the function of each part or unit in which the energy of sound waves is changed to electrical energy and then to mechanical energy. Use an actual cutting head or the manufacturer's drawings and specifications.

### Interpreting Results

1. List the parts of a cutting head and the steps required to produce a sound disc.

### Procedure:  Electrical Recording and Reproduction

1. Study the operation of a tape recorder.

2. Set up and start the recording unit.

3. Prepare a table for Recording Results and record the sounds produced according to the conditions which follow.

4. Vary the distances of the sound from the speaker to the microphone. Start at 6 in. and then move to 12 in., 2 ft. and 4 ft. (approximately).

5. Change the position of the microphone with respect to the speaker from directly in front of the speaker to the right and left sides, and then above and below the speaker.

6. Vary the input into the recording head from a minimum value to a maximum value.

7. Repeat steps 2–6 with any other type of recorder. Record the results in a similar table.

8. Play back the recorded sounds. Note the variations in fidelity and loudness for each different combination of distance, location, and electrical input in the prepared table.

## Recording Results

|   | Distance to Microphone | Location | Quality (Fidelity) | Quantity | Input to Recording Head | Quality (Fidelity) | Quantity |
|---|---|---|---|---|---|---|---|
| A | 6 in. | In front |  |  | Min. |  |  |
|   |  | R, L, A, B |  |  | Max. |  |  |
| B | 12 in. | In front |  |  | Min. |  |  |
|   |  | R, L, A, B |  |  | Max. |  |  |
| C | 2 ft. | In front |  |  | Min. |  |  |
|   |  | R, L, A, B |  |  | Max. |  |  |
| D | 4 ft. | In front |  |  | Min. |  |  |
|   |  | R, L, A, B |  |  | Max. |  |  |

## Interpreting Results

1. State the effect of varying the distance from the speaker to the microphone on the quality and quantity of the reproduced sound.
2. Make a similar statement about the effect on fidelity and loudness due to the locations of the speaker and the microphone.
3. Compare the effect of variations of the power input into the recording head with the quality and quantity of sound reproduced.
4. State the recording conditions which produce the best fidelity and loudness.

## Procedure: Light Recording and Reproduction

1. Use a magnifying glass to examine a section of sound film having a sound track to see the visual pattern made by sound.
2. Study either a model or a sectional or schematic drawing of projection and sound transcribing equipment to learn how sound can be transformed from one form of energy to another.
3. Note the major stages through which the sound on the film track is transformed until sound waves are produced.

## Interpreting Results

1. Briefly explain the steps and principles involved in changing the original sound energy so it can be recorded on film and, finally, reproduced as sound.

## EXPERIMENT C. TRANSMITTING AND REPRODUCING SOUND (RADIO)

### Procedure

1. Connect a carbon-type microphone to a recording unit.

2. Record speech and musical sounds. Vary the distance and position of the microphone from the sounds.

3. Prepare a table for Recording Results. Play back the sounds reproduced and note variations in tone quality and quantity.

4. Select other types of microphones and repeat the procedure.

   NOTE: If the microphones are not available, study the manufacturer's drawings and specifications for the construction details and operating characteristics of various types of microphones.

**Recording Results**

| | Microphone Type | Distance Microphone to Sound Source | Position of Microphone | Quality of Sound | Quantity of Sound |
|---|---|---|---|---|---|
| **A** | | 6 in. | In front | | |
| | | | R, L, A, B | | |
| | | 2 ft. | In front | | |
| | | | R, L, A, B | | |
| | | 4 ft. | In front | | |
| | | | R, L, A, B | | |
| **B** | | 6 in. | In front | | |
| | | | R, L, A, B | | |
| | | 2 ft. | In front | | |
| | | | R, L, A, B | | |
| | | 4 ft. | In front | | |
| | | | R, L, A, B | | |
| **C** | | 6 in. | In front | | |
| | | | R, L, A, B | | |
| | | 2 ft. | In front | | |
| | | | R, L, A, B | | |
| | | 4 ft. | In front | | |
| | | | R, L, A, B | | |

**Interpreting Results**

1. Name two places and/or conditions under which each of three different types of microphones may best be used.

2. Select from any three microphones the one having the best qualities of fidelity and volume control.

## EXPERIMENT D. SOUND REPRODUCTION

**Procedure**

1. Connect a simple, older model of magnetic loudspeaker to a sound source.

2. Note the action of the rigid wire and the paper cone speaker.

3. Increase the power volume and listen for the effect on the sound fidelity.

4. Repeat steps 1–3 using a permanent magnet dynamic speaker and an electro-magnet dynamic loudspeaker.

5. Add a sound amplifier to the circuit. Note any change in the sound output.

### Interpreting Results

1. Based on the experiment, identify whether the electromagnet loudspeaker or the permanent magnet dynamic loudspeaker produces the better sound fidelity.

2. Select the type of loudspeaker which is best suited to reproducing regular and loud sounds accurately and with the greatest volume.

3. What function does a sound amplifier serve? Is the fidelity improved or impaired with the use of a sound amplifier? Explain.

## PRACTICAL PROBLEMS ON TRANSMITTING, RECORDING, AND REPRODUCING SOUND

### Transmitting and Reproducing Sound

Complete statements 1 to 5 by adding a word, phrase, or quantity to make each one correct.

1. Bell's first telephone consisted of a (an) _____ and a (an) _____ .

2. The essential parts of this early telephone included: _____ , _____ , _____ , and the electrical connections.

3. Any variation in air pressure due to sound waves changes the _____ of a diaphragm in a modern telephone.

4. Variations in pressure on the carbon granules in the telephone transmitter determine the _____ which flows.

5. Transoceanic communication combines sound waves with faster moving _____ waves which travel at _____ .

6. Prepare a table similar to the one shown at the top of page 457. Name the five lettered parts of the receiver and transmitter illustrated in figure 36-11 and give the function of each.

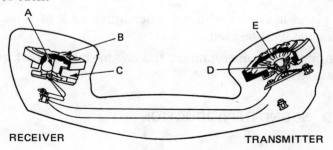

RECEIVER                    TRANSMITTER

**Fig. 36-11**

| | Part Name | Function |
|---|---|---|
| A | | |
| B | | |
| C | | |
| D | | |
| E | | |

## Sound Recording and Reproduction

For statements 1 to 10, determine which are true (T) and which are false (F).

1. Edison first recorded sound by applying the principle of resonance.

2. The function of the cutting head and needle used to mechanically record sound is to increase the frequency and amplitude of sound.

3. Good fidelity refers to the faithful reproduction of an original sound.

4. The vibrations recorded on a spiral groove on a record are duplicated during playback to produce reproductions of the original sounds.

5. Sound energy is changed to heat energy on the sound film track.

6. The light energy produced by a sound track is converted first to electrical energy and then back to sound energy.

7. The tape recorder works on the principle that a sound pattern can be induced in an unmagnetized tape as it passes over an air gap in the recording head.

8. The pulsating magnetic field of a recording head has no effect on inducing a sound pattern on a tape.

9. Magnetic tape recorders have poor sound fidelity.

10. The recording tape can be used only a limited number of times before the recorded sound is erased automatically.

## Sound Transmission and Reproduction

Select the word, phrase, or quantity to complete statements 1 to 5.

1. Radio waves travel at the rate of (1090 ft./sec.) (1200 ft./sec.) (186 000 miles/sec.).

2. Microphones convert mechanical variations in air pressure to electrical current having (equal) (greater) (smaller) frequency.

3. Throat and lip microphones react to (all vibrations of air pressure) (selected vibrations).

4. Microphones are sensitive to frequency ranges of (16 to 20 000) (50 to 5000) (30 to 10 000) (10 to 80 000) cycles per second to insure good fidelity.

5. The sensitivity of a microphone (decreases) (increases) (is unaffected) as the distance from the source of sound increases.

Column I lists four different types of microphones. Select the description or operation in Column II which best describes each type.

| Column I | Column II |
|---|---|
| 6.  Carbon Microphone | a. A coil of wire moving in a strong radial magnetic field induces a voltage in the coil which is an electrical reproduction of sound waves. |
| 7.  Crystal Microphone | |
| 8.  Capacitor Microphone | b. Vibrations of a diaphragm are transmitted to a large surface area to set great quantities of air in motion. |
| 9.  Dynamic Microphone | |

c. The vibration of the diaphragm varies the space between the backplate and the diaphragm, thus changing the voltage to produce an electrical equivalent of the sound waves.

d. Operates on the principle that the electrical pressure exerted by a quantity of carbon granules varies as the air pressure exerted upon them varies.

e. The greater the strength of the fixed magnetic field, the greater is the movement of the paper cone.

f. The vibrations of crystal elements produce an alternating voltage having the same frequency and intensity as the sound waves.

## Sound Reproduction

Select the correct word or phrase to complete statements 1 to 5.

1. The strength of the permanent magnet of (a permanent magnet dynamic) (an electrodynamic) loudspeaker becomes weaker in time.

2. The (permanent magnet dynamic) (electrodynamic) loudspeaker is better adapted to loud sounds.

3. The (magnetic) (permanent magnet) (electrodynamic) loudspeaker has a wide range of power output and constant strength.

4. The (microphone) (horn) (loudspeaker) transforms sound energy at high pressure and low velocity to low pressure and high velocity.

5. The (sound horn) (microphone) (loudspeaker) (sound amplifier) is used to increase the current and voltage output of a receiver.

6. Name the lettered parts of the three types of loudspeakers shown in figure 36-12. Describe the function of each part.

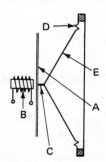
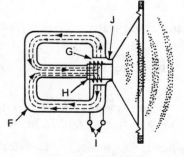
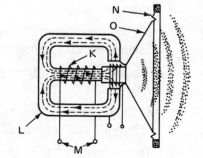

Fig. 36-12

|  | Part Name | Function |
|---|---|---|
|  |  | **Magnetic Loudspeaker** |
| A |  |  |
| B |  |  |
| C |  |  |
| D |  |  |
| E |  |  |
|  |  | **Permanent Magnet Dynamic Loudspeaker** |
| F |  |  |
| G |  |  |
| H |  |  |
| I |  |  |
| J |  |  |
|  |  | **Electromagnet Dynamic Loudspeaker** |
| K |  |  |
| L |  |  |
| M |  |  |
| N |  |  |
| O |  |  |

7. Explain briefly how new developments which increase the range of receivers and transmitters in electrical communication affect the fidelity and quantity of reproduced sounds.

8. List three new devices for receiving, transmitting, or reproducing sound. Give a use of each device.

# Achievement Review
## SOUND ENERGY

### SOUND PROPERTIES, PRODUCTION, AND ACOUSTICS

For statements 1 to 10, determine which are true (T) and which are false (F).

1. Sound may be transmitted through a vacuum.

2. Sound travels at the same rate through solids, liquids, and gases.

3. The higher the temperature, the faster sound travels (and conversely).

4. Compression refers to the low-pressure part of a single sound wave.

5. A compression and a rarefaction are part of a single sound wave.

6. Pitch indicates the difference between high and low tones.

7. The number of vibrations per second (vps) needed to produce a tone is the frequency of vibration.

8. Irregular frequencies produce musical sounds.

9. The amplitude of an original vibration determines the quantity (loudness and softness) of a sound.

10. Forced vibrations result when a vibrating object produces vibrations of the same frequency in another object.

Select the correct word or phrase to complete statements 11 to 18.

11. When the natural and the applied frequencies of vibration are equal, (softness) (a series of beats) (resonance) occurs.

12. The correct resonance length of a tube open at one end and closed at the other is equal to (one-fourth) (one full) (two full) wavelength(s).

13. The best resonance for an open tube occurs with a resonance tube which is (one-half) (three-fourths) (one-fourth) the wavelength of the sound it reinforces.

14. The range of human hearing is between (20 000 and 80 000) (16 and 20 000) (20 000 and 200 000) vps.

15. Instantaneous moments of loudness and softness resulting from two different vibration frequencies produce (beats) (compressions only) (sympathetic vibrations).

16. (Discord) (Harmony) (Noise) is produced when the tone frequencies have simple ratios with respect to each other.

17. Excessive reverberation in a room may be overcome by using a (high reflective) (sound-absorbing) material.

18. Refraction refers to the (changing of direction) (absorption) (resonance) of sound waves.

19. Describe briefly two ways in which adjoining rooms may be constructed so that minimum sound travel occurs.

20. Compute the missing wavelengths or broadcasting frequencies for radio stations A, B, C, and D in the table. Use 186 000 miles per second as the speed of radio waves.

| Radio Station | Broadcast Frequency (Kilocycles) | Wavelength (Miles) |
|---|---|---|
| A | 1488 | |
| B | 93 | |
| C | | 1/3 |
| D | | 1/16 |

21. Determine the missing values in the table for sounds A, B, C, and D. Round off the answers, where needed, to one decimal place.

| Sound | Medium | Frequency (vps) | Temperature | Wavelength (ft.) |
|---|---|---|---|---|
| A | Air | 20 | 75°C | |
| B | Water | 20 | | 238.2 |
| C | Steel | | 200°C | 70 |
| D | Steel | 176 | 176°F | |

22. Find the reverberation time of a laboratory having the specifications given in the table.

| Volume of Room | Materials and Coeff. of Absorption | | | Reverberation Time |
|---|---|---|---|---|
| | Marble | Hair Felt | Plaster | |
| 5000 cu. ft. | 1% | 50% | 3% | |
| | Area of Each Material (Sq. Ft.) | | | |
| | 100 | 420 | 300 | |

23. Compute the reverberation time if hair felt is used to cover the marble and plaster surfaces indicated in the table in question 22.

## TRANSMITTING, RECORDING, AND REPRODUCING SOUND

Select the correct word or phrase to complete statements 1 to 7.

1. The electrical currents sent out by the telephone (amplifier) (receiver) (transmitter) have the same frequency in vibrations per second as the original sound.

2. The telephone (microphone) (transmitter) (receiver) reproduces sound waves at the same frequency in vibrations per second as the original sound.

3. The telephone changes (sound) (mechanical) (electrical) energy to (electrical) (mechanical) energy and back to form new sounds which are reproductions of the original sound.

4. The cutting needle on a sound recording head vibrates at (the same) (a greater) (a lesser) frequency and amplitude as the original sound.

5. When reproducing sound, the sound track on some films converts (light) (mechanical) (heat) energy into (electrical) (light) (mechanical) energy and then converts this energy back to (sound) (heat) (electrical) (mechanical) energy.

6. Microphones convert (mechanical) (electrical) energy to the equivalent of the original (mechanical) (electrical) (sound) energy.

7. The loudspeaker changes variations in (mechanical) (electrical) energy to the equivalent of the original (mechanical) (electrical) (sound) waves.

8. Give three reasons why a sound recorder is preferred to a disc recorder.

For statements 9 to 14, determine which are true (T) and which are false (F).

9. The carbon microphone operates on the principle that variations in the pressure exerted on a diaphragm cause a corresponding variation in the current flow.

10. Crystal microphones are used in high quality public address systems where a lightweight, rugged, and easily maintained system is needed.

11. The preferred microphone for good frequency response and direction is the capacitor microphone.

12. Sounds are reproduced in the loudspeaker by converting electrical energy into vibrations of a flexible cone which, in turn, sets large quantities of air in motion.

13. The magnetic loudspeaker provides a more constant strength and wider range of power output than does the electrodynamic loudspeaker.

14. Sound amplifiers increase the current, voltage input, and quantity of reproduction of a receiver or a microphone.

# Section 8
# Electronics

## Unit 37  Basic Electronics in Industry and Communications

Electronics is the study of the behavior and motion of free electrons passing through conductor mediums other than metal conductors (such as wires and cables). These other mediums include vacuum tubes, tubes containing various gases, photo-tubes, and semiconductor materials.

The field of electronics is important because the use of electronic devices and components produces equipment having several desirable characteristics: fast operation, lack of vibration, minor or reduced maintenance, precise automatic control, quiet operation, consistent high quality results, and amplification of weak control signals into large amounts of power.

An understanding of the principles of electronics is of great importance because of the multitude of applications of electronic equipment to industry and the field of communications.

## BASIC PRINCIPLES OF ELECTRONICS AND ELECTRONIC TUBES

### Edison Effect

The principle underlying the operation of electronic tubes was discovered by Thomas Edison in the early 1880s. While experimenting with incandescent lamps, Edison sealed a fine wire and a separate filament inside a glass lamp bulb. The fine wire was connected to a battery-ammeter circuit, figure 37-1. As the filament glowed and the lamp lighted, Edison noticed that current flowed through the space between the wire and the filament even though the wire was not connected to the filament. This phenomenon became known as the *Edison Effect*.

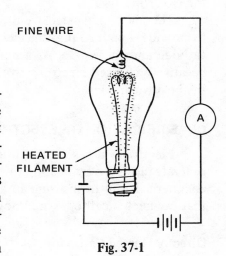

**Fig. 37-1**

The discovery of the Edison effect was important because it showed that free electrons are the ones that can be harnessed to do useful work. It is known that the easier outer electrons are freed from the atoms of any substance, the better that sub-stance is as an electrical conductor. Thus, in the atoms of very good conductors such as copper and silver, the outermost orbit has a single, easily exchanged electron.

In contrast, materials such as mica, rubber, and porcelain are better insulators because the electrons in the outer orbit of the atoms of these materials are not easily detached.

### Electron Flow

Electron flow can be represented by a stream of electrons moving from a region of a negative charge to one having a positive charge. In other words, these free electrons move through a conductor to the positive end where they are replaced by electrons from the negative end. As a result, there is a movement of electrons around the circuit.

In electronic tubes containing either a vacuum or a gas, electrical impulses can be transmitted through the conductor at a speed of 186 000 miles per second.

### Parts of the Basic Electronic Tube

One of the simplest of the many forms of electronic tubes is the vacuum tube. This device has four principal parts: a cathode, an anode or plate, an envelope, and termi-nals, figure 37-2. The *envelope* is the outer enclosure which houses and protects the anode and cathode. The *terminals* provide a connection for the tube in a circuit. Within the envelope, the *cathode* gives off electrons and the *anode* or plate receives the electrons.

The simple tube shown in figure 37-2 is known as a *diode.* Other tubes may be provided with one or more grids between the cathode and the anode which serve to regulate the electron flow.

The air is exhausted from the glass envelope of the tube to provide a path so that the electrons can travel between the cathode and the anode. As a positive charge from a circuit is applied at the anode, the negatively charged electrons from the cathode move across the vacuum in the tube to the positive anode.

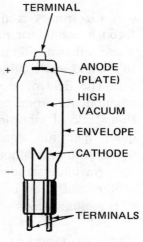

Fig. 37-2

## CLASSIFICATION OF ELECTRONIC TUBES

Electronic tubes are classified according to the type of cathode used. Many tubes generate electrons by heating the cathode. This method is widely used and is a practical commercial way of insuring an adequate and uniform supply of electrons. Those tubes that use heated cathodes are known as *thermionic* tubes.

### Directly Heated Cathode

A directly heated cathode can be made from a high-resistance material such as the tungsten wire used in incandescent bulbs. A current passed through this resistance pro-duces the necessary heat at the cathode. This type of cathode is used where the speed of heating the cathode is important.

Schematic drawings are used to represent electrical circuits. Electronic tubes are shown on these drawings by symbols such as the one shown in figure 37-3 for the directly heated cathode.

In vacuum tubes, the directly heated cathode emits more electrons than does an indirectly heated cathode. Thus, directly heated cathode tubes can be used to supply large currents.

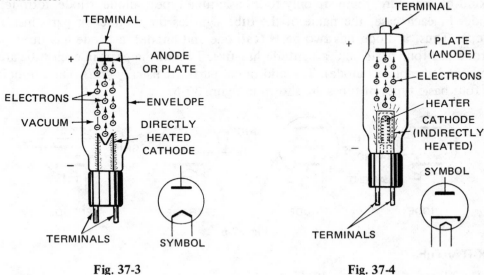

Fig. 37-3    Fig. 37-4

### Indirectly Heated Cathode

In some tubes, the cathode is heated by an electric heating coil known as a heater or filament. This heater is not connected to the cathode but is close to it. This coil heats the cathode to the proper temperature so that it can emit electrons.

Tubes having indirectly heated cathodes are used where a uniform flow of electrons is required at low anode voltages and in circuits requiring low-current power supplies. The symbol for this type of tube is given in figure 37-4.

### The Cold Cathode

Another type of tube is known as the *cold cathode tube*. In this case, the electrons must jump the gap between the anode and cathode. A high positive voltage at the anode actually pulls electrons from the cathode to the anode. One of the most common of the cold cathode tubes is the *ignitron tube*. The electronic emission in this type of tube is started by an igniter. Once the emission begins, the electron flow itself keeps the process going.

The cutaway drawing in figure 37-5 shows the simple construction and the schematic symbol for the ignitron tube. As shown, the igniter

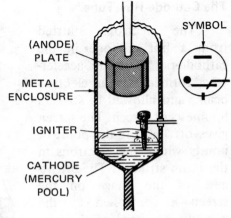

Fig. 37-5

rod (a graphite anode) dips into a mercury pool (the cathode). The anode must be made of a material which mercury cannot wet. One such material is graphite. This fact is important because the current must pass from the igniter to the mercury pool if the tube is to function as a cold cathode tube.

## BASIC TYPES OF ELECTRONIC TUBES:

There are many different forms of vacuum tubes. Regardless of their shape, size, and kind of enclosure, there are only four basic tube types: diode, triode, tetrode, and pentode. In each case, the name of the tube signifies the number of parts inside the envelope. Thus, a diode has two parts (cathode and anode), a triode has three parts, a tetrode has four parts, and a pentode has five. Two of the parts in each tube are the cathode and the plate (anode). The additional parts are usually *grids*. The symbols for these four basic types of tubes are shown in figure 37-6.

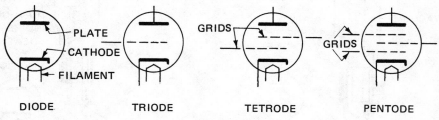

Fig. 37-6

### The X-Ray Tube

The electronic emission of X-ray tubes is produced by creating an extremely high voltage (in the order of millions of volts) at the anode which then attracts electrons with a tremendous force to a target. As the electrons strike the anode, the anode atoms are greatly disturbed. As the atoms return to their normal energy states, they give off radiations known as X rays.

X rays are applied to industrial processes by combining them with phototubes to measure the thickness and internal characteristics (such as fractures or inclusions of impurities) of steel, glass, or numerous other materials. X rays are very penetrating and can pass through most materials, including human tissue.

### The Cathode-Ray Tube

The electrons emitted from a hot cathode in the cathode-ray tube are accelerated, formed into an electron beam, and allowed to strike a luminescent screen. The screen gives off light at the point or points where the electrons in the beam strike it. The brightness of the image on the screen is controlled by the negative potential of the control grid. The simplified drawing of a typical cathode-ray tube is shown in figure 37-7A.

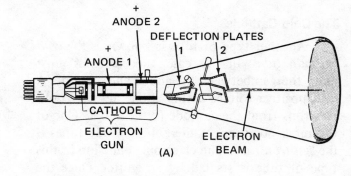

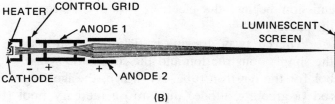

Fig. 37-7

*Forming and Deflecting an Electron Beam.* Once the electron beam is formed in the cathode-ray tube, it can be guided within either an electrostatic or electromagnetic field. As a result, the beam can be directed instantly to strike any point on the field screen. In other words, the electron beam can be made to follow the variations of a directing field so that a visible trace appears on the screen.

The concentration of electrons into a beam takes place in the *electron gun* section of the tube. Since anode 1, figure 37-7B, has a positive potential with respect to the cathode, the electrons passing through the control grid are accelerated.

Anode 2 is even more positive than anode 1; thus, the electrons are further concentrated to form a beam.

The electron beam is then deflected by two sets of plates which are at right angles to each other. The first set of plates (1) deflects the electron beam in a horizontal direction; the second set of plates (2) deflects the electron beam in a vertical direction.

The vertical deflection as shown on the screen represents the magnitude of the quantity being measured. The horizontal deflection across the screen usually represents time. The polarity of the voltage can be changed so that the spot on the screen is swept across the screen in a linear motion and then is returned to its original position. This action is commonly seen on the screens of oscilloscopes (which use cathode-ray tubes).

### The Phototube

Another type of electronic tube is the *phototube*. This device consists of an anode and a cathode enclosed in a glass envelope from which the air is evacuated. The cathode is coated with a light-sensitive material. When light strikes this material, electrons are emitted instantaneously from the cathode. Figure 37-8 shows the features of a phototube and the symbol used to indicate the tube on circuit diagrams.

The phototube can be used to control any type of motion in industrial production processes. In addition, the phototube can be used in a circuit to count, select, and sort objects on an assembly line by controlling the current and automatically turning it on and off, figure 37-9.

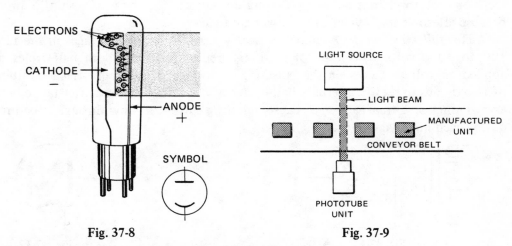

Fig. 37-8                              Fig. 37-9

## FUNCTIONS OF ELECTRONIC TUBES AND CIRCUITS

Electronic tubes have two primary functions. They act as check valves because the current flows in one direction only (the cathode alone emits electrons). Electronic tubes also function as control valves. The tube can be made to control the amount of current flowing through it. The name *valve* was first applied to electronic tubes by the British scientist Fleming when he observed that the current flowed in only one direction.

Electronic tubes are combined in circuits with other devices such as resistors, capacitors, coils or inductances, transformers, and switches. Three basic electronic circuits are of great importance: (1) the rectifier circuit, (2) the amplifier circuit, and (3) the oscillator circuit.

### Rectification

*Rectification* refers to the changing of alternating current to direct current. In this process, a device is needed which permits the current to flow in one direction only. One common tube which can be used to achieve rectification is the diode tube considered previously. Recall

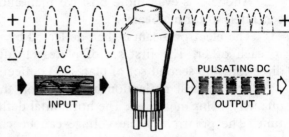

Fig. 37-10

that the diode has two elements, the cathode and the plate (anode). Both of these elements are enclosed in a glass envelope from which the air is removed. In some cases, the tube is refilled with an inert gas which improves the electron flow.

The cathode of the diode is heated until electrons are emitted. These electrons are attracted by the positive charge on the anode (plate).

*Rectifier Circuits.*  In simplest terms, a rectifier circuit takes alternating current from a power line and transforms it to direct current which is then used by other circuit components and devices.

Large rectifiers are used to provide the direct current required for large-scale electrochemical processes. Smaller, portable rectifier circuits supply the necessary current for mining operations.

*Half-wave and Full-wave Rectifier Circuits.*  The *half-wave rectifier circuit* uses a single rectifier unit to change ac to a pulsating dc output. This type of output is produced because alternate half-cycles of ac power are used.

The *full-wave rectifier circuit* is widely used to supply voltages in the range of 100 to 5000 volts. This type of rectifier circuit transforms both half-cycles of the applied ac voltage to obtain the pulsating dc voltage. This circuit uses a combination of two diodes to rectify alternate half-cycles of the voltage and a center-tapped secondary winding of a transformer to cause pulsating current to flow through a load resistor during each half-cycle of ac.

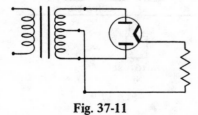

Fig. 37-11

Fig. 37-12

A full-wave rectifier circuit is shown in figure 37-11. Figure 37-12 is the schematic symbol for a full-wave rectifier having two anodes and a common cathode. The symbol shows that these elements are contained in the same glass envelope.

## Amplification

Amplification is achieved by the use of the triode, tetrode, or pentode vacuum tube. The function of these tubes is shown graphically in figure 37-13. Amplifier tubes and circuits are used principally to change small ac voltages into large ac voltages. The process of amplification can be explained by considering the signal voltage of a radio antenna. A minute antenna voltage, such as a few millionths of a volt, is amplified by the average radio set into a signal capable of operating a loudspeaker.

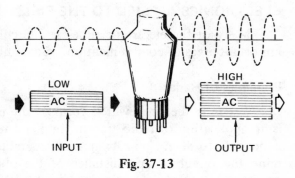

Fig. 37-13

Amplifiers are divided into three major groups, depending on the frequency range of the signals they amplify.

— Audio amplifiers are used in radio receivers, sonar devices, intercom equipment, and other sound reproduction systems to amplify frequencies within the audible range from 15 Hz (cycles/second) to 15 000 Hz.

— Video amplifiers are used in radar, fire control equipment, and television systems to amplify frequencies ranging from 30 Hz to more than 6 000 000 Hz.

— Radio frequency amplifiers are used to amplify a narrow band of frequencies ranging from 30 000 Hz to several billion Hz.

*Amplifier Circuits.*  Amplifier circuits take small voltage changes and feeble signals and enlarge them into large voltages and strong signals that can be used with oscilloscopes, loudspeakers, and similar equipment.

Electronic devices are widely used in industry to enlarge or amplify faint lines or movements to control many manufacturing processes. For example, a minute change in the light from a phototube positioned to follow a fine line printed on a paper roll in a printing plant can be amplified to control the speed of the presses and the quality of the printing. In another application, it is possible to amplify flaws in a metal strip passing over the light directed on a phototube. As a result, marking devices in strip mills can detect, mark, and remove imperfect sections.

## Oscillation

All radio frequency and audio frequency electronic tube oscillator circuits operate on the principle that it is possible to feed back voltage pulses from the plate to the grid circuit of the tube. Oscillations are produced by a grid connected to a parallel resonant circuit. This same circuit also controls the pulses fed back to the grid and the frequency of oscillation.

*Oscillator Circuits.*  *Oscillator circuits* are used to generate ac voltages of any specified frequency to carry radio signals from one point to another and to test other electronic circuits.

Broadcasting operations at high frequencies depend on electronic tubes to change the current from one frequency to another. In power transmission (where the electrical energy is rated in kilowatts), frequency converters are used to tie a power line at one frequency (such as 25 hertz) to another line at a different frequency (60 hertz).

## ELECTRONICS APPLIED TO THE FIELD OF COMMUNICATION

Many of the electronic tubes and circuits covered to this point are applied in two important areas of communication: radar and television transmitting and receiving.

### Principles of Radar

*Radar* is the abbreviated form of the phrase *radio detection and ranging.* Radar is an application of radio transmitting and receiving circuits. With the use of radar, it is possible to detect objects at a considerable distance from the equipment and determine the direction and distance of the objects. Radar is a valuable aid in military operations, meteorology, air and sea navigation, and astronomy.

*Reflection of Radio Wave Pulses.* The operation of a radar unit is based on the principle of the reflection of rapid pulses of radio waves, figure 37-14. A short-wave radio transmitter sends out radio waves in a concentrated beam. This beam is focused by a *parabolic reflector.* When the radio beam hits an object it is reflected back to the same reflector which acts as an antenna. The reflector receives the reflected beam and feeds it to a radio receiver to be amplified. The amplified wave is then passed through a cathode-ray tube in an oscilloscope. The light pattern traced on the

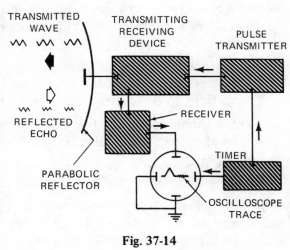

**Fig. 37-14**

face of the cathode-ray tube provides a visible picture which may be used to measure the distance and direction of an object and its speed.

The outgoing signal from a radar antenna is transmitted as a series of instantaneous pulsations lasting only a few millionths of a second each. The radio transmitter thus provides a large power output which is fed into a directive antenna to produce a narrow, accurate beam.

One important feature about radio waves is that they travel at a speed of 186 000 miles per second. At this rate, the beam can strike a target a mile away in 1/186 000 of a second. The beam is reflected back the same distance and the total time is less than 1/100 000 second. This means that the feeble refelcted wave must strike the radar antenna when there is no powerful outgoing signal. It is for this reason that the outgoing pulsations last only a few millionths of a second.

*Recording Reflected Waves on the Oscilloscope.* The electron beam in some oscilloscopes (scopes) moves across the face in a horizontal line while a radio wave pulsation is sent. If no radio waves are reflected, the line on the oscilloscope face remains straight and horizontal. However, if there is an incoming signal, this signal is amplified and the beam registers as a *blip,* or a deviation from the horizontal line, figure 37-15 (page 471).

When the transmitted pulsating radar waves hit an object and are reflected, the amplified signal produces blips which are projected on the scope. The frequency with which these blips are seen on a calibrated scale on the scope determines the speed of the object.

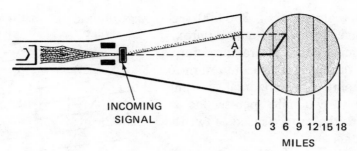

INCOMING
SIGNAL

0  3  6  9  12 15 18
MILES

**Fig. 37-15**

As the object detected approaches the radar antenna, the blip appears nearer to the starting point of the line on the oscilloscope. Conversely, if the blip moves to the end of the scope scale, this indicates that the object is moving away from the antenna. The size of the object also affects the blip appearing on the scope. The larger the object, the greater is the strength of the reflected wave and the greater is the height of the blip on the screen.

By rotating the antenna of the radar installation while the radio signal is transmitted, objects can be detected in a full 360° sweep of the area. The electron beam of the scope is electrically timed for direction with the radar antenna so that the reflected beam is displayed according to the angle of the antenna.

Radar is essential to national security, airport control, and meteorological investigations because of the tremendous speeds at which the radio waves travel and because these waves are not affected by adverse weather conditions, or by the time of the day or night.

### Principles of Television

Television is an entire field of study in itself. However, the student should be familiar with the fundamental principles governing the operation of the television scanning process, transmitter, and receiver.

*Scanning.* Every scene that is televised is broken up into a series of signals. These signals start at the top of the picture on the left side, move to the right and, at the end of the line, move back to the beginning but are displaced a very short distance downward from the initial line. The average television picture consists of more than 525 lines of scanning.

The scanning process is similar to reading. Starting at the left side of a printed page, the eye moves to the right as it scans letters and words. Before the end of the line is reached, the eye shifts to the beginning of the second line. In this shifting step between lines, no reading takes place.

The same action occurs in television scanning. An electron beam starts at the top left of a picture and moves toward the right. Any variation in light intensity in the scene being picked up by the television camera is translated into electrical impulses which are transmitted to the receiver. At the receiver, these impulses are again changed into variations in light intensity to reproduce the original scene.

As the electron beam continues scanning and reaches the end of one line, it is again brought back to the left side. Now, however, the beam is raised up another line (because the image is inverted in the television camera), figure 37-16 (page 472). The scanning process continues at such a rapid pace that the electron beam scans an entire image (consisting of 360 000 sections) 30 times every second. Approximately 60 microseconds are required for a beam to scan the entire image and return to the starting point. The horizontal scanning speed is almost two miles per second as compared with the retrace speed of twenty miles per second.

Television scanning is *interlaced* as shown in figure 37-16. Note that alternate lines are scanned by the electron beam in field 1. At the bottom of the picture, the

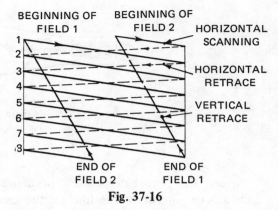

**Fig. 37-16**

beam shifts vertically to begin scanning field 2. After scanning field 2, the beam begins the entire scanning process again.

***Synchronizing and Blanking Signals.*** The scanning of the electron beam is started and stopped by the *timing generator*. This device controls the synchronization of the signals, the scanning process, and the blanking signals which bring about the required momentary lapses in scanning.

The electron beam in the picture tube of the television receiver is synchronized with the signals sent out on a carrier beam by transmitting the synchronizing and blanking signals at the same instant as the video signals.

***The Television Transmitter.*** The *television transmitter,* figure 37-17, converts the light waves from a scene being televised into electrical currents known as *video signals.* The frequencies of these signals are amplified and are impressed on a carrier beam together with synchronizing and blanking pulses from the timing generator. After further amplification, this combined signal is transmitted by an antenna.

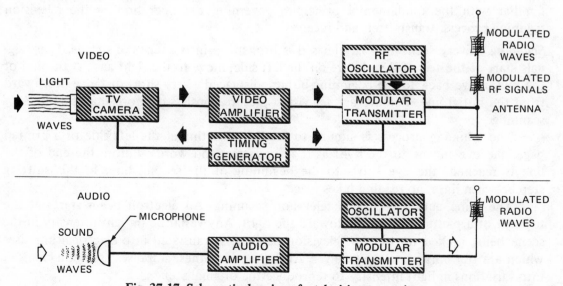

**Fig. 37-17 Schematic drawing of a television transmitter**

At the same time, the audio signals are transmitted over a separate transmitter. The sound waves (ranging from about 30 to 30 000 cycles/second) are first amplified in the *audio amplifier*. The amplified sound is then impressed on a radio frequency

(RF) carrier and is transmitted by the same antenna used for the video signal. The carrier frequency is the same for both the audio and the video signals.

The AM video signal and the FM sound signal are coupled to the antenna through a *diplexer*, figure 37-18. This device prevents any undesirable interaction between the video and sound portions of the transmitter by isolating the two sections.

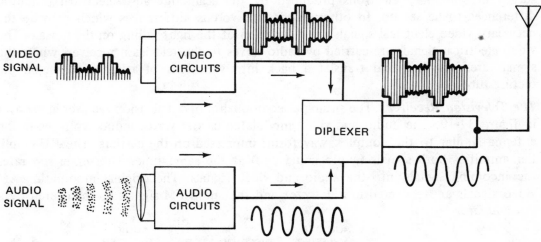

**Fig. 37-18  Coupling AM-FM signals**

The scenes broadcast in television are scanned with regard to space and time. The scene is divided into very small parts to accomplish the scanning. These parts are transmitted one at a time at a tremendous speed so that an entire scene is scanned in 1/30 of a second. Each of these small parts is then reassembled in the same sequence in the receiver to reproduce the original scene on the picture tube.

*The Television Camera Tube.*  One of the most important components in the transmission system is the television camera tube. The light and dark areas of the scene being televised provide variations in light energy. These variations are translated in the television camera tube into electrical pulses.

The camera tube is a sophisticated device consisting basically of a camera, a cathode-ray tube, and a photoelectric cell. The television camera consists of a system of optical lenses. The glass of the camera tube facing the lenses is especially clear so that an undistorted image can be formed inside the tube on a thin sheet of pure mica. The mica sheet is covered with millions of tiny globules of silver which are insulated from each other. A fine layer of cesium oxide sensitizes each globule so that as light falls on a globule it loses electrons.

The intensity of the light reaching each globule determines the number of electrons which are released. The light striking the globules distributed over the mica sheet (called the *photoelectric mosaic*) translates the light pattern into a similar distribution of electrical charges. The back of the mica plate is coated with colloidal graphite so it becomes a signal plate.

The cathode-ray section of the camera tube contains an *electron gun* and *deflection plates* or *deflection coils*. A sweep circuit connected to the deflection coils controls the electron beam which passes over and scans the photoelectric mosaic. The high-velocity electrons of the beam strike the globules of the mosaic and cause them to emit other electrons known as *secondary electrons*. The number of secondary electrons emitted from each globule depends on the photoelectric action of the light on each globule as it is scanned by the electron beam. Variations in darkness and lightness in the mosaic determine the number of free electrons emitted.

The inner surface of the camera tube contains a metalized ring which is connected electrically to a conducting layer. This ring is called a *collector anode* because the surges of secondary electrons produced by the scanning cause electrical potential differences to be set up. In other words, the voltage differences which provide the necessary video electrical signals are a function of the light falling on the mosaic. The video electrical signals are carried by radio waves to the television receivers where the signals are recovered and translated back into variations of a light pattern on the picture tube.

*The Television Receiver.* The principal components of a television receiver are shown in figure 37-19. The audio and video modulated carrier waves induce voltages in the antenna similar to the complex waveforms impressed on the carriers. The RF amplifier amplifies both signals (audio and video) at the same time. The mixer separates the incoming signals into the audio and video signals. The original modulating AM video signals are recovered in a *detector* and the audio FM signals are recovered by a *discriminator*.

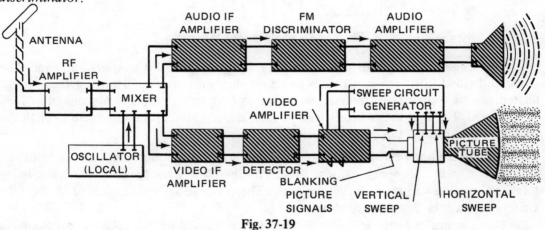

Fig. 37-19

The original modulating signals then pass through a *video amplifier* and an *audio amplifier*, respectively. The video signals (including the blanking and synchronizing pulses) are impressed on the picture tube and the sweep circuit generator. The video signals cause the electron beam to reproduce the televised picture on the luminescent screen of the picture tube. At the same time, the audio signals are reproduced through an audio amplifier and a loudspeaker.

## Color Television Transmitters and Receivers

*Color Television Transmitter.* A color television transmitter is similar to a black and white transmitter. The color television receiver has an AM section to process the video portion of the signal and an FM section to process the audio portion of the signal. Other circuits are added to the video section to process the complex color signal.

Three video input signals from the television camera are received by the color transmitter, figure 37-20. These signals contain the red (R), green (G), and blue (B) color content of the scene being televised. When these signals are applied to a matrix circuit, the circuit adds the signals in different proportions to produce three separate outputs: the Y, Q, and I signals.

The brightness or *luminance* variations of the picture are represented by the Y signal. The color variations are represented by the Q and I signals which form the *chrominance* (C) signal. The C signal contains all of the color information necessary to reproduce the televised picture.

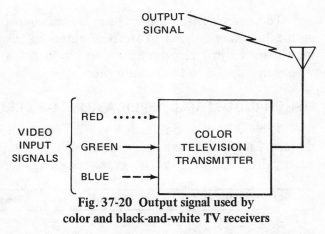

Fig. 37-20 Output signal used by color and black-and-white TV receivers

An *adder circuit* in the transmitter receives and combines the chrominance (C) and luminance (Y) signals into a single video signal. A group of timing circuits generates horizontal and vertical sync pulses, as well as blanking and equalizing pulses. All of these pulses are received by the adder and are inserted at the correct intervals in the video signal.

One color sync burst (to provide one additional sync pulse) is added to the video signal. The *color sync bursts* are added to demodulate the Q and I signals in the television receiver. All necessary information for the complete color video signal is provided at the output of the adder circuit. The remaining circuits in the color transmitter are similar to those for the black and white television transmitter.

*Color Television Receiver.* All of the circuits of the black and white receiver are contained in a color television receiver. However, additional circuits are required in the color receiver to recover the original red, blue, and green signals from the modulated carrier. Once the total video signal (S) is detected, it is amplified by a video preamplifier stage. The components of the video signal (S) are then separated by three filter networks, figure 37-21.

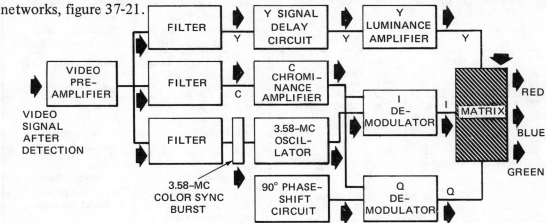

Fig. 37-21 Basic circuits of a color television receiver

The chrominance signal is separated from the video signal and is then amplified and applied to both an I demodulator and a Q demodulator. The luminance signal (Y) is coupled through a delay circuit and is amplified. The signal is then sent to a matrix circuit. The delay circuit insures that the luminance signal and the chrominance (I and Q) portion of the signal reach the matrix at the same time.

The demodulated I and Q signals are also applied to the matrix at the same time as the luminance signal (Y). These three signals (I, Q, and Y) are combined in the matrix in the proper proportions to produce the original red, blue, and green signals, which are applied to the picture tube.

## OTHER INDUSTRIAL APPLICATIONS OF ELECTRONIC TUBES AND DEVICES

Table 37-1 lists only a few typical applications of electronic tubes and electrical devices for the measurement, control, or protection of matter, machines, electrical energy, light energy, heat energy, sound energy, and color energy.

| Electronic Device | Function | Electronic Principle | Typical Applications |
|---|---|---|---|
| Pressure Indicator | Measurement of gases | Measurement of variations in capacitance circuit containing the pressure element | Engine design<br>Ballistics<br>Chemical production |
| Stroboscope | Measurement—Machines and mechanics (rotating and reciprocating) | Control of current pulses through electronic lamp by a thyratron tube | Machine design<br>Manufacturing<br>Rotating machines and equipment |
| Electronic Wheatstone bridge | Measurement of electrical resistance | High-resistance measurement using an electronic tube in one or two arms of a Wheatstone bridge | Design of electrical equipment<br>Electrical laboratory practice |
| Oscilloscope | Measurement of electrical waveforms | Cathode-ray tube used in combination with amplifiers | Electrical laboratory testing<br>Vibration studies |
| Photometer | Measurement of light intensity | Photoelectric current generated in a phototube or photocell | Illumination design<br>Process control<br>Photography |
| Thermocouple with phototube relay and meter | Temperature control | Thermocouple actuating a phototube relay through a metering device and control relay | Machine and metals technology<br>Metal manufacturing<br>Chemical production |
| Sound-level meter | Measurement of sound and acoustics | Electronic amplification of sound vibrations | Acoustical design<br>Materials testing<br>Machine design |
| Color comparator and colorimeter | Measurement of color | Generation and amplification of photocolor in phototube or photocell | Textile industries<br>Printing and ink making<br>Paint and pigment industries |
| Facsimile Transmitter | Transmission of graphic materials | Changes varying dc amplitude output into frequency variations of a high-power radio frequency carrier wave | Intercontinental communications<br>Graphic industries<br>Industrial electronics |

Table 37-1

---

### SUMMARY

- Electronics is the study of electrons moving in gases, a vacuum, or semiconductor materials.

- The Edison Effect consists of the release of electrons from a negatively charged surface and a resulting flow of electrons in space from negative to positive.

- The four principal parts of electronic tubes are the cathode, anode or plate, envelope, and terminals. The four basic types of electronic tubes include the diode, triode, tetrode, and pentode.

- Electronic tubes are classified into three groups: those having directly heated cathodes, those having indirectly heated cathodes, and cold cathode tubes.

- The electrons emitted in a cathode-ray tube are accelerated, concentrated into an electron beam in an electrostatic or electromagnetic field, and controlled by variations in a directing field. The beam produces a visible trace on the screen of the tube.

- Electronic tubes serve two prime functions: as check valves permitting current to flow in one direction only, and as control valves to regulate the amount of current flow.

- Rectification is the changing of an alternating current to a direct current.

- Amplifier tubes and circuits amplify small ac voltages and feeble signals into large ac voltages and signals. Amplifier circuits are grouped into three major categories: audio amplifiers, video amplifiers, and radio frequency amplifiers.

- Oscillator circuits generate ac voltages of any specified frequency by controlling the pulses fed back and the frequency of oscillation.

- Radar involves the instantaneous transmission of strong radio waves and the amplification and recording of the reflected waves according to a pattern traced by an electron beam on the screen of an oscilloscope.

- Television involves the scanning of a televised scene by an electron beam which transforms variations in light intensity into a series of electrical currents (signals).
  - These video signals are combined with synchronizing and blanking pulses, amplified, impressed on a carrier signal, and finally are transmitted at the same instant as the audio signals.

- A televised picture is reproduced on a television receiver from the carrier waves received at the antenna. The incoming audio and video signals are separated, amplified, and changed back to light energy by the electron beam in the picture tube to reproduce the televised scene.

- The color television transmitter receives red, blue, and green video input signals and combines these into a single luminance and chrominance signal for transmission.

- A color television receiver receives, amplifies, separates, and demodulates the luminance and chrominance signals. These signals are delivered to the matrix circuit which combines the transmitted color signals in the proper proportions to recover the original signals (R, G, and B) on a picture tube.

## ASSIGNMENT UNIT 37 BASIC ELECTRONICS IN INDUSTRY AND COMMUNICATIONS

### PRACTICAL STUDIES OF ELECTRONIC TUBES AND DEVICES

**Equipment and Materials Required**

| Experiment A<br>Electronic Tubes and Devices | Experiment B<br>Applications |
|---|---|
| Diode tube<br>Ignitron tube<br>Cathode-ray tube<br>Manufacturers' catalogs and descriptions of tubes<br>Electronic eye apparatus with light beam<br>Small sheets of materials such as heavy paper, sheet metal, plastic | Manufacturers' drawings and catalogs of electronic products used to measure or control light, heat, and electrical, mechanical, sound, and color energy |

### EXPERIMENT A. TYPES AND FUNCTIONS OF ELECTRONIC TUBES AND DEVICES

**Procedure: The Diode Tube**

1. Examine a cutaway section or a cross-sectional drawing of a diode tube.
2. Name each of the parts observed and describe the function of each part.

**Procedure: The Ignitron Tube**

1. Examine a cutaway section or a cross-sectional drawing of an ignitron tube.
2. Name only those parts of the tube that serve a different function from any part of a diode listed and described in the first part of this experiment. Explain the function of each new part.

**Procedure: The Cathode-ray Tube**

1. Examine a cathode-ray tube and a drawing showing the internal construction of a cathode-ray tube.
2. Review the manufacturers' description and specifications of the cathode-ray tube.
3. List the parts in the electron gun section of the crt (cathode-ray tube) used to deflect the electron beam.

**Procedure: The Phototube**

1. Connect an electric eye device to an electrical outlet.
2. Add an electric counter to the circuit.
3. Direct a beam of light into the electronic eye at a distance from it.
4. Pass a sheet of heavy paper through the light beam. Note what effect this action has on the circuit and the counter.

5. Repeat step 4 with a small piece of sheet metal, plastic, or any other material. Note in each case the speed with which the device operates.

### Interpreting Results

1. What effect does any interruption of the light beam have on the phototube of the electric eye?

2. Name a practical application of such a device (A) in the home, (B) on the farm, and (C) in business or industry.

## EXPERIMENT B.  INDUSTRIAL APPLICATIONS OF ELECTRONIC TUBES AND DEVICES

### Procedure

1. Study the literature available from manufacturers of electronic equipment.

2. Prepare a table like the one illustrated and furnish the information required to complete the table. List an application (other than those given in this unit) of an electronic device which is used to measure or control: (A) machines, (B) electrical energy, (C) light energy, (D) heat energy, (E) sound energy, and (F) color energy.

|   | Electronic Device | Function | Electronic Principle |   | Typical Applications |
|---|---|---|---|---|---|
| A |  |  |  | a. |  |
|   |  |  |  | b. |  |
| B |  |  |  | a. |  |
|   |  |  |  | b. |  |
| C |  |  |  | a. |  |
|   |  |  |  | b. |  |
| D |  |  |  | a. |  |
|   |  |  |  | b. |  |
| E |  |  |  | a. |  |
|   |  |  |  | b. |  |
| F |  |  |  | a. |  |
|   |  |  |  | b. |  |

## PRACTICAL PROBLEMS WITH ELECTRONIC TUBES AND DEVICES

### Principles of Electronics and Electronic Tubes

Select the correct word or phrase to make statements 1 to 4 correct.

1. An Edison Effect is produced when a current is passed through a filament in an enclosure with the result that a current is induced in another wire that (is) (is not) separated from the filament.

2. The easier it is to free the outer electrons from any substance, the (poorer) (better) the substance is as an insulator.

3. Electron flow may be represented as a stream of electrons which move from a (negative) (positive) to (negative) (positive) terminal.

4. Electrical impulses are transmitted in a conductor at the rate of (1090 ft./sec.) (186 000 miles/sec.) (15 in./min.).

5. Name five characteristics of electronic tubes which make their use desirable.

6. Make a simple sketch of a tube having a directly heated cathode. Label the parts.

7. Make a simple sketch of a tube having either a cold cathode or an indirectly heated cathode. Label the parts.

## Basic Types of Electronic Tubes

1. Sketch the symbols that are used to indicate (a) an indirectly heated cathode tube, (b) a triode tube, and (c) a pentode tube. Label the parts which are represented in each symbol.

Match the tubes listed in Column I with the function described in Column II.

| Column I | Column II |
|---|---|
| 2. X-ray tube | a. Uses an igniter to start electronic emission. |
| 3. Phototube | b. Provides a heater for the cathode to boil out the electrons. |
| 4. Cathode-ray tube | c. Emits electrons instantaneously from the surface of a light-sensitive material on the cathode. |
| | d. Accelerates electrons within an enclosure, forms them into an electron beam, and controls the beam to produce a light pattern on the tube screen. |
| | e. Produces an electronic emission by drawing electrons at extremely high voltages on an anode to a target with terrific impact. As the anode atoms return to normal energy levels, they give off radiations. |

## Functions of Electronic Tubes and Circuits

Add a word or words to complete statements 1 to 9.

1. The two primary functions of electronic tubes are to act as _____ and _____ valves.

2. _____ flows in vacuum tubes in one direction.

3. Three common electrical devices used with electronic tubes in circuits are: _____ , _____ , and _____ .

4. Two of the three basic electronic circuits are: _____ and _____ .

5. Rectification is the _____ .

6. Amplifier tubes and circuits are used primarily _____ .

7. Audio amplifiers are used to amplify frequencies within the audible range from _____ to _____ .

8. _____ amplify signals with frequency ranges from 30 Hz to over 6 000 000 Hz.

9. _____ are used to generate ac voltages of any specified frequency.

## Electronics Applied to the Field of Communication

For statements 1 to 10, determine which are true (T) and which are false (F).

1. Radar is used to detect targets and determine their direction and distance from the radar equipment.

2. Radar depends on the ability of the target to absorb sound.

3. The parabolic reflector is used in transmitting and receiving radio waves for radar equipment.

4. Outgoing radar signals are transmitted as a series of long pulsations, each pulsation lasting a few minutes.

5. The waves used in radar installations travel at the speed of light.

6. The feeble reflected wave must strike the radar antenna at an instant when there is no outgoing signal.

7. Incoming signals in radar are amplified so that an electronic beam can trace the direction, distance, and path of an object on an oscilloscope screen.

8. The closer an object, the smaller is the deviation on the oscilloscope screen.

9. The angle of the radar antenna is synchronized with the oscilloscope so that the reflected beam can be located regardless of the position of the object.

10. The speed of an object can not be approximated on the oscilloscope.

11. Explain briefly (a) what scanning is and (b) how it is accomplished.

Select the correct word or words to complete statements 12 to 18.

12. In television scanning (electrons) (an electron beam) move(s) across the picture to translate light intensity into electrical impulses.

13. (Video electrical) (Audio) impulses of the picture are transmitted to the receiver.

14. The horizontal scanning speed is (faster) (slower) than the retrace speed.

15. Synchronizing and blanking signals are sent out (slightly ahead of) (after a momentary lapse of) (at the same time as) the video signals.

16. Video signals are (strong enough) (amplified) to be impressed on an antenna and transmitted into space.

17. The carrier frequency for both the audio and video signals (is the same) (varies).

18. A whole scene is scanned once each (minute) (half minute) (second) (1/30 second).

19. Explain briefly the purpose of the collector ring in the television camera tube.

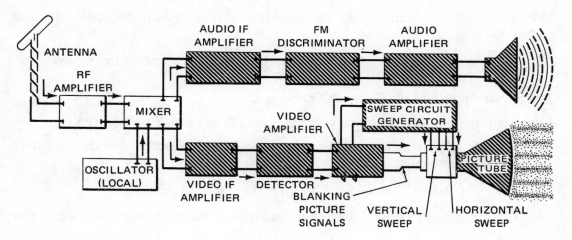

**Fig. 37-22**

20. The schematic drawing in figure 37-22 shows the components used in the audio and video stages of a television receiver. Explain briefly the function of each component (a to k).

a. Antenna
b. RF amplifier
c. Mixer
d. Audio IF amplifier
e. Audio amplifier

g. Video IF amplifier
h. Detector
i. Video amplifier
j. Sweep circuit generator
k. Picture tube

21. Describe briefly the function of a diplexer.

22. Explain how the color content signals of a televised scene are processed in a color television transmitter.

23. Name the two final circuits that (a) the luminance (Y) signal and (b) the chrominance (C) signal must pass through to reproduce the original three color signals.

# Section 9
# Nuclear Energy

## Unit 38  Production and Control of Nuclear Particles

*Nuclear physics* deals with the study of the nucleus or core of the atom. All matter is known to contain varying amounts of energy. This energy is stored in the different particles which comprise matter. Matter was described in a previous unit as being composed of molecules which, in turn, consist of atoms. As the electron theory was discussed, the atom was described as containing even finer particles, such as the proton, neutron, and electron.

### RELATIONSHIP BETWEEN MATTER AND NUCLEAR ENERGY

In the infancy of nuclear physics Albert Einstein (in 1905) expressed the quantitative relationship between matter and energy by the formula:

$$E = mc^2$$

where E = quantity of energy
m = quantity of matter
c = velocity of light

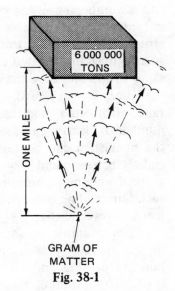

**Fig. 38-1**

Einstein developed the concept that in the process of transforming one gram (about 1/30th of an ounce) of matter to energy, 6 000 000 mile-tons of energy are released. In other words, the energy released from one gram of matter is equal to moving 6 000 000 tons a distance of one mile, figure 38-1.

Such an energy release occurs because in certain nuclear reactions the weight of the element(s) resulting from the reaction is less than the original element undergoing change.

### STRUCTURE OF THE ATOM

The atoms of all known elements are made from certain basic particles. The principal particles are the proton, neutron, and electron.

### Weight and Charge of Particles in Atoms

Certain values of relative weight and relative electric charge have been assigned to each of the particles. The *proton* was given a relative weight (*mass number*) of one (1) and has a positive electrical charge (+1). The *neutron* also was given a relative weight (mass number) of (1), but has a zero charge (0). As compared to the proton and neutron, the *electron* weighs so little that it was given a mass number of zero (0), but it has a negative unit electrical charge (−1). Another particle in the atom is the positron which has a weight of almost zero and a (+1) electrical charge.

It is assumed that all electrons are identical, all protons are alike, all neutrons are the same, and all positrons are alike. Therefore, the arrangement and number of these particles in each atom determine the way the atom behaves. An atom is neutral because the number of protons in the *nucleus* (+ charges) equals the number of electrons in the nucleus (− charge). An *atomic number* assigned to each different type of atom indicates how many protons the atom possesses. The atomic number is also equal to the number of electrons in the complete neutral atom.

### Nuclear Structure

The nucleus or core of the hydrogen atom consists of a single proton. The nucleus of every other atom consists of one or more neutrons and protons. Revolving in the paths (orbits) about each atomic nucleus are electrons located in *shells* at different distances from the nucleus.

Once the atomic numbers and weights (mass numbers) are known for each different atom, it is possible to diagram the structure of specific atoms. When drawing such a diagram, no two electrons have the same orbit, but they may be in the same shell because they revolve at the same distance from the nucleus. The maximum number of electrons in the first shell is two; the next shell can contain a maximum of eight electrons; and beyond this, there is a fixed maximum number of electrons in each succeeding concentric shell for as many as seven shells.

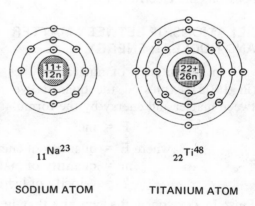

$_{11}Na^{23}$        $_{22}Ti^{48}$

SODIUM ATOM     TITANIUM ATOM

**Fig. 38-2**

For example, as shown in figure 38-2, the neutral sodium atom is represented by the symbol $_{11}Na^{23}$. Since the atom has an atomic number of 11, its nucleus contains 11 protons and 12 neutrons to equal a total mass number of 23. The planetary shells of the atom contain, respectively, 2, 8, and 1 electrons. The shells of the neutral titanium atom ($_{22}Ti^{48}$) contain, respectively, 2, 8, 8, and 4 electrons.

### Isotopes

It is known that all atoms of the same element are not always identical. Although the atomic numbers of the atoms are the same, many atoms of the same element may have a different atomic weight. An *isotope* is an atom of the same element having a different number of neutrons. Isotopes of the same element, therefore, differ in their weight.

For example, two well-known isotopes of uranium are $U^{235}$ and $U^{238}$, having atomic weights (mass numbers) of 235 and 238, respectively. The isotopes of hydrogen are ordinary hydrogen ($_1H^1$), deuterium ($_1H^2$), and tritium ($_1H^3$).

## CHARACTERISTICS OF RADIOACTIVE MATERIALS

### Alpha and Beta Particles and Gamma Rays

Radioactive materials emit alpha particles, beta particles, and gamma rays. *Alpha particles* are actually helium nuclei. These particles have a positive charge and travel at a speed of a few thousand miles per second. However, they can be stopped by a thin sheet of paper. *Beta particles* are electrons and also travel at great speeds. It is possible to stop beta particles by using a great number of sheets of paper. However, *gamma rays,* which are similar to X rays but of shorter wavelength, can penetrate thick plates of lead.

Figure 38-3 shows that the alpha and beta particles have opposite charges and are deflected by a magnetic field; gamma rays have no charge.

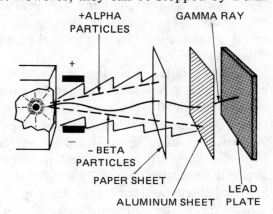

Fig. 38-3

### Devices for Observing Radioactivity

*The Wilson Cloud Chamber.* There are several methods of detecting alpha and beta particles and gamma rays. The *Wilson cloud chamber* is one of the most important of the devices used, figure 38-4. This instrument forms ions around which drops of moisture condense to cause trails of fog in the chamber. An alpha particle makes a heavy track in the fog. The lighter, high-speed beta electron produces a light beaded track having a zigzag pattern due to many collisions. The particle tracks form visible traces that can be photographed.

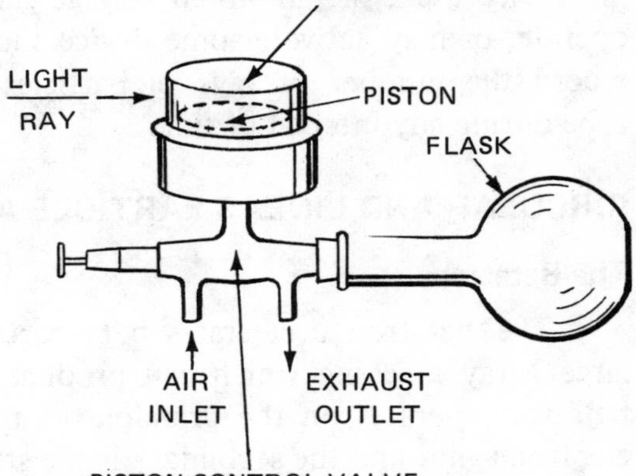

Fig. 38-4 Wilson cloud chamber

The charge of the particles can be determined by placing magnets on each side of the cloud chamber. The alpha particles are attracted to the negative pole and the beta particles are attracted to the positive pole of the magnets. The uncharged gamma rays are not attracted by the poles. The gamma rays strike atoms in the chamber and knock out high-speed electrons from these atoms.

While radium emissions were first used to smash atoms, other machines now develop rays far more efficiently and effectively. The fragments known as alpha particles can be produced mechanically by a number of devices including high-voltage Van de Graaff accelerators and other linear and circular particle accelerators.

Charged particles can be accelerated to extremely high velocities by driving the particles in a circular path. A number of accelerators constructed in a circular form provide particles having maximum energy. Three commonly used circular particle accelerators are the *betatron*, the *cyclotron*, and the *synchrotron*. These accelerators are described later in this unit.

### The Geiger-Mueller Counter Tube

The Geiger-Mueller tube, also known as a counter tube, is a device which can detect X rays, gamma rays, beta particles, and alpha particles. The counter tube operates on the principle that each radiating particle or ray produces ionization in neutral atoms along its path.

The counter tube includes a metal cylinder and a fine wire, such as tungsten wire. Both of these components are en-

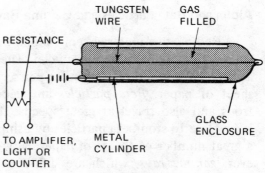

Fig. 38-5

closed in a glass envelope containing a gas at low pressure, figure 38-5. The tungsten wire is the positive terminal and the metal cylinder is the negative terminal in an electrical circuit which is connected to some type of amplifying device. The voltage between the tungsten wire and the metal cylinder is adjusted so that it almost reaches the point of discharge (but without doing so). If any ray or ionizing particle enters the tube, a momentary current flows through the tube and there is a voltage drop across the resistor. The voltage drop of the resistor is amplified. Depending upon the detecting device used, the amplified voltage causes a lamp to flash, a mechanical counter to operate, or may activate some device such as a loudspeaker. It is possible to count or record the number of rays or particles passing through the Geiger-Mueller counter tube during any interval of time.

## CIRCULAR AND LINEAR PARTICLE ACCELERATORS

### The Betatron

The betatron accelerates beta particles (electrons). In effect, the betatron is a large X-ray machine which can produce particle energies in the 300 MeV range. The betatron operates on the principle of the transformer. Thus, the primary is a large electromagnet and the secondary is the stream of electrons which are to be accelerated.

A hollow, evacuated, curved tube (known as a doughnut) encloses the electrons. As the electrons are driven in a circular path, they acquire energy. Once they reach the necessary energy level, these electrons are directed to a plate at high velocity. The impact of the electrons on the plate produces high-energy rays. The resulting X rays are used for research applications in industry and medicine, and for military and nuclear power research.

### Atom Smashing with the Cyclotron

The American physicist E. O. Lawrence developed the cyclotron in 1932. Many kinds of charged particles can be used in this circular accelerator. The *cyclotron* produces ions having tremendous speeds and energies. As these ions emerge from the accelerator they can be directed to strike a target of phosphorous, gold, sodium, iodine, or any other element. As a result, the target is made radioactive. Thus, the cyclotron substitutes other materials for radium as the radioactive source resulting in lower costs and greater control of the process.

The cyclotron consists of three major parts as shown in figure 38-6: (1) an electromagnet, (2) two D-shaped electrodes connected to a high ac voltage, and (3) a vacuumtight chamber enclosing the electrodes. The vacuum chamber is located between the poles of the electromagnet.

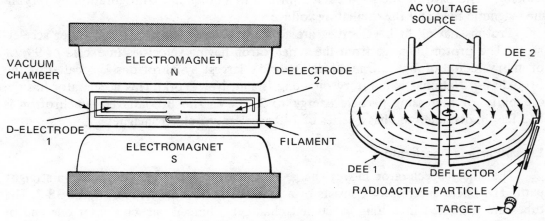

**Fig. 38-6  Cyclotron**

The operation of the cyclotron begins when protons or deuterons are fed into the space between the two D-electrodes (dees). The charged particles are drawn across the gap between the electrodes when an ac voltage is applied. In general, the voltage used has a value to 20 000 volts at a frequency of 15 000 000 cycles per second. The magnetic field of one of the electrodes causes the particles to move in a circular path. At the instant the particles complete a semicircle, the ac voltage reverses causing the particles to be attracted to the other electrode where they also move in a circular path. As the particles return to the first electrode, the voltage is reversed. Thus, the particles continue to move in a circular path. Each time the particles cross the gap between the electrodes, they are accelerated.

As the particles gain speed, the circle of motion increases as the particles spiral outward. The speed of the particles can be determined by recording the number of times they travel around the cyclotron and by noting the voltage. Finally, as the particles emerge from the cyclotron, they strike a target. The action of the particles striking the target at great speeds causes the atoms of the target to emit gamma rays and other particles. In turn, these particles may bombard other atoms. In a conventional cyclotron, the energy of the accelerated particles may reach 15 MeV.

### The Synchro-cyclotron

As the velocity of the accelerated particles increases, so too does their mass. This increased mass slows down the particles. In this state, the particles do not reach the gap between the electrodes at the time the oscillating current reverses. This difficulty is overcome in an accelerator known as the *synchro-cyclotron*. This device is capable of increasing the energy of the particles to the 800-MeV range.

The synchro-cyclotron varies the frequency of the ac voltage. As the mass of the particles increases, the acceleration decreases, and the frequency of the oscillator is reduced. The changes in the field voltage are adjusted to coincide with the slower acceleration of the more massive particles.

### The Synchrotron

The *synchrotron* is another type of particle accelerator. In this device, the particles are also accelerated by causing them to move in a circular path with increasing velocity. The path of the particles is confined to a doughnut configuration by varying the magnetic field and the oscillating voltage.

Protons having high velocities are fed into the synchrotron from other accelerators. The protons emerge from the synchrotron having velocities in excess of 99.9% of the velocity of light. One of the world's largest synchrotrons is located at the Brookhaven National Laboratory on Long Island, New York. This accelerator has the capability of increasing particle energy to 33 BeV. The Brookhaven synchrotron is 843 feet in diameter and has a series of 240 accelerating electromagnets.

### Linear Particle Accelerators

The linear accelerator moves charged particles to high velocities along a straight path. The linear accelerator consists of a series of hollow *drift tubes,* figure 38-7. The tubes are mounted in a long evacuated chamber. Charged particles enter one end of the chamber and are accelerated by a high-frequency alternating voltage applied to the drift tubes. At any instant, alternate drift tubes have opposite charges.

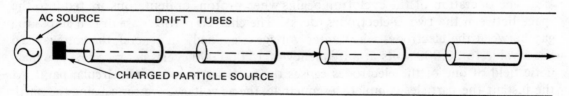

**Fig. 38-7 Simplified diagram of linear particle accelerator**

The alternating voltage is timed so that the particle is repelled by the drift tube it is leaving and is attracted by the drift tube it is approaching. Thus, the particle is accelerated each time it crosses the gap between two drift tubes. The increasing acceleration of the particles requires that the drift tubes be successively longer. This increased length insures that the particles are between tubes when the voltage is reversed.

Circular accelerators are limited in their ability to accelerate electrons because of electromagnetic radiation losses. Linear accelerators are not subject to this difficulty. There are several advantages to the use of a linear accelerator:

— the device is more accessible

— the beam of charged particles is more easily controlled

— the accelerator can provide a higher frequency of impacts on the target material.

The largest operating linear accelerator is two miles long and is operated by Stanford University near Palo Alto, California. However, even larger circular and linear accelerators are in the design or construction stages. American accelerators are being planned which will provide particles with energies in the 600 to 1000 BeV range.

## NUCLEAR FISSION

*Nuclear fission* is the process of splitting apart the nucleus of an atom. Tremendous quantities of energy bind the nuclear particles together. Splitting the nucleus of the atom causes the release of this energy.

Nuclear fission occurs as a result of bombarding the nucleus of an atom with a variety of high-speed particles. In experiments conducted in 1919, Sir Ernest Rutherford used the natural emissions of radioactive elements such as radium to study the behavior of atoms as they are bombarded by particles.

### Neutrons as Atom Smashers

To split the nucleus of an atom, any particle directed toward the nucleus of the atom must first pass through the cloud of electrons surrounding the nucleus. In early experiments, the fragments used to strike the nucleus were emitted naturally from radioactive materials or were charged particles accelerated in a Van de Graaf generator or a cyclotron. The process of using an extremely limited number of charged particles randomly striking the nuclei of atoms was a very inefficient method of splitting nuclei.

It was Enrico Fermi who disclosed in the 1930s that the neutron is a powerful atom — smashing bullet because of its weight and its lack of charge. Neutrons are obtained by bombarding certain elements with protons, alpha particles, and/or other charged particles.

### Fission of Uranium

The fission of the uranium nucleus following bombardment by neutrons produces a tremendous liberation of energy. Fission results when a neutron is captured by the nucleus.

When an atom of uranium 235 is split, barium and krypton atoms are formed. Since barium and krypton do not have as many neutrons as $U^{235}$, neutrons are released in the process, figure 38-8.

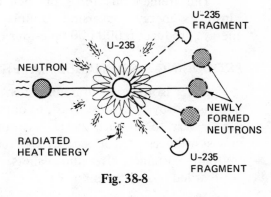

Fig. 38-8

The neutrons, in turn, bombard other $U^{235}$ atoms resulting in the release of other neutrons which then bombard still more atoms of $U^{235}$. This process is known as a *chain reaction.*

### Critical Size

An atomic explosion does not take place until the mass of the fissionable material is such that the chain reaction initiated by the release of the neutrons proceeds uncontrolled. The point at which such a reaction takes place because of the size of the fissionable mass is called the *critical size.*

This value differs from the critical size of an atomic reactor or pile. The critical size of a pile is the point at which it is possible to free a few neutrons to induce a controlled reaction. The critical size of a pile depends on the type of fissionable materials used and their shape, size, and control. The basic difference between a nuclear explosion and the operation of a reactor is in the control of the fission of $U^{235}$.

## THE ATOMIC BOMB

A nuclear warhead (atomic bomb) contains two pieces of either plutonium or $U^{235}$. Both of these pieces are slightly larger than half the critical size. While the two pieces are separated, each piece continues to lose neutrons to the surroundings and nothing happens.

Whenever the two pieces are brought together with tremendous force, the combined mass is greater than the critical size. The neutrons no longer are lost to the surroundings; instead they are emitted at tremendous speeds from the splitting nuclei and set off a chain reaction in an infinitesimal part of a second (less than 1/1 000 000th second). The sudden release of energy causes an extremely high temperature which, in turn, results in sudden expansion and blast effects. The radiation produced consists of destructive gamma rays.

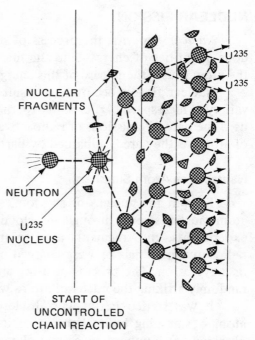

START OF
UNCONTROLLED
CHAIN REACTION

Fig. 38-9

The tremendous force of such a nuclear device can be visualized by recalling Einstein's energy conversion formula; that is, one gram of matter, transformed into energy, produces 6 000 000 mile-tons of energy.

## PRODUCING FISSIONABLE MATERIALS

Other isotopes of uranium (such as $U^{234}$ and $U^{238}$) do not undergo fission when bombarded by neutrons. Thus, to obtain sufficient quantities of fissionable material, it is necessary to produce the $U^{235}$ isotope. When neutrons are absorbed by $U^{238}$, several new substances are formed: the isotope $U^{239}$ and two new elements, neptunium and plutonium (named after the planets). Plutonium also undergoes nuclear fission when bombarded by neutrons.

### Separating $U^{235}$ From $U^{238}$

There are four ways of separating $U^{235}$ from $U^{238}$: (1) thermal diffusion, (2) gaseous diffusion, (3) centrifugal action, and (4) an electromagnetic method, figure 38-10.

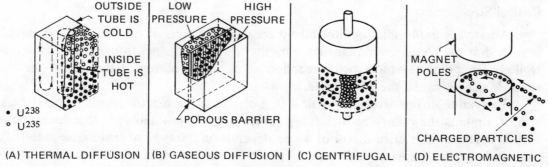

(A) THERMAL DIFFUSION | (B) GASEOUS DIFFUSION | (C) CENTRIFUGAL | (D) ELECTROMAGNETIC

Fig. 38-10

Figure 38-10A shows that as fluid uranium circulates in the *thermal diffusion* method, the lighter $U^{235}$ isotope concentrates at the top of the system where it can be removed. In the *gaseous diffusion* method, figure 38-10B, the lighter $U^{235}$ in a gaseous state passes readily through the porous barrier and is separated from the heavier $U^{238}$. In the third method, figure 38-10C, the uranium again is in a gaseous state. Thus, as it is revolved in a centrifugal device, the lighter $U^{235}$ isotope concentrates at the center of the device. In the *electromagnetic method* shown in figure 38-10D, the uranium particles enter a strong magnetic field where the lighter $U^{235}$ particles are deflected more than the heavier $U^{238}$ particles.

### The Nuclear Reactor (Atomic Pile)

Every neutron that produces a nuclear fission must create another neutron which then produces the fission of another nucleus and so on to cause a continuing chain reaction. As the nuclei of atom after atom are bombarded by neutrons and the nuclei split, tremendous amounts of energy are released in the form of heat and radiation consisting of gamma rays and high-speed alpha and beta particles.

*Producing Plutonium.* As stated previously, the fission reaction in a nuclear reactor produces two new elements. Thus, when atoms of $U^{238}$ are bombarded by neutrons, the nuclei each emit an electron. A new element known as neptunium is produced, figure 38-11. The neptunium atom is unstable and radioactive. Therefore, it releases an electron and changes to the element plutonium. Plutonium is important because (1) it can be produced artificially in nuclear reactors under controlled conditions, (2) it is more abundant than $U^{235}$ and is less expensive to produce, and (3) it acts much like $U^{235}$.

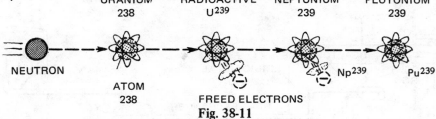

Fig. 38-11

*Function and Operation of the Nuclear Reactor.* The primary uses of the nuclear reactor are to provide a continuous release of energy and to form a new fissionable element such as plutonium. Figure 38-12 shows a nuclear reactor for producing plutonium. Within the reactor, the fission of $U^{235}$ is triggered either by cosmic rays or stray neutrons.

The neutrons produced by fission are slowed down by a graphite moderator, figure 38-13 (page 492). Having a reduced energy, the neutrons are captured by the $U^{238}$ with a resulting formation of plutonium. The moderator

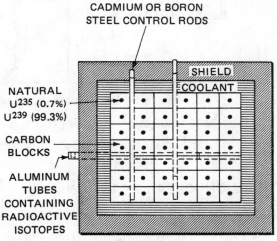

Fig. 38-12

is a material which slows down neutrons by permitting them to bounce off the atoms of the
material until both the energy and speed of the
neutrons are reduced.

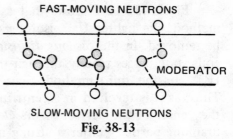

The cadmium control rods in the pile
absorb neutrons as the rods are moved into
or out of the pile, the number of neutrons
absorbed controls the rate of release of atomic
energy. The plutonium ($Pu^{239}$) and the uranium ($U^{238}$) mixture is later removed from the pile so that the plutonium may be separated chemically.

The atomic energy which is released takes the form of heat enery or radiation.
The heat energy can be harnessed to do useful work. The direct energy from the pile
is *contaminated* with harmful radioactive rays. It is for this reason that workers
in atomic energy plants must constantly guard against exposure to alpha, beta, or
gamma rays. Heavy shielding and mechanical devices for handling radioactive materials
are provided around nuclear reactors as safety precautions.

### Breeder Reactors

The *breeder reactor* produces fissionable material. A nonfissionable material is
used in the reactor. After converting (breeding), a fissionable material such as plutonium is produced and can be recovered in the reactor for use elsewhere. The operation of this reactor consists of pumping liquid fuel ($U^{238}$ enriched with $U^{235}$ or plutonium) into the core of the reactor, figure 38-14. Within the core, the $U^{235}$ acts on
a natural uranium slab ($U^{238}$) to breed it into fissionable plutonium.

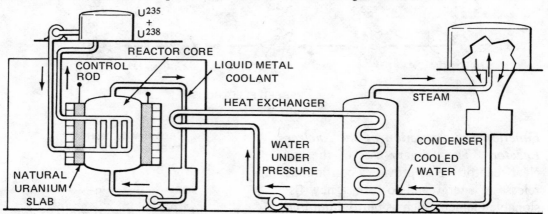

Fig. 38-14 Breeding a nonfissionable material in a breeder reactor

The heat resulting from this process is transferred to a liquid metal coolant. Eventually the heat energy produces steam which is then condensed, cooled, and recircu-
lated as water.

## THERMONUCLEAR REACTIONS

### Atomic Fusion of Deuterium and Tritium Atoms

Energy can be obtained from nuclei by another process called fusion. *Fusion*
refers to the building up of atoms using high temperatures to provide nuclei with

enough energy to overcome their mutual repulsion and join together. This process is the opposite of fission or the splitting of the atom.

The control of fusion energy is of great importance since the solution of the problems of controlling and harnessing the power from fusion reactions will conserve exhaustible natural energy resources, help eliminate pollution, and provide an inexhaustible source of electrical power.

For example, one gallon of ordinary seawater contains enough heavy hydrogen (deuterium) to produce (by fusion) the equivalent energy of 300 gallons of gasoline. In addition, the by-product of the reaction is harmless nonradioactive helium. The cost of extracting deuterium is only a fractional part of one percent of the cost of the coal necessary to obtain the equivalent amount of energy.

Fusion begins when deuterium atoms are in rapid motion at temperatures above 100 million degrees Celsius. However, it is necessary to develop and maintain temperatures above 350 million degrees Celsius over long periods of time if the acceleration of the atoms is to be held at 1500 miles per second. Under these extreme conditions of high temperature and velocity, the nuclei of the deuterium atoms collide and fuse. At the moment of fusion, the atoms release tremendous quantities of heat and light energy.

An additional difficulty of obtaining energy from the fusion process is the cooling effect which occurs as the deuteron plasma comes into contact with the container walls. To overcome this problem, deuteron plasma is confined (bottled) to a beam through the use of powerful magnetic fields created by electric current flowing through and around the plasma, figure 38-15. The electrically charged deuterium particles are responsive to a magnetic field. To prevent the plasma from escaping from the ends of the beam, the ends of the magnetic field are designed to be exceedingly strong.

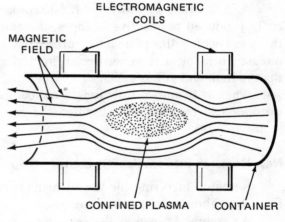

Fig. 38-15 Magnetic field confines plasma in a beam

Atomic fusion requires energy comparable to that released by atomic fission in an atom bomb to produce a temperature high enough so that the atoms of the fusion material start the fusion process. One fusion reaction producing a great deal of energy is the fusion of hydrogen atoms having a weight of 2 (deuterium) and hydrogen atoms having a weight of 3 (tritium).

The size of atomic bombs is limited because the two pieces of $U^{235}$ or plutonium must be smaller than the critical size or a chain reaction begins automatically. In comparison, there is no known critical mass consideration and no known limit to the size of hydrogen bombs. To review, atomic fusion is not dependent upon critical size but rather upon maintaining temperatures in the range of hundreds of millions of degrees Celsius over a period of time, rapid motion, and fusion materials such as deuterium and tritium (forms of hydrogen).

Although the problems involved in the fusion process are immense, the need to develop fusion energy is imperative. Only limited amounts of fissionable materials can be produced from non-fissionable resource materials in the earth's crust. When these natural sources of fission fuel are eventually used up, other sources of energy will be needed. When fusion energy is produced effectively and at a reasonable cost, it will help to satisfy ever-increasing demands for electrical, nuclear, and other forms of energy.

## APPLICATIONS OF NUCLEAR ENERGY

### Use of Radioactive Isotopes in Medicine and in Biological, Physical, and Chemical Research

For research purposes, many materials are placed in a nuclear reactor so that they can be changed into new atoms. When bombarded by atomic particles, these atoms become radioactive and are known as *radioactive isotopes*. One such artificially manufactured isotope is carbon 14. In the form of carbon dioxide, this isotope is fed to plants in an attempt to discover how plants make food from carbon dioxide, water, and sunlight. It is known that the atoms of the elements comprising air are constantly being bombarded by cosmic rays. Thus, all plants actually absorb radioactive carbon dioxide. Animals which thrive on vegetation also ingest certain amounts of radioactive carbon.

Another field in which artificial radioactive elements are used is medicine. Artificially produced radioactive isotopes of elements the body needs can be used to reach the areas where the radioactive properties may provide the necessary treatment of a disease. In medical treatment requiring radiation, radioisotopes provide better control than radium and at less expense.

When it is known how long a radioactive material requires to disintegrate, radioisotopes can be used to determine the age of trees, fossil remains, rock, and other items.

### New Research into Nuclear Structure

Research involving the use of high-energy particles continues to uncover new information about nuclear structure.

A machine known as the *cosmotron* consists of a circular doughnut-shaped tube from which the air is evacuated. Protons are injected into the tube from a four million electron-volt electrostatic accelerator. As the protons travel in the tube, they receive another pulse of energy at the rate of 800 electron-volts per revolution.

The protons receive these increments of energy until their energy level is in the range of billions of electron volts. As the protons strike a target with this tremendous energy, the nuclear fragments can be observed and recorded photographically or with various instruments and Geiger counters.

### Nuclear Energy in Industry

The potential for developing and using atomic energy in industry is endless. A device such as the nuclear reactor can serve as the furnace in which this form of energy is produced. The fuel is natural uranium in purified form plus pure carbon. It is estimated that one pound of a mixture of $U^{238}$ and $U^{235}$ can produce the equivalent heat energy of 10 tons of coal. By contrast, one pound of pure $U^{235}$ produces the equivalent heat energy of 1500 tons of coal. This energy can be harnessed to operate steam turbines, generate electricity, or provide a central heating source.

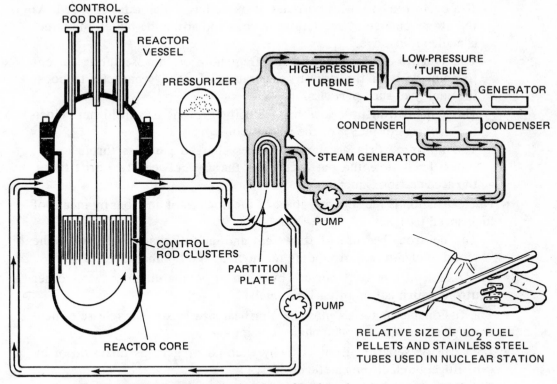

**Fig. 38-16 Representational schematic of nuclear electric generating station**

There are many unsolved problems and challenges in transforming atomic energy into electrical energy. As fundamental research continues, the whole field of nuclear energy takes on greater importance in serving modern civilization.

─────────────────── **SUMMARY** ───────────────────

- The quantitative relationship between matter and energy is expressed as $E = mc^2$.

- The atomic number indicates the number of protons in the nucleus of the atom. This number is equal to the number of satellite electrons.

- The nucleus of the atom consists of one or more neutrons and protons. Electrons circle the nucleus in one or more orbits.

- Isotopes are the atoms of the same element having a different number of neutrons and a difference in weight.

- Alpha and beta particles and gamma rays are emitted by radioactive materials. They can be detected and measured by devices such as the Wilson cloud chamber or the Geiger-Mueller counter tube.

- Linear and circular particle accelerators accelerate charged particles to extremely high velocities so that when these particles impact on a target, maximum energy is produced.

  - The betatron accelerates beta (electron) particles. The X rays produced when the beta particles strike a target are widely used in medicine, industry, and in engineering and scientific research.

- The cyclotron produces particles at exceedingly high rates of speed. As the atoms emerge at accelerated speeds and strike the target, the target is made radioactive.
- The synchro-cyclotron varies the frequency of the oscillating current to coincide with the slower acceleration of massive particles to boost the energy of such particles.
- The synchrotron processes high-velocity protons to still higher velocities in excess of 99.9% of the velocity of light.
- The linear particle accelerator moves charged particles through drift tubes. Each time the particles cross the gap between two drift tubes, the acceleration is increased.
- Nuclear fission is the splitting apart of the nucleus of an atom by means of high-speed particles.
  - The neutron, because of its weight and lack of charge, penetrates the cloud of electrons surrounding the nucleus and bombards it.
- The fission of one $U^{235}$ nucleus by neutron bombardment releases other neutrons which act on other $U^{235}$ nuclei.
- The fissionable materials must be a critical size before the release of neutrons starts an uncontrolled chain reaction.
- $U^{235}$ can be separated from $U^{238}$ by thermal or gaseous diffusion or by centrifugal or electromagnetic devices.
- The nuclear reactor provides the continuous controlled release of atomic energy to form new fissionable elements such as plutonium, or to produce radioactive isotopes.
- To produce energy from the fusion process and to sustain fusion requires that temperatures in the 350 million degree Celsius range be maintained over a period of time and that the movement of atoms such as those of deuterium and tritium be sustained at 1500 miles per second.
- Radioisotopes are widely used in the medical, physical, chemical, and other areas of scientific research.
- Nuclear energy provides a transportable medium with high energy potential.

## ASSIGNMENT UNIT 38 PRODUCTION AND CONTROL OF NUCLEAR PARTICLES

### PRACTICAL EXPERIMENTS WITH NUCLEAR PARTICLES

**Equipment and Materials Required**

| Experiment A<br>Alpha Particles | Experiment B<br>Radioactive Rays | Experiment C<br>Observational Tools |
|---|---|---|
| Luminous tape<br>Luminous surface<br>on watch<br>Microscope | Pitchblende (radioactive uranium ore)<br>Photographic paper<br>Black opaque paper<br>Small solid metal object<br>Photographic developing materials | Geiger-Mueller or other type of counter with manufacturers' specifications<br>Pitchblende (two small pieces of radioactive uranium ore)<br>Lead container<br>Carbon container<br>Wood container |

## EXPERIMENT A. IDENTIFYING ALPHA PARTICLES

### Procedure

1. Adjust and focus a microscope on either a piece of luminous tape or any other luminous surface such as a watch face.

2. Darken the space around the microscope and wait five minutes or so until the eyes are adjusted to the darkness.

3. Look through the microscope eyepiece to observe the bursts of light from the molecules of the paint as they are struck by the alpha particles of the radioactive material with which the paint is mixed.

   NOTE: The zinc sulfide molecules of the paint are struck by alpha particles which are ejected from the radioactive material at a speed of 10 000 miles per second.

## EXPERIMENT B. IDENTIFYING RADIOACTIVE RAYS

### Procedure

1. Wrap one small piece of uranium ore and a small metal object (such as a key) in a piece of sensitized photographic paper.

2. Place the package in a piece of black opaque paper or container and let it stand for a day or two.

3. Wrap another piece of sensitized photographic paper in a similar black opaque paper or container.

4. Develop both pieces of paper.

5. Note any differences between the two developed photographic sheets.

### Interpreting Results

1. Based on the results of the experiments, prove or disprove the statement that radioactive rays are emitted by certain ores.

## EXPERIMENT C. MECHANICAL DETECTION OF RADIOACTIVE RAYS

### Procedure

1. Place one piece of uranium ore or other radioactive ore in a lead container.

2. Turn the detecting device to the ON position. Note any indication of radioactive rays.

3. Remove the container cover and note any differences in the indication on the detecting device.

4. Move the Geiger counter, or other counter device, toward and then away from the ore. Listen for any difference in the sound of the instrument or for a variation in the measurement of radioactivity.

5. Add the second piece of ore. Note any differences in sound or recording on the counter.

6. Repeat steps 1–5 using the single piece of ore first and then the two pieces of ore. Place the single piece and then the two pieces of ore in a carbon box and then a wooden box.

7. Note in each case any variations in sound or recording on the counter.

## Interpreting Results

1. Which of the three materials (lead, carbon, or wood) acts as the best shield from radioactive rays? Justify the selection of one of these materials on the basis of this experiment.

2. Explain what effect distance has on the intensity of radioactive rays.

3. What effect does the mass of a radioactive material have on the quantity of radio-activity? Explain the answer in terms of the results of the experiment.

## PRACTICAL PROBLEMS ON NUCLEAR ENERGY: ITS STRUCTURE, CONTROL, AND USE

### Structure of the Nucleus

Add a word or phrase to complete statements 1 to 10.

1. The weight of the elements resulting from a nuclear reaction is _____ the weight of the original element undergoing change.

2. The transformation to energy of one gram of matter is equal to _____ of released energy.

3. A proton has a weight of _____ and a (an) _____ electrical charge.

4. A neutron has a weight of _____ and a (an) _____ electrical charge.

5. An electron has a weight of _____ and a (an) _____ electrical charge.

6. The arrangement and number of particles in the _____ of an atom determine the characteristics of the atom.

7. The nucleus of almost every atom consists of one or more _____ and _____ .

8. The electrons in the same shell are at _____ from the nucleus.

9. An atom of the same element which has a different number of neutrons is known as _____ .

10. The element whose nucleus is simply a proton is _____ .

## Production and Characteristics of Radioactive Materials and Nuclear Fission

Match the description in Column II with the appropriate term in Column I.

| Column I | Column II |
|---|---|

**Column I**

1. Cyclotron

2. Nuclear fission

3. Atom smashing

4. Nuclear bullets

5. Critical size

**Column II**

a. The quantity of a fissionable material necessary to start and continue a nuclear chain reaction.

b. The production of $U^{238}$.

c. A neutron particle capable of passing through the cloud of neutrons surrounding the nucleus of an atom.

d. A quantity of fissionable material which loses its neutrons outside its own mass.

e. Splitting the nucleus of an atom apart to release tremendous quantities of energy.

f. Bombarding the nucleus with high-speed particles.

g. Charged particles are accelerated to speed outwardly in ever-widening circles with greater and greater energy to bombard atoms on contact with the target material.

For statements 6 to 15, determine which are true (T) and which are false (F).

6. Radioactive materials emit only beta particles.

7. Gamma rays have a positive charge and longer wavelengths than X rays.

8. Alpha particles travel at speeds of thousands of miles a second, but may be stopped by a thin sheet of material.

9. The Wilson cloud chamber operates on the principle that radioactive particles and rays form ions around which moisture condenses to form trails of fog in a controlled chamber.

10. Alpha particles have a positive charge and can be separated in a Wilson cloud chamber or similar device using magnets.

11. Beta particles are attracted to a negative magnetic pole.

12. The Van de Graaff generator and the cyclotron are used to break up natural radioactive materials.

13. The Geiger-Mueller counter tube is used only to identify gamma rays.

14. A Geiger counter detects the presence of radioactive materials and the quantity of radioactivity.

15. A target consisting of almost any material can be made radioactive by bombarding it with ions traveling at tremendous speeds and with high energy.

16. Match each function in Column II with the appropriate particle accelerator in Column I.

| Column I | Column II |
|---|---|
| a. Betatron | 1. Circular and linear particle accelerators to boost the energy output of charged particles. |
| b. Cyclotron | |
| c. Synchro-cyclotron | 2. Processes high-velocity protons to a speed of 99.9% of the velocity of light. |
| d. Synchrotron | 3. Consists of an electromagnet and D-shaped electrodes encased in a vacuum chamber. |
| e. Linear particle accelerator | 4. A difference of electrical potential is built up on an insulated moving belt which is used to accelerate electrons or ions. |
| | 5. Vertical movement of charged particles to increase velocity. |
| | 6. Accelerates electron particles to produce high-energy X rays. |
| | 7. Boosts the energy level of slower moving massive charged particles. |
| | 8. Charged particles are successively accelerated as they cross the gap between drift tubes of increasing length. |

## Producing Fissionable Materials

Select the correct word or words to complete statements 1 to 5.

1. $U^{234}$ and $U^{238}$ do not undergo fission when bombarded by (protons) (electrons) (neutrons).

2. $U^{235}$ can be separated from $U^{238}$ by (thermal and gaseous diffusion) (the cyclotron) (the betatron) (a Geiger counter).

3. A chain reaction continues when every (proton) (electron) (neutron) (gamma ray) splits the nucleus of an atom and produces another identical particle.

4. When $U^{238}$ is bombarded with neutrons, the two new elements produced are: ($U^{235}$ and plutonium) (neptunium and plutonium) (barium and $U^{235}$) ($U^{235}$ and $U^{238}$).

5. The material which slows down the speed of the neutrons produced by fission in a nuclear reactor so that these neutrons lose much of their energy, is known as (a moderator) (an atomic bullet) (a shield).

6. Make a simple sketch of a nuclear reactor. Label the important parts and materials. Explain briefly how the reactor works.

## Thermonuclear Reactions

For statements 1 to 5, determine which are true (T) and which are false (F).

1. Fusion is splitting of the atom.

2. Fusion takes place when atoms are in rapid motion and have reached temperatures in the range of millions of degrees.

3. As the nuclei of atoms collide and fuse together, sufficient energy is released to maintain a high temperature and to sustain the fusion process.

4. One of the greatest thermonuclear reactions is produced by deuterium and tritium (hydrogen isotopes).

5. Atomic fusion and atomic fission both depend on a critical size to start and continue a nuclear chain reaction.

6. State three problems in the production of energy by fusion.

7. Cite three advantages to the production of heat and light energy by fusion.

## Industrial and Other Applications of Nuclear Energy

1. The cosmotron produces particles with energies in the range of billions of electron volts. What advantage does such a machine have over the cyclotron which produces atomic particles having energies of almost 500 million electron volts?

2. Read some of the latest scientific journals or articles in current publications concerning nuclear energy and radioactive particles or isotopes. Name two new applications of such particles in each of three branches of science covered in this unit.

3. State Einstein's 1905 prediction about the relationship between matter and energy. In what manner has this statement been proved in recent years?

4. Cite two reasons why science requires atom-smashing devices of greater and greater capacity.

# Section 10
# Direct Conversion of Energy

## Unit 39  Direct Conversion of Energy Systems

The principles of the direct conversion of energy from one form to another usable form have been known in many instances for more than a century. In 1832, T.J. Seebeck produced a thermopile in which an electromotive force was produced across a circuit consisting of two dissimilar metals. A few years later in 1839, A.E. Becquerel described the first *photogalvanic cell.* That same year, Sir William Grove produced the first successful fuel cell.

In 1876, selenium was used as a light-sensitive material in a solid-state device. Selenium photocells are still used today in the photoelectric eye, photographic exposure meters, and in numerous similar devices.

Each of these early applications of the principles of the direct conversion of energy had limitations due to size, difficulty of production, high comparative cost, or productivity as measured by the percent of useful output. The discovery of new materials, improved electronic circuitry, and the increased capability of industry have accelerated the speed with which new knowledge is gained and applied. Phenomenal technological breakthroughs occurred during World War II and in the years that followed. Further development of the earlier conversion devices became practical, necessary, and economical.

### NEED FOR DIRECT CONVERSION OF ENERGY TO USABLE FORMS

It is evident that technology is an evolutionary process. Progress depends on the ability of scientists, researchers, and technicians to define a problem clearly, such as the characteristics of a new material or product, and then apply available resources to reach a solution of the problem. For example, intercontinental ballistic missiles and spacecraft require a nose cone that must withstand extremely high temperatures upon reentry to the earth's atmosphere. Having detailed requirements, several industries accelerated their experimentation with high-temperature materials.

The combined effort resulted in the production (at a reasonable cost) of a ceramic material that met the requirements of excessively high temperatures while experiencing minor deterioration. The products of this space age technology in the development of high-velocity, high-temperature nose cone materials are being applied to other applications such as high-temperature furnaces and magnetohydrodynamic generator systems.

The following needs govern the present attention and research efforts concentrating on the direct conversion of energy.

- There is a growing worldwide awareness that natural energy resources on this planet are limited and must be conserved.

- The work ethic of the nation, the position of the nation as a world leader in technological developments and applications, full utilization of human resources, and human survival are factors that are inseparably interwoven with the vital questions of energy and its conversion to useful purposes.

- An efficiency plateau is being reached with common conversion equipment and processes. In nuclear plants, for example, there is a tremendous waste of heat energy from the production of the energy at the reactor to the cooling process at the heat exchanger. Improved efficiency of existing equipment means the use of costly materials, more difficult controls, and excessive expenditures. Thus, new ways are being sought to perform the same functions as present equipment through direct, economical, and energy-conserving machines and devices.

- Portable energy generating equipment is needed for inaccessible land areas. Such a generator must depend on the local natural fuel resources which may be in limited supply.

- The increased costs of accessible fossil fuels and other types of fuels places greater dependence on direct conversion processes.

- Problems of air, land, and sea pollution resulting from ever increasing per capita demands for energy require the development of equipment and systems yielding a limited amount of pollutants, if any.

- Direct conversion processes offer the potential for higher productivity through increased efficiency. Also, direct conversion processes, in combination with conventional energy conversion systems may significantly increase the output efficiency.

- Underdeveloped and developing nations require sources of power for industrial development and economic self-sufficiency. Direct energy conversion systems provide the likelihood that the energy sources available in the specific regions can be used.

- Studies of energy use among different nations show that there is a positive relationship between energy consumption and productivity, economic wealth and full employment, standards of living, the well-being of the people, and national security. Energy generation and consumption are highest among the advanced, highly industrialized nations.

Another interesting observation can be made with regard to the acceleration of technological developments in the past three centuries. Figure 39-1 (page 504) shows periods when commercially practical and useful applications were made of energy conversion systems. Note that all of the practical energy conversion systems for generating power were developed since the mid-eighteenth century.

The speed of energy system development has accelerated markedly during the period following World War II, particularly with respect to direct power conversion systems.

Direct energy converters are classified according to (1) the energy source, (2) the energy transfer mechanism, and (3) the actual process of energy conversion. Some typical direct energy converters are: (1) the fuel cell, (2) the photovoltaic cell, (3) thermionic devices, (4) thermoelectric devices, and (5) the magnetohydrodynamic (MHD) generator. The following sections of this unit investigate these converters, their advantages and disadvantages, the principles of operation on which the devices are based, and the applications of the converters.

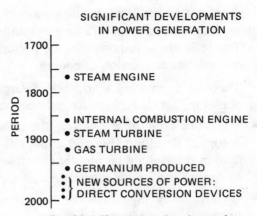

Fig. 39-1 Shortening time intervals
in the application of power technology

## FUEL CELLS: CHARACTERISTICS, PRINCIPLES, AND TYPES

The *fuel cell* is an electrochemical cell. The chemical energy of a fuel supplied continuously to the fuel cell is converted into electrical energy. The fuel cell consists of three basic parts: (1) the fuel and oxidizer, (2) electrodes, and (3) the electrolyte. There are four major types of fuel cells: (1) hydrox, (2) redox, (3) hydrocarbon, and (4) the ion-exchange membrane.

### Advantages of Fuel Cells

The following list gives the major advantages of fuel cells at their present stage of development. Fuel cells:

— contain no moving parts, operate silently and reliably, and require limited maintenance,

— do not depend on a heat cycle and have low heat rejection capability,

— have a high efficiency for their size over a wide range of applications,

— are especially adapted to electrochemical industries requiring low-voltage, direct-current devices,

— can be operated independently of the environment since they are nonpolluting,

— are adaptable to high-volume energy production,

— require low-cost fuel and an oxidant, and

— consume no energy when there is no load connected to the cell.

### Oxygen Concentration Cell

The oxygen concentration cell, figure 39-2 (page 505) is one of the simplest fuel cells. The figure shows that the cell consists of two electrodes suspended in an electrolyte. The electrolyte conducts an electric charge consisting of oxygen ions and acts as an insulator of electrons.

The operation of this type of fuel cell is based on the movement of molecules of oxygen (in air) through the porous cathode to the junction with the electrolyte. Each oxygen molecule picks up four electrons to create separate ions. The ions move through the electrolyte to the anode (fuel) where the ions release their electrons. As a result, oxygen molecules are reformed at the anode.

The electrons released by the oxygen ions move to the cathode which becomes negatively charged. The oxygen continues to enter the system at the cathode and then combines with the fuel at the anode or is exhausted from the system. A current flows when the two electrodes are joined externally with a load. The electrical energy is produced by a difference in the oxygen concentrations between the cathode and the anode. The production of energy continues as long as there is fuel at the anode to react with the incoming oxygen.

### Hydrox (Hydrogen-oxygen) Fuel Cell

The hydrox fuel cell uses an acid electrolyte. Figure 39-3 illustrates the parts of the hydrox cell and its operation. At the anode, hydrogen atoms each give up an electron ($e^-$) to the load at the same time that hydrogen ions ($H^+$) are released to the electrolyte. The hydrogen ions combine at the cathode with oxygen and the electrons from the load circuit. This combination of hydrogen and oxygen yields electrical energy and produces water (which is drained off).

The electrical energy of the hydrox cell is produced and maintained by balancing the amount of electrons moving through the circuit and the amount of ions moving through the electrolyte. The energy generated by the cell can be interrupted by stopping the current flow in the external circuit and/or stopping the flow of ions to the load circuit.

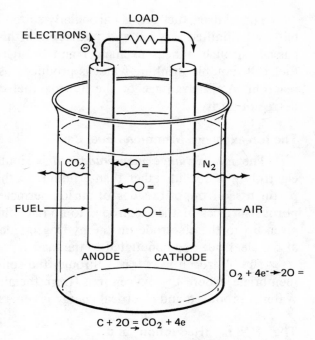

$$C + 2O \rightleftharpoons CO_2 + 4e$$

**Fig. 39-2  Oxygen concentration cell**

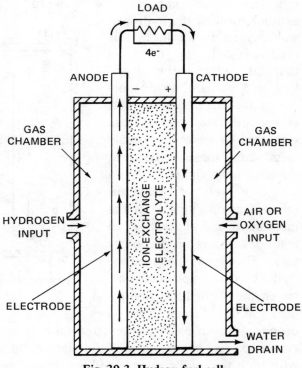

**Fig. 39-3  Hydrox fuel cell**

The design of hydrox fuel cells can be based on either acid or alkaline electrolysis. A large reaction area is provided by catalytic nickel or carbon electrodes having a controlled porosity. The hydrox fuel cell must be pressurized to control the partial pressures of both the fuel and the oxidant.

The hydrox fuel cell is particularly suited for spacecraft use because of the availability of liquified oxygen and hydrogen within the craft. The liquified gases can be passed through a heat exchanger and brought to the operating temperature of the fuel cell. Potable drinking water is produced as a valuable by-product of the fuel cell reaction. A disadvantage of the hydrox fuel cell is the need for gases having a high degree of purity.

### The Ion-exchange Membrane Fuel Cell

The *ion-exchange membrane cell* is similar to the hydrox cell except that the electrolyte is a solid rather than a liquid. In this cell, the hydrogen and air enter compartments on opposite sides of an ion-permeable membrane. The gases penetrate the porous surfaces of the electrodes to contact the membrane surfaces. The electrons are given up to the electrode on the hydrogen side of the cell. The electrons are collected at the electrode and conducted to the load.

The hydrogen ions travel through the solid electrolyte to the other surface of the membrane, where the oxygen ions (from the air) combine with the returning electrons. Water is produced and electrical energy is generated as a result of these processes.

### The Redox (Reduction and Oxidation) Fuel Cell

The *redox fuel cell* is another form of the hydrogen and oxygen fuel cell. In this type of cell, the reactions that take place between the fuel and oxidant do not occur at the electrodes. Rather, two electrolyte solutions are separated by an ion-permeable membrane located between the electrodes, figure 39-4.

One set of reactions takes place on one side of the membrane; and another set of reactions occurs on the other side. Thus, an intermediate (reduction) reaction occurs between the gasified fuel (hydrogen) and liquid electrolyte (A)

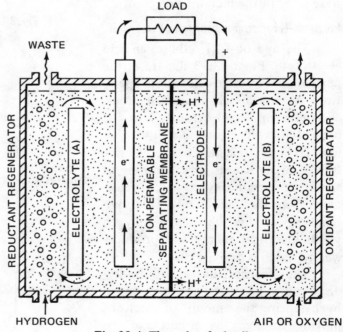

Fig. 39-4 The redox fuel cell

on one side. An oxidation reaction takes place on the other side of the membrane between oxygen or air and a liquid electrolyte (B). The liquid electrolyte is in contact with both the solid electrode and the separating membrane. The migration of hydrogen ions ($H^+$) across the permeable membrane produces electrical energy.

### The Hydrocarbon Fuel Cell

The hydrocarbon fuel cell uses a hydrocarbon as the fuel. The system operates at temperatures above 500°C. Molten carbohydrates are used as the electrolyte which is held in a sponge-like matrix of magnesium hydroxide. Metallic electrodes make direct contact with the solid electrolyte matrix.

When gasoline (or other hydrocarbon fuel) is cracked (reduced to other substances) inside the cell, hydrogen and carbon monoxide are produced. These gases diffuse into the matrix at one electrode and react with ions in the electrolyte to produce water and carbon dioxide. Electrons are released at the same time to the electrode. At the other electrode, the oxygen or air picks up the electrons to produce ions which then move through the electrolyte, completing the circuit.

Most hydrocarbon fuel cells are still experimental in nature because of limitations in the materials available and the size of the cell. The efficiency of fuel cells is normally defined as a ratio of the output of electrical energy to the heat of combustion of the fuel. Theoretically, fuel cells can operate at an efficiency of nearly 100%. In practice, the efficiency of the conversion of chemical energy into electrical energy is in the range between 70% and 90%. These values are far superior to modern generating plants which operate at an efficiency of 40% to 50%.

### The Nuclear Battery

The nuclear battery is another direct energy conversion device. This device converts the energy of a strong beta-emitting radioisotope into electrical energy. The central rod of the battery, figure 39-5, is coated with a radioactive material such as strontium 90 which has a strong emission of beta particles. The rod is part of a vacuum-sealed container.

The electrons and beta particles from the radioactive material cross the vacuum gap from the central rod to the outer metal sleeve which serves as a conductor. Electric current flows when the sleeve is connected to a load.

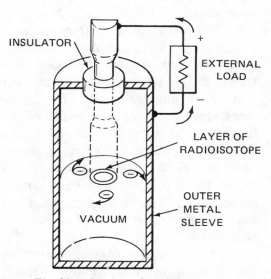

Fig. 39-5  Section of nuclear battery

The advantage of the nuclear battery is that nuclear electrons are not affected by space charge effects. Nuclear batteries are capable of applying voltages in the $10^4$ to $10^5$ volt range. Unfortunately, at the present stage of development, the electron flow output is limited to the microampere range.

## SOLAR ENERGY AND THE SOLAR BATTERY (PHOTOVOLTAIC CONVERTER)

The sun, as the major source of natural energy, produces thermal energy at an estimated continuous rate of 177 trillion ($177 \times 10^{12}$) kilowatts. Other sources of natural energy include water and the wind; natural stored energy sources consist of fossil fuels, and nuclear or cosmic energy that can be released through the processes of fission or fusion.

Light energy has been described as existing in small packets of energy called photons. The amount of energy in a photon is proportional to the wavelength of the light. Bright sunlight, which is a combination of all colors and wavelengths, produces approximately 130 watts of energy per square foot.

Solar battery cells operating under conditions of full sunlight produce an open-circuit voltage (potential difference) of about 0.6 volt. Electric current flows when a circuit is connected to the solar cell terminals. Solar cells are used to provide low voltages for those applications where space, weight, and dependence on a continuous natural source of energy are factors.

Scientists and other technologists have tried over the centuries to develop devices that will harness solar energy to produce other forms of useful energy. In the 1950s, a breakthrough occurred at the Bell Telephone Laboratories when three scientists (D. M. Chapin, C. S. Fuller, and G. L. Pearson) improved upon earlier photovoltaic cells. Their research produced a silicon photovoltaic cell with an operating efficiency of 11%. Subsequent research and experimentation have increased the efficiency of the silicon photovoltaic cell so that it is now a practical energy source providing power to the receiving and transmitting systems of communications satellites and radar installations.

## Semiconductor Properties and Functions

The photovoltaic cell or converter is commonly called a *solar battery,* figure 39-6. The solar battery is constructed of semiconducting materials which are neither conductors nor insulators. It was shown in an earlier unit that in a conductor loosely bound electrons move about easily because of the lower resistivity of the material to the flow of electricity. For an insulator, however, the electrons are tightly bound to the atoms of the material and cannot move easily. Thus, an insulating material has a high resistivity to the flow of electrons.

The properties of semiconductors are affected by temperature changes. When the semiconductor is heated, more

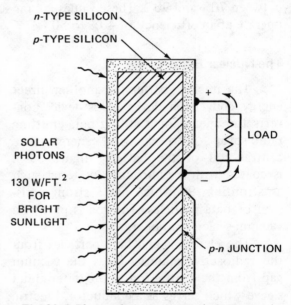

Fig. 39-6 Photovoltaic converter (solar battery)

electrons are freed from the atoms of the material to move about. In this manner, the semiconductor becomes a better conductor. Cooling the semiconductor means that fewer electrons are free to leave their orbits. Cooling causes the semiconductor to become a better insulator.

The importance of semiconductors is due to their ability to control electric current by combining positive (*p*-type) and negative (*n*-type) semiconductor materials. Semiconductor devices depend on the properties at the junction of *n*- and *p*-type materials. One of the most important properties at this junction is that ordinary current flows in one direction only through a *p-n* junction. The permanent electrical field at the *p-n* junction keeps the electrons in the negative (*n*-) side and the holes (lack of electrons) in the positive (*p*-) side. In the solar battery, electron-hole pairs are produced by photons striking the surface of the battery. The electrical field at the junction acts on the electron-hole pairs so that the holes move to the *p*-side and the electrons move to the *n*-side. A voltage is created between the ends of the semiconductor device by the displacement of freed charges.

Figure 39-7A shows a semiconductor device where part of the device carries current by electrons (*n*-type material). The remainder of the device carries current by means of the holes in the *p*-type material. When a potential difference is applied across the ends of the semiconductor device, (the *p*-end is positive), the holes in the *p*-region migrate to the left and the holes disappear, figure 39-7B. Similarly, the electrons in the *n*-region move to the right and disappear. Under these conditions, the current through the device is negligible.

If the connections to the device are changed so that the *p*-end is positive and the *n*-end is negative, figure 37-C, a semiconductor junction is formed that readily permits current to flow from the *p*-region to the *n*-region. However, current cannot flow in the reverse direction. The holes migrate to the right and the electrons move to the left to produce a positive current flowing from positive (+) to negative (−). Both the electrons and holes meet at a junction between the *p*-region to the *n*-region and recombine at that point.

A hole is a missing electron. When electrons and holes come together and combine, they disappear into the regular

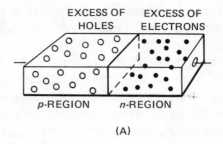

(A)

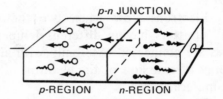

HOLES AND ELECTRONS DISAPPEAR

(B)

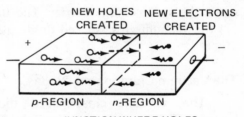

*p-n* JUNCTION WHERE HOLES
AND ELECTRONS RECOMBINE

(C)

Fig. 39-7

structure of the crystal and can no longer act as current carriers. This phenomenon accounts for the fact that current can flow through a semiconductor from the *p*-region to the *n*-region, with very little flow in the opposite direction.

The importance of this principle is that semiconductors which are sensitive to heat and light energy create free electrons and holes because the energy causes more electrons to leave their orbits. Silicon and germanium (first produced at the Bell) Telephone Laboratories) are excellent semiconducting materials with a desirable crystalline structure. Electrons can be set free readily in these materials either by thermal energy or as a result of photons of light energy striking the material.

In the solar battery, a number of wafers (cells) of semiconductor material are combined and connected in series to produce electrical energy. The battery is energized by solar energy.

## THERMIONIC CONVERTERS OF ELECTRICAL ENERGY

A typical *thermionic converter* is a low-voltage, high-current generator, figure 39-8 (page 510). Such a converter operates on the principle that as the temperature of a metal is increased, the electrons in the metal receive enough energy to leave the surface. The higher the temperature, the greater is the number of electrons emitted by the metal and the greater is the energy of these electrons. Electrons flow from an electrode (cathode) at a high temperature, to an electrode (anode) at a lower temperature. The anode is known as the *collector*. The electrons are emitted by a *thermionic emitter*.

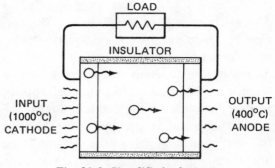

Fig. 39-8 Simplified schematic of thermionic converter.

Electrons flow when two thermionic emitters having different temperatures are connected externally with an electrical conductor to a load, figure 39-8. The energy of emission of the electrons depends upon the temperature of the emitters and the materials used.

The boiling off of the electrons in thermionic converters from the heated cathode to the anode collector produces an electrostatic space charge of negative electrons forming near the cathode. This space charge blocks the flow of other electrons and repels them. Different types of thermionic converters have been developed to overcome the space charge barrier. The three main types of thermionic converters include (1) the vacuum close-spaced diode, (2) the magnetic triode, and (3) the cesium plasma diode.

### The Vacuum Close-spaced Diode

The vacuum close-spaced diode has an efficiency of approximately 15% and is the lightest in weight of the energy converters. The space charge barrier effects are overcome by placing the anode and cathode of the diode extremely close together in a vacuum tube, figure 39-9.

One disadvantage of the vacuum close-spaced diode is the high temperature required for cathode operation. This high temperature requirement means that the tube life is shortened.

### Magnetic Triode

In the device known as the magnetic triode, the space charge effect is reduced by crossing electric and magnetic fields. To accomplish this action, an accelerator is placed above a hot cathode and a cold anode which are on the same plane. Figure 39-10 shows that a battery inserted between the accelerator

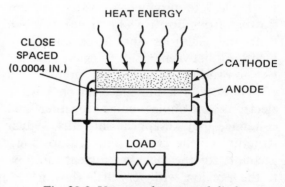

Fig. 39-9 Vacuum close-spaced diode

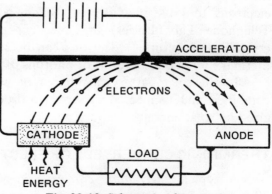

Fig. 39-10 Schematic of magnetic triode components and operation

and the cathode produces an electric field between the cathode and the anode. The entire device is placed between the poles of a large magnet to superimpose an external magnetic field.

In operation, an electron is emitted from the cathode and is deflected as it speeds toward the accelerator to the surface of the anode. During acceleration, the action of the electric field produces a gain in electron kinetic energy. The electron returns the energy to the field when it drops back to the anode. Thus, the energy of each electron arriving at the anode is the same as the energy of each electron at the cathode. Although the magnetic triode does not produce more power than the vacuum close-spaced diode (due to increased electron losses), it has the advantage of being able to operate at a lower temperature.

### Cesium Plasma Diode

In the cesium plasma diode, figure 39-11, the space charge barrier is overcome by the use of an ionized gas (cesium). The positive ions in a diode filled with cesium plasma neutralize the negative space charge to allow the electrons to flow toward the anode.

Cesium plasma diodes operate at an efficiency of nearly 30%. However, this device has the following disadvantages: (1) cesium vapors are corrosive, (2) the device is difficult to produce, and (3) the cathode must operate at exceedingly high temperatures. Such operating conditions shorten the operating life of the device and result in increased costs.

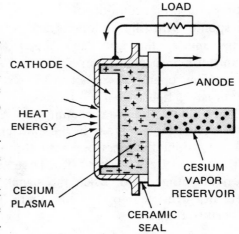

Fig. 39-11  Schematic of cesium plasma diode

### TRANSISTORS

Commercial production of the element germanium expanded the development and potential of the transistor. In turn, the transistor enabled tremendous advances to be made in science and technology, particularly with respect to the properties of solids, lasers, light amplifiers and modulators, and the complete fabrication of electronic circuits on paper-thin wafers. Because of these developments, integrated circuits that combine the functions of multiple transistors and numerous other devices are possible.

Common applications of transistors include radio, television, telephony, electronic watches and instruments, automated machine tool control and production machines, computer components for the instantaneous gathering, storing, and processing of information, and multimedia communications devices. Transistors serve the same functions as vacuum tubes in that they amplify current, provide electronic signaling, and control and combine electrical pulses.

Important advantages of transistors include greater reliability, smaller size, less power consumption than vacuum tubes, and simplicity of design. Adaptations of the transistor to integrated circuits have revolutionized the design and production of electronic devices for unlimited applications.

A transistor consists of a crystal of one type semiconductor (*p* or *n* material) sandwiched between two slices of the opposite type of semiconductor material (*n* or *p*), figure 39-12. Figures 39-12A and B are known as NPN and PNP transistors, respectively. The semiconductors in each transistor combination are known as the collector, the base, and the emitter. The functions of these components correspond with the functions of the anode, the grid, and the emitter in a vacuum tube.

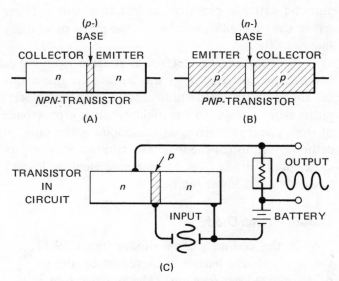

Fig. 39-12 NPN transistor serving as an amplifier

The operation of an NPN transistor, figure 39-12C, is similar to that of a triode vacuum tube. A positive signal applied to the base of the transistor causes the electrons in the emitter to be attracted to the base. Since the base area is very thin, the electrons rush through to the collector where a positive voltage is maintained by a battery. The large current causes a voltage drop across the resistor. On the other hand, when the signal to the base is negative, electrons in the emitter are not attracted to it. Thus, there is no appreciable current flow with the negative signal to the base.

### Integrated Circuits and Miniaturization

The development of the transistor was an intermediate stage in the research leading to miniaturization of equipment, devices, components, and parts. The transistor combines the functions of many vacuum tubes and is a more reliable less expensive device. It was possible to reduce the size, weight, and cost of devices that formerly required several different vacuum tubes by substituting one solid-state semiconductor device. Further reductions in the size of electronic components and equipment resulted from the development of integrated circuits.

An *integrated circuit* device of only a millimeter square in size (about 0.040 in.$^2$) may contain over 100 transistors. Integrated circuits are produced by depositing tiny globules (a few molecules thick) of *n*- and *p*-type semiconductor materials in a predetermined pattern on a small board. Varying patterns of these *n*- and *p*-type materials form the transistors, diodes and resistors of the integrated circuit. The development of transistors and integrated circuits was the key to the miniaturization of complicated equipment and was a primary contribution to the success of the American space program.

## MAGNETOHYDRODYNAMIC GENERATION OF ENERGY

### The Magnetohydrodynamic (MHD) Propulsion Engine

The term *magnetohydrodynamic* (MHD) is descriptive of the scientific phenomena involved in this method of converting one form of energy to another usable form.

MHD devices are designed and produced for propulsion and power generation processes. In an MHD propulsion engine, an electromagnetic field causes a gas (*plasma*) to flow. The gas is *pinched* or accelerated magnetically to an exceedingly high velocity.

A thrust is produced by directing the high-velocity gas through a continuously expanding nozzle. This action produces an impulse which is opposite in direction to the exhaust velocity of the engine. The rate of the pinch process sustains the required level of thrust. Among the advantages of an MHD propulsion engine are its high exhaust velocity, high specific impulse, operation at a comparatively low temperature, controllable and variable thrust level, and instantaneous starting and stopping.

### The Magnetohydrodynamic (MHD) Generator

The operation of the MHD generator is almost the reverse that of the MHD plasma pinch propulsion engine. In the MHD generator an electromotive force is created by forcing a fast-flowing plasma through a magnetic field which is at right angles to the direction of plasma flow. A conducting stream of partly ionized gas containing a number of free electrons is used as the plasma. However, the main body of the gas is nonionized. By ionizing a portion of the gas and heating it to a very high temperature, the gas becomes a conductor. Because the high temperature approaches the maximum capacity that present-day materials are able to withstand, the gas is seeded with salts such as those of cesium or potassium. The salts ionize at oxygen combustion temperatures in the 2500°K to 3000°K range. To generate the hot gaseous ionized fluid plasma, seeding takes place in a burner similar to a rocket engine.

In addition to the magnetic field, the plasma flows past a battery of electrodes, figure 39-13. As the gas moves through the electrical field, the electrons are deflected and move (between collisions with the gas particles) to one of the electrodes. This movement of the electrons from the cathode through the load to the anode and back to the gas stream causes a flow of electricity.

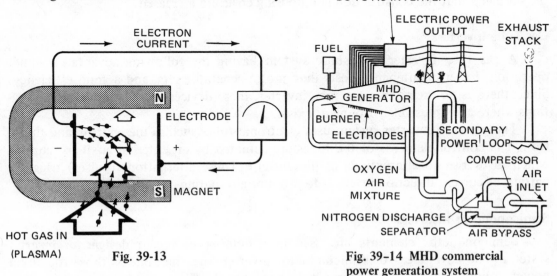

Fig. 39-13

Fig. 39-14 MHD commercial power generation system

The voltage generated by an MHD unit is directly proportional to the intensity of the magnetic field, the velocity of the gas, and the distance between the electrodes. The dc output of the MHD requires the use of an inverter to produce alternating current.

Among the current problems in MHD generator development are the high temperatures necessary; air pollution resulting from the seeding operation; the physical limitations of materials required in the system; and the initial design, development, testing, and installation costs. The MHD generator has a tremendous potential for producing electrical energy in the range of megawatts at a reasonable cost. It is also expected that an MHD generator can convert heat losses in nuclear reactor power plant operations to electrical energy at a high level of efficiency.

## INVERTERS: CONVERSION OF DIRECT CURRENT TO ALTERNATING CURRENT

The greatest use of electrical energy is in the form of alternating current. Since the direct conversion devices covered in this unit produce direct current, it is necessary to use a conversion device (inverter) to change the dc to ac. To be useful commercially, this conversion process must be accomplished economically, with high productivity and low maintenance costs. Three common types of inverters are described in the following paragraphs:  (1) the motor-generator set, (2) the gas tube inverter (using a mercury-arc rectifier), and (3) semiconductor power inverters.

### The Motor-generator Set

The motor-generator set has been used successfully as an inverter for many years. A dc motor drives an ac generator to convert direct current to alternating current. The motor-generator set has the advantage of being able to withstand overloads and to provide reliable service over a long period of time (years). The disadvantages of the motor-generator set include the fact that the motor-generator set is a dynamic system with the result that there is no output unless the unit is rotating; the rotor and stator parts and the housing bearings require continuous maintenance; and the weight and size of the set mean that the construction and housing costs are increased.

### Gas Tube Inverters

A gas tube inverter is a static system having the following advantages: quiet operation, lower maintenance costs than motor-generator sets, and a good efficiency. Since there is no arcing in a gas tube inverter, these devices are adaptable to installations where an explosive environment exists.

The gas tube inverter system uses electronic tubes such as the ignitron and thyratron. The control elements of the tubes are actuated by an external auxiliary current which gives rise to alternations in the current. These alternations (pulses) trigger a pattern of current direction reversals to produce alternating current.

### Semiconductor Inverters

Semiconductor elements are used in a number of recent designs of inverter systems. The advantages of semiconductor inverters are the same as those for direct power generators:  limited noise; high operational efficiency; lower maintenance, weight, and volume requirements; and reliability under adverse operating conditions.

However, static inverters are limited in their operation because of limited thermal energy storage capacity. As a result, inverters must be designed to operate at their maximum output. The system must be brought up to the highest possible voltage (without overload) because the inverter efficiency increases in proportion to the increase in the operating voltage. Any overloads which produce thermal energy may cause failure of the semiconductor devices.

Inverters are used in military, commercial, and industrial applications for power supplies, speed controls, and motor design and control. Transistors and a controlled rectifier element known as a Trinistar have been used in electrical power inverters for high power applications.

### The Trinistar and High-power Generation

The Trinistar is a semiconductor having the characteristics essential to high-power operation under an instantaneous switching mode. Energy losses in semiconductors are extremely low under minimum and maximum conduction conditions. In other words, the *all-on* or *all-off* switching time must be as fast as possible to produce a high efficiency with a minimum of dissipated energy. The semiconductor performs this instantaneous high speed switching function without deterioration when a heat dissipating element is included in the circuit to hold the operating temperature within fixed limits.

The Trinistar can function as an inverter or a rectifier. An inverter requires a pair of Trinistar units:  one unit supplies the positive half of the ac wave; the other unit supplies the negative half of the wave. The waveforms produced may be either gradual or instantaneous alternations (sine wave or square wave).

The functional element of the Trinistar consists of a silicon wafer. A control element is attached to the wafer to cause it to change from a high impedance state to a low impedance state and to serve as an electronically controlled switch. The control element of the Trinistar can turn the unit on. However, the circuit can be turned off only when a zero current is reached.  This requires continuous monitoring and control of the circuit when the device is operating.

### THERMOELECTRIC GENERATORS

The phenomenon of converting heat energy directly to electrical energy was applied as early as 1821 by T.J. Seebeck.  As stated earlier, Seebeck learned that when two dissimilar metals are joined together and heated, a difference in electrical potential is produced between the hot and cold ends. One widely used application of this effect is the thermo-

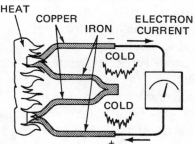

Fig. 39-15

couple. The minute quantity of electric current that is generated by the device is metered through a galvanometer. The movement of the galvanometer needle across the face of a temperature-calibrated chart provides a direct temperature reading.

Recent developments in the study and manufacture of thermoelectric semiconductor materials have resulted in the application of these materials to devices having increased power output capacity and improved efficiencies. In turn, these thermoelectric devices have led to a rapid expansion of thermoelectric power generator design, development and production. The advantages of thermoelectric power generators and similar devices are due to the lack of moving parts, the compactness and ruggedness of the units, the low noise levels, and the limited maintenance necessary.

The design of guidance and stability mechanisms in space vechicles may be simplified because the gyroscopic forces which are produced by rotating parts are eliminated with the use of thermoelectric devices.

The thermocouples in the thermoelectric generator are made of semiconducting materials which possess thermoelectric properties. These materials convert heat to electricity more efficiently than previously used metals.

### Principles Applied to Thermoelectric Generators

The ability to adjust the number of free electrons in semiconductor materials is of great significance in thermoelectric equipment. In principle, the output voltage of a thermoelectric material is inversely proportional to the number of free electrons in that material. In addition, the conductivity of the semiconductor material is directly proportional to the number of free electrons.

The optimum efficiency with more power output is achieved by adjusting the electron density until a value of $10^{19}$ free electrons/cm$^3$ is reached. This value is between high voltage and high electrical conductivity. At this point, the electrical yield is 175 microvolts per degree Celsius. Power losses result from either high voltage and low current conditions or low voltage and high current conditions.

### Operation of Thermoelectric Converters

In the positive (*p*-type) semiconductor, the positive holes carry the current. When the negative semiconductor (*n*-type) is heated, the holes (+) accumulate at the cold end. The electrons flow from the cold (+) end to the hot (−) end and in the external circuit from (−) to (+). *n*- and *p*-type semiconductors perform this same heat-to-electricity conversion process in thermoelectric converters. When heat is supplied to a negative (electron) *n*-type semiconductor, the electrons at the heated end gain both kinetic energy and velocity. As a result there is a flow of *hot electrons*

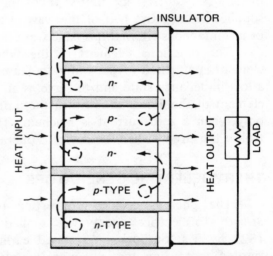

Fig. 39-16 Concept of thermoelectric converter

toward the cold end where they are accumulated. By closing an external circuit, these electrons flow out of the semiconductor and through the circuit from negative (−) to positive (+). Both types of semiconductors are connected in a thermoelectric generator because of the limited output of a single semiconductor. Adequate working currents and voltages can be produced by alternately stacking *p*- and *n*-type semiconductors around a central heat source, figure 39-16.

The positive and negative electrical charges are evenly distributed in a uniformly heated thermoelectric material. When heat is applied to one surface, the positively charged ions remain within the semiconductor and the negatively charged ions move toward the cooler end. This movement of ions produces a potential difference between the hot and cold ends of the device. The potential difference causes an electric current to flow through an external circuit.

### Disadvantages and General Applications of Thermoelectric Converters

The separate thermocouple semiconductors used in a thermoelectric device must have low contact resistance. If the device operates above 300°C, the materials are subject to corrosion from air. There is also a destructive effect on the materials if they are subjected to shock and expansion. Design features such as shielding against corrosion and spring loading to compensate for shock are incorporated into these devices.

On the positive side, thermoelectric power generators are used to supply the electricity required for the pumping and control systems along natural gas, oil, and other fuel transmission lines. Instrumentation on satellites is powered by thermoelectric converters. Thermoelectric generators may be developed as components within a nuclear power generating system. Again, the wasted heat of the fission reaction may be used in combination with thermoelectric generators to generate more electricity and improve the conversion efficiency of the power system.

### Thermoelectric Generators Powered by Radioisotopes

The electrical energy output of heated thermocouples is inefficient when compared to conventional power generating systems. However, there are some applications that require a limited power source operating under conditions of adverse weather, remote geographical area, nonavailability of fuels, and unattended operation over long periods of time.

Thermoelectric generators using the isotopes polonium$^{210}$, plutonium$^{238}$, strontium$^{90}$, and curium$^{242}$ are in operation. These generators provide power for communication and weather satellites, space probes, navigational buoys and lights, terrestrial exploration, offshore energy operations, and other scientific, military, and industrial applications. The power output of radioisotope thermoelectric generators ranges from 0.001 watts to 750 watts with a generator life ranging from 90 days to five years.

### Operation of a Radioisotope (Thermoelectric) Generator

The radioisotope generator consists of a protective outer shell lined with a radioactive shield (if required), figure 39-17. The fuel capsule containing radioactive material is at the core of the unit. A series of thermocouples is located around the generator from the core container to the outer shell. The thermocouples are connected to terminals.

The radioactive particles are trapped inside the fuel cell where their kinetic energy is converted into a continuous supply of heat energy for a predetermined period of time. As the fuel decays spontaneously, particles which produce heat upon absorption are emitted. The thermocouples convert the heat energy directly into electricity. This power source is tapped for use at the terminals.

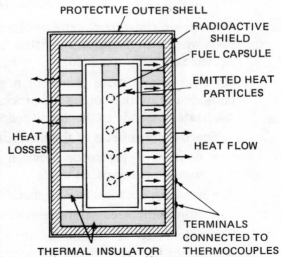

Fig. 39-17 Radioisotope thermoelectric generator

---------------------------------- **SUMMARY** ----------------------------------

- A great deal of present-day research concentrates on the applications of scientific phenomena to new systems for the direct conversion of energy as a means of conserving natural resources and improving productivity.

- Some of the forces lending urgency to the developments of direct energy conversion systems include: the reaching of efficiency plateaus with conventional power generating equipment, increased fuel and operational costs, environmental safeguards, economic and national security, increasingly higher standards of living, and the needs of developing nations.

- The fuel cell converts the chemical energy of fuels into electrical energy.
  - The three basic parts of a fuel cell are the fuel and oxidizer, electrodes, and the electrolyte.
  - The hydrox, redox, hydrocarbon, and ion-exchange membrane cell are the major types of fuel cells.

- The nuclear battery converts electrons and beta particles (having high speeds and energy levels) emitted by a radioisotope material to electrical energy.

- The amount of light energy transmitted by a photon is proportional to the wavelength of the light.
  - Solar cells have a low-voltage output and are used in installations where the factors of space, weight, and dependence for operation on a natural source of energy are important.
  - Solar batteries are constructed of positive and negative semiconductors:
    - *p*-type semiconductors are rich in holes (lack of electrons).
    - *n*-type semiconductors have an excess of electrons.
    - Electrical energy is produced by combining and connecting a number of solar cells in series and energizing the device with solar energy.

- Thermionic converters are low-voltage, high-current generators. Electrons flow when two thermionic emitters having different temperatures are connected externally to a load.

- An electrostatic space charge of negative electrons is produced near the cathode during the boiling off process when electrons are attracted from the heated cathode to the anode collector.
  - Space charge effects of thermionic converters are reduced by:
    - control of the space between the cathode and the anode in the vacuum close-spaced diode.
    - crossing electrical and magnetic fields in the magnetic triode.
    - using an ionized gas in the cesium plasma diode.

- Transistors have revolutionized the design, functions, and performance of equipment, devices, components, and parts used in the industrial, commercial, health, and all other occupational sectors of society.
  - A transistor consists of a layer of one type of semiconductor material sandwiched between two layers of the opposite type of semiconductor material.

- Transistors perform functions similar to those of vacuum tubes:  including amplification, rectification, detection, mixing, wave shaping, and oscillation.
- Integrated circuits are minute (miniaturized) circuits produced by depositing tiny globules of *n*- and *p*- materials on a small board. The materials and circuitry form transistors, diodes, resistors, and other components. The resulting integrated circuit has the operating characteristics of its component circuits.
- The MHD propulsion engine produces a thrust which is opposite in direction to the exhaust velocity. The engine uses a gas plasma which flows under the action of an electromagnetic field. The plasma is pinched to accelerate it to an exceedingly high velocity.
- The MHD generator operates by forcing a heated partially ionized plasma at a high rate of speed (1) through a magnetic field which must be at right angles to the flow and (2) past a battery of electrodes.
- Inverters are needed to change the form of electrical energy from dc to ac.
  - The motor-generator set, gas tube inverters, and semiconductor inverters are three common types of inverters.
  - The Trinistar converter functions as an inverter or a rectifier.
- The output voltage of a thermoelectric generator is inversely proportional to the number of free electrons in the material. Conductivity is directly proportional to the number of free electrons.
  - Working currents and voltages are produced in thermoelectric converters by stacking *p*- and *n*-type semiconductors around a source of heat.
  - Radioisotope thermoelectric generators produce electrical energy by converting the heat energy of the decaying radioisotope fuel; the electrical energy produced is tapped at the terminals of the thermocouples.

## UNIT 39  DIRECT CONVERSION OF ENERGY SYSTEMS

### PRACTICAL PROBLEMS IN THE APPLICATION OF DIRECT ENERGY CONVERSION SYSTEMS

#### Need for Direct Energy Conversion to Other Usable Forms

1. Make four statements to justify the need for accelerating the development of practical direct energy conversion systems.

#### Fuel Cells:  Principles, Characteristics, Types

1. Compare the potential and actual range of efficiency of fuel cells with one nonelectrochemical system for generating electrical energy.
2. Cite four advantages of fuel cells over other systems of producing electrical energy.

Secure a manufacturer's technical manual having illustrations, specifications, and descriptions of one type of fuel cell.

3. Make a simple schematic drawing of the fuel cell.

4. Name and state the function of each major part of the cell.

5. Review the fuel cell specifications. Give the electrical energy output rating(s) and operating characteristics of (a) one fuel cell or (b) the output range of the series of fuel cells produced by the manufacturer.

6. Identify two business, industrial, or scientific applications of the selected fuel cell.

## Solar Energy and the Solar Battery

Secure a technical manual for a solar battery. Check that the manual provides construction details, describes the operation of the battery, and gives applications.

1. Explain briefly the characteristics of purified silicon, arsenic, and boron in terms of their semiconductor properties.

2. State the phenomena which explain the properties of *n*-type and *p*-type semiconductors.

3. Trace the steps in the production of (a) a single crystal *n*-type ingot, (b) a thin *n*-type wafer, and (c) a *p*-type skin covering the *n*-type wafer.

4. State the function of the *p-n* junction.

## Thermionic Converters of Electrical Energy

1. Explain how a thermionic generator functions.

2. Identify three basic types of thermionic converters.

Secure a manufacturer's technical manual illustrating and describing different types of thermionic converters.

3. State the type and give the manufacturer's trade name for one type of converter.

4. Describe how the thermionic converter works.

5. Give one practical application of the thermionic converter.

## Transistors

1. State three functions that are served by transistors.

2. Give three advantages of transistors as compared to vacuum tubes.

3. Describe briefly the construction of a transistor.

4. Make a simple schematic drawing of an NPN transistor which is to serve as an amplifier.

   a. Label the parts of the NPN transistor according to their function.

   b. Explain how the NPN transistor operates.

5. Define an integrated circuit.

6. State three values to the miniaturization of electronic equipment.

## Magnetohydrodynamic (MHD) Generation of Electrical Energy

1. Contrast the operating principles of the MHD plasma propulsion engine with those of the MHD generator.

Secure an engineering or manufacturer's technical manual describing a proposed or currently operating MHD generator in an electrical generating system.

2. Give two of the stated advantages of the MHD generator over any other piece of electrical power generating equipment.

3. Describe briefly the function of the MHD generator.

4. Check the specifications and give the estimated output or operating efficiency of the MHD generator.

### Inverters: Conversion of Dc to Ac

1. Discuss the functions of an inverter.

2. Select one of the three basic types of inverters and describe the function of this type of inverter.

3. Name a new semiconductor power inverter.

   a. Give three advantages of this type of inverter as compared to a rotating machine type.

   b. Give two applications of the semiconductor inverter listed.

### Thermoelectric Generators

Study a current technical journal article or manual which illustrates and describes a thermoelectric converter powered by radioisotopes.

1. Give the name of the converter and state its use.

2. Indicate the specifications of the device with regard to:

   a. The type of radioactive fuel used

   b. the electrical power generating capacity, or range, and

   c. generator life.

3. Describe how the thermoelectric generator functions.

# Section 11
# Applications of Physical Science to Consumer Needs, Career Planning, and Development

## Unit 40  The Physical Sciences Applied to Consumer Needs

Every individual is directly affected both as a consumer and a producer by the interrelationship of science and all materials, processes, products, and services. All living and nonliving bodies, in turn, are known and described according to laws, principles, concepts, and phenomena that are founded in one or more of the sciences.

There is a basic body of general science that enriches the life of each person as a consumer. In addition, there are specialized areas of science that are directly related to shop and laboratory practices; business, office and distributive occupations; home economics-related employment opportunities; the health industries; the public service sector; and in fact, to every field of occupational endeavor.

As stated previously, the units in this text were determined by both occupational and job analyses and careful assessment of everyday consumer needs. The preceding sections have covered essential laws and principles of physics and their applications through experimentation and other practical problems in both consumer and occupational settings.

The level of understanding and application of these fundamentals of physics increases as the worker advances through job levels to the craftsman and technician levels. Employment at semiprofessional, professional, and scientific levels, as well as other highly specialized technical occupations, requires a considerably greater comprehension of science spread over many of the branches. Correspondingly, the level of mathematical competency required increases in difficulty as more complex problems are encountered.

A series of summarizing statements concerning the laws of physics with examples of everyday applications are given in this unit. These statements show the interlocking relationships between the scope and sequence of all units in *Fundamentals of Applied Physics*.

The concepts, principles, and laws of physics are applied in a general way in the Achievement Review which follows. The applications and test items serve as supplementary examples which relate physics to consumer situations. The sequence of this unit follows that of the text itself, beginning with Section I, Science, Matter and Measurement.

## APPLICATIONS OF SCIENCE:  MATTER AND MEASUREMENT

Scientific investigations and achievements are based on careful analysis, systematic planning, use of the scientific method, and demonstration and experimentation. All of these steps lead to the establishment of conclusions and generalizations which become principles and laws.

Physics deals with matter. Matter occupies space, is perceptible to the senses, and has weight and mass. Matter is measured according to universal standardized units of measurement in the British system, the modified European MKS system of metric measurement, and the evolving International Standards (SI) systems.

### Consumer Applications of Matter and Measurement

The scientific approach and systematic planning are techniques an individual can apply to any area of study and at any level, time, or place. Defining the dimensions of a problem, gathering objective evidence, analyzing the data and experiences observed, and making recommendations based on the best tested knowledge provides a systems approach to objective problem solving.

It has been shown that all mechanisms, equipment, and parts within the home, or in business, industry, the community, or the world that can be seen or manipulated by one of the senses consist of matter. Knowing the properties, structure, and conditions relating to matter, the consumer has a base for comparing the value of goods, products and services. The basic knowledge of matter allows the consumer to interpret technical data and other specifications which are a necessary and important part of purchasing and maintaining refrigeration, air conditioning (heating and cooling), and communications systems, as well as power equipment, transportation, and other equipment.

## APPLICATIONS OF SCIENCE:  MECHANICS, MACHINES, AND WAVE MOTION

Concepts treated in the science of mechanics relate both to the statics (position) and dynamics (motion) of matter in space. Physical phenomena in statics deal with bodies at rest. The science of dynamics covers bodies in motion and studies the causes of motion.

A machine transforms an applied (output) force or torque into another output force which is different in magnitude, direction, or both from the original force. The ratio between the input and output forces is identified as the mechanical advantage. In most physics problems, ideal (theoretical) circumstances are used and the mechanical advantage is theoretical (TMA). The theoretical condition can be contrasted with the actual mechanical advantage which exists under normal operating conditions where friction and other forces reduce the efficiency of machines.

All machines are a combination of two basic types of machines:  the lever and the inclined plane. The different classes of levers, the inclined plane and wedge, the wheel and axle, the screw thread, pulleys, and gears are classified as simple machines.

These simple machines are all integral parts of more complex machines. The principles of mechanical power transmission, friction, lubrication, fluids at rest and in motion, and atmospheric pressure are all important factors in the design and operation of mechanical devices and machines.

## Consumer Applications of Mechanics, Machines, and Wave Motion

The automobile is an excellent example of the application of mechanics to consumer goods. The foot pedals of the automobile and other mechanical linkages such as levers multiply and transmit force to the brake shoes, acting through a complementary hydraulic system. The screw as an application of the inclined plane is used to fasten, adjust, and measure components. As a fastener, the screw secures door hinges, safety belts and miscellaneous accessories. Under the hood, the carburetor, alternator, cylinder head block and many, many other parts are joined together or held by screws.

In the automobile radio, the movement of the station selector is produced by a simple application of the wheel and axle. Mechanical power is transmitted from the drive shaft to the fan, alternator, and air-conditioning compressor by a series of pulleys and V-belts. Other power is transmitted from the engine through a drive shaft and compound gearing in a differential to produce motion in the wheels. The design of all of these parts and units is based on a consideration of friction and lubrication for different materials under differing conditions.

The principles of fluids at rest and in motion form the basis for the air intake and carburetion and air-conditioning systems. Similarly, the principles of hydraulics are applied in the brake and automatic transmission systems and to the various fluid pumps which move liquids in the systems.

## APPLICATIONS OF SCIENCE: MAGNETISM AND ELECTRICAL ENERGY

Magnetism is considered to be a form of energy by which usable electricity is produced. Magnetism, the magnetic properties of materials, polarity, magnetic fields, shielding and so on, are fundamental to the whole area of electromagnetism and other electrical phenomena.

## Consumer Applications of Magnetism and Electricity

The study of electromotive force, the control of current and the determination of values for current, voltage, and resistance provide a basis for understanding common applications such as electrical installations and maintenance; lighting systems, and motor circuits and their operation.

The conversion of chemical energy to electrical energy is a basic principle applied to the typical automobile battery. The most widely used form of electricity, alternating current, is generated by changing the energy of fossil fuels such as oil, coal, and natural gas; falling water and other energy sources; or nuclear energy, into electricity. Power generating stations, high-tension heavy-duty transmission lines, distributing stations, step-down transformers at the pole for controlling voltage, and the watthour meter for recording and measuring electrical energy are everywhere in evidence.

This electrical energy output finds multiple daily uses by the consumer to operate door chimes, radios, and other audio-visual sound equipment, heating and lighting controls, motors for temperature control, and refrigeration equipment to preserve foods. The consumer depends on electrical energy for healthful living and personal and national security.

## APPLICATIONS OF SCIENCE: HEAT ENERGY AND HEAT ENGINES

Heat as another form of energy is a measure of molecular motion. Every particle within a solid, liquid, or gas is capable of motion. Thus, a body of matter possesses internal (kinetic) energy. Within the body, the speed of the motion of particles increases or decreases with temperature.

Another fundamental quantity of measurement is temperature. Temperature can be expressed in degrees Celsius, Kelvin, or Fahrenheit. Energy lost from the visible motion of objects that does not reappear as visible potential energy, may appear as a rise in temperature. Heat is also determined by the number of particles in the mass of the body, the average activity of these particles, and the specific heat of the material.

The design and operation of parts, components, and mechanisms depends upon several factors: varying coefficients of expansion for different materials; conduction, convection, and radiation as methods of transferring heat; and possible changes of physical state.

### Consumer Applications of Heat Energy and Heat Engines

Steam engines, steam turbines, the internal combustion gasoline engine, diesel engines, gas turbines, rocket engines, and other propulsion systems depend on heat energy. Engines whose operation depends on fluid flow, such as steam and gas turbines, and jet and rocket engines, must be designed so that at each successive stage the stator and rotor blades compensate for specific temperature changes in the fluid, rate of fluid flow, or the expansion or compression of the fluid to meet specific heat requirements. For machines, all moving parts must be machined to tolerances which permit mating surfaces to move in the proper relationship to each other under varying conditions of expansion and contraction due to temperature changes.

Most importantly, heat energy is essential for human survival.

## APPLICATIONS OF SCIENCE: LIGHT ENERGY

Light, which travels in free space at the speed of 186 000 miles per second, is responsible for all that an individual perceives through the sense of sight. The propagation of light and the formation of shadows are important in estimating distances, direction, and shape. Illumination refers to the intensity of a light source; that is, the density of luminous flux falling on a surface.

Light was treated as an electromagnet system of waves with only a small section as a visible spectrum having seven colors. Objects continuously emit radiant heat energy which is proportional to the temperature of each object. When heat is applied and the thermal equilibrium of the object is changed, radiation emitted from the object may become visible and affect the sense of sight. As the radiant energy increases, its color usually changes from a dull red to a white hot color within the visible region. Color depends on the wavelengths of light and how the light is absorbed and reflected by objects.

All objects visible to the eye depend on light that is emitted, reflected, refracted, or dispersed from the object. Mirrors are used to enlarge, decrease, and control light. Polarization is another example of the manner in which wave motion oscillations are controlled to produce desired patterns. Laser devices amplify light and produce mono-chromatic, coherent, and collimated light beams. These characteristics make lasers important for consumer, industrial, military, scientific and other applications.

## Consumer Applications of the Light Energy

Vegetation and plant growth, essential for human survival, requires the light energy of the sun. This same energy is the power source for solar batteries essential to the operation of telecommunications satellites and space exploration vehicles.

Illumination is based on scientific principles which relate intensity, area of illumination, and the distance from the light source. The nature and degree of illumination is usually governed by the degree of precision involved. Precise instrument-making and fine measurement operations require a greater illumination on the workpiece and area than is required by workers performing rough operations requiring little skill.

Cameras and projection devices are examples of equipment used by consumers which are concerned with light, degree of illumination, color, and lenses for their operation.

In the laboratory, materials are heated as a manner of identification. Each element can be heated to a gaseous or vapor state. In this state, each element has a particular wavelength and spectrum line. The spectroscope and recording spectrometer apply this principle to analyze and establish the quantity of a chemical element in a substance. In the recording spectrometer the color waves are converted to electrical currents which are then read directly on a recording device.

On highways and high-speed expressways, color and light are used in safety devices such as white and yellow lines, light reflectors, and high-intensity lamps. In the home, color-coded equipment helps maintain a safe environment and identify electrical circuits. Color is used on the exterior of houses to assist in temperature control as well as for esthetic purposes. Certainly, all of interior decoration, art, and design are influenced by the control of light and the application of color.

## APPLICATIONS OF SCIENCE: SOUND ENERGY

Sound is produced by vibrating objects. The vibrations move in longitudinal waves along the length of the wave, accompanied by pressure variations. The medium (or air) in the path of a sound wave becomes more or less dense depending on the condensations and rarefactions produced. The physiological sensation of sound is produced by changes in pressure which cause the eardrum to vibrate at frequencies within a range of 20 hertz to about 20 000 hertz. These frequencies corresponding to wavelengths ranging from 54 feet down to two-thirds of an inch. Ultrasonic waves with high frequencies above 20 000 Hz can be detected by electromechanical devices.

Pitch, frequency, wavelength, velocity, amplitude, intensity, interference, and resonance are properties that are considered in the production, recording, application, and acoustical treatment of sound.

## Consumer Applications of Sound Energy

Oceanographers probing for sources of marine food, charting ocean floors, or searching for mineral and oil deposits depend on the sound waves generated by a sonar device and returning to establish time intervals (to measure distance) between the vessel and other objects or formations.

Orchestras, theatrical productions, and large audience programs depend on the treatment of sound waves by the use of construction materials and design features which compensate for the reflection, refraction, and absorption of certain sound waves. Designers and engineers are responsible for the design of domestic, commercial, and industrial structures to provide structurally sound and physically desirable environments.

The telephone, tape and disc recorders, audio-visual films, and amplification systems are used to transmit, record and produce sound. In other applications of sound waves, materials are formed and cleaned with ultrasonic devices.

## APPLICATIONS OF SCIENCE: ELECTRONICS

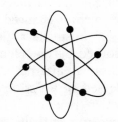

The science of electronics is an area of physics that deals with the emission, behavior, effects, and applications of electrons. Two areas where electronics is applied to an important degree are communications and industrial control equipment. Solid-state components, printed circuitry, and other developments by which comparatively large electronic components are being replaced with smaller, more efficient units have resulted in microminiaturized circuits and equipment. Such equipment provides greater efficiency and, in most instances, leads to reduced operating and maintenance costs.

### Consumer Applications of Electronics

Telecommunication satellites increase international communication in the sharing of technical, medical, and scientific experiences, and educational and other knowledge on a worldwide scale. Earth orbiting experimental stations with sophisticated electronic devices provide a base for observations of the earth that are important to energy conservation, location of food resources, protection of the environment, and national security. Other electronic equipment carried into space for interplanetary exploration provides otherwise unobtainable samples and evidence of environmental conditions on other planets and solar system relationships. Electronically guided intercontinental missiles are part of the national defense systems of all countries.

Closer to the individual, electronically controlled operations in nuclear power plants and other energy conversion power plants result in the generation of essential electrical energy. Almost every house contains one or more of the following electronic devices and systems: tape recorders, television, electronic sensing devices to provide protection and to operate motors and switches, and other automatic electronic equipment used for food preparation, heating, and lighting systems.

On the highway, speed and traffic are radar controlled. Electronic systems serve the consumer in banking, transportation, and health care. Electronic communication devices provide almost instantaneous information that is programmed, stored, and retrieved; scanning and probing equipment are essential for airport security; and quality control equipment monitors the production of food, textiles and fibers and the materials of industry that serve as the raw materials for limitless goods and products.

Solid-state automotive ignition systems are provided by automobile manufacturers because of the advantages to the consumer. Solid-state components eliminate the points and the ignition condenser and increase the life of spark plugs. These, in turn, provide greater reliability and decrease maintenance and replacement costs and services.

The continually expanding applications of electronics to the exploration of space, production, sales, distribution, agriculture, banking, health, and other personal services indicated the increasing daily dependence on electronics.

## APPLICATIONS OF SCIENCE: NUCLEAR ENERGY

Nuclear forces and electromagnetic and gravitational forces comprise the known forces in nature. Isotopes of certain elements that contain unstable nuclei emit radiation (rays) and decay under the action of a nuclear force. The emission of alpha, beta, or gamma particles within the nucleus changes the character of the nucleus so that one element is changed to another element (*transmutation*).

In nuclear fission, the neutrons absorbed by one element give the nucleus additional internal energy. The neutrons released in a fission reaction can generate a chain reaction. In the element $U^{235}$, one neutron initially causes the fission of one uranium nucleus, releasing two or three neutrons. These neutrons, in turn, may cause the fission of additional uranium nuclei and the process multiplies into a chain reaction.

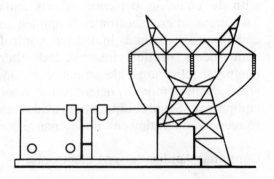

The mass that disappears in this fission reaction is changed to energy that appears as kinetic energy. The importance of fission and a resulting chain reaction is in the tremendous amount of energy that is released ($E = m \cdot c^2$). The fission of one gram of uranium is multiplied 3 000 000 times to produce the same heat energy as 3 000 kilograms (3.3 tons) of coal as a fossil fuel. Nuclear energy depends on maintaining a critical mass that will produce a self-sustaining and controllable chain reaction. The energy released as heat energy by the nuclear reaction is transformed into other usable forms.

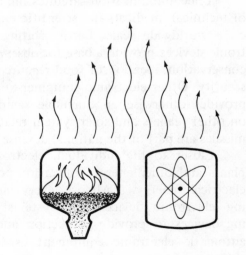

### Consumer Applications of Nuclear Energy

The awesome destructive force of nuclear energy has been demonstrated by atomic bombs, the detonation of nuclear devices in underground silos, and the surface firing of intercontinental ballistic missiles. Radioactive fallout from nuclear bombs, radioactive decay from unstable elements and their isotopes, and the waste products of nuclear reactions constitute a threat to health and human survival unless adequate safeguards are provided in their design, production, use, and disposal.

In other applications, radioactive materials are used in geology and anthropology to establish the characteristics and dates of rocks, fossilized remains of earlier creatures, the history of the earth, and the evolution of biological organisms. Exploration for untapped sources of energy requires equipment that uses nuclear energy as the power source.

In the field of health services, radioactive tracers are used in the detection of physical disorders and in the treatment of cancer and other diseases. Increasing consumer dependence on electric energy within the home, in industry for the manufacture of goods and products, and in agriculture for essential food and fiber production is being met in part by the energy output of nuclear reactor power plants.

## APPLICATIONS OF SCIENCE: THE DIRECT CONVERSION OF ENERGY

Developmental work is being conducted to design practical components and mechanisms to convert energy directly from one source to another. The advantage of direct conversion is apparent. By eliminating or decreasing multiple conversion processes, it is possible to simplify the design of equipment, reduce construction and maintenance costs, and manufacture equipment having a long service life, low weight, high reliability and increased efficiency with less environmental pollution.

Solar batteries, nuclear batteries, and magnetohydrodynamic (MHD) generators are examples of direct energy conversion units where the conversion is accomplished without reciprocating or rotating machines.

### Consumer Applications of the Direct Conversion of Energy

Improvements in the direct conversion of the radiant energy of the sun into electric energy through the use of the photovoltaic cell has made possible the numerous types of communications satellites orbiting the earth. The solar batteries of these satellites provide adequate power to relay conversations via radio and television around the globe. The solar batteries on the surface of the satellite convert sunlight into electrical energy to operate the electronic receiving and transmitting units. Rural telephone systems using solar batteries have demonstrated the feasibility of this energy conversion source in producing the necessary power to operate the system.

Electrochemical fuel cells that convert chemical energy directly into electrical energy eliminate the thermal cycle, have no moving parts, and operate with a higher efficiency. Electrical energy produced by fuel cells powers the electronic life-support systems in spacecraft.

Thermoelectric converters are used in deep-sea explorations, lunar landings, floating weather stations, navigational lights and buoys, ocean bottom beacons, and in various other land, sea, and space services.

Thermionic converters are low-voltage, high-current generators which require elevated temperatures at the cathode and a lower temperature at the anode. Converters of this type are used for power generation in space vehicles.

Magnetohydrodynamic (MHD) generators combine electromagnetic and fluid dynamic principles. Both the MHD and thermionic converters may be used with steam and gas turbines and nuclear reactors to extract higher levels of energy with a resulting increase in the potential efficiency of the energy generating equipment. The extreme heat losses incurred in changing steam back to water in the condensers of the power plants may be cycled through an MHD generator unit. These units are ideally suited to power stations where large quantities of electricity must be produced.

## UNIT 40  ACHIEVEMENT REVIEW OF CONSUMER APPLICATIONS OF SCIENCE

### CONSUMER APPLICATIONS OF SCIENCE TO MATTER AND MEASUREMENT

Use a refrigerator, air-conditioning unit, telephone, television set, or stereophonic system as a technical project and complete problems 1–5.

1. Secure the manufacturer's specifications for the equipment selected.
2. List five different materials used in the construction of the equipment.
3. State three properties for each of two of the materials listed.
4. Name three units of measurement used in the manufacture or operation of the equipment.
5. Prepare a table and show four measurements in (a) the BGS and (b) equivalent measurements in the SI system.

### CONSUMER APPLICATIONS OF MECHANICS AND MACHINES

Study a home appliance with mechanical moving parts. Also refer to the exploded views in the manufacturer's technical manual to identify the parts of the appliance and show the operation.

1. Name the appliance.
2. Study the construction of the appliance. Then list two parts that are applications of (a) the screw (inclined plane), (b) a lever, and (c) either a mechanical power transmission system or the wheel and axle.
3. Make a simple sketch to illustrate the principles of mechanics which apply to the parts listed in 2(a), 2(b), and 2(c).

Select a piece of equipment used in the home, a gasoline station, or a repair shop that operates according to the principles of fluids at rest and in motion. Analyze the functions of the various parts and components of this equipment.

4. Name the piece of equipment.
5. Describe in scientific terms which relate to mechanics and machines how the equipment functions.
6. Give an example of mechanical advantage as applied to any part or component of the equipment.
7. List five parts, mechanisms, or machines that represent additional applications of the principles of mechanics and machines.

## CONSUMER APPLICATIONS OF MAGNETISM AND ELECTRICAL ENERGY

1. Identify two devices in an airplane that depend on magnetism for their operation.

Secure and study a technical brochure from a utility company which describes the electrical power generation and distribution system.

2. Make a series of simple statements which give the steps in transforming the fuel energy into electrical energy. (State the nature of the fuel.)

3. Determine and state the electrical energy capacity of the system and the average output for a stated period of time.

4. Explain briefly the functions of (a) the system of transmission, (b) a power distribution substation, and (c) the further distribution of electrical energy from the utility company pole to the consumer.

Locate a motor on an electrical appliance. Check the nameplate or technical manual, if available, for the motor specifications.

5. State briefly the functions and characteristics of (a) the motor, (b) the electromagnetic coils, and (c) the need for overload protection.

6. Compare a primary cell used in a flashlight with a secondary cell as a power source.

## CONSUMER APPLICATIONS OF HEAT ENERGY AND HEAT MACHINES

Secure a table listing the coefficients of linear expansion.

1. Compare the linear expansion of three different solid materials, each of which is two meters long, for a 200°C rise in temperature.

2. State what effect a variation in the linear expansion of two mating revolving parts has on the tolerance to which the parts are machined.

Examine a refrigeration or cooling unit and, if possible, the specification brochure.

3. Name the unit according to its type and function.

4. Determine what features were incorporated in the design to compensate for the heat transfer by conduction, convection, and radiation.

5. State what function insulation serves in the unit.

6. Name three different types of heat engines that are used for (a) land transportation and (b) the generation of electrical energy (power). (c) Name one principle type of heat engine used for exploration in space.

7. Explain the differences between the operation of a diesel engine and an internal combustion gasoline engine.

## CONSUMER APPLICATIONS OF LIGHT ENERGY AND COLOR

Examine a series of colored photographs in a technical manual, magazine, or other publication. Use a magnifying glass.

1. Explain how the colored object seen in the photograph is made distinguishable in the printing process.

2. Describe briefly how color in a photograph is reproduced in printing.

3. Scan either a newspaper, magazine, or other printed matter that reports scientific developments in space exploration and investigation. Explain how a knowledge of the electromagnetic spectrum is related to the scientific investigation.

4. List two applications of lenses in the projection and control of images for the purposes of medical health services.

5. Examine a photographic camera. Explain the relationship of the object distance, lens opening, and image size.

## CONSUMER APPLICATIONS OF SOUND ENERGY

Secure a cutaway section of a telephone receiver and transmitter or an illustration of such a device. Study its construction and operation.

1. Make a simple line drawing to show schematically how (a) the transmitter is actuated by sound, (b) the vibrations are transmitted to the receiver, and (c) energy is converted in the receiver back to sounds.

Select a construction material that is used for acoustical treatment.

2. Name the material and describe its composition.

3. List two sound properties that are treated by the acoustical material.

4. Name two pieces of equipment or accessories that are used for (a) recording (transmitting) sounds, (b) storing sounds, (c) reproducing sounds, and (d) amplifying sounds.

5. Examine a combination microphone, recorder, and loudspeaker system. Explain how sound energy is transformed so that it can be transmitted, stored, reproduced, and amplified with this equipment.

## CONSUMER APPLICATIONS OF ELECTRONICS

Identify and name one piece of electronic equipment used in industry or in the health services field that affects each person.

1. State the function of the equipment.

2. List three electronic terms associated with the parts or circuitry and give the function of each.

3. Describe briefly how the industrial electronics equipment works.

4. State two electronic developments that have improved either the audio or visual signals in television reception, provided better coverage of televised programs, or decreased maintenance costs.

Study a current scientific article or report on any one of the following: (a) telecommunication satellite system, (b) earth-orbiting experimental station, (c) supersonic aircraft, (d) interplanetary spacecraft, (e) earth-probing energy exploration equipment, or (f) electronic equipment related to new scientific endeavors.

5. Identify two pieces of electronic equipment for the selected system or vehicle that are used for scientific exploration purposes.

6. State three different electronic terms that apply to the parts or circuitry and the function of each term.

7. Explain briefly how different forms of energy are converted, transformed, or transmitted in the system, equipment, or vehicle.

8. Discuss the impact of the new scientific knowledge and technical experiences on the individual as a consumer.

## CONSUMER APPLICATIONS OF NUCLEAR ENERGY

Secure an illustrated descriptive technical report on the production of electrical energy from an actual or proposed power generating station using a nuclear reactor.

1. Use a simple line drawing to trace the energy conversion processes. Begin with the nuclear energy fuel source for the system and end with the generation of the electrical energy.

2. Determine the percentage of the total electrical energy output of the power station that is provided by nuclear energy.

3. State two possible forms of pollution and the precautions that are taken to safeguard employees and the public.

4. Give (a) one advantage and (b) one disadvantage of producing electrical energy from nuclear energy.

5. Give one example of the use of radioisotopes or radioactive fuel in (a) health and medicine, (b) biology, and (c) communications.

## CONSUMER APPLICATIONS OF THE DIRECT CONVERSION OF ENERGY

1. Outline three pressures on society that are producing accelerated developments in the direct conversion of energy.

Locate a technical report containing illustrations of manned spacecraft or telecommunications satellites for which fuel cells or some other type of sophisticated direct electrical generators were developed as power sources.

2. Identify two different developments and state the purpose of each.

3. Explain the principles of operation of one direct electrical converter and state briefly how the converter operates.

4. Briefly state two advantages to the design of a nuclear reactor for electric power generation which includes an MHD generator.

# Unit 41 Science Essential to Careers, Career Planning, and Development

## WORK, THE WORK FORCE, AND CAREERS

### The Work Ethic as a Foundation for Human Resource Development

Any nation's most priceless resource is its human potential. Economic wealth and national security are derived from the freedom of people to serve and to grow, using individual talents to work at an occupation that is fulfilling and productive. The development of the human potential and social and economic wealth all center around a concept of work. This is known as the work ethic.

- Work is essential to human existence. A worker produces goods or performs services that are necessary to a fellow person or society.
- Work is a privilege. It is the vehicle through which each person finds his or her mission in life.
- Work provides human relationships whereby one person helps another to reach a a state of effective citizenship and wholesome living.
- Work recognizes that unemployment, underemployment — idleness or separation from meaningful work — is not conducive to the human condition.
- Work is the foundation for human wealth: economic wealth and spiritual, moral, and cultural wealth.
- Work involves the individual in activities that are essential to each person, consistent with the needs of society.
- Work provides self-fulfillment so that each person can find meaning and purpose, ideals and values, and happiness in being part of society.
- Work produces a recognition of human integrity in relation to society.

Work-oriented career education and manpower development and training are designed to reach the individual continuously throughout his or her life span. The pyramiding of in-school and on-the-job employment training and experiences develops the capability in the individual to deal at his or her level with life problems, with environmental problems, and with social problems.

### Distribution of Manpower in the Work Force: Careers Based on the Economic Sectors of Society

An analysis of the world of work reveals that it includes a series of *occupational constellations* which represent the economic sectors, figure 41-1. For example, the agriculture sector of the economy comprises a major occupational

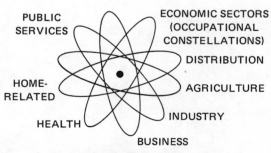

Fig. 41-1 World-of-work galaxy

constellation. Other sectors relate to marketing and distribution, home economics-related occupations, allied health occupations, trades and industry, and public services occupations.

These economic sectors or occupational constellations include major occupational *clusters.* For instance, there are no less than 40 major occupational clusters in the trades and industry sector alone, figure 41-2. The building industry, the textile design and clothing fabrication industry, the automotive industry and the personal service industries are all samples of the 40 major occupational clusters.

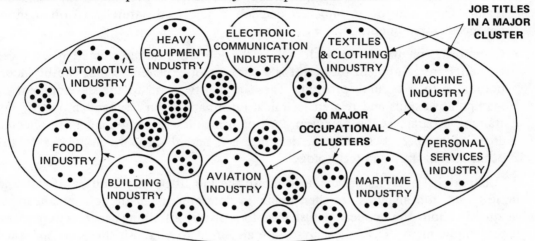

Fig. 41-2 Occupational constellation for the trades and industry economic sector

## Career Ladder Possibilities Within an Occupational Cluster

An examination of each occupational cluster in a sector reveals that it includes all job titles which relate to the occupation. For example, in the building industry occupations cluster, there is a whole series of jobs in carpentry; there are a number of jobs in masonry and the trowel trades; other jobs are available in construction electrical work, in plumbing and pipefitting, in sheet metal fabrication, air conditioning and refrigeration, and in painting and decorating. Within each of these fields the jobs are further classified, according to the skill required. The jobs range from those at the entry level to those at an intermediate level to those on the advanced level. Job mobility means vertical advancement to successively higher levels of jobs requiring greater skill, technical knowledge, and responsibility; or horizontal movement into a related occupational field; or movement out of the cluster by retraining for an entirely new economic sector.

## SCIENCE APPLIED TO CAREERS AND CAREER PLANNING

A study of the characteristics of people in the labor market shows that their interests and abilities range over a wide spectrum. Some individuals are best able to perform tasks at the lower levels. Others can master higher level skills, related sciences and technical knowledge to serve at semiskilled and skilled levels. Still others are interested in professional level positions and can satisfactorily complete a higher education program and perform at this level.

Labor market analysts group jobs according to established criteria into levels such as laborer — unskilled or semiskilled; skilled craftsman or technician; foreman, middle level supervisor, and manager; semiprofessional; and professional.

### Criteria for Identifying Jobs for a Particular Career Level

The job titles in an occupational cluster for a particular career level should fulfill the following conditions:

- There are a number of different job titles in the cluster to provide a reasonable spread of job opportunities for the individual who completes a training program for that level.

- The jobs must be closely associated with each other so that the worker can shift readily from one job to another; one job should lead naturally to other more advanced employment.

- Common elements are to be found in the job titles in the cluster. When these elements are combined, the result is a package of experiences that can be incorporated into a training program. These elements relate to the development of manipulative skills and related technical information. An inseparable part of any training program is the common core of principles and phenomena in selected areas of mathematics, science, drafting, art, and design, which relates to the occupational skills and technology.

Occupational analyses are made to group the job titles on the different skill and training levels. Job analyses are then made to determine from the job specifications what qualifications the applicant must have to profit by a training program in preparation for employment. The job analyses are also used to establish those occupational skills that must be developed in school, on the job, or by a combination of institutional and in-plant education and training. Different types of training programs with measurable performance objectives may be established from these analyses.

### A Model of a Career Development Education Ladder

The terms awareness, orientation, exploration, and occupational preparation are all a part of the career development ladder that begins in the elementary school and extends through high school and postsecondary institutions. These terms describe a progressive series of learning experiences which may be grouped around a general common core and a specialized vocational core. Integrating the general and special cores provides the foundation for effective career planning for a total educational and manpower training program.

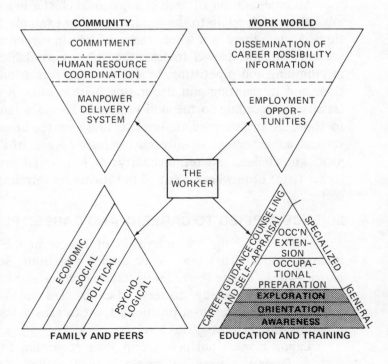

Fig. 41-3  Forces influencing career decisions

### Science Essential to Careers: Awareness, Orientation, Exploration and Occupational Preparation

In this context, science is used daily to interpret and to understand the nature, functions, properties, and applications of matter in all forms, sizes, shapes, and relationships.

The initial awareness to science and its importance to the living world is nurtured throughout the early school years so that a better understanding of its application to work and citizenship responsibilities is developed. Then, orientation to science is provided by showing how science is related to work and career planning and development.

In the junior high school years, the emphasis in science changes to one of exploration. Here, in addition to specific science courses, science is introduced as a part of the practical arts such as industrial arts, home arts and crafts, business arts and agricultural arts. These practical arts provide general education experiences. The experiences should relate broadly across each economic sector. For example, many curriculum plans for the exploratory phase of industrial arts education broadly classify the learning experiences into manufacturing, construction, communications, and power and transportation. These broad-based experiences explore the student's interest in employment possibilities which extend over the forty major occupational clusters in the trades and industry sector.

Science, therefore, is an integral part of general education. Science is also a part of specialized occupational education, when it is used to prepare individuals for entrance into, advancement in, or upgrading in employment. Physics is regarded as the foundation science upon which all other sciences dealing with living and nonliving things depend. Physics investigates basic concepts and relationships among matter, energy, and space. The domain of physics usually encompasses a study of matter and its measurement, mechanics and machines, magnetism and electricity, electronics, heat, light and sound energy, and nuclear energy.

## SCIENCE NECESSARY TO CONTINUING CAREER DEVELOPMENT

The determination to prepare for a vocation, or retrain for different employment, or enter into an upgrading program requires a team input. Professional assistance in career planning includes the screening of applicants, guidance, counseling, referral into a planned performance-based training program, placement, on-the-job followup, and continued education and training to remain occupationally competent or to achieve occupational mobility.

Additional assistance in career planning can be obtained from community agencies that provide employment and human resource development services; vocational guidance and counseling personnel; vocational school counselors and others. Publications such as the *Occupational Outlook Handbook,* a labor market analysis, manpower demand reports, occupational visual aids, examples of school-based and industry-based models for career planning and development and other business/industry/labor materials, provide assistance in planning for a career.

### Physics as a Fundamental Science for Careers at all Occupational Levels

Physics — as well as mathematics, materials of industry, blueprint reading, and other subjects related to an occupation — is part of the technical language of society and the world of work. A simple term like friction is used by scientists all over the

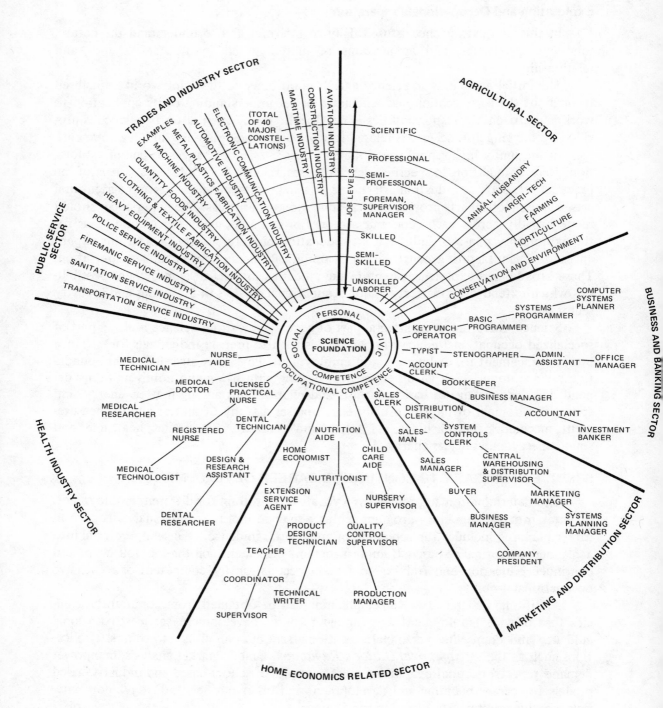

**Fig. 41-4 Science as a core in career planning, training, placement, and advancement in all economic sectors at all job cluster levels**

world (in the appropriate language translations, of course) to indicate that there are natural laws which relate to a body at rest and in motion, and to indicate that there are complementary concepts such as force, resistance, magnitude, direction, resultant forces, and equilibrium. Each of these terms has a special relationship to friction and must be understood and applied in everyday situations.

In any study of physics or other branch of science, a knowledge of related mathematics must parallel physical science principles. Mathematics competency is needed to solve problems and to describe physical quantities and conditions accurately. In summary, a study of physics requires a critical investigative spirit; an accurate interpretation of terms, laws, concepts, and phenomena; and the ability to apply this technical knowledge.

The table of contents of this text provides an example of those basic principles of science which are essential to the development of occupational competence. These basic principles were selected through the use of occupational and job analyses of a score of occupational clusters. Essential areas of science were evaluated to establish those having the principles and applications most commonly found. In addition, principles that are important to the solution of everyday problems and those needed to make science functional were studied.

Using science as a core or foundation, figure 41-4 illustrates the various levels of employment in each major economic sector. As a person moves from one level to the next higher job level, the skill requirements and the scientific knowledge required become more advanced. The important point is that every job level requires the mastery of basic science principles and the ability to apply these principles in a practical way to meet everyday citizenship needs.

## UNIT 41 ACHIEVEMENT REVIEW OF SCIENCE ESSENTIAL TO CAREERS, CAREER PLANNING, AND DEVELOPMENT

### WORK, THE WORK FORCE, AND CAREERS

1. Make four brief work ethic statements which are the foundation for human resource development.
2. Describe a labor market area.
3. Secure a labor market report which contains information on the *Characteristics of the Work Force.* Prepare a chart similar to the one shown.

| Selected Characteristics of the Work Force | | | | | | | | | | |
|---|---|---|---|---|---|---|---|---|---|---|
| | | Employed | | | Unemployed | | | 5-year Projection | | |
| | Total Number | Total | Male | Female | Total | Male | Female | Total Work Force | Employed | Unemployed |
| Labor Market Area Work Force | | | | | | | | | | |
| Major Categories | | | | | | | | | | |
| 1 | | | | | | | | | | |
| 2 | | | | | | | | | | |
| 3 | | | | | | | | | | |
| 4 | | | | | | | | | | |

    a. Determine the total number of people who are in the work force of the labor market at the time of the report.

    b. Indicate how many of the total work force are (1) employed, (2) unemployed.

    c. Record separately the number of males and the number of females who are (1) employed, (2) unemployed.

    d. Select and list four major categories of job titles according to the classification system used in the report.

    e. Record the labor market information given for each major category as called for in parts a–d. Assume a projected work increase at an annual rate of 2 1/2% of the total, the same uniform rate of increase across each of the four major categories, and that the unemployment rate at the end of five years is 5% for the total and for each category.

    f. Project the labor market data (as reported for parts a–e) total for the work force, employment and unemployment for the four categories, five years from now.

    g. Make a general statement, based on labor market data, on the employment potential (persons in the work force at a particular time) in a selected occupational level that is classified as technical.

4. Identify five economic sectors in your labor market area in which trained personnel are employed.

5. Secure a copy of the U.S. Department of Labor publication *Occupational Outlook Handbook.*

    a. Select one major economic sector and name it.

    b. Name one major occupational cluster in this economic sector.

    c. Identify three fields of employment in the cluster.

    d. Group five job titles in one of the fields on one level of employment. State the level of employment.

    e. State both the level of education and the nature of training required for entry employment in any of the five jobs listed in d.

## SCIENCE APPLIED IN CAREER PLANNING

1. Make a chart to show briefly how the need for applied and/or pure science differs among the following four job classification levels:

    a. Unskilled worker

    b. Skilled craftsperson

    c. Semiprofessional (technologist)

    d. Professional (engineer)

2. Make a brief statement about the importance of science to career development in each of the following cases:

    a. An apprenticeship for an apprenticable trade.

    b. On-the-job training for any economic sector and at any level.

c. An in-school full-time vocational education program in preparation for employment.

d. Career extension training to be upgraded to the next job level of responsibility.

e. Developing an awareness of, becoming oriented to, and exploring the world of work.

## SCIENCE NECESSARY TO CONTINUING CAREER DEVELOPMENT

1. Prepare a partial labor market demand table of eight career opportunities. This information may be obtained from Department of Commerce reports or from continuing labor market job orders placed through the area State Employment Office. Then record the following data:

   a. Levels of employment and distribution of job titles at one or more levels.

   b. The number of job openings (employment demand).

   c. Analyze one of the clusters of job titles. Then state five major topics or units of instruction in physics that are needed by the consumer and a worker at the occupational level.

2. Identify another cluster of job titles in one of the engineering or scientific fields.

   a. Indicate three branches of science that must be included in any preparation for employment in the selected field.

   b. Make a statement to indicate any differences in the depth of study of the sciences for the field selected as compared with the scope, depth, and purposes served by this text.

# Glossary

**Absolute Temperature.** A temperature reading on the Kelvin (Absolute) scale. The degree absolute represents the same temperature change as the degree Celsius. The Kelvin temperature (K) is related to the Celsius temperature (C) by the formula: $^{\circ}K = {^{\circ}C} + 273$.

**Absolute Zero.** A temperature of $0^{\circ}K$, equivalent to $-273^{\circ}C$, believed to be the lowest possible temperature that can be attained.

**Acceleration.** The rate at which the velocity of a body changes with time. Changes in velocity may be in direction, magnitude, or both.

**Acceleration of Gravity (g).** The acceleration of a freely falling body near the earth's surface.

$g = 9.8 \text{ m/s}^2$ (metric system)

$g = 32 \text{ ft./s}^2$ (British system)

**Acoustics.** The science of the production, transmission, and effects of sound. Also, the properties of a room which aid or retard sound waves.

**Adhesion.** The attractive force between molecules of different materials.

**Alpha Particle.** The nucleus of a helium atom having two neutrons and two protons. The alpha particle is emitted by certain isotopes in their radioactive decay.

**Alternating Current.** Electric current which reverses its direction at regular intervals of time.

**Ampere.** The basic unit of electric current, both in the SI metric system and the United States (British) system.

**Ampere-turns.** The product of current flow (amperes) and the number of turns in a current-carrying coil.

**Amplifier.** An electronic circuit which increases the quantity of electric current flow.

**Amplitude.** The distance from a normal position of an object to the final position, as the object vibrates.

**Amplitude Modulation (AM).** Changes of a constant frequency carrier wave caused by variations in the amplitude of the wave.

**Aneroid Barometer.** A barometer actuated by an expanding and contracting mechanism instead of by a liquid.

**Angle of Incidence.** The angle formed by an incident light ray and a perpendicular to the surface.

**Anode.** The positive element or electrode in an electronic tube, or the positive terminal in an electrolytic cell.

**Antenna.** An aerial which receives radio signals and may have various forms of modulation signals.

**Archimedes' Principle.** The force buoying up an object in a fluid (liquid or gas), equal to the weight of fluid displaced.

**Atmospheric Refraction.** The effect of the atmosphere in bending light rays.

**Atom.** The smallest particle of an element containing all of the properties of that element.

**Atomic Mass Unit (U).** A measurement unit which expresses atomic and nuclear masses. $U = 1.66 \times 10^{-27} \text{ kg}$

**Atomic Number.** A fixed (constant) number assigned to each element to indicate the number of protons in the nucleus of the atom.

**Atomic Pile (Also, Atomic Reactor).** A combination of radioactive and other materials which provide a continuous release of atomic energy to form fissionable atoms.

**Atomic Weight (See Atomic Mass Unit).** A relationship between the weight or mass of an atom of one element and that of an oxygen atom (fixed at 16).

**Audio Frequencies.** The range of sound frequencies audible to the ear.

**Audio Signal.** Those sound waves within the audible range which are converted to radio waves and transmitted.

**Avoirdupois.** A series of weight units based on 16 drams to the ounce and 16 ounces to the pound.

**Back Emf.** A counter electromotive force (emf) which is induced in the rotating coils of an electric motor and is opposite in direction to the external voltage applied to the coils.

**Barometer.** An instrument or device for measuring atmospheric pressure.

**Base Units (SI).** Seven base units are included in SI metrics. These units are: length (meter), mass (kilogram), time (second), electric current (ampere), temperature (kelvin), luminous intensity (candela), and the amount of substance of a system (mole).

**Beats.** The instantaneous moments of silence and loudness caused by the interference of two regular sound waves of slightly different frequencies.

**Bernoulli's Principle.** Any change in the velocity of a fluid caused by a constriction produces an opposite change in the pressure.

**Beta Rays.** Particles with negative charges (electrons) which are emitted by radioactive materials.

**Betatron.** A machine that accelerates beta particles and directs them at a high velocity to a plate.

**Binding Energy of a Nucleus.** The energy that is equal to the difference between the mass of a nucleus and the sum of the masses of its individual constituent nucleons.

**Boiling Point.** The temperature at which a liquid boils under normal atmospheric pressure (76 centimeters of mercury).

**Boyle's Law.** The volume of a gas at a constant temperature is inversely proportional to its pressure.

**Breeder Reactor.** A reactor that produces nonfissionable material and converts it into a fissionable material for use in other places.

**British Thermal Unit (Btu).** A unit of measurement of heat energy in the British system. A Btu is the amount of heat required to raise the temperature of one pound of water one Fahrenheit degree.

**Calorie.** A unit of measurement of heat energy in the metric system which equals the amount of heat required to raise the temperature of one gram of water one Celsius degree.

**Candela (Cd).** The basic unit of luminous intensity in the SI metric system and the United States (British) system.

**Candlepower.** A unit of measurement of illumination equal to the light energy of one standard candle.

**Capacitance.** The ability of a capacitor to hold a definite quantity of electric charge for a unit electrical pressure.

**Capacitor.** Electrical device for storing electrical energy in the form of an electrical field.

**Capillary Action.** The effect of molecular forces on liquids which cause some liquids to rise in tubes and other liquids to be depressed.

**Carrier Wave.** An electromagnetic wave used for transmitting signals.

**Cathode.** The negative electrode of an electronic tube which gives off electrons, or the negative terminal of an electrolytic cell.

**Cathode Rays.** High-velocity electrons emitted from the cathode of a cathode ray tube.

**Cathode Ray Tube.** An electronic tube in which the electrons are emitted, formed into an electron beam, and directed onto the fluorescent screen of the tube.

**Celsius Degree (°C).** Formerly called degrees centigrade in the metric system. Celsius has been adopted as the basic unit of temperature in the SI metric system. A measurement of temperature equal to 1/100 of the distance between the boiling point and the freezing point of water at standard pressure.

**Celsius Scale.** A system of temperature measurement in which 0° represents the freezing point of water and 100° the boiling point. Each 1/100th part of the scale is equal to one degree Celsius (1°C).

**Center of Gravity.** The place where the resultant of the gravitational forces on the body acts.

**Centrifugal Force.** The force which tends to move a rotating body from a circular motion to a straight-line motion.

**Centripetal Force.** The force which tends to hold a rotating body together, as opposed to centrifugal force.

**Chain Reaction (Controlled).** The process of splitting apart nuclei of a fissionable material under controlled conditions.

**Charles' Law.** The volume of a sample of gas, at constant pressure, is directly proportional to its absolute temperature. $V/T$ = constant at that pressure regardless of changes in V or. T, respectively.

**Chromatic Aberration.** The inability of a lens to focus light of different colors at one point.

**Circuit Breaker.** A protective device which automatically opens to prevent overloading of a circuit.

**Coefficient of Friction.** A numerical value equal to the force required to overcome friction divided by the weight of the moving body acting upon a horizontal surface.

**Coefficient of Linear Expansion.** A numerical value for each different material which indicates its expansion for each degree of temperature change.

**Coefficient of Volume Expansion.** A numerical value indicating the increase in volume for each degree of temperature change.

**Cohesion.** The attractive force between like molecules.

**Cold-cathode Tube.** An electronic tube in which the electrons jump the gap between the anode and the cathode because of the high positive voltage at the anode.

**Collimated.** A quality of a light beam (from a laser) that has an extremely limited angular divergence.

**Complementary Colors.** Those colors of the spectrum which produce white when combined.

**Components (Force).** Those forces (two or more) which replace a single force as they act together.

**Compression (Sound).** That section of a wavelength in which the medium is compressed.

**Concave Lens.** A lens with concave faces which causes parallel rays to diverge or spread out.

**Conduction of Heat.** The transfer of heat energy from one molecule to the next or from one body to another.

**Conductivity.** The ability of a material to conduct heat energy (thermal conductivity) or electrical energy (electrical conductivity).

**Convection.** The transfer of heat by the movement of matter.

**Cosmic Rays.** Ever present, high-energy waves which reach the earth from space.

**Coulomb (C).** The coulomb in SI metrics is the quantity of electricity transported in one second by a current of one ampere. The coulomb is a derived unit. $1/C = 6.3 \times 10^{18}$ electrons.

**Counter Tube.** A tube (such as the Geiger-Mueller tube) which is used to detect or measure the presence of alpha and beta particles and gamma rays.

**Critical Angle.** The largest angle of incidence of a light ray in a medium which causes the ray to be refracted and not reflected.

**Customary Units.** Units defined by the National Bureau of Standards based upon the yard and pound as commonly used in the United States.

**Deci (d).** An SI metrics prefix signifying a value of one-tenth.

**Deka (da).** An SI metrics prefix signifying a value of ten times.

**Density.** The mass of matter per unit of volume.

**Derived Units (SI).** A series of measurement units in SI metrics which are derived from the seven base units. Derived units extend the use of the base units to meet specific conditions and practical applications in all industrial sectors.

**Detector (Audio).** Electrical equipment used to identify and recover audio signals.

**Deuterons.** The nuclei of heavy hydrogen atoms (with an atomic weight of two) used in atomic fusion.

**Diffraction.** The ability of waves to bend around obstacles in their path.

**Diffraction Grating.** A series of parallel slits that produces a spectrum through the interference of light that is diffracted by the slits.

**Diffuse Reflection.** Reflection from many angles, usually produced from rough surfaces.

**Diode Tube.** An electronic tube consisting of two parts: a cathode and a plate (or anode).

**Direct Current.** An electrical current which flows in one direction.

**Dispersion.** The process of separating white light into a many-colored band.

**Displacement.** The ability of a floating object to displace its own weight.

**Domain Theory of Magnetism.** An electron revolving about the nucleus of an atom imparts a magnetic property to the atomic structure. Each electron, spinning on its own axis, acts as a minute permanent magnet. The atom in each magnetic domain is polarized parallel to the crystal axis. A material becomes permanently magnetized when the domain boundaries remain extended and aligned after the magnetizing force is removed.

**Ductility.** That property of a material which permits it to be drawn into finer and different shapes.

**Dynamic Microphone.** A moving-coil type microphone in which an induced coil voltage is an electrical reproduction of a desired sound wave.

**Efficiency.** A mathematical ratio of useful work output to total work input.

**Elasticity.** A property of matter that allows it to return to an original form after being moved out of shape (deformed).

**Elastic Limit.** The maximum deformation a solid under stress can withstand without being permanently changed.

**Electric Current (A).** A flow of an electrical charge from one place to another as measured in ampere (A) units. An ampere is equal to the flow of one coulomb per second.

**Electric Field (E) Magnitude.** A point at which an electric force acts on a charged particle of +1 coulomb placed in the electric field. The volt/m and 1 N/C are units of electric field.

**Electrodes.** The metal or carbon plates through which current enters or leaves an electric cell.

**Electrolyte.** The liquid in an electric cell which conducts electricity.

**Electromagnet.** A type of magnet produced by passing electric current through a current-carrying conductor.

**Electromagnetic Induction.** Generating an electric current by cutting the lines of force in a magnetic field.

**Electromagnetic Spectrum.** The radiations of all known electromagnetic waves ranging from radio waves on one end of the scale to cosmic rays on the other end.

**Electromotive Force (Emf).** The potential difference across a source of electrical energy when it is not connected to an external circuit and no current flows.

**Electron.** The unit of negative electric charge.

**Electron Gun.** That portion of an electronic tube which forms electrons into a beam.

**Electron Volt.** A unit of energy. The electron volt is equal to the energy gained by an electron in passing from a low potential point to a point one volt higher in potential. One electron volt = $1.6 \times 10^{-19}$ J.

**Electroscope.** One of the simplest and oldest known devices for detecting the presence of small electric charges.

**Energy (E).** A quantitative property of matter indicating its capacity to perform work or change some aspect of the physical world. The joule (J) is the metric unit of energy; foot-pounds (ft.-lb.) is the British unit. Energy is broadly categorized as kinetic, potential, and rest energy.

**Equilibrant.** That force in a vector diagram which is equal and opposite to the resultant (balances two or more other forces).

**Equilibrium.** A condition of balance or rest in which the resultant of the forces acting on a body is zero.

**Evaporation.** A process of transforming a liquid into a vapor.

**Fahrenheit Scale.** A widely used temperature scale on which the freezing point of water is recorded as 32° and the boiling point is 212° (there are 180 divisions, each one representing one degree, between the two points).

**Farad (F).** A derived unit of electrical capacitance in SI metrics. One (F) represents the capacitance of a capacitor between whose plates a difference of potential of one volt appears when the capacitor is charged by a quantity of electricity equal to one coulomb. 1F = 1C/V.

**Field, Magnetic.** The area through which magnetic forces act.

**Fluid.** Matter which is in a liquid or gaseous state.

**Fluorescence.** A type of light produced by radiations from ultraviolet light rays.

**Focal Length.** The distance from an optical device to the place where the light rays focus in a point.

**Foot-candle.** A measurement of the illumination at a distance one foot from a light source of one candlepower.

**Foot-pound (Ft.-lb.).** The unit of work and energy in the British system.

**Force (F).** Any influence that can cause a body to be moved. The newton (N) is the unit of force in the metric system; the foot-pound (ft.-lb.) is the unit of force in the British system.

**Forced Vibrations.** Those vibrations produced in an object when it is forced to vibrate at a frequency other than its natural frequency.

**Fraunhofer Lines.** The dark lines that appear in the spectrum of sunlight.

**Frequency Modulation (FM).** The variations of a carrier wave resulting from changes in vibrations per second of the wave.

**Friction.** The resistive forces opposing the motion of two bodies in contact with each other. *Static friction* deals with frictional resistance that must be overcome by a stationary body to be set in motion. *Sliding friction* is the frictional resistance of a body in motion. *Rolling friction* is the friction experienced by a circular body. rolling over a smooth flat surface.

**Fuel Cell.** A type of battery that remains charged as long as there is a continuous flow of the initial reactant chemicals.

**Fusion.** The process of changing from a solid to a liquid state. Also, atomic fusion, in which lighter atoms are combined to produce heavier atoms with a release of energy.

**Galvanometer.** A device for detecting small quantities of electric current.

**Gamma Rays.** Electromagnetic radiations produced from a radioactive source.

**Geiger-Mueller Tube.** A tube used to detect ionized particles and rays; the tube is activated by such rays and particles.

**Giga (G).** An SI metrics prefix denoting a value of one billion times ($10^9$).

**Gram (g).** A derived SI metrics unit of mass and weight. One gram equals 1/1000 kilogram of water at its maximum density.

**Gravitation.** Every body in the universe attracts every other body with a force that is directly proportional to the mass of each body and inversely proportional to the square of the distance separating the bodies.

**Ground State.** The lowest energy level of the atom.

**Harmonics.** Overtones whose frequencies are whole-number multiples of the fundamental frequency.

**Heat.** A quantitative measure of energy. The addition of heat to a body of matter causes an increase in internal energy. Removal of heat from a body causes the initial energy to decrease.

The kilocalorie (kcal) unit of measurement in the metric system represents the amount of heat required to change the temperature of one kilogram of water by one degree Celsius. The British unit of heat energy is the (Btu) or British Thermal Unit.

**Heat Engine.** A device that converts internal (heat) energy into mechanical energy and whose behavior is governed by the laws of thermodynamics.

**Heat of Combustion.** The amount of heat released by the complete burning of a specific quantity of matter.

**Heat of Fusion.** The amount of heat needed to melt a definite mass of matter without changing its temperature.

**Heat of Vaporization.** The amount of heat needed to vaporize a definite mass of a liquid without changing its temperature.

**Henry (H).** A derived unit of electric inductance in SI metrics. The henry (H) represents the inductance of a closed circuit in which an electromotive force of one volt is produced when the electric current in the circuit varies uniformly at a rate of one ampere per second. $1H = 1V \cdot s/A$.

**Hertz (Hz).** An SI metrics physical quantity representing a frequency of one cycle per second.

**Hooke's Law.** The amount of stretch of certain materials (within a definite range) is proportional to the change in the amount of applied force.

**Horsepower.** A unit of measurement of power equal to 550 foot-pounds per second or 33 000 foot-pounds per minute, or 746 W.

**Humidity.** A condition indicating the presence of water vapor in a given space or area.

**Iconoscope.** An electronic tube used as a camera tube in television.

**Impulse of a Force.** The product of the force and the time during which it acts. The change in the momentum of a body that is free to move, due to a force acting on it, is equal to the impulse generated by the force.

**Incandescence.** The glowing condition produced when a material is heated to a high temperature without burning.

**Index of Refraction.** A numerical value expressing the ratio of the speed of light through air to the speed of light through another medium.

**Inductance.** The opposition of a coil to the flow of a varying current.

**Inertia.** The tendency of a body to resist change in state, whether in motion or at rest.

**Infrared Waves.** A grouping of light waves whose wavelengths are longer than those of ordinary light and shorter than radio waves.

**Interference.** The interaction of different waves having the same nature. *Constructive interference* produces a composite wave with greater amplitude than either of the original waves. *Destructive interference* results in a composite wave with less amplitude than either of the original waves.

**Interferometer.** A device for measuring small dimensions which operates on the principle of light interference.

**International System of Units (SI).** SI metrics, as used in this text, is an up-to-date international system of weights and measures SI is a coherent system. There are seven base units with established names, symbols, and precise definitions.

**Ion.** A particle, atom, or group of atoms that is electrically charged.

**Ionization.** The process of breaking up a molecule having a neutral charge into a number of parts each having an electrical charge.

**Isotopes.** Forms of a chemical element in which the atoms have different atomic weights. Some isotopes of an ordinarily non-radioactive element may be radioactive (radioisotopes), and are used as tracers.

**Jet Propulsion.** That type of propulsion developed from the thrust of hot exhaust gases.

**Joule (J).** An SI metrics unit of work or energy. One joule (J) is equal to $10^7$ ergs, or approximately 0.739 0 gram calories, or 0.737 5 foot-pounds. $1 J = 1 kg \cdot m^2/s^2$.

**Kelvin Degree (°K).** The unit of temperature measurement in SI metrics. A degree on the Kelvin absolute scale is equal to the °C.

**Kilo (k).** An SI metrics prefix denoting one thousand times $(10^3)$.

**Kilocalorie (Kcal).** The unit of heat in the metric system. 1 kcal = 4185 J.

**Kilogram (kg).** The base unit of mass in SI metrics.

**Krypton$_{86}$.** An inert, colorless, gaseous element that provides the foundation for defining the meter in SI metrics.

**Laser.** A device for producing a narrow, monochromatic, coherent beam of light. The term is the abbreviation of Light Amplification by Stimulation Emission of Radiation.

**Lens.** A regularly formed object which can produce an image of an object placed before it. Converging lenses bring together parallel light rays to a single focal point. Diverging lenses diverge parallel rays outwardly.

**Linear Particle Accelerator.** An accelerator that moves charged particles through a series of successively longer drift tubes. Particles are accelerated each time they cross the gap between two drift tubes by the repelling of the particles by the tube they are leaving and the attraction of the tube they are entering.

**Lines of Force.** Denote the direction a body would move if released at a specific point. The concentration of the lines of force is proportional to the magnitude of the force.

**Liquid Propellant Rocket Engine.** An engine that carries its own supply of fuel and oxygen. It includes a propellant fuel system, a control system, and a rocket thrust chamber.

**Liter (l).** A metric system unit of volume or capacity. One liter equals one cubic decimeter ($dm^3$) or 1000 grams.

**Manometer.** A common instrument for measuring fluid pressure; has a U-shape.

**Mass (m).** A quantitative measure of the inertia of a body at rest. The greater the resistance of a body to being set in motion, the greater is its mass. The kilogram (kg) is the metric unit of mass; the slug is the British system unit of mass.

**Mechanical Advantage.** The number of times the resistance force is greater than the effort force.

**Mega (M).** An SI metrics prefix which denotes one million times ($10^6$).

**Meter (m).** A base unit of length in SI metrics. One meter equals the wavelength of light which is 1 650 763.73 wavelengths of the colored line produced by $krypton_{86}$.

**Metrication.** A widely accepted term which applies to any process or program of conversion to the International System of Units, identified in this text as SI metrics.

**Metric System.** A designation for the French and European metric systems using the meter as the basic unit of measure.

**Mircrowaves.** Waves ranging in length from 'almost one-half to one and a half inches (or from one to three centimeters, approximately).

**Mixer Tube.** An electronic tube in which frequencies of a circuit are mixed.

**Modulation.** The varying of the frequency of radio waves by other outside audio waves.

**Mole (mol).** A base unit in SI metrics representing the amount of substance of a system.

**Moments (Principle).** The sum of the clockwise and counterclockwise moments is equal when a body is balanced (in equilibrium).

**Momentum Arm of a Force.** The perpendicular distance from a pivot point to the line of action of the force.

**Momentum, Linear.** The product of the mass and velocity of a body.

**Monochromatic.** A type of one-color light widely used in measurement, produced by a laser.

**Neutron.** An uncharged particle found in the nucleus of an atom.

**Newton (N).** A derived unit of force in SI metrics. $1 N = 1 kg.m/s^2$.

**Nuclear Fission.** The splitting of the nucleus of an atom to two or more nuclei; this process releases nuclear energy.

**Nuclear Fusion.** The building up of atoms when the atoms are in rapid motion at extremely high temperatures.

**Nucleus.** The core of every atom consisting of one or more neutrons and protons.

**Ohm ($\Omega$).** A unit of electrical resistance. $1 \Omega = 1 V/A$.

**Ohm's Law.** One of the fundamental laws of electricity which states that current varies directly as its electromotive force and inversely as its resistance.

**Oscillograph.** The pattern or graph produced on the inside face (screen) of the oscilloscope.

**Oscilloscope.** An electronic cathode ray tube having one face coated on the inside for tracing electronic patterns.

**Parallelogram of Forces.** A mathematical method of representing the direction and quantity of the forces acting on a point in a parallelogram.

**Pascal (Pa).** A derived unit of pressure in SI metrics. One pascal is equal to one newton per square meter. $1 \text{ Pa} = 1 \text{ N/m}^2$.

**Pascal's Law.** A law stating that the pressure on a confined fluid (at rest) is transmitted equally in all directions.

**Period.** The time required for a body undergoing simple harmonic motion to make one complete oscillation, or for one complete wave to pass a specific point.

**Permeability.** A measure of the magnetic properties of a medium.

**Photoelectric Effect.** The emission of electrons from a metal surface when it is struck by light.

**Photometer.** A measuring device for determining light intensities.

**Plasma.** A gas composed of electrically charged particles whose behavior depends largely on electromagnetic forces.

**Polarization (Light).** The control of light vibrations using polarized glasses, plates, or shields which permit only certain light waves to pass through.

**Potential Energy.** A condition or position of energy to do work.

**Pound, Avoirdupois.** A unit of weight in the customary system. One pound avoirdupois equals 16 avoirdupois ounces, 0.453 592 37 kg, or 7000 g.

**Power.** The rate at which work is done. The watt (W) is the unit of power in the metric system and ft.-lb./s is the unit of power in the British system.

$$1 \text{ W} = 1 \text{J/s}$$
$$1 \text{ horsepower} = 746 \text{ W} = 550 \text{ ft.-lb./s}$$

**Primary Colors.** The primary colors are red, green, and blue. When these colors are mixed in the proper proportions, all other colors are produced. In painting, the primary colors are red, blue, and yellow.

**Quart.** A United States liquid unit equivalent to 57.75 cubic inches.

**Radar.** A device for transmitting radio impulses and recording their returning echoes as they strike an object and are reflected.

**Radian.** A unit of angular measure equal to $57.30°$. There are $2\pi$ radians in a full circle.

**Radiation.** The transfer of energy from one place to another as electromagnetic waves; this transfer requires no material medium.

**Radioactivity.** The emission of alpha and beta particles and gamma rays by a radioactive material.

**Rarefaction.** A portion of a sound wave produced by a rarified medium, or the region of minimum density in the sound wave.

**Refraction.** The bending of light rays as they pass from a medium of one density to a medium of another density.

**Resistance.** A measure of the degree to which the passage of electric current is impeded by a body of matter. The ohm ($\Omega$) is the unit of measure of resistance. $1\Omega = 1\text{V/A}$.

**Resonance.** A condition of response where the impressed frequency of one body is equal to the natural frequency of the other body which causes it to move.

**Resultant.** A single force having the same effect as two or more forces acting together at a point.

**Reverberation Time.** The time (seconds) required for a sound to "die down" after it is produced.

**Scanning.** A continuous process in television through which any variation in light intensity of a portion of a televised scene produces variations of electric video impulses.

**SI.** A standard abbreviation for the International System of Units of measurement.

**Significant Digit (Places).** The decimal place or digit that is necessary to accurately define a quantity or value.

**Slug.** The unit of mass in the British system. One slug weighs 32 pounds.

**Sound.** Longitudinal wave phenomenon consisting of successive compressions and rarefactions of the medium through which the sound travels.

**Specific Gravity.** A numerical value indicating the ratio of the weight of an object and an equal volume of water.

**Specific Heat.** The heat required to raise one pound of a specific material one degree Fahrenheit, or one gram one degree Celsius.

**Spectroscope.** An instrument used to identify different elements by color matching against a spectrum (colors arranged according to wavelength).

**Spectrum Analysis.** A technique of analyzing various substances by observing the spectrum pattern of each substance.

**Superposition.** The amplitude of two or more waves of the same nature traveling past a given point at the same time. The amplitude is equal to the sum of the amplitudes of the individual waves.

**Supersonic Jet Engine.** A gas turbine having a converging diffuser inlet, a gas generator section, and a large exit burning chamber with a diverging nozzle.

**Surface Tension.** A force exerted by the surface of a liquid which tends to pull it together.

**Sympathetic Vibrations.** The vibrations set up in one body by vibrations of the same frequency in another body.

**Television.** The transmitting of images by radio waves and their reproduction by separating and changing the video signals back to light energy to reproduce the scene on the picture tube.

**Tesla (T).** The unit of a magnetic field, equal to $1N/A \cdot m$.

**Thermal Conductivity.** The measure of the ability of a material to conduct heat.

**Thermionic Emission.** The emission of electrons from hot bodies in an electronic tube.

**Thermocouple.** A device consisting of two dissimilar metals joined together so that as they are heated, the electromotive force (differential) produced may be converted into a temperature measurement on another instrument.

**Torque.** The product of the magnitude of a force about a specific axis and the perpendicular distance from the line of action of the force to the axis.

**Tracer.** An isotope containing minute amounts of a radioactive material which reveals its presence in a chemical or other scientific process.

**Transistor.** A semiconductor device with similar characteristics as an electronic tube, having the advantage of reduced size, simplicity, limited maintenance, and reduced cost.

**Transverse Waves.** Waves vibrating at right angles or transversely to the direction in which they travel.

**Turbofan Jet Engine.** A gas turbine consisting of a gas generator section, a large fan wheel and inlet duct, and a turbine wheel added ahead of a converging exhaust nozzle.

**Turboshaft Engine.** A gas turbine consisting of a gas generator section, a reduction gear box to produce rotary shaft motion, and turbine wheels preceding the exhaust section to turn the drive shaft.

**Ultraviolet Rays.** Rays having shorter wavelengths than those of visible light and longer than those of X rays.

**Uranium.** An element with isotopes having atomic weights of 233, 234, 235, and 238.

**Vector.** A line on a vector diagram to represent both the direction and quantity of a force.

**Velocity (v).** A quantitative measure (vector quantity) of the speed and direction of a moving body.

**Venturi Action (Bernoulli's Principle).** The velocity of a fluid in motion in a constricted tube is greatest where the pressure is the least, and conversely.

**Video Signals.** Radio signals which are transmitted and impressed on a carrier at frequencies ranging from 30 to four million cycles per second in order to transmit a television scene.

**Volt (V).** The unit of electrical potential differences. $1V = 1J/C$.

**Watt (W).** A unit of power. One watt equals one joule per second, or 1/746 horsepower.

**Wavelength.** The distance between two corresponding points on two successive waves in a wave train.

**Work.** Work is a measure of the amount of change produced by a force acting upon an object. The amount of work is equal to the magnitude of a force multiplied by the distance through which it acts. The joule (J) is the unit of work in the metric system; the foot-pound (ft.-lb.) is the unit in the British system.

**X-rays.** Electromagnetic radiations of shorter wavelengths than visible light which are produced by bombarding a target with fast-moving electrons.

# Appendix

## TABLE I   STANDARD TABLES OF METRIC UNITS OF MEASURE

| | Unit | Value in Meters | Symbol or Abbreviation |
|---|---|---|---|
| **Linear Measure** | micron | 0.000 001 | $\mu$ |
| | millimeter | 0.001 | mm |
| | centimeter | 0.01 | cm |
| | decimeter | 0.1 | dm |
| | meter (unit) | 1.0 | m |
| | dekameter | 10.0 | dam |
| | hectometer | 100.0 | hm |
| | kilometer | 1 000.00 | km |
| | myriameter | 10 000.00 | mym |
| | megameter | 1 000 000.00 | Mm |

| | Unit | Value in Square Meters | Symbol or Abbreviation |
|---|---|---|---|
| **Surface Measure** | square millimeter | 0.000 001 | $mm^2$ |
| | square centimeter | 0.000 1 | $cm^2$ |
| | square decimeter | 0.01 | $dm^2$ |
| | square meter (centiare) | 1.0 | $m^2$ |
| | square dekameter (are) | 100.0 | $dam^2$ |
| | hectare | 10 000.0 | $ha^2$ |
| | square kilometer | 1 000 000.0 | $km^2$ |

| | Unit | Value in Liters | Symbol or Abbreviation |
|---|---|---|---|
| **Volume** | milliliter | 0.001 | $m\ell$ |
| | centiliter | 0.01 | $c\ell$ |
| | deciliter | 0.1 | $d\ell$ |
| | liter (unit) | 1.0 | $\ell$ |
| | dekaliter | 10.0 | $da\ell$ |
| | hectoliter | 100.0 | $h\ell$ |
| | kiloliter | 1 000.0 | $k\ell$ |

| | Unit | Value in Grams | Symbol or Abbreviation |
|---|---|---|---|
| **Mass** | microgram | 0.000 001 | $\mu g$ |
| | milligram | 0.001 | mg |
| | centigram | 0.01 | cg |
| | decigram | 0.1 | dg |
| | gram (unit) | 1.0 | g |
| | dekagram | 10.0 | dag |
| | hectogram | 100.0 | hg |
| | kilogram | 1 000.0 | kg |
| | myriagram | 10 000.0 | myg |
| | quintal | 100 000.0 | q |
| | ton | 1 000 000.0 | t |

| | Unit | Value in Cubic Meters | Symbol or Abbreviation |
|---|---|---|---|
| **Cubic Measure** | cubic micron | $10^{-18}$ | $\mu^3$ |
| | cubic millimeter | $10^{-9}$ | $mm^3$ |
| | cubic centimeter | $10^{-6}$ | $cm^3$ |
| | cubic decimeter | $10^{-3}$ | $dm^3$ |
| | cubic meter | 1 | $m^3$ |
| | cubic dekameter | $10^3$ | $dam^3$ |
| | cubic hectometer | $10^6$ | $hm^3$ |
| | cubic kilometer | $10^9$ | $km^3$ |

## TABLE II   CONVERSION OF ENGLISH AND METRIC UNITS OF MEASURE

### Linear Measure

| Unit | Inches to millimeters | Millimeters to inches | Feet to meters | Meters to feet | Yards to meters | Meters to yards | Miles to kilometers | Kilometers to miles |
|---|---|---|---|---|---|---|---|---|
| 1 | 25.40 | 0.03937 | 0.3048 | 3.281 | 0.9144 | 1.094 | 1.609 | 0.6214 |
| 2 | 50.80 | 0.07874 | 0.6096 | 6.562 | 1.829 | 2.187 | 3.219 | 1.243 |
| 3 | 76.20 | 0.1181 | 0.9144 | 9.842 | 2.743 | 3.281 | 4.828 | 1.864 |
| 4 | 101.60 | 0.1575 | 1.219 | 13.12 | 3.658 | 4.374 | 6.437 | 2.485 |
| 5 | 127.00 | 0.1968 | 1.524 | 16.40 | 4.572 | 5.468 | 8.047 | 3.107 |
| 6 | 152.40 | 0.2362 | 1.829 | 19.68 | 5.486 | 6.562 | 9.656 | 3.728 |
| 7 | 177.80 | 0.2756 | 2.134 | 22.97 | 6.401 | 7.655 | 11.27 | 4.350 |
| 8 | 203.20 | 0.3150 | 2.438 | 26.25 | 7.315 | 8.749 | 12.87 | 4.971 |
| 9 | 228.60 | 0.3543 | 2.743 | 29.53 | 8.230 | 9.842 | 14.48 | 5.592 |

Example   1 in. = 25.40 mm,   1 m = 3.281 ft.,   1 km = 0.6214 mi.

### Surface Measure

| Unit | Square inches to square centimeters | Square centimeters to square inches | Square feet to square meters | Square meters to square feet | Square yards to square meters | Square meters to square yards | Acres to hectares | Hectares to acres | Square miles to square kilometers | Square kilometers to square miles |
|---|---|---|---|---|---|---|---|---|---|---|
| 1 | 6.452 | 0.1550 | 0.0929 | 10.76 | 0.8361 | 1.196 | 0.4047 | 2.471 | 2.59 | 0.3861 |
| 2 | 12.90 | 0.31 | 0.1859 | 21.53 | 1.672 | 2.392 | 0.8094 | 4.942 | 5.18 | 0.7722 |
| 3 | 19.356 | 0.465 | 0.2787 | 32.29 | 2.508 | 3.588 | 1.214 | 7.413 | 7.77 | 1.158 |
| 4 | 25.81 | 0.62 | 0.3716 | 43.06 | 3.345 | 4.784 | 1.619 | 9.884 | 10.36 | 1.544 |
| 5 | 32.26 | 0.775 | 0.4645 | 53.82 | 4.181 | 5.98 | 2.023 | 12.355 | 12.95 | 1.931 |
| 6 | 38.71 | 0.93 | 0.5574 | 64.58 · | 5.017 | 7.176 | 2.428 | 14.826 | 15.54 | 2.317 |
| 7 | 45.16 | 1.085 | 0.6503 | 75.35 | 5.853 | 8.372 | 2.833 | 17.297 | 18.13 | 2.703 |
| 8 | 51.61 | 1.24 | 0.7432 | 86.11 | 6.689 | 9.568 | 3.237 | 19.768 | 20.72 | 3.089 |
| 9 | 58.08 | 1.395 | 0.8361 | 96.87 | 7.525 | 10.764 | 3.642 | 22.239 | 23.31 | 3.475 |

Example   1 sq. in. = 6.452 sq. cm,   1 sq. m = 1.196 sq. yd.,   1 sq. mi. = 2.59 sq. km

### Cubic Measure

| Unit | Cubic inches to cubic centimeters | Cubic centimeters to cubic inches | Cubic feet to cubic meters | Cubic meters to cubic feet | Cubic yards to cubic meters | Cubic meters to cubic yards | Gallons to cubic feet | Cubic feet to gallons |
|---|---|---|---|---|---|---|---|---|
| 1 | 16.39 | 0.06102 | 0.02832 | 35.31 | 0.7646 | 1.308 | 0.1337 | 7.481 |
| 2 | 32.77 | 0.1220 | 0.05663 | 70.63 | 1.529 | 2.616 | 0.2674 | 14.96 |
| 3 | 49.16 | 0.1831 | 0.08495 | 105.9 | 2.294 | 3.924 | 0.4010 | 22.44 |
| 4 | 65.55 | 0.2441 | 0.1133 | 141.3 | 3.058 | 5.232 | 0.5347 | 29.92 |
| 5 | 81.94 | 0.3051 | 0.1416 | 176.6 | 3.823 | 6.540 | 0.6684 | 37.40 |
| 6 | 98.32 | 0.3661 | 0.1699 | 211.9 | 4.587 | 7.848 | 0.8021 | 44.88 |
| 7 | 114.7 | 0.4272 | 0.1982 | 247.2 | 5.352 | 9.156 | 0.9358 | 52.36 |
| 8 | 131.1 | 0.4882 | 0.2265 | 282.5 | 6.116 | 10.46 | 1.069 | 59.84 |
| 9 | 147.5 | 0.5492 | 0.2549 | 371.8 | 6.881 | 11.77 | 1.203 | 67.32 |

Example   1 cu. cm = 0.06102 cu. in.,   1 gal. = 0.1337 cu. ft.

*231 cu in / Gal.*

*128 oz / gal*

### Volume or Capacity Measure

| Unit | Liquid ounces to cubic centimeters | Cubic centimeters to liquid ounces | Pints to liters | Liters to pints | Quarts to liters | Liters to quarts | Gallons to liters | Liters to gallons | Bushels to hectoliters | Hectoliters to bushels |
|---|---|---|---|---|---|---|---|---|---|---|
| 1 | 29.57 | 0.03381 | 0.4732 | 2.113 | 0.9463 | 1.057 | 3.785 | 0.2642 | 0.3524 | 2.838 |
| 2 | 59.15 | 0.06763 | 0.9463 | 4.227 | 1.893 | 2.113 | 7.571 | 0.5284 | 0.7048 | 5.676 |
| 3 | 88.72 | 0.1014 | 1.420 | 6.340 | 2.839 | 3.785 | 11.36 | 0.7925 | 1.057 | 8.513 |
| 4 | 118.3 | 0.1353 | 1.893 | 8.454 | 3.170 | 4.227 | 15.14 | 1.057 | 1.410 | 11.35 |
| 5 | 147.9 | 0.1691 | 2.366 | 10.57 | 4.732 | 5.284 | 18.93 | 1.321 | 1.762 | 14.19 |
| 6 | 177.4 | 0.2029 | 2.839 | 12.68 | 5.678 | 6.340 | 22.71 | 1.585 | 2.114 | 17.03 |
| 7 | 207.0 | 0.2367 | 3.312 | 14.79 | 6.624 | 7.397 | 26.50 | 1.849 | 2.467 | 19.86 |
| 8 | 236.6 | 0.2705 | 3.785 | 16.91 | 7.571 | 8.454 | 30.28 | 2.113 | 2.819 | 22.70 |
| 9 | 266.2 | 0.3043 | 4.259 | 19.02 | 8.517 | 9.510 | 34.07 | 2.378 | 3.171 | 25.54 |

Example   1 ℓ = 2.113 pt.,   1 gal. = 3.785 ℓ

*0.554 L-oz / in³*

$1 ft.^3 = 28.25 \ell$

Appendix ■ 555

## TABLE III  CONVERSION FACTORS

| | | |
|---|---|---|
| **Length** | 1 meter (m) = 39.4 in. <br> = 3.28 ft. <br> 1 centimeter (cm) = 0.394 in. <br> 1 kilometer (km) = 0.621 mi. | 1 foot (ft.) = 0.305 m <br> 1 inch (in.) = 0.0833 ft. <br> = 2.54 cm. <br> 1 mile (mi.) = 1.61 km |
| **Area** | $1\,m^2 = 10^4\,cm^2$ <br> $= 1.55 \times 10^3\,in.^2$ <br> $= 10.76\,ft.^2$ <br> $1\,cm^2 = 10^{-4}\,m^2$ <br> $= 0.155\,in.$ | $1\,ft.^2 = 9.29 \times 10^{-2}\,m^2$ <br> $= 929\,cm^2$ |
| **Volume** | $1\,m^3 = 10^6\,cm^3$ <br> $= 35.3\,ft.^3$ <br> $= 6.10 \times 10^4\,in.^3$ | $1\,ft.^3 = 2.83 \times 10^{-2}\,m^3$ <br> $= 28.3\,liters$ <br> $= 7.48\,gal.$ <br> 1 U.S. gal. = $0.134\,ft.^3$ <br> $= 3.79 \times 10^{-3}\,m^3$ |
| **Mass** | 1 kilogram (kg) = 0.0685 slug <br> = 2.21 lb. (weight/kilogram) | 1 slug = 14.57 kg <br> = 32.17 lb. (weight/slug) <br> 1 lb. mass = 454 g <br> = 0.454 kg |
| **Velocity** | 1 m/s = 3.28 ft./s <br> = 3.60 km/hr. <br> = 2.24 mi./hr. <br> 1 km/hr. = 0.278 m/s <br> = 0.913 ft./s <br> = 0.621 mi./hr. | 1 ft./s = 0.305 m/s <br> = 0.682 mi./hr. <br> = 1.10 km/hr. <br> 1 mi./hr. = 1.47 ft./s <br> = 0.447 m/s <br> = 1.61 km/hr. <br> 60 mi./hr. = 88 ft./s |
| **Force** | 1 newton (N) = 0.225 lb. <br> = 3.60 oz. <br> $= 10^5$ dynes | 1 pound (lb.) = 4.45 N <br> $= 4.45 \times 10^5$ dynes |
| **Pressure** | $1\,N/m^2 = 2.09 \times 10^{-2}\,lb./ft.^2$ <br> $= 1.45 \times 10^{-4}\,lb./in.^2$ <br> 1 atm. $= 1.013 \times 10^5\,N/m^2$ <br> $= 14.7\,lb./in.^2$ | $1\,lb./in.^2 = 144\,lb./ft.^2$ <br> $= 6.90 \times 10^3\,N/m^2$ |
| **Energy** | 1 joule (J) = 0.738 ft.-lb. <br> $= 2.39 \times 10^{-4}$ kcal <br> $= 6.24 \times 10^{18}$ eV <br> 1 kilocalorie (kcal) = 4185 J <br> = 3.97 Btu <br> = 3077 ft.-lb. <br> 1 electron volt (eV) $= 10^{-6}$ MeV <br> $= 10^{-9}$ GeV <br> $= 1.60 \times 10^{-19}$ J <br> $= 1.18 \times 10^{-19}$ ft.-lb. | 1 foot-pound (ft.-lb.) = 1.36 J <br> $= 1.29 \times 10^{-3}$ Btu <br> $= 3.25 \times 10^{-4}$ kcal <br> 1 Btu = 778 ft.-lb. <br> = 0.252 kcal |
| **Power** | 1 watt (W) = 1 J/s = 0.738 ft.-lb./s <br> 1 kilowatt (kW) = 1.34 h.p. <br> 1 horsepower (h.p.) = 746 W = 550 ft.-lb./s <br> 1 refrigeration ton = 12 000 Btu/hr. | |
| **Temperature** | $T_K = T_C + 273°$ <br> $T_R = T_F + 460°$ | $T_C = 5/9\,(T_F - 32°)$ <br> $T_F = 9/5\,T_C + 32°$ |
| **Time** | 1 day = $1.44 \times 10^3$ min. = $8.64 \times 10^4$ s <br> 1 year = $8.76 \times 10^3$ hr. = $5.26 \times 10^6$ min. = $3.15 \times 10^7$ s | |
| **Angle** | 1 radian (rad) = 57°18′ = 57.30° <br> 1 rad/s = 9.55 rev./min. | 1° = 0.01745 rad <br> 1 rev./min. (rpm) = 0.1047 rad/s |

## TABLE IV  SYMBOLS AND DERIVED UNITS OF PHYSICAL QUANTITIES

| Quantity | Symbol | Derived Units* |
|---|---|---|
| Acceleration | a | meters/second$^2$ |
| Angular acceleration | $\alpha$ | radians/second$^2$ |
| Angular displacement | $\theta$ | radian |
| Angular momentum | L | kilogram-meters$^2$/second |
| Angular velocity | $\omega$ | radians/second |
| Area | A | meter$^2$ |
| Capacitance | C | farad |
| Charge | q | coulomb |
| Conductivity | $\sigma$ | ohm-meter$^{-1}$ |
| Current (electrons) | i | ampere |
| Current density | j | amperes/meter$^2$ |
| Density, mass | $\rho$ | kilograms/meter$^3$ |
| Displacement | r, d | meter |
| Electric dipole moment | p | coulomb-meter |
| Electric field intensity | E | volts/meter |
| Electric flux | $\Phi_E$ | volt-meter |
| Electric potential | V | volt |
| Electromotive force | $\mathcal{E}$, emf | volt |
| Energy:  heat | Q | joule |
|     internal | U | joule |
|     kinetic | K | joule |
|     potential | U | joule |
|     total | E | joule |
| Force | F | newton |
| Frequency | v | cycles/second |
| Gravitational field intensity | $\gamma$ | newtons/kilogram |
| Inductance | L | henry |
| Length | L, l | meter |
| Linear momentum | p | kilogram-meters/second |
| Magnetic dipole moment | $\mu$ | ampere-meter$^2$ |
| Magnetic flux | $\Phi_B$ | weber |
| Magnetic induction | B | webers/meter$^2$ |
| Mass | m | kilogram |
| Period | T | second |
| Power | P | watt |
| Pressure | P | newtons/meter$^2$ |
| Resistance | R | ohm |
| Resistivity | $\rho$ | ohm-meter |
| Rotational inertia | I | kilogram-meter$^2$ |
| Temperature | T | degree Kelvin |
| Time | t | second |
| Torque | $\tau$ | newton-meter |
| Velocity | v | meters/second |
| Voltage | V | volt |
| Volume | V | meter$^3$ |
| Wavelength | $\lambda$ | meter |
| Work | W | joule |

* The derived units are stated in the MKS system (meter-kilogram-second-coulomb)

## TABLE V   COMMON PREFIXES FOR METRIC UNITS

| Metric Unit Prefix | Symbol | Power of Ten | Example | | |
|---|---|---|---|---|---|
| pico- | p | $10^{-12}$ | 1 picofarad | 1 pf | $= 10^{-12}$ farad |
| nano- | n | $10^{-9}$ | 1 nanosecond | 1 ns | $= 10^{-9}$ second |
| micro- | $\mu$ | $10^{-6}$ | 1 microvolt | 1 $\mu$V | $= 10^{-6}$ volt |
| milli- | m | $10^{-3}$ | 1 milligram | 1 mg | $= 10^{-3}$ gram |
| centi- | c | $10^{-2}$ | 1 centimeter | 1 cm | $= 10^{-2}$ meter |
| kilo- | k | $10^{3}$ | 1 kilowatt | 1 kW | $= 10^{3}$ watts |
| mega- | M | $10^{6}$ | 1 megajoule | 1 MJ | $= 10^{6}$ joules |
| giga- | G | $10^{9}$ | 1 gigaelectronvolt | 1 GeV | $= 10^{9}$ electron volts |

## TABLE VI   UNITS OF MASS AND WEIGHT AND CONVERSION FACTORS

| System | Units | | Acceleration of Gravity (g) | Conversion Factors | |
|---|---|---|---|---|---|
| | Mass | Weight | | Mass (m) (Given: Weight) | Weight (w) (Given: Mass) |
| Metric | kilogram (kg) | Newton (N) | $9.8 \text{ m/s}^2$ | $m \text{ (kg)} = \dfrac{w(N)}{9.8 \text{ m/s}^2}$ | $w \text{ (N)} = m \text{ (kg)} \times 9.8 \text{ m/s}^2$ |
| | 1 kg = 0.0685 slug | 1 N = 0.225 lb. | | | |
| British | slug | pound (lb.) | $32 \text{ ft./s}^2$ | $m \text{ (slugs)} = \dfrac{w(lb.)}{32 \text{ ft./s}^2}$ | $w \text{ (lb.)} = m \text{ (slugs)} \times 32 \text{ ft./s}^2$ |
| | 1 slug = 14.6 kg | 1 lb. = 4.45 N | | | |

## TABLE VII   COMPARATIVE DENSITIES OF GASES USED IN INDUSTRY

| | Gas | Density $(\text{lb./ft.}^3)$ | Specific Weight |
|---|---|---|---|
| Lighter than air | Hydrogen | 0.005 | 0.07 |
| | Methane (natural gas) | 0.045 | 0.55 |
| | Ammonia | 0.045 | 0.60 |
| | Acetylene | 0.068 | 0.91 |
| | Carbon monoxide | 0.073 | 0.98 |
| | Nitrogen | 0.074 | 0.98 |
| | Air (approx.) | 0.075 | 1.00 |
| Heavier than air | Formaldehyde | 0.077 | 1.03 |
| | Oxygen | 0.084 | 1.12 |
| | Ethane | 0.085 | 1.05 |
| | Hydrogen sulphide | 0.096 | 1.28 |
| | Carbon dioxide | 0.116 | 1.61 |
| | Propane (bottled cooking gas) | 0.117 | 1.56 |
| | Nitrogen peroxide | 0.128 | 1.71 |
| | Isobutane | 0.151 | 2.01 |
| | Butane | 0.155 | 2.05 |
| | Sulphur dioxide (refrigerant) | 0.183 | 2.44 |
| | Chlorine | 0.183 | 2.44 |
| | Gasoline vapors (octane) | 0.290 | 3.86 |

### TABLE VIII   MASS AND WEIGHT DENSITIES OF COMMON SUBSTANCES*

| Substance | Weight Density lb./ft.$^3$ | Mass Density | | |
|---|---|---|---|---|
| | | kg/m$^3$ | g/cm$^3$ | slugs/ft.$^3$ |
| Air | $8 \times 10^{-2}$ | 1.3 | $1.3 \times 10^{-3}$ | $2.5 \times 10^{-3}$ |
| Alcohol (ethyl) | 48 | $7.9 \times 10^2$ | 0.79 | 1.5 |
| Aluminum | $1.7 \times 10^2$ | $2.7 \times 10^3$ | 2.7 | 5.3 |
| Carbon dioxide | 0.12 | 2.0 | $2.0 \times 10^{-3}$ | $3.8 \times 10^{-3}$ |
| Concrete | $1.4 \times 10^2$ | $2.3 \times 10^3$ | 2.3 | 4.5 |
| Gasoline | 42 | $6.8 \times 10^2$ | 0.68 | 1.3 |
| Gold | $1.2 \times 10^3$ | $1.9 \times 10^4$ | 19 | 38 |
| Hardwood (oak) | 45 | $7.2 \times 10^2$ | 0.72 | 1.4 |
| Helium | $1.1 \times 10^{-2}$ | 0.18 | $1.8 \times 10^{-4}$ | $3.5 \times 10^{-4}$ |
| Hydrogen | $5.4 \times 10^{-2}$ | 0.09 | $9 \times 10^{-5}$ | $1.7 \times 10^{-3}$ |
| Ice | 58 | $9.2 \times 10^2$ | 0.92 | 1.8 |
| Iron | $4.8 \times 10^2$ | $7.8 \times 10^3$ | 7.8 | 15 |
| Lead | $7 \times 10^2$ | $1.1 \times 10^4$ | 11 | 22 |
| Mercury | $8.3 \times 10^2$ | $1.4 \times 10^4$ | 14 | 26 |
| Nickel | $5.5 \times 10^2$ | $8.9 \times 10^3$ | 8.9 | 17 |
| Nitrogen | $7.7 \times 10^{-2}$ | 1.3 | $1.3 \times 10^{-3}$ | $2.4 \times 10^{-3}$ |
| Oxygen | $9 \times 10^{-2}$ | 1.4 | $1.4 \times 10^{-3}$ | $2.8 \times 10^{-3}$ |
| Softwood (balsa) | 8 | $1.3 \times 10^2$ | 0.13 | 0.25 |
| Water, pure | 62 | $1.00 \times 10^3$ | 1.00 | 1.94 |
| Water, sea | 64 | $1.03 \times 10^3$ | 1.03 | 2.00 |

* At atmospheric pressure and room temperature

### TABLE IX   EFFECT OF ALTITUDE ON PRESSURE

| Altitude (Ft. above sea level) | Pressure (Inches of mercury) |
|---|---|
| Sea level | 29.92 |
| 1000 | 28.86 |
| 2000 | 27.82 |
| 3000 | 26.81 |
| 4000 | 25.84 |
| 5000 | 24.89 |
| 6000 | 23.98 |
| 7000 | 23.09 |
| 8000 | 22.22 |
| 9000 | 21.38 |
| 10 000 | 20.53 |
| 15 000 | 16.88 |
| 20 000 | 13.75 |

## TABLE X COEFFICIENTS OF FRICTION FOR DIFFERENT SOLIDS

| Combinations of Materials | Degree of Lubrication | Coefficient of Friction* |
|---|---|---|
| Bronze on bronze | None | 0.20 |
| Bronze on cast iron | None | 0.21 |
| Bronze on cast iron | Slight | 0.16 |
| Cast iron on cast iron | Slight | 0.15 |
| Cast iron on wrought iron | None | 0.18 |
| Wrought iron on wrought iron | None | 0.44 |
| Cast iron on hardwood | None | 0.49 |
| Cast iron on hardwood | Slight | 0.19 |
| Wrought iron on hardwood | Complete | 0.08 |
| Leather on hardwood | None | 0.33 |
| Leather on cast iron | None | 0.56 |
| Smooth surfaces | Complete | 0.04 |
| Metal to metal (rolling) | — | 0.002 |

\* The coefficients given are for pressures of 14 to 20 pounds per square inch.

## TABLE XI COEFFICIENTS OF STATIC AND SLIDING FRICTION FOR SELECTED MATERIALS

| Materials in Contact | Coefficient of Sliding Friction, $\mu$ | *Coefficient of Static Friction, $\mu_s$ |
|---|---|---|
| Wood on wood | 0.3 | 0.5 |
| Wood on stone | 0.4 | 0.5 |
| Steel on steel (smooth) | 0.09 | 0.15 |
| Metal on metal (lubricated) | 0.03 | 0.03 |
| Leather on wood | 0.4 | 0.5 |
| Rubber tire on dry concrete | 0.7 | 1.0 |
| Rubber tire on wet concrete | 0.5 | 0.7 |

\* Approximate coefficients for selected materials in contact

### TABLE XII  HEAT VALUES OF GASES, LIQUIDS, AND SOLIDS

| | Solids, Liquids, and Gases | Coefficient of Linear Expansion (1°C) | Specific Heat Capacity* | Heat of Fusion | Heat of Vaporization (kcal/kg) | Melting Point | Boiling Point |
|---|---|---|---|---|---|---|---|
| | | | | | | (Degrees C) | |
| Gases | Air | | 0.24 | | | | |
| | Ammonia | | 0.52 | 84. | 327. | -78 | -33 |
| | Carbon dioxide | | 0.22 | 45. | 85. | | |
| | Hydrogen | | 3.40 | 14. | 108. | -259 | -253 |
| | Nitrogen | | 0.25 | 6.1 | 48. | -210 | -196 |
| | Oxygen | | 0.22 | 3.3 | 51. | -218 | -183 |
| | Steam | | 0.48 | | | | |
| Liquids | Ethyl alcohol | | 0.58 | 25. | 204. | -130 | 78 |
| | Mercury | | 0.03 | 2.8 | 65. | -39 | 357 |
| | Water | | 1.00 | 80. | 540. | 0 | 100 |
| Solids | Aluminum | 0.000 024 | 0.21 | 77. | | 660 | 1800 |
| | Brass | 0.000 019 | 0.09 | | | 1050 | |
| | Copper | 0.000 017 | 0.09 | 42. | | 1083 | 2300 |
| | Glass (reg.) | 0.000 009 | 0.20 | | | | |
| | Glass (pyrex) | 0.000 003 | | | | | |
| | Ice | 0.000 05 | 0.51 | 80. | | 0 | |
| | Iron | 0.000 011 | 0.11 | 5.5 | | 1535 | 3000 |
| | Lead | 0.000 029 | 0.03 | 5.9 | | 327 | 1620 |
| | Platinum | 0.000 009 | 0.03 | 27. | | 1774 | 4300 |
| | Quartz | 0.000 000 5 | 0.19 | | | | |
| | Silver | 0.000 019 | 0.06 | 21. | | 960 | 1950 |
| | Sulphur | 0.000 064 | 0.17 | 13.2 | | 113 | 445 |
| | Zinc | 0.000 026 | 0.09 | 28. | | 419 | 907 |

\* kcal/kg°C and Btu/lb.°F

### TABLE XIII  PHYSICAL PROPERTIES OF IMPORTANT PURE METALS

| Element | Chemical Symbol | Density (lb. per cu. in.) | Melting Point (°F) | Coefficient of Linear Expansion (Millionths of an inch per °F) | Electrical Resistivity (Millionths of Ohms per cm³) |
|---|---|---|---|---|---|
| Aluminum | Al | 0.097 51 | 1220° | 13.3 | 2.655 |
| Chromium | Cr | 0.260 | 3430° | 3.4 | 13.0 |
| Cobalt | Co | 0.32 | 2720° | 6.8 | 6.24 |
| Copper | Cu | 0.324 | 1980° | 9.2 | 1.673 |
| Gold | Au | 0.698 | 1945° | 7.9 | 2.19 |
| Iron | Fe | 0.284 | 2800° | 6.5 | 9.71 |
| Lead | Pb | 0.4097 | 621° | 16.3 | 20.65 |
| Magnesium | Mg | 0.0628 | 1200° | 14.0 | 4.46 |
| Mercury | Hg | 0.4896 | -38° | | 94.1 |
| Nickel | Ni | 0.322 | 2650° | 7.4 | 6.84 |
| Platinum | Pt | 0.7750 | 3225° | 4.9 | 9.28 |
| Silver | Ag | 0.3790 | 1761° | 10.9 | 1.59 |
| Tin | Sn | 0.2637 | 450° | 13.0 | 11.5 |
| Tungsten | W | 0.697 | 6200° | 2.4 | 5.5 |
| Vanadium | V | 0.217 | 3200° | 4.3 | 26.0 |
| Zinc | Z | 0.258 | 787° | 22.1 | 5.916 |

### TABLE XIV   TEMPERING AND HEAT COLORS

| | Color | Degrees | |
|---|---|---|---|
| | | Fahrenheit | Celsius |
| **Temper Colors** | Faint straw | 400 | 205 |
| | Straw | 440 | 225 |
| | Deep straw | 475 | 245 |
| | Bronze | 520 | 270 |
| | Peacock blue | 540 | 280 |
| | Full blue | 590 | 310 |
| | Light blue | 640 | 340 |
| **Heat Colors** | Faint red | 930 | 500 |
| | Blood red | 1075 | 580 |
| | Dark cherry | 1175 | 635 |
| | Medium cherry | 1275 | 690 |
| | Cherry | 1375 | 745 |
| | Bright cherry | 1450 | 790 |
| | Salmon | 1550 | 840 |
| | Dark orange | 1680 | 890 |
| | Orange | 1725 | 940 |
| | Lemon | 1830 | 1000 |
| | Light yellow | 1975 | 1080 |
| | White | 2200 | 1200 |

### TABLE XV   SOUND ABSORPTION QUALITIES OF COMMON MATERIALS

| Material | Sound Absorbed (%) |
|---|---|
| Curtains or draperies (heavy) | 50 to 100 |
| Acoustic plasters | 40 to 80 |
| Ordinary plasters | 2 to 5 |
| Celotex (1/2-in.) | 50 |
| Hair felt (2-in.) | 40 |
| Painted wood | 30 |
| Carpets | 10 to 30 |
| Wood or brick | 2 to 5 |
| Concrete | 2 |
| Marble | 1 |

### TABLE XVI   VELOCITY OF SOUND AT 0°C THROUGH COMMON MEDIUMS

| Solids, Liquids, Gases | Velocity of Sound | |
|---|---|---|
| | Feet per Second (ft./s) | Meters per Second (m/s) |
| Glass, pyrex | 16 960 | 5170 |
| Iron | 16 730 | 5100 |
| Water | 4757 | 1450 |
| Hydrogen | 4167 | 1270 |
| Lead | 4000 | 1200 |
| Helium | 3165 | 965 |
| Air | 1087 | 331.5 |
| Carbon dioxide | 846 | 258 |
| Chlorine | 676 | 206 |

# Acknowledgments

*Source Editor:* Marjorie A. Bruce

*Project Editor:* Mitchell T. Baer

*Editorial Assistant:* Mary V. Miller

# Index

## A

Absorption
  experiment with, 439
  and insulation, 429
  of sound waves, 429
  total, 430
Acceleration, 101-102
  and force, 103
  measurement, 69
Accelerators, circular and linear
    particle, 486-488
Adhesion
  as cause of friction, 168
  definition, 9, 19
  experiment with, 22-23
  industrial applications, 20
Air, weight of, 28-29, 190
Air pump, 193, 198
Alnico, uses of, 245
Alpha particles, 485-486, 497
Altimeter, 190
American National Metric Council
    (ANMC), 55-57
American Standards Association,
    microphone rating by, 448
AM-FM signals, coupling of, 473
Ammeter, basic ranges of, 265
Ammonia gas, 28-29
Ampere, definition, 66-67
Ampere-hours, 281
Ampere turns, definition, 258
Amplification, definition, 469
Amplifiers
  audio, 469, 472-473, 474
  radio frequency, 469
  RF, 474
  sound, 450-451
  video, 469, 474
Amplitude, definition, 220, 424
Angles, 67
ANMC. See American National
    Metric Council
Anode, 464, 465
  collector, 474
Archimedes' principle, 182
Area, 44, 45, 68
Aristotle, experiments of, 99
Armature, 307
Astronomy, definition, 3
Atmospheric pressure. See Pressure,
    atmospheric
Atomic bonb, 490
Atomic number, 484
Atomic pile. See Nuclear reactor
Atoms, 235-236
  neutrons as smashers of, 489
  structure of, 483-485

## B

Balance, dynamic and static, 89-91
  experiment on, 95-96
Barometer, 28, 190-191
  Torricelli, experiment with,
    196-197
Base ten scientific notation system,
    57, 58-59
Bath system of lubrication, 171
Battery
  care of, 283-284
  dry cell, 280
  nuclear, 507
  principle and uses of, 279
  solar, 507-509
  storage, 280-281
  testing of, 282-283
Battery voltages, experiment with,
    273-274
Bearings, and friction, 170
Beats, production of, 427
Becquerel, A.E., 502
Bell, Alexander Graham, 431, 444
Bernoulli's principle, 208
  and fluids, 183
Beta particles, 485-486
Betatron, construction of, 486
BGS. See British gravitational system
    of measurement
Biology, definition, 3
Block and tackle. See Pulley,
    mechanical advantage
Boiling point, and vapor pressure, 344
Botany, definition, 3
Boyle, Robert, 191
Boyle's law, 192-193, 203
  experiment with, 198-199
  industrial applications, 195
Brake thermal efficiency, 361-363
  formula for, 362
British gravitational system of
    measurement (BGS), 36-37
British thermal unit (Btu), definition,
    329
Brookhaven National Laboratory, 488
Btu. See British thermal unit
Buoyancy, 21, 182
  experiment on, 24

## C

Calorie, 328-329
Calorimeter, principle of, 330
Calorimetry, 330
Cams, 105, 124
Candela, definition, 67
Candlepower, definition, 383
Capacitor, types, 240-241

Capacity, measurement of, 281-282
Capillary action, 19-20
  experiment on, 23-24
Carbon dioxide, compressibility of, 29
Career decisions, forces influencing,
    536
Career development
  model of, 536
  science applied to, 535-537
  science necessary to continue,
    537-539
Career levels, job identification for,
    536
Cathode-ray tube, operation of, 466
Cathodes, types of, 464-466
Cells
  dry, 280
  primary, 279
    experiment with, 285
  secondary, 280-282
    experiment with, 286-287
Celsius scale, 327-328
Center of gravity, 90-91
  experiment on, 96
Centimeter-gram-second system (cgs),
    64
Centrifugal force, experiment on, 109
Centripetal force, experiment on, 109
Cesium-clock system, 43
Cesium plasma diode, 511
CGPM. See General Conference on
    Weights and Measures
Cgs. See Centimeter-gram-second-
    system
Chain reaction, in nuclear fission, 489
Charges, electrical
  attraction and repulsion of,
    236-237
  detection of, 240
  experiment with, 242
  negative, 236, 238
  positive, 236, 237-238
  static, discharging of, 238-239
  storage of, 240-241
Chemistry, definition, 3
Chrominance of video picture, 475
Circuits, electrical
  amplifier, 469
  open and closed, 269-270
  oscillator, 469
  parallel, 270-271
  rectifier, 468
  series, 270
  short, 270
Circuits, fluid, 205-207
Circuits, magnetic, 258
Clusters, occupational, 535

Coal, as energy source, 290
Coefficient of friction, 169
Cohesion, 9, 19, 168, 342
    experiment on, 22-23
    industrial applications, 20
Coils, electromagnetic, 258
Color
    analyzing materials through,
        406-408
    effect of wavelength frequency on,
        404-405
    experiment with, 412-413, 414-
        415
    Newton's experiments on, 403-
        404
    printing, 409
    producing by absorption and
        reflection, 405-406
    use in industry and science, 408
Color inspection, for temperature
    measurement, 326
Color sync bursts, in video signals,
    475
Commutator, in direct-current
    generator, 294
Compass, development of, 245
Component. See Force, component
Composition of forces method. See
    Parallelogram method
Compressibility of gases, 27-29
    experiment on, 30-31
Compression and force, 81
Computer, analog, 106
Condensation, 343
    pulse, 215
Condenser, in single-cylinder engine,
    353
Conduction, definition, 339-340
Conductivity, 15
Conductors, 239
    current-carrying, 305-307
Constellations, occupational, 534
Consumer applications of science
    direct conversion of energy, 529-
        530
    electronics, 527
    heat energy and engines, 525
    light energy, 526
    magnetism and electricity, 524
    matter and measurement, 523
    mechanics, machines and wave
        motion, 524
    nuclear energy, 528-529
    sound energy, 526-527
Contamination of nuclear reactor, 492
Convection, 340
Conversion, 38, 44, 60, 70-73
    of energy, 529
Converters
    thermionic, 509-511
    thermoelectric, 516-517

Cosmotron, 494
Coulomb, 264-265
Coulomb's law, 249
Counter tube. See Geiger-Mueller
    tube
Critical angle of incidence, 378
Critical size, 489
Cubes, volume of, 40-41
Current, electrical, 239-240, 264
    alternating, 272, 289, 296, 514
    controlled by resistance, 266-268
    direct, 271-272
    experiment with, 274-275
    and free electrons, 236
    insulators for, 239
    left-hand rule for, 257
    measurement of, 264-265
    thermoelectric, 325
    and voltage, 265-266
Current-carrying conductor, 256
    poles in, 257-258
Current direction, 256-257
Cyclotron, 486-487
Cylinder cycle, in gasoline engine, 354

D
Deceleration, uniform, 102
Decibel, 431
Deflection plates, in television
    camera tube, 473
Deformation due to friction, 168
Density, 8-9, 28
    experiment on, 10-11
Detector, in television receiver, 474
Deuterium, fusion of, 493
Diamagnetism, 245, 249
Diesel engine. See Engines, diesel
Diesel operated plants, 291
Diffraction of light, 379-380
Diffuse reflection, 376
Diffusion, 491
Diplexer, in television, 473
Discharge cycle, of storage battery,
    282
Discriminator, in television receiver,
    474
Displacement, as vector quantity, 92
Displacement method of measuring
    volume, 41-42
Distance, mechanical advantage of
    122, 124
Domain theory of magnetism, 246-
    248
Domains, 247-248
Dot-cross method, to indicate current
    direction, 256-257
Drift tubes, 488
Dry cell. See Cells, dry
Ductility, 15
    experiment on, 17
Dynamometer, 166

E
Edison, Thomas A., 289, 445, 463
Edison Effect, 463-464
Education and Industrial Training
    Coordinating Committee, 57
Efficiency, 81, 83-84
Efficiency, brake thermal. See
    Brake thermal efficiency
Effort, 112
Einstein, Albert, 483
Elasticity, definition, 15
Electric cell, development of, 255
Electric charges. See Charges,
    electrical
Electricity, 69, 255
    applications of, 271-272
    direct-current, 269-271
    distribution of, 296-297
    and electron motion, 235
    generation of, 289-290
    for light and heat, 310-311
    static, 235, 237-239
        experiment with, 243
    transmission of, 295-296
Electrodes, 280-281
Electrolytes, 279-282
Electromagnetic method for
    separating $U^{235}$ from $U^{238}$,
    491
Electromagnetic spectrum of wave-
    lengths, 403
Electromagnetism, 255-256
    experiments with, 260-262
    practical applications, 258-259
Electromagnets, 258-259
    in alternating-current generator,
        293
    in synchrotrons, 488
Electromotive force (emf), 265-266
Electron beam, forming and deflect-
    ing, 467
Electron flow, 264, 267-268
Electron gun, 467-473
Electron motion, 246-247, 279
Electron theory, 239-240
    description of electrons, 235-236
    importance of, 240
Electronic tubes
    basic types, 466-467
    classification of, 464-465
    Edison effect in, 463-464
    electron flow in, 464
    functions, 467-468
    industrial applications of, 476
    parts of, 464
Electronics
    applications of, 527
    applied to communication, 470-
        476
    basic principles of, 463-467

Electrons
  flow of, in electronic tubes, 464
  free, 236, 463-464
  ground state of, 390
  hot, 516
  in lasers, 388
  mass of, 484
  in metastable state, 388
  in motion, 239-240
  paired, 247
  random motion of, 388
  secondary, 473-474
  theory of, 235-236
Electroscope, 240
Element, 235
EMF. See Electromotive force
Emulsions, industrial applications, 20
Energy
  conversion to mechanical form,
    289-291
  converters, 504
  electrical, 245, 297, 298, 509-510
    experiments with, 300-301
  of fluid systems, 207
  forms of, 279
  heat, 323-325, 328-330, 352
  kinetic, 323-324
  light, 375
  magnetohydrodynamic generation
    of 512-514
  mechanical, 290-295, 353
  nuclear, 483, 494-495
  solar, 507-509
  sound, 446-447
  sources, 289
  transfer, 214
  transmission by waves, 379
Energy research, areas of, 502-503
Engines. See also Turbines, gas
  brake thermal efficiency of, 363
  diesel, 355-356
  gasoline, 354-355
    experiment with, 366
  heat, principles of, 352
  internal combustion, 354
  MHD propulsion, 513
  radial, 354
  ramjet, 360
  rocket, 361
  steam, 352-353, 361, 362-363
    experiment with, 365
  supersonic jet, 360
  turbofan jet, 359
  turbojet, 358-359
    experiment with, 367
  turboprop, 359
  turboshaft, 359
Envelope of electronic tube, 464
Equilibriant, 94
Equilibrium, 89, 90-91

Evaporation, factors affecting, 342-
  343
Exhaust nozzle, 359-360
Expansion
  by heat, 338-339
  linear, formula for, 338
Experiments
  atmospheric pressure, 196-199
  basic properties of gases, 30-31
  basic properties of liquids, 22-25
  basic properties of solids, 16-17
  centrifugal and centripetal force,
    109
  color, 412-413, 414-415
  compound machines, 160-162
  conduction, 347
  convection and convection
    currents, 347
  electrical charges, production and
    detection, 242
  electrical energy, 300-301
  electricity in motion, 273-276
  electrochemical sources of
    electrical energy, 285-287
  electromagnetic induction,
    312-314
  electromagnetism, 260-262
  electronic tubes and devices,
    478-480
  engines, 365-366
  fluid power, 210-212
  fluids and pressures, 184-186
  forces, 85-87, 173-174
  forces and balance, 95-97
  friction, 174-175
  gear trains, 150-153
  gravitation, 107-108
  heat expansion, 346
  inclined plane, 125-126
  levers, 116-119
  light, 394-398, 412-415
  magnetism, 251-253
  measurement, 48-51
  mechanical power measurement,
    172-173
  motion, 108-109
  motors, 314-315
  with nuclear particles, 496-498
  pulley and pulley systems, 159-
    160
  pyrometers, 332-333
  radiation and its control, 348-349
  radio, 454-455
  screw threads, 142-143
  sound, 433-439, 453-456
  specific heat, 333-334
  specific properties of matter, 10-12
  spectroscope, 413-414
  static charges, removal and
    reduction, 243
  steam, 367

Experiments (con't.)
  telephone, 452-453
  thermometers, 331-332
  turbines, steam, 366
  wave motion, 222
  wave pulse, 222-223
  wedge, 126-127
  wheel and axle, 132-134

F
Fahrenheit scale, 327-328
Fermi, Enrico, 489
Ferromagnetic materials, 245, 249
Filament, of incandescent lamp, 310
Fission. See Nuclear fission
Fissionable materials, production of,
  490-492
Flameout, prevention of, 357
Fleming's Right-hand Motor Rule,
  306
Flow, 203, 205, 208
Fluid circuits, 205-206
Fluid power, 203-204, 208
  experiments on, 210-212
Fluid power systems, and expansion
  compensation, 203
Fluid pressure
  application to machines, 181-182
  engineering applications, 181
  experiments on, 184-186
  and force, 179-180
  industrial applications, 195
  and size or shape of container,
    180-181
Fluid resistance, result of, 205
Fluid systems, 206-208
Fluidic control devices, 208
Fludics, application of, 208
Fluids, 180, 182, 205
  applications of, 194-195
Flux, 256-257
  conductor, 305
  field, 305
  line of, 248
  magnetic, 248-249, 258
Focal length, 383
Focal point, 381
Force
  angular, 93
  centrifugal and centripetal, 104
  compared with pressure, 180
  component, 94
  definition, 80
  effects of, 91
  effort and resistance, 81-82
  experiments on, 85-87
  kinds of, 80-81
  magnitude of, 94
  measurement, 68
  mechanical advantage of, 82-83,
    114-115, 121-122, 124, 130,
    138-139

Force (con't.)
    moment, 89, 114
    parallel, 92-93
    point of application, 94
    resultant, 91-92
    torque, 89
    transmission of, 204-205
    as vector quantity, 92
Fraunhofer lines, 407
Freely falling objects, 99-100
Freezing point, 342
Frequency
    in alternating-current generator,
        291
    of light waves, 377
    measurement, 68
    and pitch, 423-424
    of sound, 430
        experiments with, 433-436
    of waves, 219-220
Friction
    advantages and disadvantages, 167
    causes, 168, 237
    coefficient of, 169
    deformation due to, 168-169
    experiments with, 173-174
    internal, 205
    overcoming, 170-171
    types, 168-169
Fuel cells, 504-507
    types, 505-507
Fulcrum, 112, 129
Full-wave rectifier circuit, 468
Fusibility, definition, 15
Fusion, 492-494
    latent heat of, 342
Galileo, experiments of, 99-100
Gamma rays, 485-486
Gases
    change to liquid or solid state, 29
    evaporation of liquids to form, 29
    pinched, 513
    properties of, 27-29
    result of pressure on, 203
Gasoline engine. See Engines, gasoline
Gauge, venturi, 183
Gear box, 359
Gear train
    compound, 148
    practical applications of, 149
    simple, 145, 147
Gears
    driver and driven, 146
    effect on force, 147-148
    in power transmission, 165
    rotation and, 145-146
    and speed, 146-147
Geiger-Mueller tube, 486
General Conference on Weights and
    Measures (CGPM), 64-65

Generators
    alternating-current, 292-294,
        307-310
    definition, 292
    direct-current, 294-295, 305
    elementary, 291-292
    gas, 357-358
    MHD, 513-514
    thermoelectric, 515-517
Gilbert, William, 246
Gravity, 37, 100. See also Center of
        gravity; Specific gravity
    experiment on, 107-108
Gravity pressure, 180
Grove, Sir William, 502

H
Halftone, 409
Half-wave rectifier circuit, 468
Hardness, 14
    experiment on, 16
Hearing
    mechanism of human ear, 426-427
    threshold of, 431
Heat
    conversion to steam, 290
    from electricity, 311
    expansion by, 338-339
    insulation against loss of, 341
    latent, 342-343
    measurement of, 328-330
    mechanical equivalent of, 361-362
    methods of transfer, 339-341
    radiant, 341
    specific, 329
        experiment with, 333-334
    waves, 340-341
Heat energy and heat engines,
    applications of, 525
Heat engine. See Engines, heat
Helix angle of screw threads, 138-139
Hertz (Hz), 68
Holograms, 391-392
Holography. See Laser photography
Horsepower, compared with work, 83
Hydraulic press, 181-182
Hydraulic systems, 203-204
    uses of, 208
Hydrocarbon fuel cell, operation of,
    506-507
Hydroelectric plants, 291
Hydrometer, 182, 282
Hydrox fuel cell, operation of,
    505-506
Hz. See Hertz

I
Igniter rod, 465
Illumination, 384-385
    industrial requirements for, 385
    measurement of, 383

Image
    distance from lens, 383
    magnification of, 382-383
Impenetrability, 9
    experiment on, 11
Incidence, angle of, 377, 378
Inclined plane, 121-123
    experiment with, 125-126
Induction
    experiment with, 312-313
    magnetic, 295
    transferring static electricity
        through, 238-239
Inertia, 9, 102, 103, 104
    experiment on, 12
Infrared waves, 403
Inlet duct, types, 359-360
Insulation, 429
    experiments with, 437-439
Insulators, 340, 341
    properties of, 239
Intercontinental ballistic
        missiles, research on, 502
Interference
    constructive and destructive,
        218, 380
    principle of light, 410
    wave, 427
Interference bands, 410-411
Internal combustion engine. See
        Engines, internal combustion
International Bureau of Weights and
        Measures, 64
International Metric Convention, 64
International metric units, using,
        59-60
International Standards Organization
        (ISO), 55
International System of Units (SI)
    adoption of, 55
    advantages, 65
    base ten scientific notation
        system, 57
    base units, 65-67
    basic measurement units in, 36-37
    common metric units, symbols
        and formulae, 69-70
    derived units, 68-70
    development, 64-65
    supplementary units, 67
Invariant quantity in nature, 37
Inverters, 514-515
Ion-exchange membrane fuel cell,
        operation of, 506
ISO. See International Standards
        Organization
Isotopes, 484-485
    radioactive, 494

J
Joule, James P., 362

**K**

Kelvin absolute scale, 328

**L**

Lamps, types of, 310-311
Laser, 390-391
    metastable energy levels in, 389
    photography, 391-392
    properties of, 387-388
    ruby, 389-390
Law of Inverse Squares, 384
Law of Moments, 114
Law of Reflection, 377
Law of Refraction, 377-379
Lawrence, E.O., 486
Left-hand rule, for electron flow, 257
Length, measurement of, 38-39
Lenses
    experiments with, 396-397
    function of, 382-383
    types, 381-382, 409-410
Levers
    characteristics, 112
    classification of, 113
    experiments with, 116-119
    industrial applications, 115-116
    mechanical advantage of, 114-115
    types, 113-114
Leyden jar, 240-241
Light
    action of, 376
    artificial, 375
    conditions for seeing an object,
        375-376
    diffraction of, 379-380
    from electricity, 310-311
    experiments with, 412-413, 414-
        415
    incoherent emission of, 388
    interference, in optical flats,
        410
    interference in wavelets of, 379-
        380
    measurement of, 397
    monochromatic, 410
    polarized, 385-386
    reflection of, 376, 377
    refraction of, 377-379
    separation of, 403-404
    sources, 375
    speed of, 376
    wave motion theory of, 376-377
Light Amplification by Stimulated
    Emission of Radiation. **See**
    Laser
Light energy, applications of,
    525-526
Lightmeter, 384
    experiment with, 397-398
Line loss, in electrical transmission,
    296

Linear measure. **See** Length,
    measurement of
Linear particle accelerator, 488
Liquid measure, 47
Liquids
    with greater adhesion than
        cohesion, 19-20
    with greater cohesion than
        adhesion, 20
    properties, 19-21
    results of pressure on, 203
Loudness, factors affecting, 424
Loudspeakers, 447, 450
Lubrication, 171
    experiment on, 174-175
Luminance, of video picture, 475
Luminous intensity, derived units
    of, 69

**M**

Mach number, 361
Machinability, definition, 15
Machines, 112, 166
    compound, 157-158
        experiments with, 160-162
    ideal, 181
    simple, 156
Magnetic fields
    effect of interaction of, 310
    electric. **See** Electromagnetism
    experiment with, 252
    magnetic flux density, 248-249
    rotating, 308-310
    of stator, 309
Magnetic poles, 246, 249
    experiment with, 251-252
Magnetic saturation of electromagnet,
    259
Magnetic triode, 510-511
Magnetism
    domain theory of, 246-248
    and electron motion, 246-247
    first and second law of, 246
    induced, 249
    permanent, 247
    residual, 249
Magnetism and electrical energy,
    applications of, 524
Magnetohydrodynamics (MHD),
    definition, 512
Magnets, 245-246
    permanent, 245, 305, 450
Magnification, by lenses, 382-383
Magnitude. **See** Force, magnitude of
Malleability, 14
    experiment on, 16-17
Manpower, distribution of, 534-535
Maser, applications of, 387
Mass
    definition, 8
    derived units of, 68

Mass (con't.)
    fissionable, 489
    measurement, metric units of,
        46-47
    and momentum, 103
Matter
    and nuclear energy, 483
    physical properties, 7-10
        experiments on, 10-12
    states of, 7, 14
    structure of, 7
Matter and measurement, applications
    of, 523
Measurement
    of area, 39-40
    conversion from one system to
        another, 37-38, 44
    of length, 38-39
    linear, with optical flats, 409-410
    standardization, 36-37
    systems, 36-38
    of time, 42-43
    of volume, 39-42
    of weight, 42-43
Measuring devices, 137
Mechanical advantage
    of compound gear train, 148-149
    of compound machines, 157, 158
    experiment on, 86-87
    of force, 82-83
    of hydraulic press, 181-182
    of inclined plane, 121-122
    of levers, 114-115
    of pulley, 156-157
    of screw threads, 138-139
    of speed, 82-83
    of wedge, 124
    of wheel and axle, 130
    of worm and worm wheel, 157
Mechanical movements, 104-106
Mechanical power, experiment with,
    172-173
Mechanics, machines and wave motion,
    applications of, 523-524
Melting point, 342
Mercury, cohesion and adhesion of,
    20
Metallurgy, definition, 3
Meter, definition, 65-66
Meters-kilograms-seconds-amperes
    system (MKSA), 64-65
Meters-kilograms-seconds system
    (MKS), 64
Metric Conversion Act, 55
Metric system of measurement
    (MKSA), 36-37
    compared with English measure,
        45, 46, 47
    computations in, 44-47
    legislation in U.S., 64
    units, 43

Metrication, 55-57
    conversion factors, 70-73
MHD. See Magnetohydrodynamics
Micrometer, 39
Microphones, 448-449
    conversion of sound to electrical
        waves in, 447-448
Microwave Amplification Stimulated
    Emission of Radiation. See
    Maser
Miniaturization, 512
MKS. See Meters-kilograms-seconds
    system
MKSA. See Meters-kilograms-
    seconds-amperes system
Mole, definition, 67
Molecular attraction, 9
Molecular motion, 27
    experiment on, 30
Molecules, 7, 9, 235
Moment, 89-90
Moment of force. See also Torque
    experiment on, 132-133
    of wheel and axle, 129-130
Momentum, factors controlling, 103
Motion
    common concepts of, 100-102
    experiments on, 108-109
    fluid, 207
    laws of, 102-104
    molecular, and heat, 323-324
    positive rotary, 145
    rectilinear, 105, 136
    rotary, 105, 136
    transmitting, 105
Motors, electric
    alternating-current, 307-310
    direct-current, 305
    experiments with, 314-315
    single-phase and three-phase
        induction, 308
Movement, mechanical advantage of,
    115
Musical tones, 427-428

N

National Bureau of Standards, 57, 64
National Metric Conversion Board, 56
National Metric Study of the National
    Bureau of Standards, 57
NBS. See National Bureau of
    Standards
Neutrons, 236
    as atom smasher, 489
    relative weight of, 484
Newcomen, Thomas, 352
Newton, Sir Isaac, 403-404
    law of gravitation, 100
    laws of motion, 102-104
        applied to gas turbines, 356-357
        applied to rocket engines, 361

Nonfissionable material, conversion
    of, 492
Nonlegislative consensus, 55
Nonmagnetic materials, 245, 249-250
Nuclear energy, applications of, 528
Nuclear fission, 489
Nuclear reactor, 491-492
Nucleus, atomic
    research into, 494
    structure of, 236, 484
Nylon, development of, 1-2

O

Occupational Outlook Handbook,
    537
Ohm, George S., 268
Ohm, definition, 267
Ohmmeter, use of, 267
Ohm's law, 268-269, 271
Oil, miscible, 20
Opaque objects, light action on, 376
Opposition, 205. See also Resistance;
    Reluctance; Friction; Fluid
    resistance
Optical comparator, 411
Optical flats, 409-410
Oscillation, production of, 469
Oxygen concentration cell, operation
    of, 504-505

P

Parabolic reflector, 470
Parallelogram method of vector
    representation, 93-94
    experiment on, 97
Paramagnetism, 245, 249
Pascal's Law, 180-181, 182
Permalloy, uses of, 245
Permeability of magnetic substances,
    249
Photoelectric mosaic, in television
    camera tube, 473
Photogalvanic cell, 502
Photons, emitted by lasers, 390
Phototube, 467
Photovoltaic cell. See Battery, solar
Physical change, definition, 7
Physics, 3, 537-539
Pinch process, 513
Pitch, 423
    experiments with, 433-435
Plasma, 513
Plutonium, production of, 491-492
Pneumatic systems, 203, 204-205,
    208
Point of application of forces, 94
Polarity. See Magnetic poles
Polarization, 280, 385, 386
    applications of, 386-387
Polaroid
    materials, 386-387

Polaroid (con't.)
    experiment with, 398
    process, 386
Population inversion in lasers, 389
Porosity, 9
    experiment on, 12
Power
    continuous transmission, 165
    coupling devices, 165
    definition, 83, 164
    intermittent, 165
    measurement, 68, 166-167
    mechanical transmission of,
        164-165
    of sound, 431
    sources, 164
    standard unit of measure, 83
Power systems, principle governing,
    204
Power transmission, through fluid
    system, 204
Pressure. See also Pressure,
    atmospheric; Pressure, fluid
    and compressibility of gases, 27-28
    experiment on, 30-31
    measurement, 68
Pressure, atmospheric, 28
    barometers and, 190-191
    basic principles of, 191-193
    Boyle's Law, 192-193
    and fluid movement, 194-195
    industrial applications, 195
    measurement of, 190-191
Pressure, fluid, 180-181
    Pascal's law, 180
Pressure system, of lubrication, 171
Principles of light interference, 410
Printing, color, 409
Prism, effect on light, 403
Problem solving, systematic approach
    to, 4-5
Prony brake, 166-167
    experiment with, 172-173
Propellants, in rocket engines, 361
Proton, 236, 484
Pulley
    experiments with, 159-160
    mechanical advantage, 156-157
    in power transmission, 164
    types, 156
Pulses, 215-218
    experiment with, 222-223
    superposition of, 217-218
Pumps
    fluid, 195-196
    force, 194
    lift, 194
Pyrometer
    advantage of 325
    optical, 326-327
        experiment with, 332-333
    radiation, 341
    uses of, 327

**R**

Radar, 470-471
Radian. See Angles
Radiation, 340-341
Radio Detection and Ranging. See
    Radar
Radio wave pulses, reflection and
    recording of, 470-471
Radioactive materials, 485-486
    experiments with, 497-498
Radioactivity, devices for observing,
    485-486
Radioisotopes, generators powered
    by, 517
Rarefaction pulse, 215
Rectification, definition, 468
Redox fuel cell, operation of, 506
Reflection, 376-377
    experiment on, 394-395
    index of, 377
    Law of, 377
    of sound waves, 428
Refraction
    application of principles of,
        381-383
    experiment on, 395-396
    index of, 378
    Law of, 377-378
    of sound waves, 428
Regular reflection, 376
Reluctance, in magnetic circuits, 205
Reservoir, for fluid systems, 207
Resistance
    applications of, 267-268
    in conductors, 266
    contact, 517
    and cross-sectional area, 267
    definition, 266
    experiments, 210-211, 275-276
    to flow, 205
    of fluid systems, 206
    in insulators, 266
    and length of material, 266-267
    and levers, 112
    measurement of, 267
    parallel circuit and, 206, 270-271
    path of least, 249-250
    series circuit and, 206, 270
    of series-parallel system, 207
    and temperature, 267
    thermal, 208
Resistors, 267-268
Resolving power, definition, 380
Resonance, 425-426. See also
        Vibrations
    experiments with, 436-437
Resonant cavity, in lasers, 389
Resultant, definition, 91
Reverberation, 428
Roemer, Olaf, 376
Rotation, direction of, 306

Rotor, 308, 310
Rules, steel, 38-39
Rutherford, Sir Ernest, 489

**S**

Saturation point, 343
Scalar, 91-92
Scanning, in television, 471-472
Science, importance of, 1-3
Scientific laws, 4
Scientific method, 3-4
Screw threads
    applications, 137-138, 141
    characteristics of, 139-140
    experiments with, 141-143
    as fastening devices, 137-138
    force transmitted with, 138-139
    motion transmitted with, 136-137
    terms, 136
    uses of, 136-139
Second, definition, 66
Seebeck, T.J., 502, 515
Semiconductor junction diode laser.
        See Laser
Semiconductors
    for high-power generation, 515
    importance of, 508-509
    in inverters, 514-515
Service entrance of three-wire system,
        296
Shadows and light, 376
Sheave, 156
Shielding for magnetic materials,
        249-250
    experiment with, 252-253
SI. See International System of Units
SI metrics. See International System
        of Units
Single-cylinder engine. See Engines,
        steam
Siphon, and atmospheric pressure,
        193
    experiment with, 197-198
Slip ring, in alternating current
        generator, 292
Slope of inclined plane, 121
Solar battery. See Battery, solar
Solar energy. See Energy, solar
Solar spectrum. See Visible spectrum
Solids
    computing volume of, in metric
        system, 44
    measurement of, 41-42
    molecular arrangement in, 14
    properties, 14-15
        experiments on, 16-17
Sound
    acoustics of, 428-430
    amplifiers, 450-451
    amplitude of, 424
    definition, 421

    experiments with, 452-456
    fidelity of, 448
    frequency of, 423-424
    horns, 451
    intensity, 431
    as interpreted by human ear, 426
    loudness of, 424
    magnetic recorder, 446-447
    measurement of, 430
    pitch of, 423
    principles of, applied to music,
        427-428
    recording, 446
    reproduction, 445-451
    speed of, 421-422
    transmission by microphones,
        447-449
    in vacuum, 421
    waves, 422-423, 444
Sound energy, applications of, 526
Sound waves. See Waves, sound
Specific gravity, 21, 182
    experiment on, 24-25
Specific weight of gases, 29
Spectrometer, recording 407-408
Spectrophotomer. See Spectroscope
Spectroscope, 406, 407, 408
    experiment with, 413-414
Spectrum lines, 407
Speed, 101-102
    change in, using wheel and axle,
        130-131
    control of, experiment on, 133-
        134
    mechanical advantage of, 82-83,
        115, 148-149
Square measure, 40
Stability. See Equilibrium
Static electricity. See Electricity,
        static
Stator, 308-309
Steady-state, in fluid systems, 208
Steam, conversion to mechanical
        energy, 290-291
Steam engine. See Engines, steam
Steam generating plants, 290-291
Steam turbine, operation of, 290-
        291
Sublimation, 344
Substations, for step-down trans-
        formers, 296
Surface tension, 19
    experiment on, 23
Switch, for electrical circuits, 270
Sympathetic vibrations. See
        Vibrations
Synchro-cyclotron, 487-488
Systeme International d'Unites (SI).
        See International System of
        Units

**T**

Tape recorder, 446-447
Tape reproducer, 447
Tapes, steel, 38-39
Telephone, 444-445
Television
  camera tube, 473-474
  color receivers, 475-476
  color transmitters, 474-475
  principles of, 471-474
  receiver, 474
  signals, 472
  transmitter, 472-473
Temperature
  absolute, 328
  effect on gas pressure, 28
  measuring instruments for, 324-327
  and resistance, 267
  systems of measurement of, 327-328
Tenacity, definition, 15
Tension, and force, 80
Terminals, of electronic tube, 464
Thermionic converter. **See**
  Converters, thermionic
Thermionic emitter, 510
Thermocouples, 325-326
  in thermoelectric generators, 515-516
Thermograph, 325
Thermometers, 324-325
  experiment with, 331-332
  principles of, 324
Thermonuclear reactions, 492-494
Three-wire system of electricity
  distribution, 296-297
Threshold of hearing, 431
Time, 43
  derived units of, 68-69
Time constant, of fluid systems, 208
Torque. **See also** Force
  in current-carrying conductor, 307
  formula for, 82
  of lever, 114
Torricelli, Evangelista, 190
Torricelli barometer. **See** Barometer
Toughness, definition, 14
Transformers, 295-296
Transistors, 511-512
Translucent materials, and light, 376
Transmitter, 444-445
Transparent Materials,
  and light, 376
Tritium, fusion of, 493
Tuning fork, 422-423
Turbines, gas
  applications of, 360
  brake thermal efficiency of, 363
  laws of motion applied to, 356-357
  operation of, 356-357
  types of, 358-360
Turbines, steam, 353-354
  experiment with, 366

**U**

Ultraviolet waves, 403

Uncharged body, 236
Unit pole, 249
Uranium
  fission of, 489
  isotopes of, 490-491
U.S. Constitution, and regulation
  of weights and measures, 64
U.S. Metric Board, 55, 57

**V**

Vacuum, sound in, 421
Vacuum close-spaced diode, 510
Van de Graaff accelerators, 485
Vapor, definition, 310
Vapor pressure, and boiling point,
  344
Vaporization, 343-345. **See also**
  Evaporation
Vector
  definition, 91
  graphic representation, 93-94
  quantities, 92
Velocity
  measurement, 69
  and momentum, 103
  of sound, 430
  as vector quantity, 92
Vibrations, 424-425
Video input signals, in color
  television, 474-475
Viscosity, 20-21
  of fluids, 205
Visible spectrum, 403-404
Volatile liquids, storage of, 29
Volt, 265
Voltage
  applications of, 265-266
  of dry cell, 280
  experiment with, 274-275
  induced, 292
  measurement, 266
  overload protection for, 266
  in parallel circuit, 270-271
  in series circuit, 270
  stepped-up, 295
Volume
  changing units of, 42
  computation in metric system,
    44-46
  computed in liquid measure, 46
  definition, 8
  derived units of, 68
  measurement of, 39, 40-42, 46

**W**

Water, 20
  composition of, 235
  conversion to mechanical energy,
    291
Watt, James, 352
Watthour, 297-298

Wave motion, 214
  experiment with, 222
  pulse of, 215-217
Wave motion theory
  applied to lenses, 381-382
  of light, 376-377
Wave train, 218. **See also** Pulses
Waveform, 218
  in alternating-current generator,
    293
Wavefronts, of light waves, 379
Wavelength, 219-220
  of daylight, 410
  electromagnetic spectrum of, 403
  of light, 376-377
  of sound, 430
  visible spectrum of, 403
Waves
  amplitude, 220
  electromagnetic, 214
  frequency of, 219-220
  interference of, 218
  longitudinal, 215
  mechanical, 214
  period of, 219
  periodic, 218, 219-220
  in phase, 219
  radio, 470
  sound, 422-423, 428, 429-430
  transverse, 214-215
  velocity of, 215, 219
Wedge
  compared with inclined plane,
    122-123
  experiment with, 126-127
  mechanical advantage of, 124
  uses, 123-124
Weight, 8, 42
Wetting action. **See also** Adhesion
  of liquids, 19
Wheel and axle, 129-130
  changing speeds using, 130-131
  experiments on, 132-134
  gears as application of, 147
  industrial applications, 131
Wilson cloud chamber, 485-486
Work, 80, 83
  measurement, 68
  mechanical and heat energy, 362
Work ethic, and human resource
  development, 534
World-of-work galaxy, 534
Worm and worm wheel, 157, 158

**X**

X-ray tube, 466
X-rays, 466

**Y**

Yardsticks, 38-39

**Z**

Zero, absolute, 328
Zoology, definition, 3